Gold and Silver Nanoparticles

Micro and Nano Technologies

Gold and Silver Nanoparticles
Synthesis and Applications

Edited by
Suban K. Sahoo
Department of Chemistry, Sardar Vallabhbhai National
Institute of Technology (SVNIT), India

M. Reza Hormozi-Nezhad
Department of Chemistry, Sharif University
of Technology, Iran

ELSEVIER

Elsevier
Radarweg 29, PO Box 211, 1000 AE Amsterdam, Netherlands
The Boulevard, Langford Lane, Kidlington, Oxford OX5 1GB, United Kingdom
50 Hampshire Street, 5th Floor, Cambridge, MA 02139, United States

Notices

Knowledge and best practice in this field are constantly changing. As new research and experience broaden our understanding, changes in research methods, professional practices, or medical treatment may become necessary.

Practitioners and researchers must always rely on their own experience and knowledge in evaluating and using any information, methods, compounds, or experiments described herein. In using such information or methods they should be mindful of their own safety and the safety of others, including parties for whom they have a professional responsibility.

To the fullest extent of the law, neither the Publisher nor the authors, contributors, or editors, assume any liability for any injury and/or damage to persons or property as a matter of products liability, negligence or otherwise, or from any use or operation of any methods, products, instructions, or ideas contained in the material herein.

ISBN: 978-0-323-99454-5

For Information on all Elsevier publications visit our website at
https://www.elsevier.com/books-and-journals

Publisher: Matthew Deans
Acquisitions Editor: Ana Claudia A Garcia
Editorial Project Manager: Czarina Mae S Osuyos
Production Project Manager: Erragounta Saibabu Rao
Cover Designer: Greg Harris

Typeset by Aptara, New Delhi, India
Transferred to Digital Printing 2023

Working together
to grow libraries in
developing countries

www.elsevier.com • www.bookaid.org

Contents

Chapter 3: Implementation of gold and silver nanoparticles in sensing and bioengineering 83

Geetika Bhardwaj, Navneet Kaur, Narinder Singh

Chapter 4: Glucose biosensing with gold and silver nanoparticles for real-time applications...109

R. Balamurugan, S. Siva Shalini, M.P. Harikrishnan, S. Velmathi, A. Chandra Bose

Chapter 5: Atomically precise gold and silver nanoclusters: Synthesis and applications...137

Rajanee Nakum, Raj Kumar Joshi, Suban K. Sahoo

Chapter 8: Silver and gold nanoparticles: Potential cancer theranostic applications, recent development, challenges, and future perspectives........................247

Swapnali Londhe, Shagufta Haque, Chitta Ranjan Patra

Chapter 9: Recent progress in gold and silver nanoparticle mediated drug delivery to breast cancers..291

Parth Malik, Gajendra Kumar Inwati, Rachna Gupta, Tapan Kumar Mukherjee

Contributors

Samira Abbasi-Moayed   Department of Chemistry, Sharif University of Technology, Tehran, Iran

Tripti Ahuja   Department of Chemistry, Indian Institute of Technology Delhi, Hauz Khas, New Delhi, India

Nidhi Andhariya   Thapar Institute of Engineering and Technology, Patiala, Punjab, India

R. Balamurugan   Nanomaterials Laboratory, Department of Physics, National Institute of Technology, Tiruchirappalli, Tamil Nadu, India

Geetika Bhardwaj   Department of Chemistry, Centre for Advanced Studies in Chemistry, Panjab University, Chandigarh, India

A. Chandra Bose   Nanomaterials Laboratory, Department of Physics, National Institute of Technology, Tiruchirappalli, Tamil Nadu, India

Cristina Buzea   IIPB Medicine Corporation, Owen Sound, ON, Canada

Bhupendra Kumar Chudasama   Thapar Institute of Engineering and Technology, Patiala, Punjab, India

Daiane Dias   Laboratório de Eletroespectro Analítica (LEEA), Escola de Química e Alimentos, Universidade Federal do Rio Grande, Rio Grande, Brazil

Subrata Dutta   Department of Chemistry, Sardar Vallabhbhai National Institute of Technology, Surat, Gujarat, India

Nafiseh Fahimi-Kashani   Department of Chemistry, Isfahan University of Technology, Isfahan, Iran

Mahdi Ghamsari   Department of Chemistry, Sharif University of Technology, Tehran, Iran

Forough Ghasemi   Department of Nanotechnology, Agricultural Biotechnology Research Institute of Iran (ABRII), Agricultural Research, Education, and Extension Organization (AREEO), Karaj, Iran

Rachna Gupta   Department of Biotechnology, Vishwa-Bharati Shantiniketan, Bolpur, West Bengal, India

Shagufta Haque   Department of Applied Biology, CSIR-Indian Institute of Chemical Technology, Hyderabad, Telangana, India; Academy of Scientific and Innovative Research (AcSIR), Ghaziabad, Uttar Pradesh, India

M.P. Harikrishnan   Nanomaterials Laboratory, Department of Physics, National Institute of Technology, Tiruchirappalli, Tamil Nadu, India

Mônika Grazielle Heinemann   Laboratório de Eletroespectro Analítica (LEEA), Escola de Química e Alimentos, Universidade Federal do Rio Grande, Rio Grande, Brazil

M. Reza Hormozi-Nezhad   Department of Chemistry, Sharif University of Technology, Tehran, Iran; Institute for Nanoscience and Nanotechnology, Sharif University of Technology, Tehran, Iran

Gajendra Kumar Inwati   Department of Physics, University of the Free State, Bloemfontein, ZA, South Africa

Zahra Jafar-Nezhad Ivrigh   Department of Chemistry, Sharif University of Technology, Tehran, Iran

Ritambhara Jangir   Department of Chemistry, Sardar Vallabhbhai National Institute of Technology, Surat, Gujarat, India

Raj Kumar Joshi   Department of Chemistry, Malaviya National Institute of Technology (MNIT), Jaipur, India

Navneet Kaur   Department of Chemistry, Centre for Advanced Studies in Chemistry, Panjab University, Chandigarh, India

Swapnali Londhe   Department of Applied Biology, CSIR-Indian Institute of Chemical Technology, Hyderabad, Telangana, India; Academy of Scientific and Innovative Research (AcSIR), Ghaziabad, Uttar Pradesh, India

Parth Malik   School of Chemical Sciences, Central University of Gujarat, Gandhinagar, Gujarat, India

Maneet   Department of Chemistry, National Institute of Technology Hamirpur, Hamirpur, India

Tapan Kumar Mukherjee   Institute of Biotechnology, Amity University, Noida, Uttar Pradesh, India

Rajanee Nakum   Department of Chemistry, Sardar Vallabhbhai National Institute of Technology, Surat, Gujarat, India

Afsaneh Orouji   Department of Chemistry, Sharif University of Technology, Tehran, Iran

Ivan Pacheco   IIPB Medicine Corporation, Owen Sound, ON, Canada; Department of Pathology, Grey Bruce Health Services, Owen Sound, ON, Canada; Department of Pathology and Laboratory Medicine, Schülich School of Medicine & Dentistry, Western University, London, ON, Canada

Chitta Ranjan Patra   Department of Applied Biology, CSIR-Indian Institute of Chemical Technology, Hyderabad, Telangana, India; Academy of Scientific and Innovative Research (AcSIR), Ghaziabad, Uttar Pradesh, India

Jai Prakash   Department of Chemistry, National Institute of Technology Hamirpur, Hamirpur, India

Caroline Pires Ruas   Laboratório de Eletroespectro Analítica (LEEA), Escola de Química e Alimentos, Universidade Federal do Rio Grande, Rio Grande, Brazil

Suban K. Sahoo   Department of Chemistry, Sardar Vallabhbhai National Institute of Technology, Surat, Gujarat, India

Samriti   Department of Chemistry, National Institute of Technology Hamirpur, Hamirpur, India

S. Siva Shalini   Nanomaterials Laboratory, Department of Physics, National Institute of Technology, Tiruchirappalli, Tamil Nadu, India

Narinder Singh   Department of Chemistry, Indian Institute of Technology Ropar (IIT Ropar), Rupnagar, Punjab, India

Anjana K. Vala   Department of Life Sciences, Maharaja Krishnakumarsinhji Bhavnagar University, Bhavnagar, Gujarat, India

S. Velmathi   Organic and Polymer Synthesis Laboratory, Department of Chemistry, National Institute of Technology, Tiruchirappalli, Tamil Nadu, India

Plasmonic noble metal (Ag and Au) nanoparticles: From basics to colorimetric sensing applications

Nafiseh Fahimi-Kashani[a], Afsaneh Orouji[b], Mahdi Ghamsari[b], Suban K. Sahoo[d], M. Reza Hormozi-Nezhad[b,c]

[a]Department of Chemistry, Isfahan University of Technology, Isfahan, Iran [b]Department of Chemistry, Sharif University of Technology, Tehran, Iran [c]Institute for Nanoscience and Nanotechnology, Sharif University of Technology, Tehran, Iran [d]Department of Chemistry, Sardar Vallabhbhai National Institute of Technology, Surat, Gujarat, India

1.1 Plasmonics: A basic background

Plasmonics, a prominent branch of nanophotonics deals with the interaction of electromagnetic radiation with the conduction electrons of a metal-dielectric interface on the nanometer scale. Indeed, when light impinges upon the surface of a metallic film or nanostructure, the alternating electric field causes the delocalized electrons to collectively oscillate at a resonant frequency. These rapid oscillations of the electron density in a conducting medium like a metal lattice are so-called plasma oscillations or Langmuir waves [1]. Resembling photons which delineates a quantum of light, the quantization of these electron oscillations underlies the term Plasmon describing a quantum of plasma oscillations [2]. Contrary to bulk plasmons, surface plasmons are strongly localized to the surface since electromagnetic waves can only penetrate a metal's surface to a shallow depth. Surface plasmons are predominantly responsible for the interaction with the incident light and controlling the metals' optical properties [3]. The optical properties of gold (Au) and silver (Ag) nanoparticles (NPs) are mainly justified in light of the topics explored in plasmonics and the following sections attempt to bring up the main concepts of this field and elaborate on their implications in the chromogenic properties of these NPs.

1.1.1 Surface plasmon resonance (SPR) versus localized surface plasmon resonance (LSPR)

SPR is deemed the most remarkable optical property of metals and is briefly defined as the collective oscillations of the delocalized electrons in a metallic surface stimulated by the incident light under resonance conditions. When the frequency of the incident light matches

Gold and Silver Nanoparticles.
DOI: https://doi.org/10.1016/B978-0-323-99454-5.00005-6

the oscillation frequency of the surface plasmons, the resonance condition is achieved and the energy transfer occurs between the light's electric field and the surface plasmons, leading to a dramatic decrease in the reflected light intensity (Fig. 1.1A and B). On a planar (2D) surface, surface plasmons promulgating along the metal-dielectric interface are referred to as surface plasmon polaritons, and this phenomenon is known as SPR [3,4]. However, in metallic NPs with sizes comparable to or smaller than the wavelength of the incident light, the surface plasmons are confined in the particle and are called localized surface plasmons. By irradiating the NPs with light, the oscillating electric field polarizes the entire electron cloud of conduction band, resulting in the accumulation of surface charges alternatively on the particle's ends. In the meantime, the coulombic attraction between the electrons and the nuclei generates a restoring force, causing the electron cloud to oscillate coherently (Fig. 1.1C). This phenomenon is known as localized surface plasmon resonance (LSPR). As a consequence of this process, the light beam is partially absorbed in the plasmon resonance frequency and

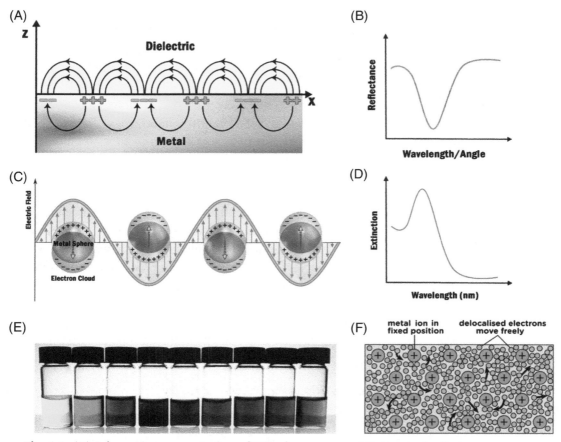

Fig. 1.1: (A) Schematic representation of SPR phenomenon, (B) A typical reflectance spectrum associated with SPR, (C) Schematic representation of LSPR phenomenon, (D) A typical extinction spectrum associated with LSPR (E) Color variations images of Au NPs' dispersions, (Note, the first vial from the right is $HAuCl_4$ solution). [8] (F) Schematic representation of the Drude model.

an absorption peak that is, LSPR peak appears in the particle's optical absorption profile (Fig. 1.1D). That's why a wide range of brilliant colors is observed for the dispersions of different Au and Ag NPs possessing different structural features (Fig. 1.1E) [5–7].

1.1.2 Dielectric functions of metals: Drude model

In advance of shedding light upon different subtle aspects of the LSPR phenomenon, it is crucial to acquire a more detailed knowledge of the Drude Model and the metals' dielectric function. In fact, this model can rationalize the optical properties of metals over a wide frequency range. In this model, a gas of electrons moves on a fixed background of ions and then oscillates when an electromagnetic field is applied (Fig. 1.1F) [7,9,10]. This behavior can be delineated using the Drude dielectric function (ε_{Drude}) for a free electron gas, where the electrons' oscillation is damped via collisions happening with a characteristic collision frequency (γ):

$$\varepsilon_{Drude}(\omega) = 1 - \frac{\omega_p^2}{\omega^2 + i\gamma\omega} \tag{1.1}$$

In this equation, (γ_b) is the bulk damping constant related to the Fermi velocity (v_F), ω_p and l_m are the plasma frequency and the mean free path respectively. These factors are defined as:

$$\omega_p = \sqrt{\frac{ne^2}{\varepsilon_0 m}} \tag{1.2}$$

$$\gamma_b = \frac{v_F}{l_m} \tag{1.3}$$

where n is the conduction electron density, e is the electron charge, ε_0 is the permittivity of the free space, and m is the electron mass. In this theory, ω_p is deemed a pivotal factor in determining the optical behavior of metals. In particular, the incident light is transmitted providing that its frequency exceeds ω_p, whereas it is reflected if its frequency is lower than ω_p. To illustrate, in the former case, the metal's free electrons are not so fast to screen the electric field of the light, but in the latter they are.

After all, the dielectric function of the metal is composed of two real (ε_1) and imaginary (ε_2) parts controlling the absorption and reflection behaviors respectively.

$$\varepsilon(\omega) = \varepsilon_1(\omega) + i\varepsilon_2(\omega) \tag{1.4}$$

In the case of metal NPs, the position of the LSPR peak is determined by the real part, whereas the imaginary part affects dephasing. Based on Eqs 1.1 and 1.4, these two components are as follows:

$$\varepsilon_1(\omega) = Re((\varepsilon)) = 1 - \frac{\omega_p^2}{\omega^2 + \gamma^2} \tag{1.5}$$

$$\varepsilon_2(\omega) = Im((\varepsilon)) = \frac{\omega_p^2 \gamma}{\omega(\omega^2 + \gamma^2)} \tag{1.6}$$

In the vast majority of metals, ω_p lies in the UV region and that is why most metals show an almost total reflecting behavior in the visible region. However, when $\omega < \omega_p$, the real part is a negative value, the imaginative part is small, and as a result, the excitation of the surface plasmons can occur [7]. Note that the frequency-dependence of damping in noble metals can be strongly altered by interband transitions [10]. These transitions take place from the lower-energy electrons compared to the conduction electrons, and the effect of this phenomenon only becomes significant when the energy of the incident light is high enough. Some examples of interband transitions are 3.9 eV for Ag, 2.4 eV for Au, and 2.1 eV for Cu. The interband transitions can be taken into account by adding an additional term ($\varepsilon^{ib}(\omega)$) to the metal's dielectric function:

$$\varepsilon(\omega) = \varepsilon^{ib}(\omega) + 1 - \frac{\omega_p^2}{\omega\left(\omega + i\gamma\left(l_{eff}\right)\right)} \tag{1.7}$$

1.1.3 Extinction and scattering cross-sections of plasmonic NPs

1.1.3.1 Mie theory for spherical NPs

When dealing with the visible spectrum of plasmonic NPs, it is of great importance to justify how the intensity of light incident on the NPs is diminished through absorption and scattering. In 1908, Gustav Mie reached an analytical solution to Maxwell's equations describing the scattering and absorption (extinction) of light by the spherical NPs possessing a smaller diameter than the wavelength of light [5–7,11]. Consequently, the following equations were attained for scattering, extinction, and absorption cross-sections of a uniform conducting sphere irradiated by a plane wave:

$$\sigma_{sca} = \frac{2\pi}{|k|^2} \sum_{L=1}^{\infty} (2L+1)\left(|a_L|^2 + |b_L|^2\right) \tag{1.8}$$

$$\sigma_{ext} = \frac{2\pi}{|k|^2} \sum_{L=1}^{\infty} (2L+1)\left[Re\left(a_L + b_L\right)\right] \tag{1.9}$$

$$\sigma_{abs} = \sigma_{ext} - \sigma_{sca} \tag{1.10}$$

Here, k is the incoming wavevector, L is an integer representing the dipole, quadrupole, and higher multipole of the scattering and a_L and b_L are the two parameters calculated according to the following expressions using the Riccati-Bessel functions ψ_L and χ_L:

$$a_L = \frac{m\psi_L(mx)\psi_L'(x) - \psi_L'(mx)\psi_L(x)}{m\psi_L(mx)\chi_L'(x) - \psi_L'(mx)\chi_L(x)} \tag{1.11}$$

$$b_L = \frac{\psi_L(mx)\psi_L'(x) - m\psi_L'(mx)\psi_L(x)}{\psi_L(mx)\chi_L'(x) - m\psi_L'(mx)\chi_L(x)} \tag{1.12}$$

In the Eq. 1.12, $x = k_m r$, where r is the radius of the particle and k_m is the wavenumber in the medium. Furthermore, $m = \dfrac{\tilde{n}}{n_m}$, where $\tilde{n} = n_R + i n_I$ is the complex refractive index of the metal, and n_m is the real refractive index of the surrounding medium [12].

Supposing that the NP is far smaller the wavelength (i.e. $x \ll 1$), the approximation of Riccati-Bessel functions by power series, results in the simplified forms of Eqs. 1.10 and 1.11:

$$a_1 \approx -\frac{i2x^3}{3}\frac{m^2 - 1}{m^2 + 2} \tag{1.13}$$

$$b_1 \approx 0 \tag{1.14}$$

This approximation is the dipolar approximation that is known as the quasistatic or Rayleigh limit as well. In this way, it is assumed that the electronic polarization is exactly in phase with the excitation field and the electrons are displaced as a whole. Thus, the charge distribution in the particle can be treated as if it were a static distribution. Herein, by keeping the terms up to x^3, the higher-order a_L and b_L will be zero. Substituting $m = \dfrac{n_R + i n_I}{n_m}$ into Eq. 1.12, gives the following equation:

$$a_1 = -i\frac{2x^3}{3}\frac{n_R^2 - n_I^2 + i2n_R n_I - n_m^2}{n_R^2 - n_I^2 + i2n_R n_I + 2n_m^2} \tag{1.15}$$

Next, introduction of the complex metal dielectric functions ($\varepsilon_1 = n_R^2 - n_I^2$, $\varepsilon_2 = 2n_R n_I$), and the medium's dielectric function ($\varepsilon_m = n_m^2$) into the above equation leads to:

$$a_1 = \frac{2x^3}{3}\frac{-i\varepsilon_1^2 - i\varepsilon_1\varepsilon_m + 3\varepsilon_2\varepsilon_m - i\varepsilon_2^2 + i2\varepsilon_m^2}{\left(\varepsilon_1 + 2\varepsilon_m\right)^2 + \left(\varepsilon_2\right)^2} \tag{1.16}$$

Finally, the substitution of Eq. 1.15 into Eqs. 1.8 and 1.9, yields the following expressions for the extinction and scattering cross-sections, widely reported for NP plasmon resonances:

$$\sigma_{ext} = \frac{18\pi\varepsilon_m^{3/2}V}{\lambda}\frac{\varepsilon_2(\lambda)}{\left[\varepsilon_1(\lambda) + 2\varepsilon_m\right]^2 + \varepsilon_2(\lambda)^2} \tag{1.17}$$

$$\sigma_{sca} = \frac{32\pi^4\varepsilon_m^2 V^2}{\lambda^4}\frac{\left(\varepsilon_1 - \varepsilon_m\right)^2 + \left(\varepsilon_2\right)^2}{\left(\varepsilon_1 + 2\varepsilon_m\right)^2 + \left(\varepsilon_2\right)^2} \tag{1.18}$$

The foregoing equations not only can provide a clear insight into the LSPR of very small particles, but can also be applied to larger spherical particles with acceptable accuracy [13]. According to Eq. 1.13, σ_{ext} is maximized providing that ε_2 approaches zero and $\varepsilon_1 = -2\varepsilon_m$ so that the resonance condition is established. These two conditions are met solely by a few metals including Au, Ag, and Cu. Alternatively, in the case of small particles ($R \ll \lambda$), merely absorption is in play and the scattering is negligible $\left(\dfrac{\sigma_{sca}}{\sigma_{abs}} \propto \left(\dfrac{R}{\lambda} \right)^3 \right)$ [6,10]. The extinction behavior of different spherical Au NPs with different diameters have been illustrated in (Fig. 1.2A) which further elucidates the Mie theory.

1.1.3.2 Generalized Mie theory for spheroidal NPs (Gans theory)

The formula calculated in the previous section for Mie theory is only applicable to spherical NPs. Hence, a generalized form of this theory was developed by Richard Gans representing the LSPR behavior of small-sized spheroidal NPs with any aspect ratio [6,14]. The following equation is the peer of Eq. 1.16 for spheres, albeit it yields the absorption cross-section of a prolate spheroidal NP:

$$\sigma_{abs} = \frac{\omega}{3c} \varepsilon_m^{3/2} V \sum_j \frac{\left(1/P_j^2 \right) \varepsilon_2}{\left\{ \varepsilon_1 + [(1 - P_j)]\varepsilon_m \right\}^2 + \varepsilon_2^2} \tag{1.19}$$

In this expression, the sum over j takes the three dimensions of the particle into account. That is, depolarization factors (P_j) comprises P_A, P_B, and P_C, for each axis of the particle, where $A > B = C$ for a prolate spheroid. These factors change the values of ε_1 and ε_2 and, in turn, exert influence on LSPR peak frequencies. The calculation of depolarization factors is as follows:

$$P_A = \frac{1 - e^2}{e^2} \left[\frac{1}{2e} \ln \left(\frac{1+e}{1-e} \right) - 1 \right] \tag{1.20}$$

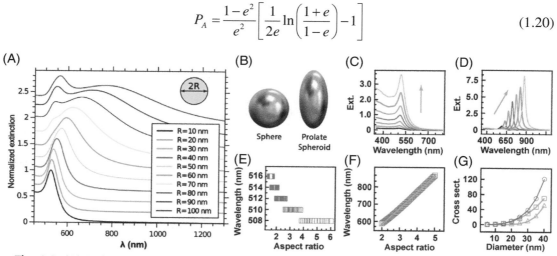

Fig. 1.2: (A) Extinction spectra of spherical Au NPs with different diameters [18] (B) Schematic illustration of sphere and prolate spheroid (C and D) Simulated extinction spectra of prolate spheroids (d = 5 nm) with different aspect ratios (1.5–6) using Gans theory under transverse and longitudinal excitations respectively. (E–F) Dependence of the transverse and longitudinal LSPR wavelengths on the aspect ratio respectively. (G) Extinction (circles), scattering (triangles), and absorption (squares) cross-sections versus the diameter calculated based on Gans theory [19].

$$P_B = P_C = \frac{1 - P_A}{2} \tag{1.21}$$

where e is the following parameter, calculated using the aspect ratio R;

$$e = \left[1 - \left(\frac{B}{A}\right)^2\right]^{1/2} = \left[1 - \frac{1}{R^2}\right]^{1/2} \tag{1.22}$$

Eq. 1.19 well describes the extinction profile of prolate spheroids [15,16] (Fig. 1.2B). On the one hand, they signify two LSPR peaks, one originating from the contribution of x and y directions is referred to as transverse mode, and the other one emanating from the contribution of the z-axis is so-called the longitudinal mode (Fig. 1.2C and D). On the other hand, it clearly indicates how the aspect ratio (R) determines the wavelength of the LSPR peak (Fig. 1.2E and F). According to this equation, $\left[\dfrac{1 - P_j}{P_j}\right]$ is a quantity that is proportional to the aspect ratio and weights ε_m. This factor is 2 for spherical particles but can be much greater for prolate spheroids. In fact, it not only justifies the red-shift of the LSPR peak by increasing the aspect ratio but substantiates why such NP exhibit higher sensitivity to the dielectric constant of the surrounding medium.

Although Mie and Gans theories can be efficiently employed for spherical NPs, concentric spherical shells, spheroids, and infinite cylinders, they are not suitable for describing nanostructures with arbitrary geometrical shapes. That is why a number of numerical approaches comprising Discrete Dipole Approximation (DDA), Multiple Multipole Method (MMP), Finite Element Method (FRM), Finite Difference Time Domain (FDTD), and T-matrix method have been developed in this regard [5,17].

1.1.4 Experimental factors controlling LSPR

The LSPR oscillation frequency and extinction cross-section of plasmonic nanostructures depend on the density of electrons, the effective electron mass, and the size and shape of the charge distribution, which are primarily determined by the dielectric constants of both metal and the surrounding medium along with compositional and geometrical characteristics of the nanostructure. The following sections specifically elaborate on different key factors controlling the LSPR behavior of plasmonic nanostructures.

1.1.4.1 Effect of the surrounding medium

According to the Drude model, the refractive index of the surrounding medium is capable of affecting the position of the LSPR peak significantly [6,20]. One can utilize the real portion of the dielectric function to support this relationship:

$$\varepsilon_1 = 1 - \frac{\omega_P^2}{\omega^2 + \gamma^2} \tag{1.23}$$

where ω_p is the plasma frequency, and γ is the damping parameter of the bulk metal. For visible and near-infrared regions, $\gamma \ll \omega_p$ and the above equation can be simplified to:

$$\varepsilon_1 = 1 - \frac{\omega_P^2}{\omega^2} \qquad (1.24)$$

By considering the resonance condition and setting $\varepsilon_1 = 2\varepsilon_m$, one can obtain the following expression in which ω_{max} is the LSPR peak frequency:

$$\omega_{max} = \frac{\omega_P}{\sqrt{2\varepsilon_m + 1}} \qquad (1.25)$$

Following the frequency-to-wavelength conversion $\left(\lambda = \frac{2\pi c}{\omega} \right)$ and the dielectric function to the refractive index ($\varepsilon_m = n^2$), the final expression is attained:

$$\lambda_{max} = \lambda_P \sqrt{2n_m^2 + 1} \qquad (1.26)$$

Here, λ_{max} and λ_p are the wavelength of the LSPR peak and the wavelength corresponding to the plasma frequency of the bulk metal respectively. As it appears, the LSPR band of plasmonic NPs gradually redshifts with increasing the refractive index of the solvent. Additionally, it has been revealed based on the experiments that the variations of the LSPR peak wavelength upon changing the medium's refractive index are approximately linear over small ranges of n (Fig. 1.3). Clearly denotes this behavior for different plasmonic nanostructures.

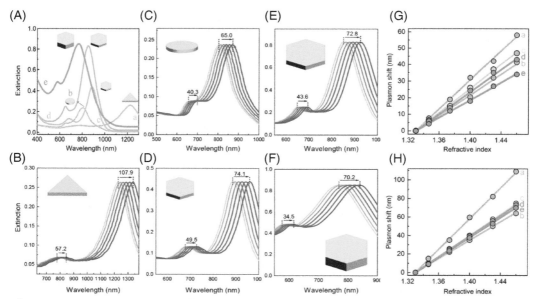

Fig. 1.3: (A) Extinction spectra of the triangular, circular, and three overgrown hexagonal Au NPs, (B–F) Extinction spectra of triangular, circular, and three overgrown hexagonal Au NPs in water-glycerol liquid mixtures of varying percentages of glycerol including 0%, 10%, 30%, 50%, 70%, and 90%, respectively. (G and H) Correlation of the LSPR shift of the quadrupolar and dipolar modes with the RI of the surrounding medium for the triangular, circular, and three overgrown hexagonal Au NPs [21].

1.1.4.2 Effect of crystal composition

One of the crucial factors which primarily determines the LSPR behavior and in turn controls different characteristics of the extinction spectrum that is, the position, bandwidth, and intensity of the resonance peaks is the chemical composition of the nanostructures. As previously stated regarding the dipole approximation in Mie theory, the resonance is established providing that the real part of the metals' dielectric function and that of the surrounding medium meet the condition $\varepsilon_1 = -2\varepsilon_m$. Considering the fact that the surrounding medium is typically air or water possessing positive dielectric functions, the real part should be necessarily negative. The imaginary part, however, that is, ε_2 may exert influence on the quality of the resonance peak as well. Therefore, it should be small at the resonance condition to limit the choices to some Nobel metals comprising Au, Ag, and Cu whose light interaction in the visible and NIR regions is considerable (Fig. 1.4A and B). Not to mention, the foregoing explanation is an implication of Mie theory which is mainly applicable to spherical NPs, whereas considering the effect of other factors especially the structure is indispensable for studying the LSPR behavior of a nanostructure.

In this context, a few criteria can be considered for examining the metals' performance comprising the plasmonic efficiency, wavelength range, and chemical stability. Some of the widely-used noble metal NPs for example, Au, Ag, and Cu are of acceptable chemical

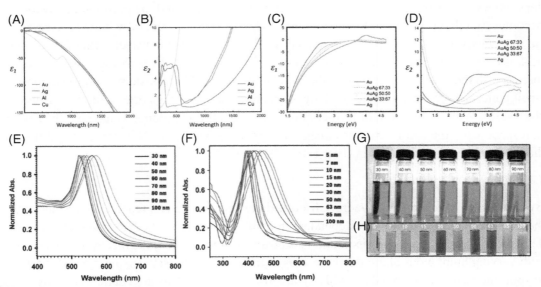

Fig. 1.4: (A and B) Real and imaginary parts of dielectric functions as a function of wavelength for Au, Ag, Al, and Cu (C and D) Real and imaginary part of dielectric functions of Au, Ag, and Au/Ag alloy thin films as a function of energy [10,27]. (E) Normalized absorption spectra of spherical Au NPs with different diameters (30–100 nm). (F) Normalized absorption spectra of spherical Ag NPs with different diameters (5–100 nm). (G) Color variations' images of spherical Au NPs with different diameters (30–90 nm). (H) Color variations' images of spherical Ag NPs with different diameters (5–100 nm) [28,29].

stability and exhibit intriguing LSPR behavior in the visible-NIR region. Generally, Ag and Au nanostructures display narrow and intense plasmonic peaks ranging from visible to NIR range, and those of Cu nanostructures range from red to NIR [10,22,23]. On the contrary, the LSPR bands of Al, Ga, In, and Mg usually appear in the UV region but yet a few cases exhibiting visible LSPR peaks have been reported either [10,24]. In addition to the NPs composed of pure metal, the combination of different metals for the creation of random-alloy or ordered-alloy NPs can be regarded as a potent tool for tailoring the LSPR behavior (Fig. 1.4C and D) show how the ratio of metals affects the real and the imaginary parts of the alloy's dielectric function [25].

1.1.4.3 Effect of crystal size

For NPs with sizes less than the incident radiation's wavelength (<25 nm for Au NPs), only the dipole term should contribute to the extinction cross section. These variations in the absorption spectra are known as inherent size effects [7,26]. In accordance with the experimental reports, it has been well established that there is a negative correlation between the absorption bandwidth and the size of NPs, meaning smaller particles ensure larger bandwidth and vice versa. In contrast, as the sizes of large NPs are comparable to the wavelength of the interacting light, the size dependency enters through the full expression of Mie's theory and the plasmon bandwidth becomes exactly proportional to the NPs' size. In contrast, as the sizes of large NPs are comparable to the wavelength of the interacting light, the size dependency enters through the full expression of Mie's theory and the plasmon bandwidth becomes exactly proportional to the NPs' size. This behavior is generally known as *extrinsic size effects*. In addition to the dependence of the bandwidth, the wavelength of the LSPR peak is dependent upon the NP size. Despite the fact that the LSPR peak red shifts as the size increases in the extrinsic size region, the NPs belonging to the intrinsic size region exhibit small patternless shifting toward both higher energies and lower energies with increasing size [7,26]. With the purpose of further elucidating the aforementioned behaviors, the extinction spectra and the color variations images of spherical Au and Ag NPs possessing different diameters have been provided in (Fig. 1.4E–H).

1.1.4.4 Effect of crystal shape

By comparing Mie and Gans theories discussed previously, one can readily understand that the LSPR behavior is drastically dependent upon the geometry of the NPs and that a trivial deformation might change the extinction spectrum entirely. In fact, the electrons can oscillate with varying amplitudes along the three axes of the anisotropic NP, causing the LSPR band to split into two or more bands. Highly symmetrical spherical NPs are characterized by single scattering peaks, while anisotropic NPs display multiple LSPR peaks on account of highly localized charge polarizations at the corners and edges [5,7]. The effect of particle geometry on the LSPR properties has been thoroughly studied in the work of Mock et al., in which the spectra of similar volumes but different morphologies of Ag NPs (spheres, triangles, and cubes) were correlated with their structure [30]. Thanks to the versatile optical properties of anisotropic NPs, a myriad of new Au and Ag nanostructures featuring sharp tips have been developed over the past few decades.

As far as Au and Ag nanorods are concerned, a low-energy LSPR band corresponding to the longitudinal resonance can be observed which red shifts as the aspect ratio increases. On the contrary, the transverse resonance peak appears at higher energies and slightly blue shifts by increasing the aspect ratio [7,19] (Fig. 1.5A). The same behavior is observed for prolate spheroids, as already discussed in Section 1.3.2. The shape of nanorods, however, is clearly different from that of spheroids in terms of the edges and corners. That is why numerical methods are usually employed for modeling the optical properties of nanorods [5,7,31].

Owing to the existence of several symmetry axes in the nanocube geometry, it is expected that the extinction spectrum exhibits several LSPR bands. For instance, one can pinpoint about eight LSPR bands (D1–D8) at 468, 427,405, 399, 398, 375, 339, and 332 nm in the extinction spectra (300–900 nm) of Ag nanocubes in water (Fig. 1.5B). The position of these peaks is dependent upon the size and as the size increases, a red-shift is observed on account of the retardation effects. In particular, as the mean size of the cubes increases from 50 to 150 nm, the LSPR peak with the lowest energy not only red-shifts from ~450 to ~750 nm, but is broadened and becomes less intense. Indeed, these behaviors are justified in light of a set of dipole resonance modes and radiation effects. To begin with, D1–D5 modes originate from

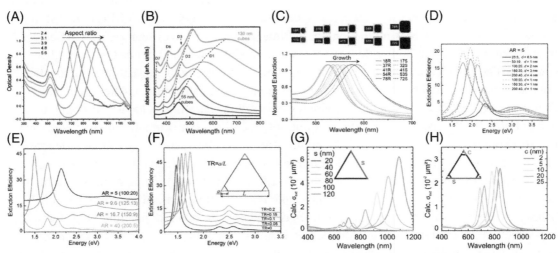

Fig. 1.5: (A) Extinction spectra of Au nanorods with different aspect ratios (2.4–5.6) [41], (B) Extinction spectra of different Ag nanocubes with different sizes (55–130 nm) [32]. (C) Extinction spectra of different Au nanocubes with different sizes and corner geometries [34]. (D–F) Extinction spectra of different Ag nanotriangles with varying volumes, aspect ratios, and different degrees of truncation respectively [36]. (G and H) Extinction spectra of different Au nanotriangles with varying length and different degrees of truncation respectively [40].

a series of dipole plasmon modes with a concentrated charge at the corners and along the edges of the nanocubes. In the case of small-sized cubes, these modes are closely clustered and the extinction peaks of D2–D5 modes are embedded in that of the D1 mode. However, as a result of size increment, the resonance wavelengths of these plasmon modes increase at different rates, leading to wider separation. Indeed, these modes differ in the number of nodes. Those with more nodes and shorter spatial oscillation periods have weaker dipole moments but resonate at higher frequencies. Moreover, modes D7 and D8 are dipole modes with charge density distributed over the faces of the cubes [32]. In contrast to Ag nanocubes exhibiting many absorption bands, the spectra of Au nanocubes only show the excitation of a single plasmonic mode analogous to spherical NPs (Fig. 1.5C). Seemingly, this LSPR band represents different dipole modes whose bands appeared unresolved due to their close proximity. Besides, as the corners sharpen and the edge length increases, this peak red-shifts gradually on account of the retardation effect [5,33,34].

Nanotriangles, also known as nanoprisms are roughly two-dimensional nanostructures possessing an extreme degree of anisotropy, favoring a high tunability of their LSPR behavior. For example, in the case of Ag nanotriangles, the edge length, the thickness, the aspect ratio, and the volume of the nanoprisms clearly control the LSPR behavior (Fig. 1.5D–F) [35–37]. Generally, the dimensions of the nanotriangles range from 50 to 200 nm in edge length and from 5 to 7 nm in thickness. Accordingly, the aspect ratio which is defined as the ratio of the edge length to the thickness ranges from 5 to 40. The extinction spectrum of Ag nanotriangles presents three SPR peaks in the range of 300–1200 nm which can be attributed to the dipolar, quadrupolar, and octupolar resonances, respectively. All of the observed LSPR bands shift to lower energies by increasing the length, volume, or aspect ratio of nanotriangles. Besides, it has been revealed that higher degrees of truncation shift the LSPR bands toward higher energies. The enhancement of the band intensity with increasing the aspect ratio at constant volume can be justified by the fact that the triangular face area increases. Alternatively, the characteristic LSPR band of Au nanotriangles possessing sharp tips is located between 650 and 750 nm. As the nanotriangles grow, it not only red-shifts but is greatly intensified, reaching 1000 nm in cases with 100 nm side length. (Fig. 1.5G) Furthermore, the higher degrees of truncation lead to the larger blue-shift of the LSPR peak (Fig. 1.5H). In a similar fashion, by increasing the thickness of nanotriangles the LSPR peak blue-shifts as well. Not to mention, due to the high sensitivity of the anisotropic nanostructures, the aforementioned LSPR behaviors can be easily altered and vary from one batch to another, depending on different experimental parameters ranging from degree of monodispersity and size-distribution to the effect of other coexisting nanostructures [38–40].

1.1.4.5 Effect of structural configuration

Apart from the homogenous NPs discussed in the previous sections, there is another class of plasmonic NPs possessing multiphasic architectures ranging from core-shell and multi-shell core-shell to heteromeric structures such as Janus and core-satellite nanostructures [42,43]. Depending on the dielectric functions of the metallic or nonmetallic materials used in the

NPs' composition and also depending on the morphological details of the studied architecture, the LSPR peaks can be adjusted over a wide frequency range. Most interestingly, some extensions of Mie theory have been developed for modeling the LSPR behavior of core-shell and multi-shell core-shell plasmonic nanostructure which can truly rationalize the effect of structural configuration [44–51].

Extension of Mie theory for the LSPR behavior of a sphere (core size R) coated with one shell (thickness d) in the Rayleigh approximation, eventually results in the following equation for the extinction cross-section of a core-shell plasmonic NP:

$$\sigma_{ext}(\lambda, R, d) = \frac{8\pi^2 (E + d^2)^3 \sqrt{\varepsilon_M}}{3\lambda} . \text{Im} \left[\frac{(\varepsilon_{sh} - \varepsilon_M)(\varepsilon_c - 2\varepsilon_{sh}) + \left(\dfrac{R}{R+d}\right)^3 (\varepsilon_c - \varepsilon_{sh})(\varepsilon_M + 2\varepsilon_{sh})}{(\varepsilon_c + 2\varepsilon_{sh})(\varepsilon_{sh} + 2\varepsilon_M) + 2\left(\dfrac{R}{R+d}\right)^3 (\varepsilon_c - \varepsilon_{sh})(\varepsilon_{sh} - \varepsilon_M)} \right]$$

Where ε_c, ε_{sh}, and ε_M are the dielectric functions of the core, the shell, and the surrounding medium respectively, and "Im" means the imaginary part of the expression in the bracket. The foregoing expression can be utilized for a metallic core coated with a nonmetallic shell and vice versa. Comparing this equation with Eq. 1.16 well indicates how the introduction of a shell impacts the LSPR behavior.

When a transparent core coated with a metallic shell is concerned, the resonance can occur in case the following conditions are met:

$$\varepsilon_{sh} \approx -\frac{R}{2d}(\varepsilon_c + 2\varepsilon_M)$$

$$\varepsilon_{sh} \approx -\frac{1}{2}\varepsilon_c$$

The first condition is fulfilled providing that the real part of the dielectric function ε_{sh} becomes more negative than that of a compact metal sphere. Since a more negative is obtained at longer wavelengths, the LSPR peak red-shifts. The smaller the shell is or the bigger the core is, the more the LSPR peak shifts. The computed extinction spectra of different Ag-coated silica clearly illustrate the dependence of LSPR peak on the core size (Fig. 1.6A). As a consequence of increasing the core size, additional extinction maxima appear as well which are attributed to the higher multipolar orders.

Conversely, when a metallic core is coated with a dielectric shell like an oxide layer, the resonance is established subject to the condition that the following equations are satisfied:

$$\varepsilon_c = -2\varepsilon_{sh} \frac{(\varepsilon_{sh} + 2\varepsilon_M) - \left(\dfrac{R}{R+d}\right)^3 (\varepsilon_{sh} - \varepsilon_M)}{(\varepsilon_{sh} + 2\varepsilon_M) + 2\left(\dfrac{R}{R+d}\right)^3 (\varepsilon_{sh} - \varepsilon_M)}$$

$$\varepsilon_c = -2\varepsilon_{sh}$$

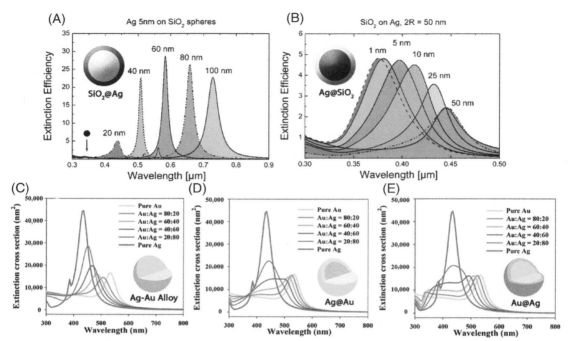

Fig. 1.6: (A) Simulated extinction efficiency spectra of Ag-coated SiO_2 nanospheres with core diameters of 20, 40, 60, 80, and 100 nm and a Ag shell thickness d = 5 nm. (B) Simulated extinction efficiency spectra of silica-coated Ag nanospheres with core diameter of 50 nm and shell thickness of 1, 5, 10, 25, and 50 nm. The dashed line corresponds to the uncoated Ag sphere and the dashed-dotted-dotted line to an uncoated Ag sphere embedded in silica [51]. (C–E) Extinction spectra of alloy and Ag@Au and Au@Ag core-shell NP with varying compositions [59].

The foregoing equations have been obtained by the Rayleigh approximation. Therefore, in large d where the sphere surpasses the limit of Rayleigh approximation, it fails and full Mie theory must be employed. Anyway, in this case, the increment of the shell thickness gives rise to the red-shift of the core's LSPR peak. Computed extinction spectra of different silica-coated Ag spheres clearly show this behavior (Fig. 1.6B).

Apart from the aforementioned core-shell designs, there are cases wherein both core and shell are metallic, such as Ag-coated Au NPs (Au@Ag NPs) and Au-coated Ag NPs (Ag@Au NPs) (Fig. 1.6C–E) [51–55]. Actually, due to the fact that Au and Ag possess almost the same physical properties, the grain boundary between core and shell is absent and instead an intermediate alloy region is observed. Therefore, since the work functions of Ag and Au are different, the electronic systems remain separated and behave independent of each other. After all, one can find numerous reports in the literature concerning synthesis of various binary, ternary, and other high-order heteromeric plasmonic NPs featuring multifarious structural configurations like islands-on-core [56], islands-on-shell [57], Janus-like [58], and so forth [42]. All the same, each of them exhibits its own unique LSPR behavior, denoting the role of structural configuration in controlling the LSPR phenomenon. Green for Ag NPs is observed as a result of interparticle plasmon coupling.

1.1.4.6 *Effect of interparticle coupling*

Another factor that strongly affects the oscillation frequency of surface plasmons in metallic NPs is the interparticle coupling in which following the approaching of individual plasmonic NPs to a distance below five times the NP radius, the plasmon resonances of individual particles couple to one another, giving rise to new plasmon modes depending on the size and geometry of the involved NPs. This phenomenon greatly resembles the dipole-dipole interaction, in which the interaction energy is proportional to the magnitudes of the dipole moments and the inverse cube of the distance [60]. This type of interparticle coupling is known as the near-field coupling in which the NPs' separation is comparable to the extent of the evanescent field at the particle surface and the Coulombic interaction between the NPs is the key determinant of this phenomenon [31]. In this regard, when a pair of metallic nano-spheres are placed in close proximity, the coupling of the lower-energy resonance correspond-ing to two longitudinally aligned dipoles engenders the strong red-shifted LSPR peak in the extinction spectrum. Furthermore, the coupled dipoles in higher energy resonances cancel each other, leading to resonance with a zero net dipole moment that does not appear in the extinction absorption. Meanwhile, depending on the NPs' size and the interparticle distance, higher-order multipoles can contribute to this interaction as well. After all, the magnitude of the plasmon coupling shift primarily depends on the interparticle gap, as well as particle geometry and the cluster size in aggregation systems [60]. The aggregation behavior of the spherical Au and Ag NPs is the best illustration of interparticle plasmon coupling (Fig. 1.7). In fact, upon the onset of aggregation, a dramatic color variation from red to blue for Au NPs

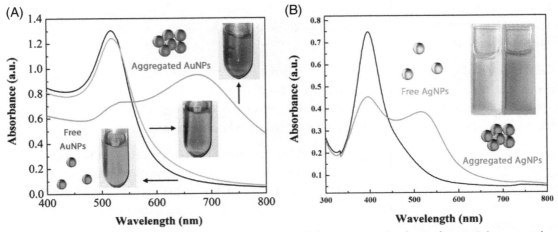

Fig. 1.7: (A) Extinction spectra and color variations of citrate-capped spherical Au NPs' aggregation upon addition of 4-mercaptophenylboronic acid [61], (B) Extinction spectra of citrate-capped Ag NPs' aggregation upon addition of Thiophanate-methyl [62].

and from yellow to brown to green for Ag NPs is observed as a result of interparticle plasmon coupling.

1.2 Surface functionalization of Au and Ag NPs

Bare nanoparticles (NPs) are rarely employed in applications. In fact, surface functionalization, also known as surface coating, is crucial for imbuing NPs with particular characteristics. Nevertheless, it is essential to carefully evaluate the potential effects of surface functionalization pertaining to their function [63].

The surface coating of NPs has a significant effect on the stability of the colloidal suspension, contributing to their aggregation state. In the absence of repulsive forces to counterbalance Van der Waals attraction forces, an unprotected colloidal solution of NPs would coagulate. Particle-particle interactions influences the optical and physical properties of plasmonic nano-materials. This phenomenon must be avoided for *in-vivo* diagnostic or therapeutic purposes, whereas it is desired for *in-vitro* quantification.

Surface functionalization is an indispensable task for specific target recognition. It is worth mentioning that surface properties of NPs influence their intrinsic binding affinity to proteins. Although formation of a protein shell on the surface of NPs improves their stability, it comes with a number of drawbacks. Due to the fact that proteins form a nonuniform layer on the surface of NPs; they interact proportionally regardless of their serum concentrations. In addition, the affinity of different proteins for binding to NPs varies, and in the case of weak interactions, the shell surrounding NPs will not have a stable composition, resulting in the exchange of surface ligands and rendering them unsuitable for use. The strong affinity of a few identical proteins, on the other hand, prevents interactions between proteins and surface cell receptors, resulting in unfavorable outcomes such as Cell activation or particle internalization. Furthermore, such protein-particle interactions can cause protein unfolding, which may alter their ability to bind with receptors.

The surface modification is also pivotal to elimination of NPs from the biological fluids. In this regard, rapid elimination rates are desired for diagnosis, while slow elimination rates are typical for therapeutic applications. In addition, surface coating enables NPs to remain invisible to the immune system, thereby preventing triggered immune reactions.

Surface coating with protective ligands would be a remedy to control NPs aggregation, target-specific interactions as well as their biocompatibility. Surface modifications are typically accomplished in two steps: First a linker molecule is attached to the surface, then a target molecule is bound. As the gold and silver NPs are the primary focus, their modification will be discussed in detail below.

As stated previously, surface modification aims to generate a protective coating to ensure the stability of AuNPs while achieving its mission pertaining to prophylaxis, detection and treatment [63,64]. Stabilization of NPs can be achieved through electrostatic sta-bilization or steric stabilization. However, covalently attaching desirable molecules to the outermost surface of water-soluble NPs is a viable way for adding functionality to the NPs.

In traditional AuNPs prepared by reducing aqueous $[AuCl_4]^-$ with sodium citrate, adsorbed citrate, chloride anions, and cations, on the surface of colloidal NPs generate an electrical double layer generated, creating coulombic repulsion between NPs. If the electric potential associated with the double layer is sufficiently high, then electrostatic repulsion will prevent particle aggregation. The AuNPs are electrostatically stabilized, which increases the possibil-ity of NP aggregation by raising the ionic strength of the dispersing medium. When adsorbed anions are replaced by a neutral adsorbate that has a higher binding affinity, the probability of agglomeration increases due to the influence of van der Waals forces. However, the lim-ited electrostatic stability of metal nanoparticles never prevents their substantial aggregation in actual complex matrices. In this regard, a significantly improved technique for preventing nanoparticles from aggregating is either based on steric effects and involves the addition of polymers and surfactants to the particles' surfaces, or makes use of covalent bonding on the NPs' surfaces. Traditional ligands in complex chemistry have been revealed to specifically stabilize metal nanoparticles. The ligand molecule and nanoparticle interact largely via cova-lent bonds. The functionalization of NPs with these ligands results in highly stable AuNPs that can be separated in solid form without aggregation. In addition, ligand exchange pro-cesses can be performed using ligands of differing strengths in terms of the ultimate solubility and chemical properties of NPs [65].

Electrostatic adsorption, chemisorption of thiol derivatives, covalent binding through bifunc-tional linkers, and particular affinity interactions have all been employed to manufacture functionalized NPs (Scheme 1.1). Applying such methods, variety of small biomolecules, proteins, nucleic acids, DNA, aptamers, enzymes, drugs, contrast agents, and other molecules can be conjugated with nanoparticles [66].

1.2.1 Noncovalent interactions

Electrostatic interactions, hydrogen bonding, hydrophobic and steric interactions have been extensively implemented to functionalize AuNPs and AgNPs a wide range of polymers, biomolecules, drug [67], carbohydrates, DNA, aptamers and proteins [66–71]. The preserva-tion of the biomolecule's original, active state is a great merit of this interaction because the capping agent, herein the biomolecule, is not subjected to harsh chemical alterations. The pH and ionic strength of the surrounding medium play critical roles in the application of this nanoparticle functionalization technique [72]. However, they are prone to disintegration from NPs either through successive washing steps or in the presence of more affine compounds to the surface of NPs.

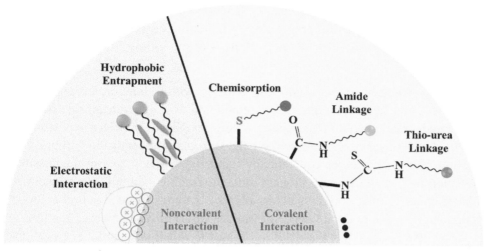

Scheme 1.1: Common functionalization strategies gold and silver NPs.

1.2.2 Covalent interactions

Regarding the synthesis route for preparation of NPs, there are variety of functional groups available on the surface of bare NPs, as stabilizing agent, namely thiols, carboxylic acids, amines, and hydroxyls.

Among them, thiol-containing compounds have exhibited a strong affinity for gold (126–146 KJ/mol) and silver nanoparticles. Due to strong electronegativity of silver and gold, these interactions are often considered covalent (1.93 and 2.4 on the Pauling scale for silver and gold respectively). Most compounds used to produce a protective monolayer over AuNP are thiol-modified. Sulfur atoms forms this bond with metals like Au and Ag by donating an electron pair to their empty shells at the interface. Formation of metal-S bonds has also been utilized for adhering oligonucleotides to the surfaces of NPs. Mirkin and Alivisatos independently introduced oligonucleotide-functionalized AuNPs in 1996, and this study has now been applied to numerous biomedical domains [73–75]. Some researchers have used the metal-S bond to attach targeting ligands like peptides to the surface of gold and silver NPs, making it easier for the particles to be taken up by cells. Cysteine residues in the peptides are responsible for the functionalization in this system [76,77].

Because direct binding of the target molecule via thiol bonding is not always achievable, nanoparticle conjugate systems often rely on bi-functionalized linkers to covalently attach biomolecules, mitigating the risk of structural alterations caused by direct interaction. This alternative strategy involves attaching the target molecule to the shell formed by the primary capping ligands like thiol-functionalized carboxylic acids (thioglycolic acid, 3-Mercaptopropionic acid, 11-Mercaptoubdecanoic acid and etc.). Such thiol-containing ligands often possess carboxylic groups that stabilize NPs through electrostatic repulsion of negative charges.

Furthermore, these groups can be used to attach other biologically relevant compounds. These modifications induce either NPs hydrophobicity or hydrophilicity for application in lipidic or aqueous media respectively while controlling particle reactivity and stability. Oligoethylene glycol, amine, carboxyl, isothiocyanate, and maleimide functional groups are frequently utilized as linkers in biomedical contexts as well.

The most popular protocol is cross-linking amines and carboxylic acids following carbodiimide activation, forming a covalent link between an amino group present on the biological molecule and a carboxylic group present on the shell, using a chemical reagent such as EDC (N-Ethyl-N'-(3-dimethylaminopropyl)-carbodiimide) (Fig. 1.8A). Even though this chemical reaction is rather well-understood, there are some issues associated with this strategy including side reactions due to bridging between particles, which leads to formation of aggregates,

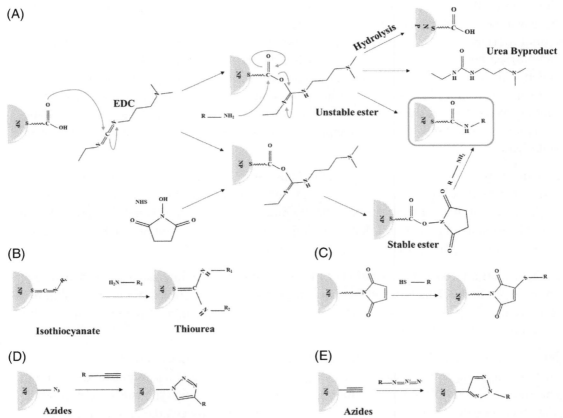

Fig. 1.8: (A) EDC carbodiimide activation chemistry in the presence and absence of NHS: EDC reacts with a carboxylic-acid group on NPs, forming an amine-reactive O-acyl iso-urea intermediate. This intermediate may react with an amine group of any target, yielding a conjugate (B) Isothiocyanate-amine coupling: thiourea functionalized NPs (C) Maleimide-thiol coupling click chemistry (D and E) cycloaddition reaction of azides and alkynes for functionalization of NPs.

fast hydrolysis, and low reaction yields. N-hydroxysuccinimide (NHS) mitigates a number of difficulties connected with this process (Fig. 1.8A). The creation of amide bonds is a way of conjugation; however, this conjugation process is unstable, allowing for minimal control over specificity, conjugation numbers, surface densities, and spatial orientation of the attached molecules [78]. More stable thio-urea linkages can be formed by the reaction of iso-thio cyanates with amines. In this case iso-thio cyanates are typically play the role of bifunctional chelator in the functionalization (Fig. 1.8B) [79].

Maleimide coupling with thiols is a more selective method that results in highly specific bonds at physiological pH (Fig. 1.8C) [80]. To bind nanoparticles to biomolecules, cycloaddition reactions involving azides and alkynes are often utilized (Fig. 1.8D and E). Using "click" chemistry and cycloaddition processes, a variety of strategies for generating enzyme-functionalized AuNP conjugates have been devised [81,82]. For NP functionalization, photoinitiated thiol-ene reactions—which can be thermal, redox, or photo-initiated—have also been studied. These reactions include the formation of an alkyl sulfide from a thiol (SH) and an alkene group [83]. Several biomolecules, including as amino acids [84–86], carbohydrates [87], and nucleic acids and proteins [88–91], have been employed in functionalization of NPs through carbodiimide activation and maleimide coupling. Because of their high affinity for the NPs' surfaces, these molecules are ideal candidates for direct binding to the NPs' surfaces via covalent or noncovalent interactions. They can simultaneously act as surfactants, catalysts, capping agents, and reducing agents in the formation and stabilization of NPs.

A wide range of biomolecules present in biomass or living organisms can also effectively function as both reducing and capping agents during NPs synthesis. Because of the biomolecules that coat their surface, the biologically produced nanoparticles are biocompatible and could be used in a number of different settings. Metal NPs have been biosynthesized using microorganisms, plant extracts, and even mammalian systems [92].

In the bacterium-assisted preparation of metal NPs enzymes, quinones, lactic acid, and sugars are involved in the reduction and functionalization of NPs. Functional groups present in bacterial cells such as $–NH_2$, $–OH$, $–SH$, and $–COOH$ perform a dual role in reduction as well as the stabilization of NPs [93].

Fungi-assisted metal NPs production and functionalization similarly rely heavily on enzymes and fungal biomolecules such as polysaccharides, peptides, amino acids, vitamins, enzymes, alkaloids, flavonoids, saponins, steroids, tannins, carboxylic acids, quinones, and other secondary metabolites [93]. Due to interactions between the amine group of proteins and their residual amino acids, controlling pH aids in capping and stabilizing the NPs. The blue-green algae and cyanobacteria pigments perform a dual role in the synthesis and stability of the NPs [94,95]. Fucoidan from brown algae has also been exploited in the synthesis of metal nanoparticles as a capping agent [96,97]. In the production of NPs, water-soluble polysaccharides derived from algal species also function as reducing and stabilizing agents [98–101].

Plant extracts contain a wide variety of metabolites, some of which may have a role in the production of nanomaterials as reducing or stabilizing agents. Many different types of NPs have been biologically synthesized using plant gums [102]. The functional groups present in plant extracts of Marchantia aid in stabilizing the surface of AgNPs. The stabilization of the AgNPs was accomplished primarily through the chemical functional groups of isothiocyanate (-N=C=S) and amide (-N=C=O) [103].

Animal systems such as wings of insects, honey, spider silk, and different components of chicken eggs have also been widely used in the biosynthesis of nanoparticles. Synthesis of AgNPs has been accomplished using the chitin-rich wings of an American cockroach [104]. Carbohydrate, vitamins, minerals, and a number of active substances found in honeybees, such as flavonoids, phenolic acids, tocopherols, catalase, superoxide dismutase, reduced glutathione, and a number of enzymes, all have the potential capability to contribute to the reduction and functionalization of NPs [105].

Paper wasps construct their paper-like nest out of wood that has been collected, chewed, and cemented together with saliva. Proteins found in paper wasp nests have the potential to be used as effective bio-reductants in the biofabrication of nanoparticles. The efficient production and stability of AgNPs is facilitated by the presence of proteins and a phenolic component in the extract [106]. The synthesis of NPs has also been reported by the use of cysteine-modified silk fibroin as a reducing and stabilizing agent. The COO^- group in silk sericin has crucial in the formation of AgNPs, and NH_2^+ stabilize the prepared NPs [107].

1.2.2.1 Determination of NP core/surface ligand stoichiometries

To ensure colloidal NPs' solubility in aqueous solutions and enable their future usage, the surface of the particles must be coated with the proper ligands. Identifying the ligand density, sometimes referred to as the level of functionalization or "coverage density," which eventually enables the calculation of core/ligand stoichiometry, is crucial and a contemporary chemistry issue. Sulfur is a common element found in the ligands that are employed for functionalization of AuNPs. Common methods for calculating the coverage density include X-ray crystallography, XPS, and surface plasmon resonance. When it comes to measuring Au/S ratios in the particle, molecular mass spectrometry (ICP-MS) is the superior and well-developed analytical method. Quantifying ligand densities and assessing the impact of ligand chain length on coverage density is now possible because of advancements in extended ICP-MS technology. As the length of the ligand chain increased, the ligand density decreased linearly. In addition to estimating the mean number of ligands bound to the surface of each AuNP, this method also provided an estimate of the total number of ligands present in a given volume of AuNP solution [108,109].

1.2.3 Polymer-functionalized nanoparticles

Polymer modification of NPs allows fine-tuning of the size, morphology, stability and assembly of NPs. Type of the polymer and NPs determine not only the synthesis route but the characteristics and potential applications of NPs. Prior to the advent of synthetic polymer chemistry, natural polymers such as gelatin and agar were widely used for preparation and functionalization of NPs. However, the wide variety of polymers available today, makes it possible to select a material with the optimized combination of properties for a given application. Characteristics such as porosity, wettability, chemical/biological stability, mechanical/thermal resistance, biocompatibility, and biodegradability can be adjusted to the needs of the application [110,111].

Several polymerization methods have and their potential for use in surface functionalization of NPs has been investigated [112–115]. For direct attachment of polymer, the reactive sites on the surface of the NP are necessary. Thus, thiol-functionalized polymers are recommended for the covalent binding to surface of metal NPs. Metal–carbon covalent bonds can also be used to attach organic groups to NPs surfaces [116–118]. Diffusion of additional polymer chains close to the free reaction sites on the NPs surface is, however, hindered by steric effects of already attached polymer chains. Thus, this method cannot ensure a high packing density of the polymers on the NPs surface. Polymer-stabilized NPs can be also prepared by the growth of bound initiators on the NPs surface. This method employs a variety of polymerization techniques, including cationic, anionic, ring-opening, and reversible deactivation radical polymerization (RDRP), in particular reversible addition fragmentation chain transfer (RAFT), nitroxide-mediated (NMP), or atom transfer radical polymerization (ATRP) [119–123]. The growth of initiators enables the formation of a densely packed surface of NPs and fabrication of multilayered, well-defined core shell NPs. In addition, the copolymerization of initiators with free monomers produces a homogeneous covering polymer, allowing the thickness of the polymer's surface layer to be adjusted [124–126]. Fig. 1.9 shows surface functionalization of metal NPs by variety of polymers through metal-sulfur, metal-carbon and metal-nitrogen bonds. Backbone of several amphiphilic polymer, as sell on the surface of NPs, contains carboxylic acid groups by which further functionalization of NPs can be achieved through EDC/NHS chemistry. Generally, regardless of the type, polymers reduce nonspecific adsorption of proteins and provides greater AuNPs stability by preventing interactions between particles by steric hindrance.

The ability of a polymer to stabilize NPs has been proposed to be quantified using a protective value. This is defined as This is the polymer mass required to stabilize 1 g of a standard red AuNPs containing 50 mg. L^{-1} gold at ionic strength of 1% sodium chloride solution [127].

Polyethylene glycol (PEG) is one of the popular surface agents that increases the colloidal stability of nanoparticles in biological settings. Furthermore, PEG serves as a versatile spacer for the bioconjugation of targeted ligands [128]. Poly (N-isopropylacrylamide) (PNIPAM) is

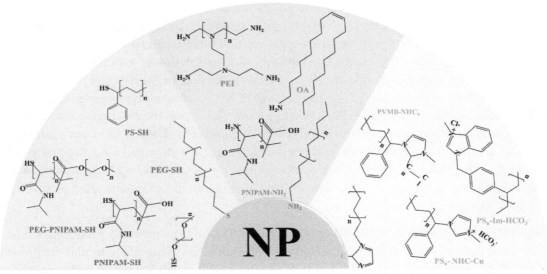

Fig. 1.9: Surface functionalization of NPs by polymers via metal-S (thiol-functionalized Polyethylene glycol (PEG-SH), Poly(N-isopropylacrylamide) (PNIPAM-SH), polystyrene (PS) and their combination), metal-N (amine-functionalized PNIPAM, polyethylene imine (PEI) and oleyl amine) and metal-C bonds (PS- N-heterocyclic carbene (PS-NHC), methylimidazolium bicarbonate terminated polystyrene (PS–Im HCO$_3^-$) and poly (vinyl benzyl N-methylbenzyl N-heterocyclic carbene) (PVMB-NHC).

another polymer which is utilized in surface modification of metal NPs. -SH, -NH$_2$, -COOH, and -H functional groups in PNIPAM allows further functionalization assembly of the PNIPAM-AuNPs [129]. Metal-C bond allows N-heterocyclic carbene (NHC) polymers to be employed as surface coating. Affinities between metals NPs and sulfur and those between NPs and carbon are distinct.in terms of length and energy of the bond. Compared to its equivalent Au-S bond, the Au-C bond is significantly stronger. Furthermore, binding through carbon has the additional benefit of facilitating a high electron transfer rate from ligand to metal via σ donation. Thus, the electron-rich surface contributes to an increase in NPs stability, which is further enhanced by the bond's inherent strength [130]. Surface functionalization of NPs has also made use of end-group functionalized polystyrene (PS) ligands [131]. Polymers such as polyethyleneimine (PEI), poly (vinyl pyrrolidone) (PVP), and poly (vinyl alcohol) (PVA) have been utilized in one-step preparation of metal NPs, silver in particular [132].

In order to lessen the toxicity of metal NPs in the biological medium without compromising their stability, biopolymers, a class of large biocompatible molecules, can be used in surface functionalization. Polysaccharides are among the most abundant biopolymers on the earth. Thanks to their high concentration of hydroxyl groups, they can be used as environmentally friendly "green templates" or "supporting materials" in the production of NPs. Polysaccharide-functionalized NPs are stable in water, and can easily be modified by chemical synthesis and physical adsorption of other species. Cellulose is the world's most common organic polymer and is made up of a linear chain of d-glucose units. Acting as mild reducing agent, it has been used for simultaneous in situ preparation and functionalization of metal NPs. The structure, impurities, and production pathway of several cellulose varieties define their distinct identities. Common forms of cellulose used to stabilize metal NPs and prevent them from agglomeration include cellulose nanocrystals (CNCs), cellulose nanofibrils (CNFs), and bacterial cellulose (BC) [133–137]. After glucose, chitosan is the most prevalent natural polysaccharide and is formed by deacetylation of chitin (Fig. 1.10A). Chitosan's unique qualities, including biocompatibility and nontoxicity, make it well-suited for a wide range of uses [138–140]. Similarly, there exist a number of other polysaccharides like hyaluronic acid, pectin, lignin and starch which have been utilized in preparation and modification of silver and gold NPs [141–148].

Poly (L-lysine) (PLL) (Fig. 1.10B) is another type of biopolymers naturally produced, cationic water-soluble and it is nontoxic to humans. Moreover, it is reported to inhibit the growth of a wide range of microorganisms, thus it has antimicrobial capability which make it a suitable coating for NPs in different applications. PLL in combination with several polymers like polyacrylic acid (PAA) (Fig. 1.10C), polydopamine (PDA) (Fig. 1.10D) and chitosan have been utilized for preparation of metal NPs [149–151].

Using long chain alkylammonium cations and surfactants, either in single-phase or in the reverse micelle synthesis of colloidal NPs, allows for the regulation of growth rate and stabilization of the generated NPs by a combination of electrostatic and steric effects. Direct gold nanorods generation with cationic alkyl ammonium surfactants like cetyltrimethylammonium bromide/chloride (CTAB/CTAC) has been acknowledged as an example of the

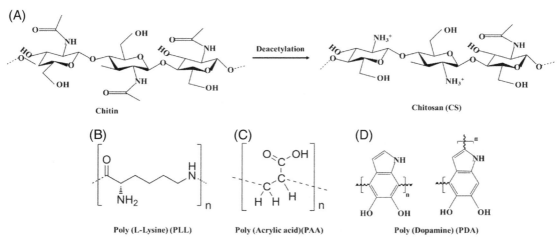

Fig. 1.10: (A) deacetylation of chitin and formation of chitosan. Chemical structure of (B) PLL, (C) PAA and (D) PDA biopolymers.

use of shape-inducing cationic surfactants [152–158]. It is worth to mention that the aspect ratio of the produced NPs can be controlled by adjusting the ratio of the surfactants utilized. Sodium dodecyl sulfate (SDS) [159], polysorbates [160–162], as reducing and stabilizing agents, have also been used in the synthesis of silver and gold NPs.

1.2.4 Silica-coated metal nanoparticles

Ancient and cutting-edge alike, noble metal/silica composites have been used to create works of art that have withstood the test of time and weather. Examples range from the Lycurgus cup, created in the fourth century, to more modest stained-glass windows whose vibrant colors have remained intact over the centuries. Many scalable synthetic approaches exist for preparing well-defined Au and Ag NPs; however, these methods are not optimal for many practical applications. Aggregation, etching, and other unintended consequences can be induced by the harsh conditions necessary for many applications, significantly undermining any attempts to adjust particle characteristics. Coating plasmonic particles with solid shells of a chemically inert material like silica has become an increasingly popular method for improving their manipulation and facilitating their integration into sensors, photovoltaics cells, catalysts, and other functional devices. The unusual stability of silica, especially in aqueous systems, is one of the main reasons for its usage as a coating material; other factors, like chemical inertness, controlled porosity, processability, and optical characteristics, also contribute. Low van der Waals interactions and significant affinity of positively charged species towards polymeric silicate layer at silica-water interface are responsible for silica's extraordinary durability. Thus, the NPs can benefit from the steric and electrostatic shielding provided by this silicate layer. In light of these benefits, silica is an excellent, low-cost material for modifying surface properties of NPs, allowing for the creation of lab-on-a-particle materials, and paving the way for the development of lab-on-a-particle NPs with tunable sizes and novel characteristics. Silica coating makes NPs biocompatible while enabling further functionalization. Multiple objectives can be achieved using silica shells, including stability of NPs, isolation of NPs from the surrounding media, modeling of novel structures, and regulation of molecule transport through the pores. There are two primary reasons attributing in silica shells' stabilizing function. The first is the accumulation of cations on the silica surface caused by the silanol groups, which creates a long-range electrostatic repulsion force between the particles. This permits the introduction of silica-coated NPs into aqueous or organic systems and has been demonstrated to effectively prevent irreversible aggregation even in liquids with extremely high ionic strengths. The second is the superior mechanical stability imposed by the coated shells, preventing morphological alterations during the formation of bimetallic NPs [163–165].

The study of metal/silica hybrids has lately seen a shift in focus due to recent breakthroughs in the science of mesoporous materials, to a high level of precision and reproducibility.

A redistribution of the electric near-field and a local rise in the refractive index result in a redshift of the LSPR when silica is used to coat plasmonic nanoparticles. Precise spacing is crucial for tuning the absorption and scattering properties of NPs due to interparticle coupling, and for regulating energy transfer from NPs to nearby fluorescent materials. In addition to metal-enhanced fluorescence, there is considerable interest in SERS-based approaches in which the core particles give near-field enhancement of molecular tags. The formation of silica shells enables the production of nanoprobes in which tags can be directly incorporated into the silica network [166,167]. In addition, mesoporous silica's high specific area is often used to boost loading with small signal molecules or medicines as well as serving as a molecular sieve to enhance the selectivity toward ions and small analyte molecules in complex biological media [168–171].

A number of classical and modern synthetic fabrication strategies have been reported for silica coating of colloidal NPs namely Stöbber synthesis, use of silane coupling agents, sodium silicate water-glass methodology, silica-coating in water-in-oil (W/O) microemulsions, silica coating of polymeric aggregates, surfactant vesicles, polymer/surfactant stabilized NPs, and assembly of silica colloids on nano- and microparticles by different physisorption strategies.

In the modified Stöbber method, condensation of silicic acids occur on NPs, as nucleation centers; tetraethyl orthosilicate (TEOS) hydrolyze and condensates in a alkaline alcohol/water mixture; forming a shell on NPs (Fig. 1.11A) [172]. This two-step process allows for a variety of morphologies of NPs while controlling shell thickness control by changing the seed/ TEOS ratio, but it typically necessitates prefunctionalization of the NPs surface with specific ligands This premodification minimizes the interfacial area between the two components which is energetically favorable in hetero NPs [173]. On the other hand, lack of colloidal stability in the conventional Stöbber reaction medium might cause aggregation problems, especially in the early stage of silica shell formation [174–177]. Coating nanoparticles with small, covalently binding molecules or polymers can mitigate these issues by reducing the interfacial energy between the metal and silica, hence increasing the nanoparticles' electro-steric stability. Amino- and mercapto- silanes, mercapto- carboxylic acids, and mercapto- sulfonate are examples of common low molecular weight surface primers [178–183].

Silane coupling plays an important role in the preparation of surface functionalization of substrates, such as glass and metals [184]. The general structure of silane coupling agents are $(RO)_3$-Si-R', where RO is a hydrolysable group such as an alkoxy group and R' is an organofunctional group. Due to hydrolysis and condensation of alkoxy groups on the surface, Silanization of the substrate takes place, providing a platform for further functionalization through its organofunctional groups. The most common type of Silane coupling agents are amino alkoxy silanes such as 3-aminopropyltrimethoxysilane (APTMS), 3-aminopropyltriethoxysilane (APTES), and 3-mercaptopropylntrimethoxysilane (MPTMS) by which further modifications can be done by a variety of coupling chemistry (Fig. 1.11B) [185].

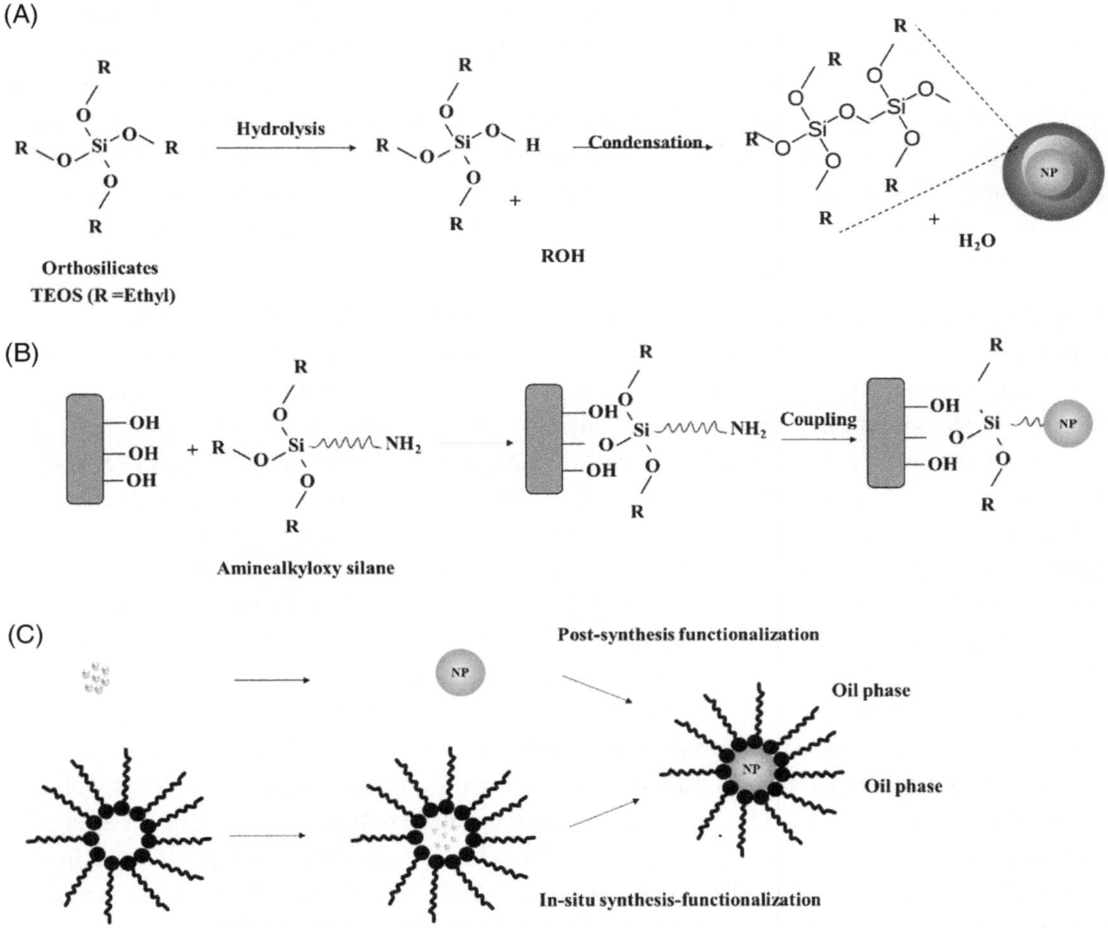

(A)

Orthosilicates
TEOS (R =Ethyl)

Hydrolysis

ROH

Condensation

H₂O

(B)

Aminealkyloxy silane

Coupling

(C)

Post-synthesis functionalization

Oil phase

Oil phase

In-situ synthesis-functionalization

Fig. 1.11: Silica-coating of NPs using (A) Stöber synthesis, (B) silane coupling agents, (C) water-in-oil (W/O) microemulsions.

Silica condensation in water-in-oil (W/O) microemulsions is an alternative to coating in typical Stöber medium. On a macroscopic scale, W/O microemulsions, also termed reverse microemulsions, are simply a homogenous combination of water, organic solvent (oil), and surfactant, but at the microscopic level, they are composed of distinct domains of water and organic solvent that are partitioned by a monolayer of surfactant. Microemulsions have been widely used as constrained reaction media due to the fact that one of these domains is typically present in the form of nanometer scale droplets. Due to their ability to efficiently dissolve inorganic salts and organometallic precursors, reverse microemulsions have proven to be ideal nanoreactors for the synthesis and silica coating of NPs. They provide sufficient flexibility to synthesize NPs with a wide range of sizes and morphologies. Numerous broad strategies for silica coating of NPs in W/O microemulsions have been reported so far [186]. One method entails coating presynthesized NPs with silica shells inside droplets, while another involves performing both the NPs synthesis and the Silanization process in situ (Fig. 1.11C). The production of NPs and shells involves different nucleation and growth pathways, making this approach sensitive to even subtle changes in processing conditions and making it challenging to regulate the relative sizes of the NPs and their corresponding silica shells. The coating uniformity, as well as the size and shape of in situ generated NPs, are all considerably influenced by the surfactant utilized in microemulsion design. Since W/O microemulsions allow for more regulation of silica nucleation than the stöbber technique, they are often used for silica coating [187,188].

Polymers and surfactants are excellent scaffolds for stabilizing NPs in the face of harsh conditions such as centrifugation and high ionic strength. This is owing to their capacity to form aggregates in aqueous solution and their propensity to adsorb at diverse surfaces and interfaces. The selection of an appropriate silica coating technique is primarily governed by the chemical affinity of the surface material for silica and the particle stability in the coating reaction media. In addition, it has been shown these surface components prevent chemical species from leaking out into the bulk solution, while protecting the NPs from etching [189–191]. While certain nanomaterials may be coated directly with silica, the majority of systems need the addition of surface primers [192]. Polymers and surfactants have been used as the "shape-inducing" agents in a number of recently published wet chemical techniques for the production of both spherical and anisotropic nanoparticles and nanocomposites [193]. Using the amphiphilic polymer poly(vinylpyrrolidone) is the most effective technique for coating silica with a monolayer of polymer (PVP). This polymer has been demonstrated to adsorb onto a broad range of NPs, ranging from spherical AuNP to Au nanoshells, as well as different sizes of AgNPs [191]. The molecular weight of the polymer has significant effects on colloidal stability and coating uniformity of NPs. Particle stability is improved by using longer PVP chains, the coating becomes less smooth though [194,195]. PVP capping is often used to make spiky and anisotropic Au and Ag NPs [196]. In addition to PVP, thiol-modified poly(ethylene glycol) and polyelectrolyte multilayers have been employed to stabilize metal NPs and silica, allowing for more complex particle morphologies [197].

Cooperative assembly of silica and AuNPs using lysine-cysteine di-block polypeptides produced strong, hollow microspheres. After combining a gold solution with the copolymer to produce inter- and intrachain disulfide and gold-thiolate linkages, the SiO_2 solution was added and the spheres were etched. Block copolymer length and polymer concentrations affected both the sphere size and the silica/Au shell thickness [198]. Functional composites may be produced by assembling small silica (or silica-coated) NPs into dense shell arrays around a central core. Au and Ag NPs have been functionalized via a variety of self-assembly processes, including layer-by-layer (LbL) wrapping. This technique may be used to control factors such as particle size, shape, content, silica wall thickness, and homogeneity.

Sequentially adsorbing polyelectrolytes with opposing charges formed a three-layer LbL film. Positively charged top layer would boost negative SiO_2 NPs adsorption. If three layers of

Fe_3O_4 nanoparticles were added between the silica NPs on either end, these nanocomposite colloidal particles might be magnetic [199–201].

Mesoporous silica is chosen over dense, compact shells for drug delivery goals and catalysis applications in which core NPs-solvent interactions are favored [202,203]. Postsynthetic etching in an alkaline media is one way to add porosity, but organic porogens like n-octadecyl trimethoxy silane and CTAB micelles may be included into the shell growth to create a porous template. All of these synthesis techniques share the need for precise regulation of the silica condensation kinetics, which is accomplished by manipulation of the reaction conditions such as pH, temperature, solvent composition, and precursor to nuclei ratio [204–206].

Shells dissolve at varying speeds in basic media or even in pure water due to the presence of silanol groups in the silica network. Temperature treatment, which might require heating the sample for several hours in the 100°C–600°C range or exposing it to reflux in boiling solvents, is a way to increase the stability of silica in water [207]. Temperature treatment, on the other hand, may make the silica shell more brittle and, in the case of mesoporous silica, can induce a reworking of the pore geometry. This process also results in the removal of surface silanol groups, which might not favorable for future silanization-based functionalization of the particles. Additionally, anisotropic nanoparticles may undergo reconfiguration if subjected to a significant temperature rise [208].

Preparation of complex hybrid NPs is another area of intense research and development. These NPs take advantage of silica shells' versatility by allowing for precise regulation of their thickness, porosity, and stability, thus facilitating the creation of new NPs with desirable characteristics. In order to create Au/SiO_2 Janus particles, 4-mercaptophenylacetic acid and polyacrylic acid-functionalized AuNPs were used as ligands. These Janus particles were used as seeds to construct ternary $Ag/Au/SiO_2$ structures during the reduction of $AgNO_3$ [209]. Using the pores of silica shells as templates for restricted metal deposition on lower length scales is a distinct synthetic route. This method was first developed for depositing noble metals into the pores of mesoporous silica materials [210].

1.3 Design principles of nanoplasmonic colorimetric assays

In designing and developing colorimetric plasmonic sensors, the most critical challenge is achieving an obvious color change strategy that depends on the target to convert the recognition events into an observable response [211,212]. As indicated above, plasmonic nanomaterials with extraordinary optical properties play a significant role in developing colorimetric strategies for visually detecting a wide range of chemical and biological species that can be observed by the unaided eye [213,214]. There are three main types of colorimetric strategies based on signal generation mechanism from plasmonic nanoparticles: (1) interparticle distance-dependent; (2) morphology/size-dependent colorimetric plasmon assay; and

(3) an ambient refractive index fluctuation. This section's objective is to describe the fundamentals of plasmonic colorimetric strategy for various sensor applications, beginning with distance-dependent, then morphology/size-dependent, and finally ambient refractive index variations. To clarify this, Scheme 1.2 provides diagrammatic representations of each tactic.

1.3.1 Interparticle plasmon coupling colorimetric assays

The unique LSPR properties of plasmonic NPs are highly dependent on the interparticle distance [215]. When analytes alter this parameter, the resonance frequency of the plasma shifts in response to changes in the surface's chemical and particle interactions; this change has an impact not only on the color of the solution but also on the wavelength of the absorption band [216, 217]. So, analyte concentrations can be measured with the naked eye in a semiquantitative way and with UV-vis spectra quantitatively at low sample volumes without requiring sophisticated sample pretreatment or instrumentation. When the distance between particles changes, there are different degrees of coupling between the plasmon fields of neighboring particles, which causes shifts in the SPR band and changes the color of the solution in an observable way. It might be a potential portable probe that can be used for colorimetric detection of different chemical and biological species [218–220]. Based on how the LSPR band changes with the distance between plasmonic nanoparticles, there are two primary approaches to designing nanoparticles for colorimetric detection: the first one is to aggregation and the second one is anti-aggregation (Scheme 1.3).

1.3.1.1 Aggregation

In analytical nanotechnology, the aggregation strategy is the most common method for developing plasmonic colorimetric sensors. In the presence of the analyte, groups of nanoparticles aggregated, resulting in a reduction in the distance between individual nanoparticles, which serves as the foundation for colorimetric assays. This reduction in the interparticle distance causes a large overlap between the plasmon fields of adjacent particles, resulting in new collective optical properties that are not observable in isolated particles. This causes the LSPR band to shift red and the solution's color to change dramatically in relation to the analyte concentration [221–223]. Indeed, when the distance between the neighboring nanoparticles is smaller than 2.5 times the size of the nanoparticles, electromagnetic coupling develops between the nanoparticles [224]. It also causes an LSPR band shift from the visible to the near-infrared area, as well as a color change for the Au nanosphere from red to purple/blue and for the Ag nanosphere from yellow to brown/red [225]. This method uses several different ways to aggregate NPs together, such as the immune-sandwich effect [226], hydrogen bonding [226], electrostatic interactions [227], DNA hybridization [228], and metal-ligand coordination [229]. To achieve the above purposes, plasmonic nanoparticles must be modified with suitable ligands interacting with analytes to make them easier to assemble and more selective [230–232]. For example, Taefi et al. demonstrated the

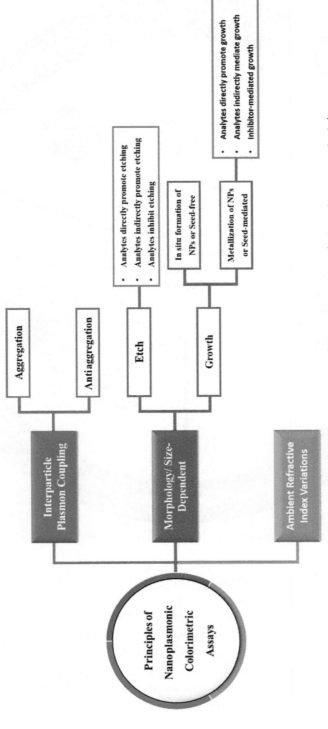

Scheme 1.2: The summary of different types of nanoplasmonic colorimetric sensor principles.

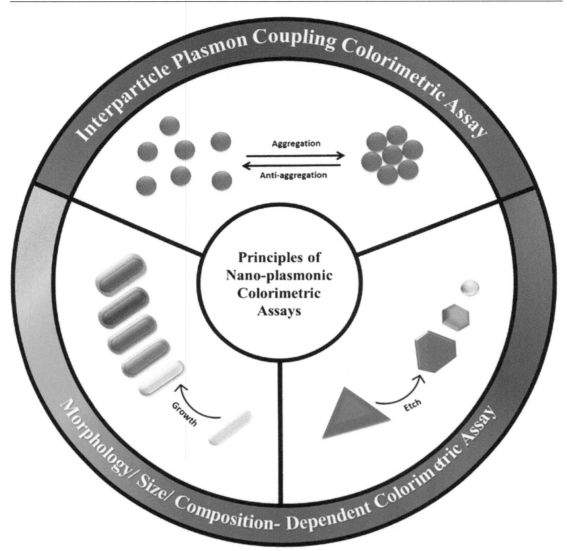

Scheme 1.3: Representation of plasmonic colorimetric strategies for various sensor applications.

colorimetric probe capable of directly monitoring pentaerythritol tetranitrate (PETN) as an explosive in an aqueous solution based on an aggregation strategy [233]. As a first step, the gold nanospheres (AuNPs) were exposed to arginine, a natural amino acid containing primary amines. Electron-deficient $-NH_2$ groups from arginine could strongly interact with the $-NO_2$ groups of PETN as electron donors. On the other hand, the $-NO_2$ group of PETN and the $-NH_2$ group of arginine molecules form hydrogen bonds. Therefore, selective aggregation of AuNPs occurred due to the donor-acceptor and hydrogen bonding interactions, resulting in an increase in absorbance at 650 nm and a decrease in absorbance at 520 nm as well as a change in color from reddish AuNPs turning blue or purple depending on the concentration of PETN, as depicted in Fig. 1.12A.

Despite the high sensitivity and flexibility of the aggregation-based colorimetric platforms, there are still some challenges, such as the lack of inexpensive ways to increase naked-eye inspection's effectiveness for ultrasensitive detection, because monitoring by the unaided eye in this sensor can usually notice a mono-color change at low color resolution. For instance, interparticle distance-dependent colorimetric sensors using AuNPs aggregation (or dispersion) commonly show a red-to-blue (or blue-to-red) single color fluctuation during various

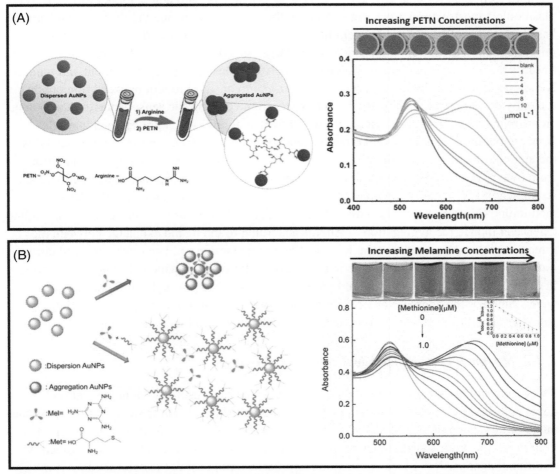

Fig. 1.12: (A) The PETN detection strategy by employing arginine-capped AuNPs. The absorption spectra of AuNPs and images of the probe, upon the addition of different concentrations of PETN, range from 1.0 to 10.0 μmol L^{-1}[1]. (B) A simplified graphic illustrating the Met assay's principle based on AuNP antiaggregation. Photographs and SPR spectrum of the Mel-AuNPs after incubation with different concentrations of Met [2].

sensing events that can be seen with the naked eye. Another issue is the "auto-aggregation" of nanoparticles, which is induced by nontarget-mediated parameters like environmental factors (e.g., temperature, pH, salt, and charged molecules), which can readily impact plasmonic nanoparticles. This phenomenon frequently results in false positives/ negatives and makes some detection events with less selectivity [234–236]. The second type of interparticle distance colorimetric probe, based on antiaggregation, has been developed to accomplish more accurate and precise detection results [237].

1.3.1.2   Antiaggregation

Another technique for changing the interparticle distance is "antiaggregation," which prevents plasmonic nanostructures from aggregating in the presence of an analyte. In antiaggregation-based sensors, an aggregation reagent first decreases the interparticle distance of NPs to promote their aggregation. Subsequently, the target analyte has a higher binding constant towards the aggregation reagent, which ultimately disables the aggregation agent and inhibits NP aggregation [238,239]. In most cases, multi-amino or -thiol functionalized compounds typically contribute to developing the antiaggregation colorimetric sensors due to their potent crosslinking and complexing capacities with target analytes, which prevent nanoparticles from aggregating together. Consequently, depending on the concentration of target analytes, the amount of LSPR peak red-shift decreases and the color varies in the opposite direction, from blue to red for Au nanospheres or from brown to yellow for Ag nanospheres [237,240]. As shown in Fig. 1.12B Zhang and coworkers developed a rapid method for the detection of methionine (Met) based on the antiaggregation of AuNPs in serum and urine [241]. Melamine (Mel) can induce the aggregation of AuNPs according to the ligand exchange between the three exocyclic amine groups (single bond NH_2) of melamine and negatively charged citrate ions and then induce a color change in the AuNPs solution from wine red to blue. However, in the presence of Met, AuNPs would prefer to bind with Met through Au-S and Au-N bands which suppress the aggregation of AuNPs; thus, Met concentration increases the solution color presented from blue to purple and finally retain the initial wine red.

1.3.2 Morphology/size-dependent colorimetric assays

Using morphology/size-dependent plasmonic colorimetric sensors as "nonaggregation" novel colorimetric platforms, great potential is shown for the colorimetric detection of a wide range of target analytes [242,243]. Nonaggregation colorimetric sensors have been reported to improve the color sensing readout by employing particular catalytic processes instead of traditional nanoparticle-based aggregation approaches. In general, the color readout is mostly dependent on the performance of the nonaggregation colorimetric sensor, which has evolved into two types of color readouts based on the naked eye's sensitivity to color changes. The first is for monocolorimetric sensors, where the visual color signal usually changes from light to dark. In contrast, the human eyes are not sensitive to changes in hue density, so it is impossible to distinguish between different samples. Multicolorimetric sensor is another type of sensor in which the color output typically fluctuates in the color tonality (rainbow presentation) to reveal brilliant color responses that are more desirable for visual recognition due to the great optical sensitivity of the naked eye [221,244]. Regarding the sensing principles for morphology/size-dependent mono and multicolorimetric sensors, there are different strategies to improve color readouts and achieve lower detection limits. Sensing principles are divided into two categories: (1) the target-induced principle is based on a direct electrochemical interaction between the target analyte and probe. (2) target-mediated principle, which indirectly operates on the target-mediated generation of a product such that the electrochemical reaction is between the product and probe indirectly [236,245]. As previously stated, the unique optical properties of plasmonic nanostructures are greatly influenced by their size, morphology, and chemical composition. Consequently, depending on how the analyte changes these parameters, both techniques of the sensing principles exhibit significant shifts in the plasmon peak and alteration of the color of light. In order to accomplish this objective, the size/morphology colorimetric nanosensors will be classified and reviewed as (i) plasmonic nanoparticle growth-dependent and (ii) plasmonic nanoparticle etching-dependent colorimetric sensors (Scheme 1.3).

1.3.2.1 Noble metal nanoparticle etch-based colorimetric sensors

On the other hand, etching refers to the redox reaction between etchants and metal nanoparticles, in which etchant, acting as an oxidizing agent, turns metal nanoparticle components (e.g., Au0) into metal ions (e.g., Au^{+1} or Au^{+3}), thereby changing the nanoparticle's structural properties. By controlling the etching process, it is possible to change the size, shape, and composition of nanoparticles while they are being etched with an etchant, and this results in a variety of LSPR that can be visible with the naked eye due to the recognition events caused by the target analyte [246–248]. By sensing strategy, etch-based plasmonic sensors can be divided into three primary categories: (1) Analytes directly promote etching, enabling the detection of analytes with a high redox potential or that function as complexing agents for noble metal ions (e.g., Au^+ and Ag^+). The simplest method is to use the analytes with powerful oxidative capabilities to etch gold or silver under specified conditions. Indeed, the high redox potential of Au and Ag makes them extremely stable and resistant to oxidation (etching). However, under certain conditions, such as the availability of suitable ligands to form the complex, the electrode potential of Au and Ag can be reduced, making them easier to oxidize and etching process faster. Lin et al., suggested a sensitive and high-resolution colorimetric approach based on a direct etching strategy for detecting benzoyl peroxide (BPO) utilizing orange sheath-like Au@Ag nanorods (NRs) as the signal amplifier and transducer-based on direct etching (Using morphology/size-dependent plasmonic colorimetric sensors as "nonaggregation" novel colorimetric platforms, great potential is shown for the colorimetric detection of a wide range of target analytes [242,243]. Nonaggregation colorimetric sensors have been reported to improve the color sensing readout by employing particular catalytic processes instead of traditional nanoparticle-based aggregation approaches. In general, the color readout is mostly dependent on the performance of the nonaggregation colorimetric sensor, which has evolved into two types of color readouts based on the naked eye's sensitivity to color changes. The first is for monocolorimetric sensors, where the visual color signal usually changes from light to dark. In contrast, the human eyes are not sensitive to changes in hue density, so it is impossible to distinguish between different samples. Multicolorimetric sensor is another type of sensor in which the color output typically fluctuates in the color tonality (rainbow presentation) to reveal brilliant color responses that are more desirable for visual recognition due to the great optical sensitivity of the naked eye [221,244]. Regarding the sensing principles for morphology/size-dependent mono and multicolorimetric sensors, there are different strategies to improve color readouts and achieve lower detection limits. Sensing principles are divided into two categories: (1) the target-induced principle is based on a direct electrochemical interaction between the target analyte and probe. (2) target-mediated principle, which indirectly operates on the target-mediated generation of a product such that the electrochemical reaction is between the product and probe indirectly [236,245]. As previously stated, the unique optical properties of plasmonic nanostructures are greatly influenced by their size, morphology, and chemical composition. Consequently, depending on how the analyte changes these parameters, both techniques of the sensing principles exhibit significant shifts in the plasmon peak and alteration of the color of light. In order to accomplish this objective, the size/morphology colorimetric nanosensors will be classified and reviewed as (i) plasmonic nanoparticle growth-dependent and (ii) plasmonic nanoparticle etching-dependent colorimetric sensors [249].

BPO may cause the oxidation of Ag nanoshells on the surface of Au@Ag NRs; the etching of Ag nanoshells alters the aspect ratio of the Au nanostructures and causes a red-shift of the longitudinal LSPR peak, resulting in a vivid multicolor transformation in under 2 minutes. In the presence of BPO, Ag nanoshells on the surface of Au@Ag NRs are oxidized to Ag+, the Ag nanoshells are etched, and the longitudinal LSPR peak is red-shifted, resulting in a multicolor alteration. The intensity of the color shift and the peak in the longitudinal LSPR is directly related to the BPO concentration; (2) analytes indirectly promote etching; in this technique, the target analyte can selectively promote the synthesis of etchant-acting products as reactive intermediates. Reactive products have a significant tendency to oxidize, which can induce etch of plasmonic nanoparticles, which can result in the etching of plasmonic nanoparticles, altering the geometry of the plasmonic nanomaterials as well as the LSPR peak and color of the detection probe. Zhang et al. proposed an indirect etching technique based on a Fenton-like reaction to etch gold nanorods (AuNRs) and applied it to the sensitive visual detection of Co^{2+} ions [250]. Co^{2+} ions produce superoxide radical ($O_2^{\bullet-}$) in the presence of bicarbonate (HCO_3^-) and hydrogen peroxide (H_2O_2) via a Fenton-like reaction. As the Co^{2+} concentration rises in the presence of $O_2^{\bullet-}$ and SCN^-, the AuNRs are etched, resulting in a shift in the LSPR peak of the GNRs to a shorter wavelength and a dramatic change in color from blue to red (Fig. 1.13); (3) analytes inhibit etching, in which some analytes protect metal nanoparticles from etching by operating as "protective agents," inhibiting etching and so maintaining the shape of the original material. The etching process can be avoided using one of two general

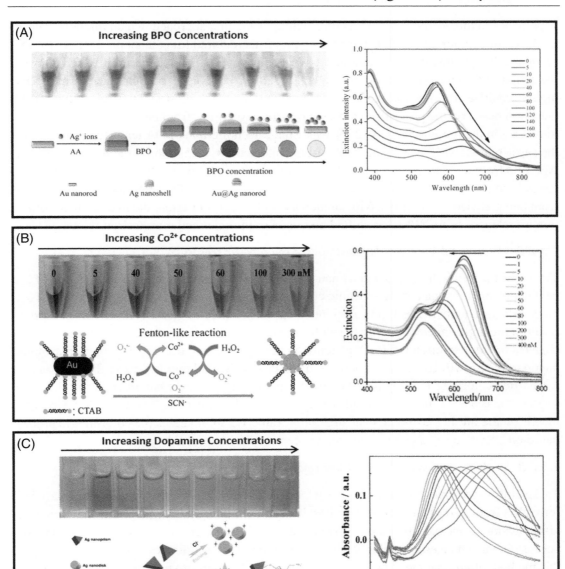

Fig. 1.13: (A) The colorimetric approach for detecting BPO by utilizing Au@AgNRs. UV-vis absorption spectra and the photos of Au@AgNRs in the presence of different concentrations of BPO [3]. (B) Visual Detection of Co2+ based on AuNRs etching. UV extinction spectra and color change pictures of AuNRs in the presence of different concentrations of Co2+[4]. (C) The principle of colorimetric detection of DA is based on AgNPRs etching prevention. Corresponding images and extinction spectra of the AgNPRs incubated with Cl− in the presence of DA at different concentrations [5].

approaches in this method: (a) preventing etching is the first approach, which might be caused by the disintegration of the etchant or the components required for etchant production; (b) the second technique involves blocking the active etching sites at the nanoparticle, which allows the action of the etchant to be circumvented [217,242,251]. Xian et al. fabricated a simple and highly sensitive colorimetric platform based on an inhibit etching strategy for dopamine (DA) detection based on the protection of Ag nanoprisms (AgNPRs) from being etched by halide ions (Cl⁻) in the presence of DA [252]. Since the catechol group of DA tends to adsorb onto the surface of AgNPRs via chemisorption-type interactions, it can inhibit the form evolution of AgNPRs from nanoprisms to nanodisks. As a result, the stability of silver atoms at corners and edges is enhanced by DA, which works as a protective agent for the AgNPRs, preventing them from being etched by chloride. With an increase in DA concentration, the maximum absorption wavelength of AgNPRs has shifted because DA provides better protection (Fig. 1.13).

1.3.2.2 Noble metal nanoparticle growth-based colorimetric sensors

Typically, the "growth" of plasmonic nanoparticles means the formation or remodeling of new nanostructures via a target-specific chemical interaction involving noble metal ions as precursors of nanoparticles [253]. Both newly proposed growth-based colorimetric sensors result in the appearance of new or the shift of existing LSPR peaks that are responsive to changes in color solution [254]. Depending on the signal generation process, these plasmonic nanoparticle growth-based colorimetric sensors can be divided into two parts. The first one is a seed-free colorimetric assay based on the formation of plasmonic nanoparticles in situ. The second one is a seed-mediated colorimetric assay based on the metallization or deposition of metals on the initial plasmonic nanostructures. Table 1.3 summarizes the colorimetric probes based on growth of NPs either seed-free or seed-mediated approaches

1.3.2.3 In situ formation of NPs/seed-free colorimetric assay

To develop a seed-free, colorimetric sensor capable of modulating direct signals, nanoparticle formation from a solution system is utilized in situ. Even though this type of monocolorimetric visual readout is inexpensive, simple to operate, sensitive enough, and stable, it can be limited because it only changes one color, leading to diminished recognition resolution. By applying this method, plasmonic metal nanoparticles can be produced in situ by some targets with a reduction capability (e.g., a reductant) via direct target-induced or indirect target-mediated principles that can react with metal ions as precursors (e.g., gold and silver ions), resulting in abrupt color changes from colorless to one color [255,256]. The fact that it only changes one color leads to diminished recognition resolution. The target-induced strategy was the most common method for fabricating this monocolorimetric sensor. The targets function as the reducing agents in a specific electrochemical reaction to form plasmonic nanoparticles from the metal ion precursors. Recent research indicates that seed-free growth-based monocolorimetric sensors relying on target-mediated methods will be challenging to construct. As a result, they play a crucial role in practical applications requiring high selectivity and sensitivity [257,258]. As illustrated

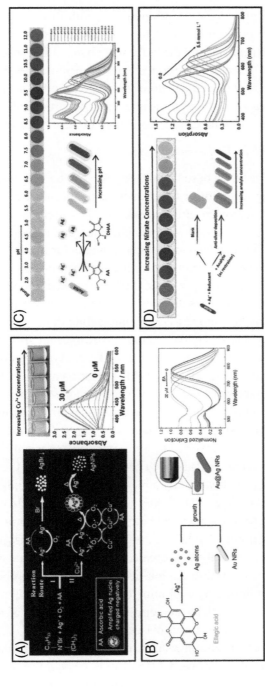

Fig. 1.14: (A) Principle for detecting Cu^{2+} through the catalyzed production of AgNPs. Absorption spectra and photos of AgNPs were obtained from various concentrations of Cu^{2+} standards [259]. (B) Diagram of the EA detection technique based on seed-mediated growth of Au@Ag core-shell nanorods. Normalized extinction spectra of AuNRs or Au@Ag core-shell NRs show the continuous blue shift of the LSPR peak when the concentration of EA increases [260]. (C) A schematic representation of the proposed multicolorimetric pH sensor's operating principle. Absorbance spectra and the images of the developed AuNRs-based colorimetric sensor in various pH levels (2.0–12.0) [261]. (D) The proposed strategy through the analyte prevents silver deposition. Absorption spectra and the corresponding images of the multicolor probe upon exposure to different concentrations of nitrate [262].

in Fig. 1.14, Yuan and coworkers developed an in-situ formation of NPs or seed-free colorimetric assay for Cu^{2+} detection by utilizing copper-catalyzed in situ synthesis of Ag NPs [259]. A milk-white suspension (AgBr) can be prepared by combining Ag^+ with Br in an aqueous solution of $AgNO_3$, cetyltrimethylammonium bromide, ascorbic acid, and bovine serum albumin. Upon adding Cu^{2+}, the reaction will terminate, and the color will change from milk-white to yellow or orange. In this way, Cu^+, as the reduction product Cu^{2+}, consumes the dissolved O_2 and inhibits O_2 from oxidizing the newly reduced Ag atoms (through ascorbic acid) back to Ag^+, hence allowing the further aggregation of Ag atoms into Ag nanoparticles. It was demonstrated that the visible color change was unique to Cu^{2+} compared to most other metal ions.

1.3.2.4 Metallization of NPs/seed-mediated colorimetric assay

On the contrary, the "seed-mediated" mode is based on the deposition of a new metal layer on the original seed substrate by reducing metal ions as precursors in the presence of the target analyte, which results in color and extinction spectrum alterations. LSPR characteristics can also be changed by employing various precursors, seeding substrates, and reducing or capping agents to create new nanostructures. This can lead to multiple visual color readouts, such as monocolorimetric assays that only change one color and multicolorimetric assays that change a rainbow of colors [236]. According to this principle, three different types of metallization-based plasmonic sensors can be recognized: (1) analytes directly promote growth in which some analytes act as reductants to produce metal nanoparticles by interacting directly with $AuCl_4^-$ and Ag^+ because of their reducing characteristics. For example, if AuNPs are used as seeds in a reaction with $AuCl_4^-$ or Ag^+ and a reducing agent, they will make gold or silver nanoshells on the surface of the AuNPs, which will change the color of the solution and the LSPR peak. Wang and coworkers developed a straightforward and sensitive technique for the detection of ellagic acid (EA) based on the direct growth of AuNRs as an optical probe to bimetallic nanoparticles with a core-shell structure (Au@Ag core-shell NRs) [260]. With the assistance of EA in an alkaline solution, Ag^+ ions can be converted to Ag atoms and deposited on the surfaces of Au nanorods (AuNRs, which act as seeds here). This process results in the generation of Au@Ag core-shell nanorods and is accompanied by a blue shift of the longitudinal LSPR band of AuNRs from the near-infrared region to the shorter wavelength (Fig. 1.14); (2) analytes indirectly mediate growth, and some enzymes can accelerate the conversion of nonreducing substrates into products with reducing ability. Under the catalysis reaction of enzymes, analytes generate a related product with a potent capacity to decrease $AuCl_4^-$ or Ag^+ as a nanoshell on the surface of nanostructures. Orouji et al. produced a multicolor colorimetric probe based on an indirect growth strategy for pH monitoring by exploiting the silver metallization on the surface of AuNRs (Fig. 1.14) [261]. Silver nanoshells are deposited on the surface of AuNRs when silver ions are reduced by ascorbic acid, which is very pH-dependent. Thus, the creation of Au@Ag core-shell NRs results in a sequence of blue shifts in the longitudinal peaks of AuNRs and sophisticated color changes at pH values between 2.0 and 12.0. Changes in the spectrum made it possible to see different colors, such as brownish pink/light green (acidic media), dark green/blue (neutral media), and purple/brown (basic media); (3) inhibitor-mediated growth, which some analytes and enzymes can catalyze reactions that decrease the growth rate of nanoshells on the surface of nanostructures by producing inhibitors that hinder the metallization of nanoparticles or directly degrading the reductant [217,242,245]. Naseri et al., designed the anti-silver deposition mechanism based on an inhibiting growth strategy in which the analyte directly or indirectly inhibits the silver coating of AuNRs to detect nitrate ions [262]. Being pretreated with the nitrate (as a restrainer agent) with ascorbic acid (a reducing agent) caused a decrease in the spectral blueshift of AuNRs, resulting in a wide range of vibrant colors ranging from reddish-orange (blank) to maroon, wine, berry/purple, dark blue, teal, green, seafoam, and mint. Therefore, the restrainer agent prevents silver deposition by acting as an oxidizer, either by reducing silver ions or by oxidizing the reducing agent (Fig. 1.14).

1.3.3 Ambient refractive index variations

The intensity and peak placement of the absorption spectra are strongly dependent on the shape, size, and composition of the plasmonic nanoparticle as well as the refractive index of the surrounding medium [263]. The other sensor type is based on the LSPR's extreme sensitivity to fluctuations in the dielectric constant of the surrounding medium or adsorbents that provide the basis for colorimetric plasmonic sensors [244]. In this sensor, plasmonic nanoparticles also function as transducers by converting minute variations in the local refractive index into LSPR spectrum shifts. Therefore, chemical reactions around metal nanostructures can explain the redshift in the LSPR spectrum [264,265]. Theoretical models and experimental data suggest that anisotropic nanostructures are more sensitive to changes in the refractive index than spherical nanoparticles [266]. Unlike AuNPs, AuNRs are extremely sensitive to changes in the bulk refractive index (AuNRs have a refractive index sensitivity of 250 nm per unit, while AuNPs have 60 nm per unit) [19,267]. The spectrum shift caused by the refractive index change at the surface of the nanoparticles has drawn attention to the capacity of LSPR sensors to detect biomolecular binding events. There is a close correlation between LSPR sensor sensitivity and plasmon properties like spectra linewidth, extinction intensity, electromagnetic field strength, and decay length [268]. The effective sensing zone of LSPR nanosensors is very limited (within tens of nanometers) since LSPR decays exponentially with distance from the nanoparticle surface, in contrast to the propagating SPR from a thin noble metal sheet, which demonstrates a long detecting distance (in microns) [269,270]. Hence, the spectral response of LSPR is sensitive only to the proximity of the environment surrounding metallic nanoparticles, making it well suited for the real-time monitoring of local refractive index changes induced by biomolecular binding events close to the nanoparticle surface [271]. Plasmonic nanosensors based on the refractive index are generally less sensitive than the other two methods. However, they can be effectively miniaturized and multiplexed, reach the limits of a single nanoparticle, reproduce the biological receptors, and work with microfluidics and in-flow experiments [244,265].

1.4 Conclusions

There is an exponential growth in the analytical applications of plasmonic Au and Ag NPs due to their unique optical and surface properties, facile synthesis, easy surface modification via noncovalent/covalent interactions and high colloidal stability. Because of the plasmonic effect with high extinction coefficient, the Au and Ag NPs are extensively applied for colorimetric assays, surface enhanced Raman spectroscopy (SERS), photocatalysis, drug delivery and biomedical applications. This chapter discussed the basics of LSPR phenomenon of Au and Ag NPs along with the different key factors that influenced the LSPR behavior of plasmonic nanostructures. The intense color and fascinating size/shape-dependent color change of functionalized Au and Ag NPs prompted the designing of numerous colorimetric assays for environmental monitoring, food quality control, and clinical and biomedical investigation. This chapter provides a details discussion on the different approaches adopted for the designing of colorimetric assays using plasmonic Au and Ag NPs, such as analyte-induced aggregation and antiaggregation of NPs, etching of NPs, analyte directed growth of NPs, morphology change of NPs etc. with suitable examples (Table 1.1 and Table 1.2). We believe this chapter will open new ideas for future research with the plasmonic Au and Ag NPs for advance analytical applications.

Table 1.1: Plasmonic metal nanoparticle based interparticle plasmon coupling colorimetric assays.

NPs	Target	Mechanism	Color change	Detection Limit	Real sample	Refs.
AuNPs	Microcystin-LR	Antiaggregation	Blue to red	0.05 nM	Water	[272]
AuNPs	Arsenic	Aggregation	Pink to red	2 $\mu g\,L^{-1}$	Water	[273]
AuNPs	Proteins-vascular endothelial growth factor (VEGF)	Antiaggregation	Purple to red	0.1 nM	-	[274]
AuNPs	Malathion	Antiaggregation	Gray to red	11.8 nM	Tap water Apple Vegetables	[237]
AuNP/CDs	Hg^{2+}	Aggregation	Wine red to blue	7.5 nM	Tap water Lake water	[275]
AuNPs	Al^{3+} Zn^{2+}	Aggregation	Wine red, purple to blue	0.51 μM 0.74 μM	Water samples	[276]
AuNPs	Co^{2+} Cr^{3+}	Aggregation	Red to blue	83.22 nM 108 nM	-	[277]
AuNPs	Cr^{3+} $Cr_2O_7^{2-}$	Aggregation	Red to blue	0.03 μM 0.07 μM	Mineral water	[278]
AgNPs	Pb^{2+}	Aggregation	-	0.64 μg/L	-	[279]
AgNPs	As^{3+}	Aggregation	Pink to blue	4 μgL^{-1}	Water Soil	[280]
Au/Ag alloy nanocomposites (Au/Ag NCs)	Hg^{2+}	Aggregation	Orange-yellow to yellowish green	4.8 nM	Tap waters Lake waters	[281]

NPs	Target	Mechanism	Color change	Detection Limit	Real sample	Refs.
Au/Ag alloy NPs (Au/AgNPs)	β-agonist ractopamine	Aggregation	Orange-yellow to green	1.5 ng/mL	Pork, swine feed and swine urine samples	[282]
gold–silver nanoparticles	Cd^{2+}	Aggregation	Orange–yellow to green	44 nM	Tap water Lake water	[283]
AuNPs	Methionine	Anti-aggregation	Blue, purple to wine red	24.5 nM	Human serum Human urine	[241]

Table 1.2: Noble metal nanoparticle etch-based colorimetric sensors.

NPs	Target	Mechanism	Color change	DL/Linear range	Real sample	Refs.
Au@Ag nanorods	Benzoyl peroxide	Direct Etching	Brown, green, plum to red	0.75 μM	Flour	[249]
AuNRs	Glucose	Indirect Etching	Red to blue	3.0 μM	Human serum	[284]
AgNPrs	Cl^-	Direct Etching	Dark-violet to red	1.3 mg L^{-1}	Tap and ground water	
Ag@Au core/shell triangular nanoplates (TNPs)	Glucose	Indirect Etching	Blue to colorless	800 μmol/L	Human urine	[285]
Au@Ag nanocages	Hg^{2+}	Direct Etching	Blue, purple to brown	10 nM	Lake and tap water	[286]
Gold nanodouble cone @ silver nanorods	Cr^{6+}	Direct Etching	Orange, pink, purple, to colorless	1.69 μM	Tap water and River water	[287]
Triangular gold nanoplate (AuNPL)	Cu^{2+} Hg^{2+}	Indirect Etching	Blue to pink	19 nM 9 nM	Tap and Pond water and food	[288]
AuNPL	Sulfide ions	Inhibition of Etching	Pink, red, purple to blue	0.1 μM.	River waters	[289]
AuNRs	CN^-	Direct Etching	Blue to pink	0.5 nM	Tap, pond, and waste water	[290]
AgNPrs	Homocysteine Cysteine Glutathione	Inhibition of Etching	Yellow to blue purple	0.041 μM 0.079 μM 0.086 μM	Human serum	[291]
TAgNPs	Hg^{2+}	Inhibition of Etching	Yellow, brown, purple to blue	0.35 nM	River water	[292]
AgNPRs	Dopamine	Inhibition of Etching	Blue, purple, red to yellow	0.16 nM	Serum	[252]
AgNPRs	Captopril	Inhibition of Etching	Yellow, orange, red, Purple, to blue	2 nM	Tablets	[293]
AgNPs	Uric acid	Inhibition of Etching	Brown to yellow	30 pM	Serum	[294]

Table 1.3: Noble metal nanoparticle growth-based colorimetric sensors.

NPs	Target	Mechanism	Color change	DL/Linear range	Real sample	Refs.
AuNRs	p-aminophenol	Direct growth	Peach, orange, yellow, green, blue to violet	0.64 µM	Tap water River water Human urine	[295]
AuNPs	IgG	Indirect Growth	Light pink to deep brown	1.4×10^{-13} M	-	[296]
AuNRs	Ellagic acid	Direct growth	Purplish red to green	40 nM	Whitening cosmetics	[260]
Au nanocages	Vitamin C	Direct growth	m blue to purplish red	24 nM	Pharmaceutical products	[297]
AgNPs	p-aminophenol	Direct growth	Yellow to brown	0.32 µM	Urine and paracetamol samples	[298]
AuNRs	Prostate-specific antigen (PSA)	Indirect Growth	Colorless, purple, blue, gray to red	3×10^{-15} g.mL^{-1}	Human serum	[299]
AuNRs	beta-galactosidase (β -gal) activity	Indirect Growth	Green to orange-red	128 pM	E. coli bacteria cells	[300]
AuNRs	Formaldehyde	Direct Growth	Pink, green to blue	6.3×10^{-11} gmL^{-1}	Tap and River samples	[301]
AuNRs	Uric acid	Direct Growth	Purple to green	0.065 µM	Human plasma and urine samples	[302]
AuNRs	Omethoate	Inhibit Growth	Red, brown, red, violet, gray, green, to pink	0.53 U/L	Food safety	[303]
AuNR	Biogenic amines-Trimethylamine (TMA)	Indirect Growth	Pink, pale yellow, green, purple to brick red	8.6×10^{-9} mol dm^{-3}	-	[304]

References

[1] L. Tonks, I. Langmuir, Note on "oscillations in ionized gases", Phys. Rev. 33 (1929) 990–990.
[2] D. Pines, D. Bohm, A collective description of electron interactions: II. Collective vs individual particle aspects of the interactions, Phys. Rev. 85 (1952) 338–353.
[3] S.A. Maier, Plasmonics: Fundamentals and Applications, Springer, Berlin, 2007.
[4] Handbook of Surface Plasmon Resonance, 2017.
[5] V. Klimov, Nanoplasmonics, CRC press, Boca Raton, 2014.
[6] K.M. Mayer, J.H. Hafner, Localized surface plasmon resonance sensors, Chem. Rev. 111 (2011) 3828–3857.
[7] L. De Sio, Active plasmonic nanomaterials, 2015.
[8] A.M. Sitarski, Development of Spectroscopic Methods for Dynamic Cellular Level Study of Biochemical Kinetics and Disease Progression, (2017) 242–242.

[9] P. Drude, Zur Elektronentheorie der Metalle, Ann. Phys. 306 (1900) 566–613.

[10] L. Wang, M. Hasanzadeh Kafshgari, M. Meunier, Optical properties and applications of plasmonic-metal nanoparticles, Adv. Funct. Mater. 2005400 (2020) 1–28.

[11] G. Mie, Beiträge zur Optik trüber Medien, speziell kolloidaler Metallösungen, Ann. Phys. 330 (1908) 377–445.

[12] C.F. Bohren, Absorption and Scattering of Light by Small Particles, John Wiley & Sons, New York, 1983.

[13] L.J.E. Anderson, K.M. Mayer, R.D. Fraleigh, Y. Yang, S. Lee, J.H. Hafner, Quantitative measurements of individual gold nanoparticle scattering cross sections, J. Phys. Chem. C 114 (2010) 11127–11132.

[14] R. Gans, Über die Form ultramikroskopischer Silberteilchen, Ann. Phys. 352 (1915) 270–284.

[15] S. Eustis, M.A. El-Sayed, Determination of the aspect ratio statistical distribution of gold nanorods in solution from a theoretical fit of the observed inhomogeneously broadened longitudinal plasmon resonance absorption spectrum, J. Appl. Phys. 100 (2006) 44324–44324.

[16] S. Link, M.B. Mohamed, M.A. El-Sayed, Simulation of the optical absorption spectra of gold nanorods as a function of their aspect ratio and the effect of the medium dielectric constant, J. Phys. Chem. B 103 (1999) 3073–3077.

[17] J. Zhao, A.O. Pinchuk, J.M. McMahon, S. Li, L.K. Ausman, A.L. Atkinson, G.C. Schatz, Methods for describing the electromagnetic properties of silver and gold nanoparticles, Acc. Chem. Res. 41 (2008) 1710–1720.

[18] T. Maurer, P.M. Adam, G. Lévêque, Coupling between plasmonic films and nanostructures: From basics to applications, Nanophotonics. 4 (2015) 363–382.

[19] H. Chen, L. Shao, Q. Li, J. Wang, Gold nanorods and their plasmonic properties, Chem. Soc. Rev. 42 (2013) 2679–2724.

[20] T.R. Jensen, M.L. Duval, K.L. Kelly, A.A. Lazarides, G.C. Schatz, R.P. Van Duyne, Nanosphere lithography: Effect of the external dielectric medium on the surface plasmon resonance spectrum of a periodic array of silver nanoparticles, J. Phys. Chem. B 103 (1999) 9846–9853.

[21] X. Luo, L. Qiao, Z. Xia, J. Yu, X. Wang, J. Huang, C. Shu, C. Wu, Y. He, Shape- and size-dependent refractive index sensing and SERS performance of gold nanoplates, Langmuir. 38 (2022) 6454–6463.

[22] E.J. Zeman, G.C. Schatz, An accurate electromagnetic theory study of surface enhancement factors for Ag, Au, Cu, Li, Na, Al, Ga, In, Zn, and Cd, J. Phys. Chem. 91 (1987) 634–643.

[23] M. Rycenga, C.M. Cobley, J. Zeng, W. Li, C.H. Moran, Q. Zhang, D. Qin, Y. Xia, Controlling the synthesis and assembly of silver nanostructures for plasmonic applications, Chem. Rev. 111 (2011) 3669–3712.

[24] S. Kim, J.M. Kim, J.E. Park, J.M. Nam, Nonnoble-metal-based plasmonic nanomaterials: Recent advances and future perspectives, Adv. Mater. 30 (2018) 1704528–1704528.

[25] M.B. Cortie, A.M. McDonagh, Synthesis and optical properties of hybrid and alloy plasmonic nanoparticles, Chem. Rev. 111 (2011) 3713–3735.

[26] F. Tam, A.L. Chen, J. Kundu, H. Wang, N.J. Halas, Mesoscopic nanoshells: Geometry-dependent plasmon resonances beyond the quasistatic limit, J. Chem. Phys. 127 (2007) 204703–204703.

[27] D. Rioux, S. Vallières, S. Besner, P. Muñoz, E. Mazur, M. Meunier, An analytic model for the dielectric function of Au, Ag, and their Alloys, Adv. Opt. Mater. 2 (2014) 176–182.

[28] S. Agnihotri, S. Mukherji, S. Mukherji, Size-controlled silver nanoparticles synthesized over the range 5–100 nm using the same protocol and their antibacterial efficacy, RSC Adv. 4 (2014) 3974–3983.

[29] P.N. Njoki, I.I.S. Lim, D. Mott, H.Y. Park, B. Khan, S. Mishra, R. Sujakumar, J. Luo, C.J. Zhong, Size correlation of optical and spectroscopic properties for gold nanoparticles, J. Phys. Chem. C 111 (2007) 14664–14669.

[30] J.J. Mock, M. Barbic, D.R. Smith, D.A. Schultz, S. Schultz, Shape effects in plasmon resonance of individual colloidal silver nanoparticles, J. Chem. Phys. 116 (2002) 6755–6759.

[31] M. Pelton, G.W. Bryant, Introduction to Metal-Nanoparticle Plasmonics, John Wiley & Sons, Hoboken, New Jersey, 2013.

[32] L. Hung, S.Y. Lee, O. McGovern, O. Rabin, I. Mayergoyz, Calculation and measurement of radiation corrections for plasmon resonances in nanoparticles, Phys. Rev. B - Condensed Matter Mater Phys 88 (2013) 1–9.

[33] M. Alsawafta, M. Wahbeh, V.V. Truong, Simulated optical properties of gold nanocubes and nanobars by discrete dipole approximation, J. Nanomater 2012 (2012) 1–8.

[34] J.E. Park, Y. Lee, J.M. Nam, Precisely shaped, uniformly formed gold nanocubes with ultrahigh reproducibility in single-particle scattering and surface-enhanced raman scattering, Nano Lett. 18 (2018) 6475–6482.

[35] C. Wu, X. Zhou, J. Wei, Localized surface plasmon resonance of silver nanotriangles synthesized by a versatile solution reaction, Nanoscale Res. Lett. 10 (2015) 1–6.

[36] P. Yang, H. Portalès, M.P. Pileni, Identification of multipolar surface plasmon resonances in triangular silver nanoprisms with very high aspect ratios using the DDA method, J. Phys. Chem. C 113 (2009) 11597–11604.

[37] Y. He, G. Shi, Surface plasmon resonances of silver triangle nanoplates: Graphic assignments of resonance modes and linear fittings of resonance peaks, J. Phys. Chem. B 109 (2005) 17503–17511.

[38] M. Gao, X. Zheng, I. Khan, H. Cai, J. Lan, J. Liu, J. Wang, J. Wu, S. Huang, S. Li, J. Kang, Resonant light absorption and plasmon tunability of lateral triangular Au nanoprisms array, Phys. Lett., Sec. A: Gen, Atomic Solid State Phys. 383 (2019) 125881–125881.

[39] S. Stangherlin, N. Cathcart, F. Sato, V. Kitaev, Gold Nanoprisms, Synthetic approaches for mastering plasmonic properties and implications for biomedical applications, ACS Appl Nano Mater. 3 (2020) 8304–8318.

[40] M.A. Huergo, C.M. Maier, M.F. Castez, C. Vericat, S. Nedev, R.C. Salvarezza, A.S. Urban, J. Feldmann, Optical nanoparticle sorting elucidates synthesis of plasmonic nanotriangles, ACS Nano. 10 (2016) 3614–3621.

[41] X. Huang, I.H. El-Sayed, W. Qian, M.A. El-Sayed, Cancer cell imaging and photothermal therapy in the near-infrared region by using gold nanorods, J. Am. Chem. Soc. 128 (2006) 2115–2120.

[42] G. Zheng, S. Mourdikoudis, Z. Zhang, Plasmonic metallic heteromeric nanostructures, Small. 16 (2020) 1–21.

[43] M. Ha, J.H. Kim, M. You, Q. Li, C. Fan, J.M. Nam, Multicomponent plasmonic nanoparticles: From heterostructured nanoparticles to colloidal composite nanostructures, Chem. Rev. 119 (2019) 12208–12278.

[44] A.L. Aden, M. Kerker, Scattering of electromagnetic waves from two concentric spheres, J. Appl. Phys. 22 (1951) 1242–1246.

[45] J.J. Mikulski, E.L. Murphy, The computation of electromagnetic scattering from concentric spherical structures, IEEE Trans. Antennas Propag. 11 (1963) 169–177.

[46] R. Bhandari, Scattering coefficients for a multilayered sphere: analytic expressions and algorithms, Appl. Opt. 24 (1985) 1960–1960.

[47] L. Kai, P. Massoli, Scattering of electromagnetic-plane waves by radially inhomogeneous spheres: A finely stratified sphere model, Appl. Opt. 33 (1994) 501–501.

[48] B.R. Johnson, Light scattering by a multilayer sphere, Appl. Opt. 35 (1996) 3286–3296.

[49] W. Yang, Improved recursive algorithm for light scattering by a multilayered sphere, Appl. Opt. 42 (2003) 1710–1710.

[50] J.R. Wait, Electromagnetic scattering from a radially inhomogeneous sphere, Appl. Sci. Res., Sec. B 10 (1962) 441–450.

[51] M. Quinten, Optical Properties of Nanoparticle Systems: Mie and Beyond, John Wiley & Sons, Weinheim, 2011.

[52] J. Sinzig, M. Quinten, Scattering and absorption by spherical multilayer particles, Appl. Phys. A 58 (1994) 157–162.

[53] R.H. Morriss, L.F. Collins, Optical properties of multilayer colloids, J. Chem. Phys. 41 (1964) 3357–3363.

[54] J. Sinzig, U. Radtke, M. Quinten, U. Kreibig, Binary clusters: homogeneous alloys and nucleus-shell structures, Zeitschrift für Physik D Atoms, Mole. Clusters. 26 (1993) 242–245.

[55] P. Mulvaney, M. Giersig, A. Henglein, Electrochemistry of multilayer colloids: Preparation and absorption spectrum of gold-coated silver particles, J. Phys. Chem. 97 (1993) 7061–7064.

[56] S. Mourdikoudis, M. Chirea, D. Zanaga, T. Altantzis, M. Mitrakas, S. Bals, L.M. Liz-Marzán, J. Pérez-Juste, I. Pastoriza-Santos, Governing the morphology of Pt-Au heteronanocrystals with improved electrocatalytic performance, Nanoscale. 7 (2015) 8739–8747.

[57] J. Zeng, M. Gong, D. Wang, M. Li, W. Xu, Z. Li, S. Li, D. Zhang, Z. Yan, Y. Yin, Direct synthesis of water-dispersible magnetic/plasmonic heteronanostructures for multimodality biomedical imaging, Nano Lett. 19 (2019) 3011–3018.

[58] Z. Tang, B.C. Yeo, S.S. Han, T.J. Lee, S.H. Bhang, W.S. Kim, T. Yu, Facile aqueous-phase synthesis of Ag–Cu–Pt–Pd quadrometallic nanoparticles, Nano Converg. 6 (2019) 1–7.

[59] R. Borah, S.W. Verbruggen, Silver-gold bimetallic alloy versus core-shell nanoparticles: Implications for plasmonic enhancement and photothermal applications, J. Phys. Chem. C 124 (2020) 12081–12094.

[60] N.J. Halas, S. Lal, W.S. Chang, S. Link, P. Nordlander, Plasmons in strongly coupled metallic nanostructures, Chem. Rev. 111 (2011) 3913–3961.

[61] Y. Zhou, H. Dong, L. Liu, M. Li, K. Xiao, M. Xu, Selective and sensitive colorimetric sensor of mercury (II) based on gold nanoparticles and 4-mercaptophenylboronic acid, Sens. Actuat. B 196 (2014) 106–111.

[62] M. Zheng, Y. Wang, C. Wang, W. Wei, S. Ma, X. Sun, J. He, Silver nanoparticles-based colorimetric array for the detection of Thiophanate-methyl, Spectrochim. Acta - Part A: Mole Biomole. Spectrosc. 198 (2018) 315–321.

[63] C.A.P. Louis, Olivier, Gold Nanoparticles for Physics, Chemistry and Biology.

[64] R. Thiruppathi, S. Mishra, M. Ganapathy, P. Padmanabhan, B. Gulyás, Nanoparticle functionalization and its potentials for molecular imaging, Adv. Sci. 4 (2017) 1600279.

[65] G. Schmid, Nanoparticles: From Theory to Application: Second Edition, 2010.

[66] J.M. Carnerero, A. Jimenez-Ruiz, P.M. Castillo, R. Prado-Gotor, Covalent and non-covalent DNA–gold-nanoparticle interactions: New avenues of research, ChemPhysChem 18 (2017) 17–33.

[67] D. Curry, A. Cameron, B. MacDonald, C. Nganou, H. Scheller, J. Marsh, S. Beale, M. Lu, Z. Shan, R. Kaliaperumal, H. Xu, M. Servos, C. Bennett, S. MacQuarrie, K.D. Oakes, M. Mkandawire, X. Zhang, Adsorption of doxorubicin on citrate-capped gold nanoparticles: Insights into engineering potent chemotherapeutic delivery systems, Nanoscale. 7 (2015) 19611–19619.

[68] C.K. Kim, P. Ghosh, C. Pagliuca, Z.-J. Zhu, S. Menichetti, V.M. Rotello, Entrapment of hydrophobic drugs in nanoparticle monolayers with efficient release into cancer cells, J. Am. Chem. Soc. 131 (2009) 1360–1361.

[69] Y. Yuan, J. Zhang, H. Zhang, X. Yang, Silver nanoparticle based label-free colorimetric immunosensor for rapid detection of neurogenin 1, Analyst. 137 (2012) 496–501.

[70] C.M. McIntosh, E.A. Esposito, A.K. Boal, J.M. Simard, C.T. Martin, V.M. Rotello, Inhibition of DNA transcription using cationic mixed monolayer protected gold clusters, J. Am. Chem. Soc. 123 (2001) 7626–7629.

[71] X. Huang, I.H. El-Sayed, W. Qian, M.A. El-Sayed, Cancer cells assemble and align gold nanorods conjugated to antibodies to produce highly enhanced, sharp, and polarized surface Raman spectra: A potential cancer diagnostic marker, Nano Lett. 7 (2007) 1591–1597.

[72] L.A. Austin, M.A. Mackey, E.C. Dreaden, M.A. El-Sayed, The optical, photothermal, and facile surface chemical properties of gold and silver nanoparticles in biodiagnostics, therapy, and drug delivery, Arch. Toxicol. 88 (2014) 1391–1417.

[73] C.A. Mirkin, R.L. Letsinger, R.C. Mucic, J.J. Storhoff, A DNA-based method for rationally assembling nanoparticles into macroscopic materials, Nature. 382 (1996) 607–609.

[74] J.-S. Lee, M.S. Han, C.A. Mirkin, Colorimetric detection of mercuric ion (Hg2+) in aqueous media using DNA-functionalized gold nanoparticles, Angew. Chem. Int. Ed. 46 (2007) 4093–4096.

[75] N.L. Rosi, D.A. Giljohann, C.S. Thaxton, A.K.R. Lytton-Jean, M.S. Han, C.A. Mirkin, Oligonucleotide-modified gold nanoparticles for intracellular gene regulation, Science. 312 (2006) 1027–1030.

[76] L.A. Austin, B. Kang, C.-W. Yen, M.A. El-Sayed, Nuclear targeted silver nanospheres perturb the cancer cell cycle differently than those of nanogold, Bioconjugate Chem. 22 (2011) 2324–2331.

[77] B. Kang, M.A. Mackey, M.A. El-Sayed, Nuclear targeting of gold nanoparticles in cancer cells induces DNA damage, causing cytokinesis arrest and apoptosis, J. Am. Chem. Soc. 132 (2010) 1517–1519.

[78] M. Valcárcel, A. López-Lorente, Gold nanoparticles in analytical chemistry, (2014).

[79] C.R. Maldonado, L. Salassa, N. Gomez-Blanco, J.C. Mareque-Rivas, Nano-functionalization of metal complexes for molecular imaging and anticancer therapy, Coord. Chem. Rev. 257 (2013) 2668–2688.

[80] M. Di Marco, S. Shamsuddin, K.A. Razak, A.A. Aziz, C. Devaux, E. Borghi, L. Levy, C. Sadun, Overview of the main methods used to combine proteins with nanosystems: Absorption, bioconjugation, and encapsulation, Int. J. Nanomed. 5 (2010) 37.

[81] M.-X. Zhang, B.-H. Huang, X.-Y. Sun, D.-W. Pang, Clickable gold nanoparticles as the building block of nanobioprobes, Langmuir. 26 (2010) 10171–10176.

[82] K. Klein, K. Loza, M. Heggen, M. Epple, An efficient method for covalent surface functionalization of ultrasmall metallic nanoparticles by surface azidation followed by copper-catalyzed azide-alkyne cycloaddition (Click Chemistry), ChemNanoMat. 7 (2021) 1330–1339.

[83] L. Taiariol, C. Chaix, C. Farre, E. Moreau, Click and bioorthogonal chemistry: The future of active targeting of nanoparticles for nanomedicines? Chem. Rev. 122 (2022) 340–384.

[84] P. Prasher, M. Sharma, H. Mudila, G. Gupta, A.K. Sharma, D. Kumar, H.A. Bakshi, P. Negi, D.N. Kapoor, D.K. Chellappan, M.M. Tambuwala, K. Dua, Emerging trends in clinical implications of bio-conjugated silver nanoparticles in drug delivery, Colloid Interface Sci. Commun. 35 (2020) 100244.

[85] R. Pardehkhorram, F.A. Alshawawreh, V.R. Gonçales, N.A. Lee, R.D. Tilley, J.J. Gooding, Functionalized gold nanorod probes: A sophisticated design of SERS immunoassay for biodetection in complex media, Anal. Chem. 93 (2021) 12954–12965.

[86] R. Panda, S. Ranjan Dash, S. Sagar Satapathy, C. Nath Kundu, J. Tripathy, Surface functionalized gold nanorods for plasmonic photothermal therapy, Mater. Today: Proc. 47 (2021) 1193–1196.

[87] A.T. Thodikayil, S. Sharma, S. Saha, Engineering carbohydrate-based particles for biomedical applications: Strategies to construct and modify, ACS Appl. Bio Mat. 4 (2021) 2907–2940.

[88] C.M. Niemeyer, B. Ceyhan, DNA-directed functionalization of colloidal gold with proteins, Angew. Chem. Int. Ed. 40 (2001) 3685–3688.

[89] D.-Q. Feng, G. Liu, W. Zheng, J. Liu, T. Chen, D. Li, A highly selective and sensitive on–off sensor for silver ions and cysteine by light scattering technique of DNA-functionalized gold nanoparticles, Chem. Commun. 47 (2011) 8557–8559.

[90] L. Nguyen, M. Dass, M.F. Ober, L.V. Besteiro, Z.M. Wang, B. Nickel, A.O. Govorov, T. Liedl, A. Heuer-Jungemann, Chiral assembly of gold–silver core–shell plasmonic nanorods on DNA origami with strong optical activity, ACS Nano. 14 (2020) 7454–7461.

[91] W. Bai, Y. Wei, Y. Zhang, L. Bao, Y. Li, Label-free and amplified electrogenerated chemiluminescence biosensing for the detection of thymine DNA glycosylase activity using DNA-functionalized gold nanoparticles triggered hybridization chain reaction, Anal. Chim. Acta 1061 (2019) 101–109.

[92] S. Srivastava, A. Bhargava, Green Nanoparticles: The Future of Nanobiotechnology, Springer, Singapore, 2022.

[93] A. Singh, P.K. Gautam, A. Verma, V. Singh, P.M. Shivapriya, S. Shivalkar, A.K. Sahoo, S.K. Samanta, Green synthesis of metallic nanoparticles as effective alternatives to treat antibiotics resistant bacterial infections: A review, Biotechnol. Rep 25 (2020) e00427.

[94] A.K. Bishoyi, C.R. Sahoo, A.P. Sahoo, R.N. Padhy, Bio-synthesis of silver nanoparticles with the brackish water blue-green alga Oscillatoria princeps and antibacterial assessment, Appl. Nanosci. 11 (2021) 389–398.

[95] A.U. Khan, M. Khan, N. Malik, M.H. Cho, M.M. Khan, Recent progress of algae and blue–green algae-assisted synthesis of gold nanoparticles for various applications, Bioprocess. Biosyst. Eng. 42 (2019) 1–15.

[96] N. González-Ballesteros, N. Flórez-Fernández, M. Torres, H. Domínguez, M. Rodríguez-Argüelles, Synthesis, process optimization and characterization of gold nanoparticles using crude fucoidan from the invasive brown seaweed Sargassum muticum, Algal Res. 58 (2021) 102377.

[97] Y.A. Yugay, R.V. Usoltseva, V.E. Silant'ev, A.E. Egorova, A.A. Karabtsov, V.V. Kumeiko, S.P. Ermakova, V.P. Bulgakov, Y.N. Shkryl, Synthesis of bioactive silver nanoparticles using alginate, fucoidan and laminaran from brown algae as a reducing and stabilizing agent, Carbohydr. Polym. 245 (2020) 116547.

[98] F. Ameen, Optimization of the synthesis of fungus-mediated bi-metallic Ag-Cu nanoparticles, Appl. Sci. 12 (2022) 1384.

[99] A. Chauhan, J. Anand, V. Parkash, N. Rai, Biogenic synthesis: A sustainable approach for nanoparticles synthesis mediated by fungi, Inorgan. Nano-Metal Chem. (2022) 1–14.

[100] A. Zielonka, M. Klimek-Ochab, Fungal synthesis of size-defined nanoparticles, Adv. Nat. Sci. 8 (2017) 043001.

[101] M.K. Zahran, H.A. Mohammed, Green synthesis of silver nanoparticles using polysaccharide extracted from Laurencia obtuse algae, Egypt J. Appl. Sci. 36 (2021) 9–16.

[102] S. Iravani, Plant gums for sustainable and eco-friendly synthesis of nanoparticles: Recent advances, Inorgan. Nano-Metal Chem. 50 (2020) 469–488.

[103] F. Fibriana, A.V. Amalia, S. Muntamah, L. Ulva, S. Aryanti, Antimicrobial activities of green synthesized silver nanoparticles from Marchantia sp. extract: Testing an alcohol-free hand sanitizer product formula, J. Microbial., Biotechnol. Food Sci. 9 (2020) 1034–1038.

[104] M. Khatami, S. Iravani, R.S. Varma, F. Mosazade, M. Darroudi, F. Borhani, Cockroach wings-promoted safe and greener synthesis of silver nanoparticles and their insecticidal activity, Bioproc. Biosyst. Eng. 42 (2019) 2007–2014.

[105] G. Bonsignore, M. Patrone, S. Martinotti, E. Ranzato, "Green" biomaterials: The promising role of honey, J. Funct. Biomat. 12 (2021) 72.

[106] R. Vishwanath, B. Negi, Conventional and green methods of synthesis of silver nanoparticles and their antimicrobial properties, Curr. Res. Green Sustain. Chem. 4 (2021) 100205.

[107] Z. Xu, L. Shi, M. Yang, L. Zhu, Preparation and biomedical applications of silk fibroin-nanoparticles composites with enhanced properties-a review, Mater. Sci. Engin.: C. 95 (2019) 302–311.

[108] H. Hinterwirth, S. Kappel, T. Waitz, T. Prohaska, W. Lindner, M. Lämmerhofer, Quantifying thiol ligand density of self-assembled monolayers on gold nanoparticles by inductively coupled plasma–mass spectrometry, ACS Nano. 7 (2013) 1129–1136.

[109] W. Jiang, D.B. Hibbert, G. Moran, R. Akter, Measurement of gold and sulfur mass fractions in l-cysteine-modified gold nanoparticles by ICP-DRC-MS after acid digestion: validation and uncertainty of results, J. Anal. At. Spectrom. 27 (2012) 1465–1473.

[110] H. Xiang, Y. Chen, Materdicine: Interdicipline of materials and medicine, VIEW. 1 (2020) 20200016.

[111] S. Barroso-Solares, P. Cimavilla-Roman, M.A. Rodriguez-Perez, J. Pinto, Non-invasive approaches for the evaluation of the functionalization of melamine foams with in-situ synthesized silver nanoparticles, Polymers. 12 (2020) 996.

[112] S. Thanneeru, K.M. Ayers, M. Anuganti, L. Zhang, C.V. Kumar, G. Ung, J. He, N-Heterocyclic carbene-ended polymers as surface ligands of plasmonic metal nanoparticles, J. Mater. Chem. C 8 (2020) 2280–2288.

[113] G. Fadillah, O.A. Saputra, T.A. Saleh, Trends in polymers functionalized nanostructures for analysis of environmental pollutants, Trends Environ. Analyt. Chem. 26 (2020) e00084.

[114] S. Sur, A. Rathore, V. Dave, K.R. Reddy, R.S. Chouhan, V. Sadhu, Recent developments in functionalized polymer nanoparticles for efficient drug delivery system, Nano-Struct. Nano-Objects. 20 (2019) 100397.

[115] W.-J. Zhang, C.-Y. Hong, C.-Y. Pan, Polymerization-induced self-assembly of functionalized block copolymer nanoparticles and their application in drug delivery, Macromol. Rapid Commun. 40 (2019) 1800279.

[116] L. Yu, N. Li, Noble metal nanoparticles-based colorimetric biosensor for visual quantification: A mini review, Chemosensors. 7 (2019) 53.

[117] M. Azharuddin, G.H. Zhu, D. Das, E. Ozgur, L. Uzun, A.P.F. Turner, H.K. Patra, A repertoire of biomedical applications of noble metal nanoparticles, Chem. Commun. 55 (2019) 6964–6996.

[118] C. Li, C. Wang, Z. Ji, N. Jiang, W. Lin, D. Li, Synthesis of thiol-terminated thermoresponsive polymers and their enhancement effect on optical limiting property of gold nanoparticles, Eur. Polym. J. 113 (2019) 404–410.

[119] J. Yan, M.H. Malakooti, Z. Lu, Z. Wang, N. Kazem, C. Pan, M.R. Bockstaller, C. Majidi, K. Matyjaszewski, Solution processable liquid metal nanodroplets by surface-initiated atom transfer radical polymerization, Nat. Nanotechnol. 14 (2019) 684–690.

[120]   M.J.A. Hore, Polymers on nanoparticles: Structure & dynamics, Soft Matter. 15 (2019) 1120–1134.

[121]   H. Liu, H.-Y. Zhao, F. Müller-Plathe, H.-J. Qian, Z.-Y. Sun, Z.-Y. Lu, Distribution of the number of polymer chains grafted on nanoparticles fabricated by grafting-to and grafting-from procedures, Macromolecules. 51 (2018) 3758–3766.

[122]   S.A. Kedzior, J.O. Zoppe, R.M. Berry, E.D. Cranston, Recent advances and an industrial perspective of cellulose nanocrystal functionalization through polymer grafting, Curr. Opin. Solid State Mater. Sci. 23 (2019) 74–91.

[123]   Y. Zhu, E. Egap, PET-RAFT polymerization catalyzed by cadmium selenide quantum dots (QDs): Grafting-from QDs photocatalysts to make polymer nanocomposites, Polym. Chem. 11 (2020) 1018–1024.

[124]   M. Madkour, A. Bumajdad, F. Al-Sagheer, To what extent do polymeric stabilizers affect nanoparticles characteristics? Adv. Colloid Interface Sci. 270 (2019) 38–53.

[125]   J.C. Cazotti, A.T. Fritz, O. Garcia-Valdez, N.M.B. Smeets, M.A. Dubé, M.F. Cunningham, Graft modification of starch nanoparticles using nitroxide-mediated polymerization and the grafting from approach, Carbohydr. Polym. 228 (2020) 115384.

[126]   K. Wieszczycka, K. Staszak, M.J. Woźniak-Budych, J. Litowczenko, B.M. Maciejewska, S. Jurga, Surface functionalization – The way for advanced applications of smart materials, Coord. Chem. Rev. 436 (2021) 213846.

[127]   H. Thiele, H. Von Levern, Synthetic protective colloids, J. Colloid Sci. 20 (1965) 679–694.

[128]   T. Špringer, M.L. Ermini, B. Špačková, J. Jabloňků, J. Homola, Enhancing sensitivity of surface plasmon resonance biosensors by functionalized gold nanoparticles: Size matters, Anal. Chem. 86 (2014) 10350–10356.

[129]   V.A. Turek, S. Cormier, B. Sierra-Martin, U.F. Keyser, T. Ding, J.J. Baumberg, The crucial role of charge in thermoresponsive-polymer-assisted reversible dis/assembly of gold nanoparticles, Adv. Opt. Mater. 6 (2018) 1701270.

[130]   L. Zhang, Z. Wei, M. Meng, G. Ung, J. He, Do polymer ligands block the catalysis of metal nanoparticles? Unexpected importance of binding motifs in improving catalytic activity, J. Mater. Chem. A 8 (2020) 15900–15908.

[131]   L. Zhang, Z. Wei, S. Thanneeru, M. Meng, M. Kruzyk, G. Ung, B. Liu, J. He, A polymer solution to prevent nanoclustering and improve the selectivity of metal nanoparticles for electrocatalytic CO2 reduction, Angew. Chem. Int. Ed. 58 (2019) 15834–15840.

[132]   C.C.S. Batista, L.J.C. Albuquerque, A. Jäger, P. Stepánek, F.C. Giacomelli, Probing protein adsorption onto polymer-stabilized silver nanocolloids towards a better understanding on the evolution and consequences of biomolecular coronas, Mat. Sci. Engin.: C 111 (2020) 110850.

[133]   S.Y.H. Abdalkarim, L.-M. Chen, H.-Y. Yu, F. Li, X. Chen, Y. Zhou, K.C. Tam, Versatile nanocellulose-based nanohybrids: A promising-new class for active packaging applications, Int. J. Biol. Macromol. 182 (2021) 1915–1930.

[134]   X. Wu, C. Lu, Z. Zhou, G. Yuan, R. Xiong, X. Zhang, Green synthesis and formation mechanism of cellulose nanocrystal-supported gold nanoparticles with enhanced catalytic performance, Environ. Sci. 1 (2014) 71–79.

[135]   N. Drogat, R. Granet, V. Sol, A. Memmi, N. Saad, C. Klein Koerkamp, P. Bressollier, P. Krausz, Antimicrobial silver nanoparticles generated on cellulose nanocrystals, J. Nanopart. Res. 13 (2011) 1557–1562.

[136]   J. Wu, Y. Zheng, W. Song, J. Luan, X. Wen, Z. Wu, X. Chen, Q. Wang, S. Guo, In situ synthesis of silver-nanoparticles/bacterial cellulose composites for slow-released antimicrobial wound dressing, Carbohydr. Polym. 102 (2014) 762–771.

[137]   K.A. Mahmoud, K.B. Male, S. Hrapovic, J.H.T. Luong, Cellulose nanocrystal/gold nanoparticle composite as a matrix for enzyme immobilization, ACS Appl. Mater. Interfaces. 1 (2009) 1383–1386.

[138]   X.-X. Dong, J.-Y. Yang, L. Luo, Y.-F. Zhang, C. Mao, Y.-M. Sun, H.-T. Lei, Y.-D. Shen, R.C. Beier, Z.-L. Xu, Portable amperometric immunosensor for histamine detection using Prussian blue-chitosan-gold nanoparticle nanocomposite films, Biosens. Bioelectron. 98 (2017) 305–309.

[139]   A.A. Menazea, E. Alzahrani, W. Alharbi, A.A. Shaltout, Bimetallic nanocomposite of gold/silver scattered in chitosan via laser ablation for electrical and antibacterial utilization, J. Electron. Mater. 51 (2022) 3811–3819.

[140] M. Sharifiaghdam, E. Shaabani, F. Asghari, R. Faridi-Majidi, Chitosan coated metallic nanoparticles with stability, antioxidant, and antibacterial properties: Potential for wound healing application, J. Appl. Polym. Sci. 139 (2022) 51766.

[141] R. Kanaoujiya, S.K. Saroj, S. Srivastava, M.K. Chaudhary, Renewable polysaccharide and biomedical application of nanomaterials, J. Nanomat. 2022 (2022) 1050211.

[142] T.A. Jacinto, C.F. Rodrigues, A.F. Moreira, S.P. Miguel, E.C. Costa, P. Ferreira, I.J. Correia, Hyaluronic acid and vitamin E polyethylene glycol succinate functionalized gold-core silica shell nanorods for cancer targeted photothermal therapy, Colloids Surf. B 188 (2020) 110778.

[143] Q. Liu, X. Yan, Q. Lai, X. Su, Bimetallic gold/silver nanoclusters-gold nanoparticles based fluorescent sensing platform via the inner filter effect for hyaluronidase activity detection, Sens. Actuators B 282 (2019) 45–51.

[144] G.V. Kumari, M.A. JothiRajan, T. Mathavan, Pectin functionalized gold nanoparticles towards singlet oxygen generation, Mater. Res. Express 5 (2018) 085027.

[145] S. Sharma, A. Jaiswal, K.N. Uttam, Colorimetric and Surface Enhanced Raman Scattering (SERS) detection of metal ions in aqueous medium using sensitive, robust and novel pectin functionalized silver nanoparticles, Anal. Lett. 53 (2020) 2355–2378.

[146] Y. Yu, S.S. Naik, Y. Oh, J. Theerthagiri, S.J. Lee, M.Y. Choi, Lignin-mediated green synthesis of functionalized gold nanoparticles via pulsed laser technique for selective colorimetric detection of lead ions in aqueous media, J. Hazard. Mater. 420 (2021) 126585.

[147] K.R. Aadil, N. Pandey, S.I. Mussatto, H. Jha, Green synthesis of silver nanoparticles using acacia lignin, their cytotoxicity, catalytic, metal ion sensing capability and antibacterial activity, J. Environ. Chem. Eng. 7 (2019) 103296.

[148] D. Kumar Ban, S. Paul, Functionalized gold and silver nanoparticles modulate amyloid fibrillation, defibrillation and cytotoxicity of lysozyme via altering protein surface character, Appl. Surf. Sci. 473 (2019) 373–385.

[149] N. Nombona, E. Antunes, W. Chidawanyika, P. Kleyi, Z. Tshentu, T. Nyokong, Synthesis, photophysics and photochemistry of phthalocyanine-ε-polylysine conjugates in the presence of metal nanoparticles against Staphylococcus aureus, J. Photochem. Photobiol. A 233 (2012) 24–33.

[150] D. Zhu, R.X. Yang, Y.-P. Tang, W. Li, Z.Y. Miao, Y. Hu, J. Chen, S. Yu, J. Wang, C.Y. Xu, Robust nanoplasmonic substrates for aptamer macroarrays with single-step detection of PDGF-BB, Biosens. Bioelectron. 85 (2016) 429–436.

[151] F. Fatima, S. Siddiqui, W.A. Khan, Nanoparticles as novel emerging therapeutic antibacterial agents in the antibiotics resistant era, Biol. Trace Elem. Res. 199 (2021) 2552–2564.

[152] M.J. Kofke, E. Wierzbinski, D.H. Waldeck, Seedless CTAB mediated growth of anisotropic nanoparticles and nanoparticle clusters on nanostructured plasmonic templates, J. Mater. Chem. C 1 (2013) 6774–6781.

[153] S.K. Meena, C. Meena, The implication of adsorption preferences of ions and surfactants on the shape control of gold nanoparticles: A microscopic, atomistic perspective, Nanoscale 13 (2021) 19549–19560.

[154] A. Knauer, D. Kuhfuss, J.M. Köhler, Electrostatic Control of Au nanorod formation in automated microsegmented flow synthesis, ACS Appl. Nano Mat. 4 (2021) 1411–1419.

[155] D. Dong, Q. Shi, D. Sikdar, Y. Zhao, Y. Liu, R. Fu, M. Premaratne, W. Cheng, Site-specific Ag coating on concave Au nano arrows by controlling the surfactant concentration, Nanoscale Horizons. 4 (2019) 940–946.

[156] C. Hamon, D. Constantin, Growth kinetics of core–shell Au/Ag nanoparticles, J. Phys. Chem. C 124 (2020) 21717–21721.

[157] Y.-C. Tsao, S. Rej, C.-Y. Chiu, M.H. Huang, Aqueous phase synthesis of Au–Ag Core–shell nanocrystals with tunable shapes and their optical and catalytic properties, J. Am. Chem. Soc. 136 (2014) 396–404.

[158] Y. Ma, W. Li, E.C. Cho, Z. Li, T. Yu, J. Zeng, Z. Xie, Y. Xia, Au@Ag core–shell nanocubes with finely tuned and well-controlled sizes, shell thicknesses, and optical properties, ACS Nano. 4 (2010) 6725–6734.

[159] C.-H. Kuo, T.-F. Chiang, L.-J. Chen, M.H. Huang, Synthesis of highly faceted pentagonal- and hexagonal-shaped gold nanoparticles with controlled sizes by sodium dodecyl sulfate, Langmuir. 20 (2004) 7820–7824.

[160] S. Wu, S.Y. Tan, C.Y. Ang, K.T. Nguyen, M. Li, Y. Zhao, An imine-based approach to prepare amine-functionalized Janus gold nanoparticles, Chem. Commun. 51 (2015) 11622–11625.

[161] A. Mandal, S. Sekar, N. Chandrasekaran, A. Mukherjee, T.P. Sastry, Synthesis, characterization and evaluation of collagen scaffolds crosslinked with aminosilane functionalized silver nanoparticles: In vitro and in vivo studies, J. Mat. Chem. B 3 (2015) 3032–3043.

[162] J. Kim, J.-M. Kim, M. Ha, J.-W. Oh, J.-M. Nam, Polysorbate- and DNA-mediated synthesis and strong, stable, and tunable near-infrared photoluminescence of plasmonic long-body nanosnowmen, ACS Nano. 15 (2021) 19853–19863.

[163] W. Zhang, M. Saliba, S.D. Stranks, Y. Sun, X. Shi, U. Wiesner, H.J. Snaith, Enhancement of perovskite-based solar cells employing core–shell metal nanoparticles, Nano Lett. 13 (2013) 4505–4510.

[164] R. Zhang, Y. Zhou, L. Peng, X. Li, S. Chen, X. Feng, Y. Guan, W. Huang, Influence of SiO2 shell thickness on power conversion efficiency in plasmonic polymer solar cells with Au nanorod@SiO2 core-shell structures, Sci. Rep. 6 (2016) 25036.

[165] S. Carretero-Palacios, A. Jiménez-Solano, H. Míguez, Plasmonic nanoparticles as light-harvesting enhancers in perovskite solar cells: A user's guide, ACS Energy Lett. 1 (2016) 323–331.

[166] M. Salehi, L. Schneider, P. Ströbel, A. Marx, J. Packeisen, S. Schlücker, Two-color SERS microscopy for protein co-localization in prostate tissue with primary antibody-protein A/G-gold nanocluster conjugates, Nanoscale. 6 (2014) 2361–2367.

[167] A.B. Serrano-Montes, D. Jimenez de Aberasturi, J. Langer, J.J. Giner-Casares, L. Scarabelli, A. Herrero, L.M. Liz-Marzán, A general method for solvent exchange of plasmonic nanoparticles and self-assembly into SERS-active monolayers, Langmuir. 31 (2015) 9205–9213.

[168] J. Liu, C. Detrembleur, M.C. De Pauw-Gillet, S. Mornet, C. Jérôme, E. Duguet, Gold nanorods coated with mesoporous silica shell as drug delivery system for remote near infrared light-activated release and potential phototherapy, Small. 11 (2015) 2323–2332.

[169] T. Tian, X. Shi, L. Cheng, Y. Luo, Z. Dong, H. Gong, L. Xu, Z. Zhong, R. Peng, Z. Liu, Graphene-based nanocomposite as an effective, multifunctional, and recyclable antibacterial agent, ACS Appl. Mater. Interfaces. 6 (2014) 8542–8548.

[170] C. Hamon, M.N. Sanz-Ortiz, E. Modin, E.H. Hill, L. Scarabelli, A. Chuvilin, L.M. Liz-Marzán, Hierarchical organization and molecular diffusion in gold nanorod/silica supercrystal nanocomposites, Nanoscale. 8 (2016) 7914–7922.

[171] G. Bodelón, V. Montes-García, V. López-Puente, E.H. Hill, C. Hamon, M.N. Sanz-Ortiz, S. Rodal-Cedeira, C. Costas, S. Celiksoy, I. Pérez-Juste, L. Scarabelli, A. La Porta, J. Pérez-Juste, I. Pastoriza-Santos, L.M. Liz-Marzán, Detection and imaging of quorum sensing in Pseudomonas aeruginosa biofilm communities by surface-enhanced resonance Raman scattering, Nat. Mater. 15 (2016) 1203–1211.

[172] T. Ribeiro, C. Baleizão, J.P.S. Farinha, Artefact-free evaluation of metal enhanced fluorescence in silica coated gold nanoparticles, Sci. Rep. 7 (2017) 2440.

[173] G. Li, Z. Tang, Noble metal nanoparticle@metal oxide core/yolk–shell nanostructures as catalysts: recent progress and perspective, Nanoscale. 6 (2014) 3995–4011.

[174] E. Mine, A. Yamada, Y. Kobayashi, M. Konno, L.M. Liz-Marzán, Direct coating of gold nanoparticles with silica by a seeded polymerization technique, J. Colloid Interface Sci. 264 (2003) 385–390.

[175] Y. Kobayashi, H. Katakami, E. Mine, D. Nagao, M. Konno, L.M. Liz-Marzán, Silica coating of silver nanoparticles using a modified Stöber method, J. Colloid Interface Sci. 283 (2005) 392–396.

[176] M. Ohmori, E. Matijević, Preparation and properties of uniform coated inorganic colloidal particles: 8. Silica on Iron, J. Colloid Interface Sci. 160 (1993) 288–292.

[177] A. Guerrero-Martínez, J. Pérez-Juste, L.M. Liz-Marzán, Recent progress on silica coating of nanoparticles and related nanomaterials, Adv. Mater. 22 (2010) 1182–1195.

[178] B. Wrzosek, J. Bukowska, Molecular Structure of 3-Amino-5-mercapto-1,2,4-triazole Self-Assembled Monolayers on Ag and Au Surfaces, J. Phys. Chem. C 111 (2007) 17397–17403.

[179] Y.J. Wong, L. Zhu, W.S. Teo, Y.W. Tan, Y. Yang, C. Wang, H. Chen, Revisiting the Stöber method: Inhomogeneity in silica shells, J. Am. Chem. Soc. 133 (2011) 11422–11425.

[180] K.W. Shah, T. Sreethawong, S.-H. Liu, S.-Y. Zhang, L.S. Tan, M.-Y. Han, Aqueous route to facile, efficient and functional silica coating of metal nanoparticles at room temperature, Nanoscale. 6 (2014) 11273–11281.

[181] M. Thiele, I. Götz, S. Trautmann, R. Müller, A. Csáki, T. Henkel, W. Fritzsche, Wet-chemical passivation of anisotropic plasmonic nanoparticles for LSPR-sensing by a silica shell, Mater. Today: Proc. 2 (2015) 33–40.

[182] Y. Zhang, X. Kong, B. Xue, Q. Zeng, X. Liu, L. Tu, K. Liu, H. Zhang, A versatile synthesis route for metal@SiO2 core–shell nanoparticles using 11-mercaptoundecanoic acid as primer, J. Mater. Chem. C 1 (2013) 6355–6363.

[183] M.K. Gangishetty, K.E. Lee, R.W.J. Scott, T.L. Kelly, Plasmonic enhancement of dye sensitized solar cells in the red-to-near-infrared region using triangular core–shell Ag@SiO2 nanoparticles, ACS Appl. Mater. Interfaces. 5 (2013) 11044–11051.

[184] C. Kumudinie, Polymer–ceramic nanocomposites: interfacial bonding agents, in: K.H.J. Buschow, R.W. Cahn, M.C. Flemings, B. Ilschner, E.J. Kramer, S. Mahajan, P. Veyssière (Eds.), Encyclopedia of Materials: Science and Technology, Elsevier, Oxford, 2001, pp. 7574–7577.

[185] D. Tosi, M. Sypabekova, A. Bekmurzayeva, C. Molardi, K. Dukenbayev, 10 - Fiber surface modifications for biosensing, in: D. Tosi, M. Sypabekova, A. Bekmurzayeva, C. Molardi, K. Dukenbayev (Eds.), Optical Fiber Biosensors, Academic Press, United States, 2022, pp. 253–282.

[186] Y. Han, J. Jiang, S.S. Lee, J.Y. Ying, Reverse microemulsion-mediated synthesis of silica-coated gold and silver nanoparticles, Langmuir. 24 (2008) 5842–5848.

[187] S. Cavaliere-Jaricot, M. Darbandi, T. Nann, Au–silica nanoparticles by "reverse" synthesis of cores in hollow silica shells, Chem. Commun. (2007) 2031–2033.

[188] Y. Xia, Y. Xiong, B. Lim, S.E. Skrabalak, Shape-controlled synthesis of metal nanocrystals: Simple chemistry meets complex physics? Angew. Chem. Int. Ed. 48 (2009) 60–103.

[189] M. Grzelczak, J. Pérez-Juste, P. Mulvaney, L.M. Liz-Marzán, Shape control in gold nanoparticle synthesis, Chem. Soc. Rev. 37 (2008) 1783–1791.

[190] I. Pastoriza-Santos, L.M. Liz-Marzán, N,N-Dimethylformamide as a reaction medium for metal nanoparticle synthesis, Adv. Funct. Mater. 19 (2009) 679–688.

[191] B. Lim, P.H.C. Camargo, Y. Xia, Mechanistic study of the synthesis of Au nanotadpoles, nanokites, and microplates by reducing aqueous HAuCl4 with Poly(vinyl pyrrolidone), Langmuir. 24 (2008) 10437–10442.

[192] C. Graf, D.L.J. Vossen, A. Imhof, A. van Blaaderen, A general method to coat colloidal particles with silica, Langmuir. 19 (2003) 6693–6700.

[193] I. Pastoriza-Santos, J. Pérez-Juste, L.M. Liz-Marzán, Silica-coating and hydrophobation of CTAB-stabilized gold nanorods, Chem. Mater. 18 (2006) 2465–2467.

[194] J. Rodríguez-Fernández, I. Pastoriza-Santos, J. Pérez-Juste, F.J. García de Abajo, L.M. Liz-Marzán, The effect of silica coating on the optical response of sub-micrometer gold spheres, J. Phys. Chem. C 111 (2007) 13361–13366.

[195] M. Sanles-Sobrido, W. Exner, L. Rodríguez-Lorenzo, B. Rodríguez-González, M.A. Correa-Duarte, R.A. Álvarez-Puebla, L.M. Liz-Marzán, Design of SERS-Encoded, submicron, hollow particles through confined growth of encapsulated metal nanoparticles, J. Am. Chem. Soc. 131 (2009) 2699–2705.

[196] N. Malikova, I. Pastoriza-Santos, M. Schierhorn, N.A. Kotov, L.M. Liz-Marzán, Layer-by-layer assembled mixed spherical and planar gold nanoparticles: Control of interparticle interactions, Langmuir. 18 (2002) 3694–3697.

[197] V. Salgueiriño-Maceira, M.A. Correa-Duarte, M. Spasova, L.M. Liz-Marzán, M. Farle, Composite silica spheres with magnetic and luminescent functionalities, Adv. Funct. Mater. 16 (2006) 509–514.

[198] M.S. Wong, J.N. Cha, K.-S. Choi, T.J. Deming, G.D. Stucky, Assembly of nanoparticles into hollow spheres using block copolypeptides, Nano Lett. 2 (2002) 583–587.

[199] V. Salgueiriño-Maceira, F. Caruso, L.M. Liz-Marzán, Coated colloids with tailored optical properties, J. Phys. Chem. B 107 (2003) 10990–10994.

[200] M. Spasova, V. Salgueiriño-Maceira, A. Schlachter, M. Hilgendorff, M. Giersig, L.M. Liz-Marzán, M. Farle, Magnetic and optical tunable microspheres with a magnetite/gold nanoparticle shell, J. Mater. Chem. 15 (2005) 2095–2098.

[201] M.A. Correa-Duarte, N. Sobal, L.M. Liz-Marzán, M. Giersig, Linear assemblies of silica-coated gold nanoparticles using carbon nanotubes as templates, Adv. Mater. 16 (2004) 2179–2184.

[202] P. Mulvaney, M. Giersig, T. Ung, L.M. Liz-Marzán, Direct observation of chemical reactions in silica-coated gold and silver nanoparticles, Adv. Mater. 9 (1997) 570–575.

[203] J. Rodríguez-Fernández, J. Pérez-Juste, P. Mulvaney, L.M. Liz-Marzán, Spatially-directed oxidation of gold nanoparticles by Au(III)–CTAB complexes, J. Phys. Chem. B 109 (2005) 14257–14261.

[204] J. Lee, J.C. Park, J.U. Bang, H. Song, Precise tuning of porosity and surface functionality in Au@SiO2 nanoreactors for high catalytic efficiency, Chem. Mater. 20 (2008) 5839–5844.

[205] I. Gorelikov, N. Matsuura, Single-step coating of mesoporous silica on cetyltrimethyl ammonium bromide-capped nanoparticles, Nano Lett. 8 (2008) 369–373.

[206] J. Kobler, K. Möller, T. Bein, Colloidal suspensions of functionalized mesoporous silica nanoparticles, ACS Nano. 2 (2008) 791–799.

[207] S. Barui, V. Cauda, Multimodal decorations of mesoporous silica nanoparticles for improved cancer therapy, Pharmaceutics 12 (2020) 527.

[208] M.N. Sanz-Ortiz, K. Sentosun, S. Bals, L.M. Liz-Marzán, Templated growth of surface enhanced raman scattering-active branched gold nanoparticles within radial mesoporous silica shells, ACS Nano. 9 (2015) 10489–10497.

[209] T. Chen, G. Chen, S. Xing, T. Wu, H. Chen, Scalable Routes to Janus Au–SiO2 and ternary Ag–Au–SiO2 nanoparticles, Chem. Mater. 22 (2010) 3826–3828.

[210] D. Zhao, J. Feng, Q. Huo, N. Melosh, G.H. Fredrickson, B.F. Chmelka, G.D. Stucky, Triblock copolymer syntheses of mesoporous silica with periodic 50 to 300 angstrom pores, Science. 279 (1998) 548–552.

[211] X. Chen, F. Wang, J.Y. Hyun, T. Wei, J. Qiang, X. Ren, I. Shin, J. Yoon, Recent progress in the development of fluorescent, luminescent and colorimetric probes for detection of reactive oxygen and nitrogen species, Chem. Soc. Rev. 45 (2016) 2976–3016.

[212] A. Bigdeli, F. Ghasemi, H. Golmohammadi, S. Abbasi-Moayed, M.A.F. Nejad, N. Fahimi-Kashani, S. Jafarinejad, M. Shahrajabian, M.R. Hormozi-Nezhad, Nanoparticle-based optical sensor arrays, Nanoscale 9 (2017) 16546–16563.

[213] L. Tang, J. Li, Plasmon-based colorimetric nanosensors for ultrasensitive molecular diagnostics, ACS Sensors 2 (2017) 857–875.

[214] K. Saha, S.S. Agasti, C. Kim, X. Li, V.M. Rotello, Gold nanoparticles in chemical and biological sensing, Chem. Rev. 112 (2012) 2739–2779.

[215] H. Jans, Q. Huo, Gold nanoparticle-enabled biological and chemical detection and analysis, Chem. Soc. Rev. 41 (2012) 2849–2866.

[216] D. Vilela, M.C. González, A. Escarpa, Sensing colorimetric approaches based on gold and silver nanoparticles aggregation: chemical creativity behind the assay. A review, Anal. Chim. Acta. 751 (2012) 24–43.

[217] L. Yu, Z. Song, J. Peng, M. Yang, H. Zhi, H. He, Progress of gold nanomaterials for colorimetric sensing based on different strategies, TrAC Trends Anal. Chem. 127 (2020) 115880.

[218] T. Lou, Z. Chen, Y. Wang, L. Chen, Blue-to-red colorimetric sensing strategy for Hg2+ and Ag+ via redox-regulated surface chemistry of gold nanoparticles, ACS Appl. Mater. Interfaces. 3 (2011) 1568–1573.

[219] B. Li, X. Li, Y. Dong, B. Wang, D. Li, Y. Shi, Y. Wu, Colorimetric sensor array based on gold nanoparticles with diverse surface charges for microorganisms identification, Anal. Chem. 89 (2017) 10639–10643.

[220] L. Chen, J. Li, L. Chen, Colorimetric detection of mercury species based on functionalized gold nanoparticles, ACS Appl. Mater. Interfaces. 6 (2014) 15897–15904.

[221] E. Mauriz, Clinical applications of visual plasmonic colorimetric sensing, Sensors 20 (2020) 6214.

[222] N. Fahimi-Kashani, M.R. Hormozi-Nezhad, Gold-nanoparticle-based colorimetric sensor array for discrimination of organophosphate pesticides, Anal. Chem. 88 (2016) 8099–8106.

[223] M.R. Mirghafouri, S. Abbasi-Moayed, F. Ghasemi, M.R. Hormozi-Nezhad, Nanoplasmonic sensor array for the detection and discrimination of pesticide residues in citrus fruits, Anal. Methods 12 (2020) 5877–5884.

[224] X. Ma, Z. Chen, P. Kannan, Z. Lin, B. Qiu, L. Guo, Gold nanorods as colorful chromogenic substrates for semiquantitative detection of nucleic acids, proteins, and small molecules with the naked eye, Anal. Chem. 88 (2016) 3227–3234.

[225] L. Polavarapu, J. Pérez-Juste, Q.-H. Xu, L.M. Liz-Marzán, Optical sensing of biological, chemical and ionic species through aggregation of plasmonic nanoparticles, J. Mater. Chem. C 2 (2014) 7460–7476.

[226] Y. He, F. Tian, J. Zhou, Q. Zhao, R. Fu, B. Jiao, Colorimetric aptasensor for ochratoxin A detection based on enzyme-induced gold nanoparticle aggregation, J. Hazard. Mater. 388 (2020) 121758.

[227] Y. Tian, Q. Liu, Y. Jiao, R. Jia, Z. Chen, Colorimetric aggregation based cadmium (II) assay by using triangular silver nanoplates functionalized with 1-amino-2-naphthol-4-sulfonate, Microchim. Acta 185 (2018) 1–6.

[228] Y. Wei, Q. Zhang, L. Jia, H. Xu, M. Liang, Effects of dietary arginine levels on growth, intestinal peptide and amino acid transporters, and gene expressions of the TOR signaling pathway in tiger puffer, Takifugu rubripes, Aquaculture. 532 (2021) 736086.

[229] Y.-X. Qi, Z.-b. Qu, Q.-X. Wang, M. Zhang, G. Shi, Nanomolar sensitive colorimetric assay for Mn2+ using cysteic acid-capped silver nanoparticles and theoretical investigation of its sensing mechanism, Anal. Chim. Acta. 980 (2017) 65–71.

[230] Y. Li, Z. Wang, L. Sun, L. Liu, C. Xu, H. Kuang, Nanoparticle-based sensors for food contaminants, TrAC Trends Anal. Chem. 113 (2019) 74–83.

[231] T. Chotchuang, W. Cheewasedtham, T.J. Jayeoye, T. Rujiralai, Colorimetric determination of fumonisin B1 based on the aggregation of cysteamine-functionalized gold nanoparticles induced by a product of its hydrolysis, Microchim. Acta. 186 (2019) 1–10.

[232] M. Iarossi, C. Schiattarella, I. Rea, L. De Stefano, R. Fittipaldi, A. Vecchione, R. Velotta, B.D. Ventura, Colorimetric immunosensor by aggregation of photochemically functionalized gold nanoparticles, ACS Omega. 3 (2018) 3805–3812.

[233] Z. Taefi, F. Ghasemi, M.R. Hormozi-Nezhad, Selective colorimetric detection of pentaerythritol tetranitrate (PETN) using arginine-mediated aggregation of gold nanoparticles, Spectrochim. Acta Part A 228 (2020) 117803.

[234] Y. Guo, W. Zhao, In situ formed nanomaterials for colorimetric and fluorescent sensing, Coord. Chem. Rev. 387 (2019) 249–261.

[235] A. Orouji, F. Ghasemi, A. Bigdeli, M.R. Hormozi-Nezhad, Providing multicolor plasmonic patterns with Au@Ag core–shell nanostructures for visual discrimination of biogenic amines, ACS Appl. Mater. Interfaces 13 (2021) 20865–20874.

[236] H. Wang, H. Rao, M. Luo, X. Xue, Z. Xue, X. Lu, Noble metal nanoparticles growth-based colorimetric strategies: From monocolorimetric to multicolorimetric sensors, Coord. Chem. Rev. 398 (2019) 113003.

[237] D. Li, S. Wang, L. Wang, H. Zhang, J. Hu, A simple colorimetric probe based on anti-aggregation of AuNPs for rapid and sensitive detection of malathion in environmental samples, Anal. Bioanal.Chem. 411 (2019) 2645–2652.

[238] F. Najafzadeh, F. Ghasemi, M.R. Hormozi-Nezhad, Anti-aggregation of gold nanoparticles for metal ion discrimination: A promising strategy to design colorimetric sensor arrays, Sens. Actuators B 270 (2018) 545–551.

[239] X. Sun, R. Liu, Q. Liu, Q. Fei, G. Feng, H. Shan, Y. Huan, Colorimetric sensing of mercury (II) ion based on anti-aggregation of gold nanoparticles in the presence of hexadecyl trimethyl ammonium bromide, Sens. Actuators B 260 (2018) 998–1003.

[240] A. Safavi, R. Ahmadi, Z. Mohammadpour, Colorimetric sensing of silver ion based on anti aggregation of gold nanoparticles, Sens. Actuators B 242 (2017) 609–615.

[241] P.-C. Huang, N. Gao, J.-F. Li, F.-Y. Wu, Colorimetric detection of methionine based on anti-aggregation of gold nanoparticles in the presence of melamine, Sens. Actuators B 255 (2018) 2779–2784.

[242] Z. Zhang, H. Wang, Z. Chen, X. Wang, J. Choo, L. Chen, Plasmonic colorimetric sensors based on etching and growth of noble metal nanoparticles: Strategies and applications, Biosens. Bioelectron. 114 (2018) 52–65.

[243] J. Cao, T. Sun, K.T. Grattan, Gold nanorod-based localized surface plasmon resonance biosensors: A review, Sens. Actuators B 195 (2014) 332–351.

[244] L. Guo, J.A. Jackman, H.-H. Yang, P. Chen, N.-J. Cho, D.-H. Kim, Strategies for enhancing the sensitivity of plasmonic nanosensors, Nano Today. 10 (2015) 213–239.

[245] J. Zeng, Y. Zhang, T. Zeng, R. Aleisa, Z. Qiu, Y. Chen, J. Huang, D. Wang, Z. Yan, Y. Yin, Anisotropic plasmonic nanostructures for colorimetric sensing, Nano Today. 32 (2020) 100855.

[246] X.-H. Chen, J. Zhu, J.-J. Li, J.-W. Zhao, A plasmonic and SERS dual-mode iodide ions detecting probe based on the etching of Ag-coated tetrapod gold nanostars, J. Nanopart. Res. 21 (2019) 1–14.

[247] H. Zhu, C. Liu, X. Liu, Z. Quan, W. Liu, Y. Liu, A multi-colorimetric immunosensor for visual detection of ochratoxin A by mimetic enzyme etching of gold nanobipyramids, Microchim. Acta. 188 (2021) 1–10.

[248] X. Yang, Y. Yu, Z. Gao, A highly sensitive plasmonic DNA assay based on triangular silver nanoprism etching, Acs Nano. 8 (2014) 4902–4907.

[249] T. Lin, M. Zhang, F. Xu, X. Wang, Z. Xu, L. Guo, Colorimetric detection of benzoyl peroxide based on the etching of silver nanoshells of Au@ Ag nanorods, Sens. Actuators B. 261 (2018) 379–384.

[250] Z. Zhang, Z. Chen, D. Pan, L. Chen, Fenton-like reaction-mediated etching of gold nanorods for visual detection of Co2+, Langmuir 31 (2015) 643–650.

[251] K. Kermanshahian, A. Yadegar, H. Ghourchian, Gold nanorods etching as a powerful signaling process for plasmonic multicolorimetric chemo-/biosensors: Strategies and applications, Coord. Chem. Rev. 442 (2021) 213934.

[252] X. Fang, H. Ren, H. Zhao, Z. Li, Ultrasensitive visual and colorimetric determination of dopamine based on the prevention of etching of silver nanoprisms by chloride, Microchim. Acta 184 (2017) 415–421.

[253] P.D. Howes, S. Rana, M.M. Stevens, Plasmonic nanomaterials for biodiagnostics, Chem. Soc. Rev. 43 (2014) 3835–3853.

[254] C.L. Nehl, J.H. Hafner, Shape-dependent plasmon resonances of gold nanoparticles, J. Mater. Chem. 18 (2008) 2415–2419.

[255] R. Wilson, The use of gold nanoparticles in diagnostics and detection, Chem. Soc. Rev. 37 (2008) 2028–2045.

[256] M.E. Stewart, C.R. Anderton, L.B. Thompson, J. Maria, S.K. Gray, J.A. Rogers, R.G. Nuzzo, Nanostructured plasmonic sensors, Chem. Rev. 108 (2008) 494–521.

[257] Q. Wang, R.-D. Li, B.-C. Yin, B.-C. Ye, Colorimetric detection of sequence-specific microRNA based on duplex-specific nuclease-assisted nanoparticle amplification, Analyst 140 (2015) 6306–6312.

[258] L. Zhang, Y. Yuan, Y. Zhang, Z. Liu, W. Xiao, J. Nie, J. Li, Equipment-free quantitative aptamer-based colorimetric assay based on target-mediated viscosity change, ACS Omega. 3 (2018) 1451–1457.

[259] X. Yuan, Y. Chen, Visual determination of Cu 2+ through copper-catalysed in situ formation of Ag nanoparticles, Analyst. 137 (2012) 4516–4523.

[260] Y. Wang, Y. Zeng, W. Fu, P. Zhang, L. Li, C. Ye, L. Yu, X. Zhu, S. Zhao, Seed-mediated growth of Au@ Ag core-shell nanorods for the detection of ellagic acid in whitening cosmetics, Anal. Chim. Acta 1002 (2018) 97–104.

[261] A. Orouji, S. Abbasi-Moayed, F. Ghasemi, M.R. Hormozi-Nezhad, A wide-range pH indicator based on colorimetric patterns of gold@ silver nanorods, Sens. Actuators B 358 (2022) 131479.

[262] A. Naseri, F. Ghasemi, Analyte-restrained silver coating of gold nanostructures: An efficient strategy to advance multicolorimetric probes, Nanotechnology. 33 (2021) 075501.

[263] K.A. Willets, R.P. Van Duyne, Localized surface plasmon resonance spectroscopy and sensing, Annu. Rev. Phys. Chem. 58 (2007) 267–297.

[264] J. Homola, Present and future of surface plasmon resonance biosensors, Anal. Bioanal.Chem. 377 (2003) 528–539.

[265] B. Sepúlveda, P.C. Angelomé, L.M. Lechuga, L.M. Liz-Marzán, LSPR-based nanobiosensors, Nano Today. 4 (2009) 244–251.

[266] K.-S. Lee, M.A. El-Sayed, Gold and silver nanoparticles in sensing and imaging: Sensitivity of plasmon response to size, shape, and metal composition, J. Phys. Chem. B 110 (2006) 19220–19225.

[267] C.-D. Chen, S.-F. Cheng, L.-K. Chau, C.C. Wang, Sensing capability of the localized surface plasmon resonance of gold nanorods, Biosens. Bioelectron. 22 (2007) 926–932.

[268] J.N. Anker, W.P. Hall, O. Lyandres, N.C. Shah, J. Zhao, R.P. Van Duyne, Biosensing with plasmonic nanosensors, nanoscience and technology, Nat. Mater., 7, A (2010) 442–453.

[269] N. Gandra, C. Portz, L. Tian, R. Tang, B. Xu, S. Achilefu, S. Singamaneni, Probing distance-dependent plasmon-enhanced near-infrared fluorescence using polyelectrolyte multilayers as dielectric spacers, Angew. Chem. Int. Ed. 53 (2014) 866–870.

[270] S.-W. Hsu, C. Ngo, A.R. Tao, Tunable and directional plasmonic coupling within semiconductor nanodisk assemblies, Nano Lett. 14 (2014) 2372–2380.

[271] M.-C. Estevez, M.A. Otte, B. Sepulveda, L.M. Lechuga, Trends and challenges of refractometric nanoplasmonic biosensors: A review, Anal. Chim. Acta. 806 (2014) 55–73.

[272] F. Wang, S. Liu, M. Lin, X. Chen, S. Lin, X. Du, H. Li, H. Ye, B. Qiu, Z. Lin, Colorimetric detection of microcystin-LR based on disassembly of orient-aggregated gold nanoparticle dimers, Biosens. Bioelectron. 68 (2015) 475–480.

[273] K. Shrivas, R. Shankar, K. Dewangan, Gold nanoparticles as a localized surface plasmon resonance based chemical sensor for on-site colorimetric detection of arsenic in water samples, Sens. Actuators B. 220 (2015) 1376–1383.

[274] D. Wu, T. Gao, L. Lei, D. Yang, X. Mao, G. Li, Colorimetric detection of proteins based on target-induced activation of aptazyme, Anal. Chim. Acta 942 (2016) 68–73.

[275] F. Wang, J. Sun, Y. Lu, X. Zhang, P. Song, Y. Liu, Dispersion-aggregation-dispersion colorimetric detection for mercury ions based on an assembly of gold nanoparticles and carbon nanodots, Analyst. 143 (2018) 4741–4746.

[276] S. Bothra, R. Kumar, S.K. Sahoo, Pyridoxal derivative functionalized gold nanoparticles for colorimetric determination of zinc (II) and aluminium (III), RSC Adv. 5 (2015) 97690–97695.

[277] C. Karami, S.Y. Mehr, E. Deymehkar, M.A. Taher, Naked eye detection of Cr3+ and Co2+ ions by gold nanoparticle modified with azomethine, Plasmon. 13 (2018) 537–544.

[278] J. Du, H. Ge, Q. Gu, H. Du, J. Fan, X. Peng, Gold nanoparticle-based nano-probe for the colorimetric sensing of Cr 3+ and Cr 2 O 7 2– by the coordination strategy, Nanoscale. 9 (2017) 19139–19144.

[279] R. Roto, B. Mellisani, A. Kuncaka, M. Mudasir, A. Suratman, Colorimetric sensing of Pb2+ ion by using ag nanoparticles in the presence of dithizone, Chemosensors. 7 (2019) 28.

[280] K. Shrivas, S. Patel, D. Sinha, S.S. Thakur, T.K. Patle, T. Kant, K. Dewangan, M.L. Satnami, J. Nirmalkar, S. Kumar, Colorimetric and smartphone-integrated paper device for on-site determination of arsenic (III) using sucrose modified gold nanoparticles as a nanoprobe, Microchim. Acta 187 (2020) 1–9.

[281] N. Bi, Y. Zhang, Y. Xi, M. Hu, W. Song, J. Xu, L. Jia, Colorimetric response of lysine-caped gold/silver alloy nanocomposites for mercury (II) ion detection, Colloids Surf. B 205 (2021) 111846.

[282] X. Hu, J. Du, J. Pan, F. Wang, D. Gong, G. Zhang, Colorimetric detection of the β-agonist ractopamine in animal feed, tissue and urine samples using gold–silver alloy nanoparticles modified with sulfanilic acid, Food Addit. Contam.: Part A. 36 (2019) 35–45.

[283] J. Du, X. Hu, G. Zhang, X. Wu, D. Gong, Colorimetric detection of cadmium in water using L-cysteine functionalized gold–silver nanoparticles, Anal. Lett. 51 (2018) 2906–2919.

[284] Q. Zhong, Y. Chen, X. Qin, Y. Wang, C. Yuan, Y. Xu, Colorimetric enzymatic determination of glucose based on etching of gold nanorods by iodine and using carbon quantum dots as peroxidase mimics, Microchim. Acta 186 (2019) 1–8.

[285] A. Liu, M. Li, J. Wang, F. Feng, Y. Zhang, Z. Qiu, Y. Chen, B.E. Meteku, C. Wen, Z. Yan, Ag@ Au core/ shell triangular nanoplates with dual enzyme-like properties for the colorimetric sensing of glucose, Chin. Chem. Lett. 31 (2020) 1133–1136.

[286] J.-K. Chen, S.-M. Zhao, J. Zhu, J.-J. Li, J.-W. Zhao, Colorimetric determination and recycling of Hg2+ based on etching-induced morphology transformation from hollow AuAg nanocages to nanoboxes, J. Alloys Compd. 828 (2020) 154392.

[287] S. Liu, X. Wang, C. Zou, J. Zhou, M. Yang, S. Zhang, D. Huo, C. Hou, Colorimetric detection of Cr6+ ions based on surface plasma resonance using the catalytic etching of gold nano-double cone@ silver nanorods, Anal. Chim. Acta 1149 (2021) 238141.

[288] Q. Wang, R. Peng, Y. Wang, S. Zhu, X. Yan, Y. Lei, Y. Sun, H. He, L. Luo, Sequential colorimetric sensing of cupric and mercuric ions by regulating the etching process of triangular gold nanoplates, Microchim. Acta 187 (2020) 1–9.

[289] Q. Wang, Y. Wang, M. Guan, S. Zhu, X. Yan, Y. Lei, X. Shen, L. Luo, H. He, A multicolor colorimetric assay for sensitive detection of sulfide ions based on anti-etching of triangular gold nanoplates, Microchem. J. 159 (2020) 105429.

[290] S. Lee, Y.-S. Nam, S.-H. Choi, Y. Lee, K.-B. Lee, Highly sensitive photometric determination of cyanide based on selective etching of gold nanorods, Microchim. Acta 183 (2016) 3035–3041.

[291] P. Li, S.M. Lee, H.Y. Kim, S. Kim, S. Park, K.S. Park, H.G. Park, Colorimetric detection of individual biothiols by tailor made reactions with silver nanoprisms, Sci. Rep. 11 (2021) 1–8.

[292] L. Li, L. Zhang, Y. Zhao, Z. Chen, Colorimetric detection of Hg (II) by measurement the color alterations from the "before" and "after" RGB images of etched triangular silver nanoplates, Microchim. Acta 185 (2018) 1–6.

[293] P. Zhang, L. Wang, J. Zeng, J. Tan, Y. Long, Y. Wang, Colorimetric captopril assay based on oxidative etching-directed morphology control of silver nanoprisms, Microchim. Acta 187 (2020) 1–8.

[294] L. Li, J. Wang, Z. Chen, Colorimetric determination of uric acid based on the suppression of oxidative etching of silver nanoparticles by chloroauric acid, Microchim. Acta 187 (2020) 1–7.

[295] T. Lin, Z. Li, Z. Song, H. Chen, L. Guo, F. Fu, Z. Wu, Visual and colorimetric detection of p-aminophenol in environmental water and human urine samples based on anisotropic growth of Ag nanoshells on Au nanorods, Talanta 148 (2016) 62–68.

[296] X.-H. Pham, E. Hahm, T.H. Kim, H.-M. Kim, S.H. Lee, Y.-S. Lee, D.H. Jeong, B.-H. Jun, Enzyme-catalyzed Ag growth on Au nanoparticle-assembled structure for highly sensitive colorimetric immunoassay, Sci. Rep. 8 (2018) 1–7.

[297] Y. Wang, P. Zhang, X. Mao, W. Fu, C. Liu, Seed-mediated growth of bimetallic nanoparticles as an effective strategy for sensitive detection of vitamin C, Sens. Actuators B. 231 (2016) 95–101.

[298] S.M. Shaban, B.-S. Moon, D.-H. Kim, Selective and sensitive colorimetric detection of p-aminophenol in human urine and paracetamol drugs based on seed-mediated growth of silver nanoparticles, Environ. Technol. Innov. 22 (2021) 101517.

[299] Y. Li, X. Ma, Z. Xu, M. Liu, Z. Lin, B. Qiu, L. Guo, G. Chen, Multicolor ELISA based on alkaline phosphatase-triggered growth of Au nanorods, Analyst. 141 (2016) 2970–2976.

[300] J. Chen, A.A. Jackson, V.M. Rotello, S.R. Nugen, Colorimetric detection of Escherichia coli based on the enzyme-induced metallization of gold nanorods, Small. 12 (2016) 2469–2475.

[301] J.-M. Lin, Y.-Q. Huang, Z.-b. Liu, C.-Q. Lin, X. Ma, J.-M. Liu, Design of an ultra-sensitive gold nanorod colorimetric sensor and its application based on formaldehyde reducing Ag+, RSC Adv. 5 (2015) 99944–99950.

[302] M. Amjadi, T. Hallaj, E. Nasirloo, In situ formation of Ag/Au nanorods as a platform to design a non-aggregation colorimetric assay for uric acid detection in biological fluids, Microchem. J. 154 (2020) 104642.

[303] Q. Zhang, Y. Yu, X. Yun, B. Luo, H. Jiang, C. Chen, S. Wang, D. Min, Multicolor colorimetric sensor for detection of omethoate based on the inhibition of the enzyme-induced metallization of gold nanorods, ACS Appl. Nano Mat. 3 (2020) 5212–5219.

[304] T. Lin, Y. Wu, Z. Li, Z. Song, L. Guo, F. Fu, Visual monitoring of food spoilage based on hydrolysis-induced silver metallization of Au nanorods, Anal. Chem. 88 (2016) 11022–11027.

CHAPTER 2

Gold and silver nanoparticles: Properties and toxicity

Cristina Buzea[a], **Ivan Pacheco**[a,b,c]

[a]*IIPB Medicine Corporation, Owen Sound, ON, Canada* [b]*Department of Pathology, Grey Bruce Health Services, Owen Sound, ON, Canada* [c]*Department of Pathology and Laboratory Medicine, Schülich School of Medicine & Dentistry, Western University, London, ON, Canada*

2.1 Gold and silver nanoparticle properties

The physico-chemical characteristics of nanoparticles are of tremendous importance in determining their properties, applications as well as toxic effects, as outlined in Fig. 2.1. Among these, their physical characteristics, such as size and composition together with their crystalline phase and morphology will decide their electro-magnetic properties, aggregation propensity, and hydrophobicity/hydrophilicity. The most relevant properties of gold and silver nanoparticles are going to be outlined below.

Nanoparticles are aggregates of a relatively small number of atoms. The smallest nanoparticles, often termed "clusters" have only a few atoms, while larger nanoparticles will be made of more than 100,000 atoms. The important thing to remember is that nanoparticles are at the confluence of the quantum mechanics (that describes effects occurring at quantum scale) and classical electrodynamics and solid-state physics (that describe bulk materials) [1,2]. The nanoscale is the intermediary between the microscopic and macroscopic worlds. As we approach the atomic scale the quantum effects will be dominant. In addition, as nanoparticle size decreases, the surface effects will exceed the bulk effects simply because nanoparticles have more and more atoms at their surface compared to their bulk. Therefore a material in nanoparticle form compared to its bulk counterpart will exhibit quite different optical, magnetic, chemical, and mechanical properties [3,4]. Surface effects are responsible for the smooth scaling of physical properties that reflect the increasing number of atoms at the nanoparticle surface compared to its core [5]. The surface effects of nanoparticles will reflect in their enhanced chemical reactivity and reduced melting point compared to bulk of the same material. The quantum size effect is related to the quantized energy spectrum of the electrons inside a nanoparticle as a result of their confinement within the physical boundaries of a nanoparticle. As we lower the size of a nanoparticle its quantum size effects will dominate their properties, resulting in altered optical, electrical and magnetic properties compared to bulk. For example, quantum effects in gold nanoparticles are responsible for their

Gold and Silver Nanoparticles.
DOI: https://doi.org/10.1016/B978-0-323-99454-5.00007-X

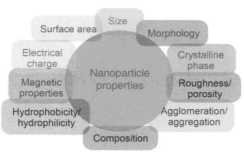

Fig. 2.1: Physico-chemical characteristics that determine a nanoparticle properties.

interesting magnetic and optical properties, as we will show later in this chapter. In general gold nanoparticles behave very different from bulk gold: they show ferromagnetic behavior compared to bulk gold which is diamagnetic; the melting temperature of small gold nanoparticles is much lower than that of bulk gold; optical properties of gold nanoparticles manifest as different colors in solution corresponding to different sized nanoparticles [6].

2.1.1 Gold and silver nanoparticle size

Gold and silver nanoparticles can be manufactured in various sizes, ranging from a few atoms or as large as particles of micron size. Usually, particles with micron size will be called microparticles, while those smaller will be under the general umbrella term of nanoparticles. Nanoparticles with sizes of up to several nanometers contain a small number of atoms (up to several thousand) and are called **clusters**. These clusters are probably the most interesting of the entire class of nanoparticles, as they have properties that differ from both molecules or atoms and bulk form, their properties being very strongly dependent on their size [7,8]. The smaller a nanoparticle, the more dominant its surface and quantum confinement effects. The size of a nanoparticle for which the transition to bulk properties occurs is not a fixed size, but a range of values and it depends on material type. A particle with a diameter of around 3 nanometers will have about the same fraction of atoms on its surface and in its core. Hence, smaller nanoparticles will have dominant surface effects while larger nanoparticles will have properties dominated by their bulk atoms [6]. Nanoparticles with sizes of a few nanometers to tens of nanometers show changes in their catalytic activity, optical activity, magnetism, and mechanical properties, such as melting temperature compared to the same material as bulk [9,6,10,11].

Very small clusters that have a "magic number" of atoms will exhibit nonmonotonic behaviors, such as local maxima or kinks in their spectra [12]. Also, small clusters of metals that are considered nonreactive in bulk form, such as noble metals, will exhibit a catalytic behavior [12]. The structure of a gold cluster depends on its size, and it can form nanowires if formed of less than 9 atoms, zigzags, polyhedral, icosahedral, cage, or tubular structures [12]. These small clusters will show catalytic properties as a result of their electronic configurations.

2.1.2 Gold and silver nanoparticle classification

Gold and silver nanoparticles can be classified according to their aspect ratio: with a short aspect ratio (such as spherical, oval, cubes, etc.) and long aspect ratio (such as fibers). Fig. 2.2 shows examples of gold and silver nanoparticles with short and long aspect ratio having various morphologies [13–16].

Short aspect ratio gold nanoparticles with various morphologies

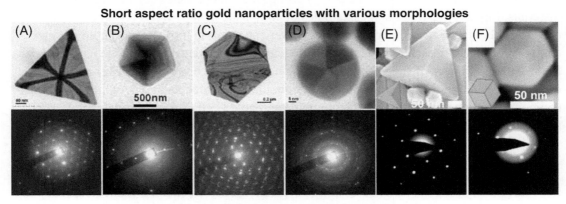

Long aspect ratio silver and gold nanoparticles

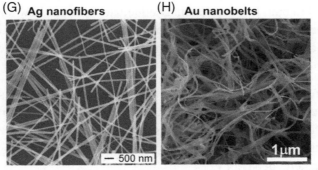

Fig. 2.2: Gold and silver nanoparticles with short and long aspect ratio having various morphologies. (A–F) Upper part: Short aspect ratio gold nanoparticles with different morphologies - transmission electron microscopy images and scanning electron microscopy images (upper part) and lower part: the selected-area electron diffraction (SAED) patterns; nanoplate with (A) triangular, (B) pentagonal, and (C) hexagonal shape; (D) spherical nanoparticle; (E) bipyramid, and (F) rhombic dodecahedron. *Images A, B, C, and D are reproduced with permission from Das S. K. et al. 2010. Small, 6, 1012-1021, John Wiley and Sons [13]. Images E and F are reproduced with permission from Personick M. L. et al. 2011. Journal of the American Chemical Society, 133, 6170-6173. Copyright (2011) American Chemical Society [14].* (G–H) Silver and gold long aspect ratio nanoparticles. Scanning electron microscopy images of (G) silver nanofibers, and (H) gold nanobelts. *Image (G) is reproduced with permission from Wiley B. et al. 2005. Langmuir, 21, 8077-8080. Copyright (2005) American Chemical Society [15]. Image (H) is reprinted with permission from Zhao N. et al. 2008. Langmuir, 24, 991-998. Copyright (2008) American Chemical Society [16].*

The morphology of a nanoparticle is important, as it controls to an important extent their physico-chemical properties. Technological advances in fabrication methods allow the synthesis of gold and silver nanoparticles with very diverse morphologies. Some of these are exemplified in Fig. 2.2, including nanoplates with triangular, pentagonal, and hexagonal shapes, spherical, or nanofibers and nanobelts [13–16]. Other morphologies not showed here may include helical nanopillars, straight nanopillars, and zigzag nanopillars.

Gold and silver nanoparticles can be part of composite nanoparticles containing two or more materials. Fig. 2.3 illustrates some examples of composite nanoparticles containing gold and silver [17–19]. Special fabrication procedures can result in intricate morphologies and compositions encompassing metals, semiconductors, and insulators within one nanoparticle. Fig. 2.3 shows images of such nanoparticles: nano-barcodes composed of silver and aluminum oxide; nano-zigzags made of nickel, silicon, titanium oxide and gold; chiral nano-hooks composed of copper, titanium, aluminum oxide and gold [17]; CdS encapsulated gold [18]; and silver nanowires coated with minute titanium oxide nanoparticles [19].

2.1.3 Magnetic properties of gold nanoparticles

Materials modify their magnetic behavior in the presence of an external magnetic field. They can be either: ferromagnetic, paramagnetic, diamagnetic, antiferromagnetic, or ferromagnetic [20].

Both silver and gold in bulk form are diamagnetic materials and show a zero net magnetic moment [20]. In the presence of an external magnetic field, diamagnetic materials show a very weak magnetic response. After the external magnetic field is removed these materials will not retain a magnetic moment.

Gold is a very interesting material because in bulk form it is chemically stable and diamagnetic, however, when in nanoparticle form with sizes below a specific threshold (few nanometers) will show ferromagnetism [21–26]. Its magnetism reverts back to the one of bulk (diamagnetism) if the nanoparticles are larger than 4 nanometers. We might add that gold thin films also show magnetic behavior [27].

In addition to the inherent magnetism of small gold clusters, their magnetic behavior can be chemically controlled by adding various ligands, and can range from diamagnetic, paramagnetic to ferromagnetic [28]. Some of these ligands are shown in Fig. 2.4. Magnetism depends strongly on gold cluster size: clusters of 25 and 38 atoms with thiol and mixed ligands are diamagnetic while nanoparticles containing 55 atoms show a ferromagnetic behavior.

The origin of magnetism in small gold nanoparticles is still under debate. Some authors attribute it to the two-dimensional character of the particle surface [23,29]. The few top surface layers of a nanoparticle exhibit a different magnetic ordering compared to the core of the nanoparticle [29]. In the case of a gold nanoparticle, its surface atoms show ferromagnetism while those in the core have a paramagnetic behavior [23]. Other authors hypothesize

Composite nanoparticles and nanofibers containing silver or gold

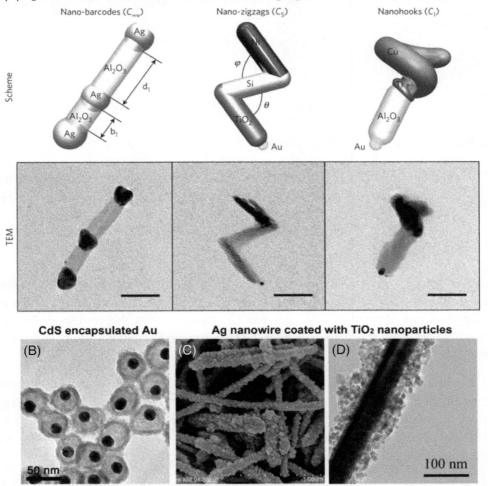

Fig. 2.3: **Composite nanoparticles and nanofibers containing silver and gold.**
(A) Nanoparticle composites with lower symmetry with morphology of nano-barcodes composed of Ag-Al2O3, nano-zigzags composed of Ni-Si-TiO$_2$-Au, nanohooks composed of Cu-Ti-Al2O3-Au. Upper row shows structure models while the lower row shows transmission electron microscopy images. (B) spherical nanocomposites made of a gold core encapsulated by CdS a shell. (C and D) Ag nanowire coated with TiO$_2$ nanoparticles. *Image (A) is reprinted from Mark A. G. et al. 2013. Nature Materials, 12, 802-807, by permission from Springer Nature [17]. Image (B) is reproduced from Xia Y. S. & Tang Z. Y. 2012. Advanced Functional Materials, 22, 2585-2593 with permission from John Wiley and Sons [18]. Images (C) and (D) are reprinted from Journal of Hazardous Materials, Volume 177, B. Cheng et al., Pages 971–977, Copyright (2010), with permission from Elsevier [19].*

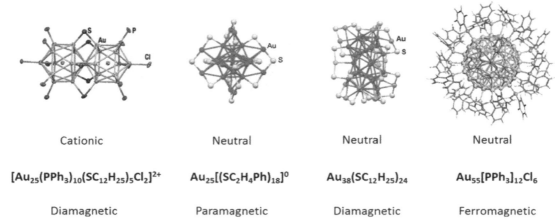

Cationic	Neutral	Neutral	Neutral
$[Au_{25}(PPh_3)_{10}(SC_{12}H_{25})_5Cl_2]^{2+}$	$Au_{25}[(SC_2H_4Ph)_{18}]^0$	$Au_{38}(SC_{12}H_{25})_{24}$	$Au_{55}[PPh_3]_{12}Cl_6$
Diamagnetic	Paramagnetic	Diamagnetic	Ferromagnetic

Fig. 2.4: Gold clusters with various ligands can exhibit diamagnetic, paramagnetic, or ferromagnetic behavior.
Image reproduced with permission from [28], John Wiley and Sons.

the origin of magnetism of small gold nanoparticles as a result of self-sustained persistent currents [30]. Some authors believe that the existence of an odd or even number of electrons affect the properties of small metallic nanoparticles, however, the observed magnetism of gold clusters with 3 nm size cannot explain the large magnetic moments of about 20 spins/particle [21].

The exposure to air modifies the magnetic moment of gold nanoparticles, namely increases the magnetic moment by about 20% [31]. The authors propose two possible mechanisms for explaining this result: (1) oxygen absorption enhances nanoparticle surface magnetization, and (2) oxygen absorption decreases shell magnetization, which is antiparallel to the core ferromagnetic magnetization, resulting in an increased magnetization.

Calculations indicate that small silver clusters might also develop magnetism [32]. The highest magnetic moment calculated was held by the Ag13 cluster, being theorized that the symmetry and lower coordination number are responsible for its value. To our knowledge, there is no experimental evidence yet for silver nanoparticles magnetism.

2.1.4 Optical properties of gold nanoparticles

Optical properties of nanoparticles reflect how they interact with electromagnetic radiation. In general the optical characteristics of nanoparticles are dependent on the composition, shape and size [1,33]. Materials can interact with light via different processes, they can absorb, scatter the light at the same frequency, re-emit absorbed light (fluorescence), and enhance the local electromagnetic field [33]. Gold and silver nanoparticles show all of the above processes.

When gold nanoparticles are exposed to an external electromagnetic radiation, their electrical field component displaces the electrons from the conduction band, resulting in uncompensated surface electrical charges [1,33]. Because they occur at the surface, these oscillations are called surface plasmons (collective excitations of the electrons from the conduction band) (Khlebtsov and Dykman, 2010). The surface plasmons have a resonance frequency dependent on size and morphology, as depicted in Fig. 2.5 for spherical gold nanoparticles and rods with various aspect ratios [34]. Fig. 2.5 (E–F) shows the optical absorption spectra for Au [34] and Ag [35] nanoparticles, respectively. Fig. 2.5 (A–D) shows the transmission electron microscopy images of Au nanoparticles measured in e) [34]. In the case of gold nanorods, both

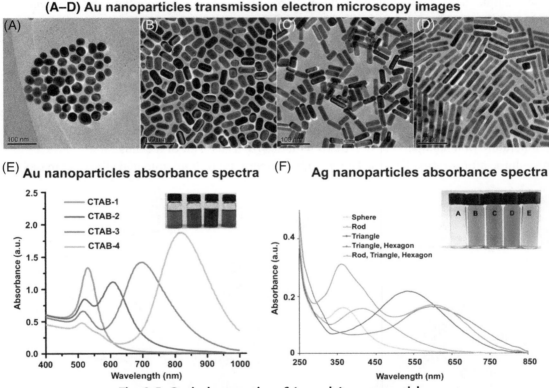

Fig. 2.5: Optical properties of Au and Ag nanoparticles.

(A–D) Transmission electron microscopy images of cetyltrimethylammonium bromide-coated Au nanoparticles with different aspect ratio; (E) Au nanoparticles absorption spectra (for nanoparticles shown in A–D); (F) Ag nanoparticles absorbance spectra for different shape of Ag nanoparticles. Insets of (E and F) are photograph of gold and silver nanoparticle suspensions, respectively, showing different colors. *Figure (A–E) are adapted from Biomaterials, Vol 31, issue 30, Qiu Y. et al, Pages 7606–7619, Copyright (2010), with permission from Elsevier [34]. Figure (F) is adapted from Ferrag, C., et al., Colloids and Surfaces B-Biointerfaces, 2021. 197. Copyright (2021), with permission from Elsevier [35].*

transversal and longitudinal oscillations contribute to an enhanced band in the near-infrared region and a reduced visible region band. In both cases, the spectra peak shifts as a function to the shape and morphology of the nanoparticles.

The fluorescence of gold nanorods exceeds that of bulk gold by 6–7 times [33]. Gold nanorods longer than 200 nm have a luminescence in the visible spectrum that can be perceived with the naked eye.

Optical properties of nanoparticles, in addition to size and shape, also depend on chirality and handedness [17]. Chirality is the property of the structure of an object whose image differs from its mirror image [36]. Handedness is the property of chiral objects which divide them into left-handed and right-handed objects [36].

For example, helical gold nanoparticles with the same size and morphology, but different chirality show different optical response. Fig. 2.6 exemplifies this. While the position of the circular dichroism (CD) spectral peak depends on the helix radius, the left-handed helical nanoparticle has a mirror-image spectrum compared to the right-handed helix.

2.1.5 Melting temperature of gold and silver nanoparticles

Thermodynamics laws as well as experimental observations indicate that the melting temperature of a nanoparticle decreases with its size as a result of a change in its surface-to-volume atoms ratio [40–46]. The melting temperature of both gold and silver decreases

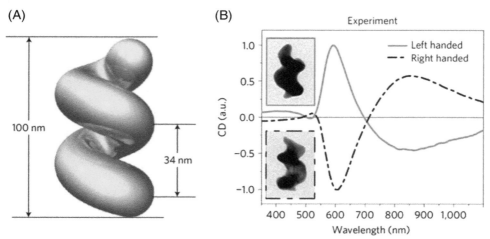

Fig. 2.6: Optical properties dependence on chirality for gold nanoparticles.
(A) Schematics of a gold nanohelix with two turns; (B) Normalized circular dichroism (CD) spectra versus wavelength of left-handed and right-handed gold helices, shown inset. *Image reprinted from Mark A. G. et al. 2013, Nature Materials 12, 802-807 by permission from Springer Nature [17], copyright (2013).*

from their bulk values T_m(bulk Au) =1337 K [47] and T_m(bulk Ag) =1233 K [38] according to Gibbs-Thomson equation for small crystallites with radius r:

$$\Delta T = \frac{4\sigma M T_{bulk}}{\Delta H_m r \rho}$$

where σ - is the surface energy of the material, M = the molar mass, T_{bulk} = the melting point of the bulk form, ΔH_m = the melting enthalpy, and ρ = the material density [38]. Fig. 2.7A–B illustrates the experimental data for the variation of the melting temperature of gold [37,48] and silver [38] with the cluster (nanoparticle) size. The melting temperature of nanoparticles versus the nanoparticle size decreases monotonically from the bulk value to lower values. The drop in the melting temperature of silver (deviation from the monotonic increase) shown in Fig. 2.7B) is due to a kink in the lattice constant corresponding to the cluster size of 12 nm [38], probably related to a phase transition. The experimental data for the melting temperature variation with nanoparticle size varies slightly from author to author, depending on their experimental conditions, fabrication of nanoparticles, etc. Fig. 2.7C) shows the evolution of the morphology for gold clusters with different configurations while D) illustrates the calculated melting point of Au clusters versus the number of atoms in the cluster for clusters with different configurations [39]. As deduced from the calculations, gold clusters with a cuboctahedral configuration will have a melting point of 645 K for a cluster made of 147 atoms and increase to 771 K for a cluster made of 923 atoms [39]. Also, as indicated by this figure, gold clusters encompassing the same number of atoms will have different melting points when in different morphological configurations, the largest difference in melting points being up to 75 K. Cuboctahedra and icosahedra gold clusters have a lower melting point compared to regular decahedra, which are lower than that of Marks decahedra and star-decahedra [39]. The larger clusters, containing more than 12,000 atoms, will have the same melting point, except those with icosahedral structure.

2.2 Release of silver nanoparticles from consumer products

Silver in nanoparticle form is increasingly utilized as an antimicrobial agent in various commercial products ranging from fabrics, to food storage plastics [49–51]. The inclusion of silver nanoparticles into various products without the existence of strong chemical bonds between nanoparticles and the existing medium results in increased likelihood of release during the use and disposal of these products. For example, silver nanoparticles are used in the textile industry for their antimicrobial properties. However nanoparticle-coated textiles may be a source of pollution and toxicity for consumers and other organisms. Fabrics coated with nanoparticles have been shown to release these nanoparticles following washing.

Several authors have studied the release of Ag nanoparticles from a variety of consumer products [49–53]. Measurements of silver nanoparticles in consumer products, such as shirts, medical masks, toothpaste, shampoo, detergent, toys, and humidifiers showed

(A)

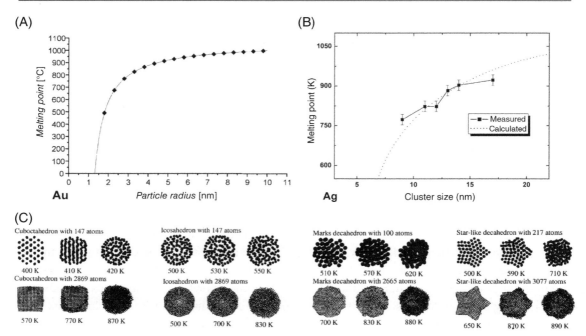

Au   *Particle radius* [nm]

(B)

Ag   Cluster size (nm)

(C)

Au clusters configuration evolution with temperature for various shapes

(D)

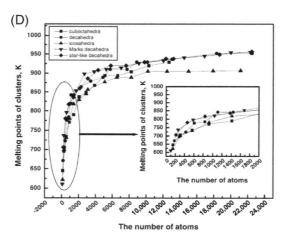

The number of atoms

Melting point of Au clusters with different configurations

Fig. 2.7: Experimental and calculated melting point of Au and Ag clusters versus the size of clusters.
(A) Experimental Melting point of Au clusters; (B) Experimental and calculated melting point of silver clusters; (C) Au clusters configurations variation with increasing temperature; (D) Calculated melting point of Au clusters versus the number of atoms in the cluster for clusters with different configurations. *Image (A) is reproduced from Schmid, G. and B. Corain (2003). European Journal of Inorganic Chemistry (17): 3081-3098, with permission from John Wiley & Sons, Inc. [37]. Image (B) is reprinted with permission from Springer Nature, Shyjumon I. et al. (2006). European Physical Journal D 37(3): 409-415. Copyright (2006) [38]. Images c-d) are adapted from Liu, H.B., et al., Surface Science, 2001. 491(1-2): p. 88-98. Copyright (2001), with permission from Elsevier [39].*

Table 2.1: The size and amount of silver nanoparticles released from nanocomposite materials into a liquid media. Data adapted with permission from Duncan T. V. & Pillai K. 2015. ACS Applied Materials & Interfaces, 7, 2-19. Copyright (2015) American Chemical Society [53].

Size (nm)	Nanocomposite Matrix	Conditions	Maximum release amount	Refs.
10 – 500	clothing	washing, water	1845 µg/g	[51]
-	Fabric, toothpaste	washing, water	46 µg/g	[50]
100 – 300	polyethylene (PE),	50°C, water and ethanol	4 µg/dm	[52]
-	Commercial plastic food storage containers	20°C, 10 days, water, acetic acid, ethanol, olive oil	30 ng/cm^2 of surface area	[49]
	Silicone	37°C, 10 days, water	250 ng/cm^2	[55]

concentrations of silver range from 1.4 µg to 270 mg Ag per product [50]. Following the washing in tap water, silver nanoparticles were released from the above products up to 45 µg/product [50]. Electron microscopy together with energy dispersive X-ray diffraction measurements demonstrated the existence of Ag nanoparticles in both the products and the wash water samples. Table 2.1 shows experimental data on the size and amount of silver nanoparticles released from consumer products into liquids as a result of diffusion, desorption and dissolution [53].

Recent experimental data also show the release of silver nanoparticles from plastics and their migration into foodstuff [49,52]. In Fig. 2.8A and B) are shown microscopy images of (A) Ag nanoparticles on the surface of a medical face mask and (B) Ag nanoparticles leached into the tap water wash of the medical face mask [50]. Fig. 2.8C and D) shows Ag nanoparticles from degraded plastic food containers that have Ag nanoparticles embedded in their structure that leached into liquid food [49]. One must specify that silver nanoparticles are toxic to microorganisms as well as to human and other organism cells, and exposure to silver nanoparticles at concentrations of 5 µg/ml exceeds asbestos toxicity [54].

2.3 Toxicity of silver and gold nanoparticles

The toxicity of nanoparticles is directly dependent on their composition, size and other physical properties, as illustrated in Fig. 2.9. Composition is imperative in deciding whether a nanoparticle is inherently toxic or not, such in the case of known carcinogens. In addition, crystallinity or the type of structure that the nanoparticle adopts, modulates its toxicity, such as asbestos [2].

In addition to nanoparticle properties, other factors that have to be taken into account are the conditions of exposure to nanoparticles and the pre-existing medical conditions of the subject. Conditions of exposure to nanoparticles are related to the concentration, frequency,

(A) **Ag nanoparticles on the surface
of a medical face mask**

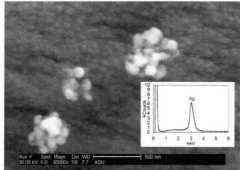

(B) **Ag nanoparticles from tap water
wash of the medical face mask**

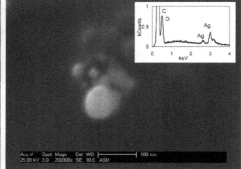

Migration of Ag nanoparticles from commercial plastic food containers

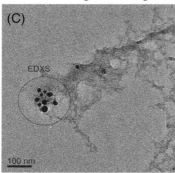

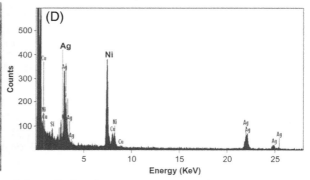

Fig. 2.8: Ag nanoparticle pollution from consumer products.

(A) Scanning electron microscopy image of Ag nanoparticles on the surface of a medical face mask coated with Ag nanoparticles with size smaller than 20 nm. (B) microscopy image of nanoparticles collected from water wash of the face mask. The inset of (A) and (B) show the energy dispersive X-ray spectrum indicating the composition as silver. (C and D) transmission electron microscopy image and (D) the energy dispersive X-ray spectrum of the circle indicated in figure (C) demonstrating Ag nanoparticles migration in liquids as a result of the degradation of food containers containing Ag nanoparticles as an antibacterial agent. *Images (A) and (B) are reprinted from Benn T. et al. 2010. Journal of Environmental Quality, 39, 1875-1882 with permission from John Wiley and Sons [50]. Images (C–D) are reprinted with permission from von Goetz N. et al. 2013. Food Additives and Contaminants Part a-Chemistry Analysis Control Exposure & Risk Assessment, 30, 612-620 by permission of Taylor & Francis Ltd, http://www.tandfonline.com [49].*

and duration of exposure. Usually, the toxicity of nanoparticles will increase for exposure to higher concentrations of nanoparticles, for more frequent exposures, and longer exposure times, as a result of higher and more sustain accumulation of nanoparticles within organisms. Pre-existing medical conditions might be exacerbated by exposure to nanoparticles and be conducive to other diseases.

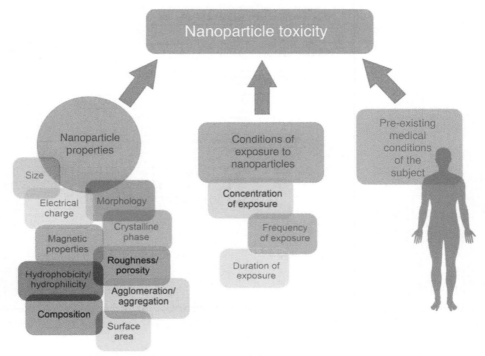

Fig. 2.9: Factors that decide the toxicity of nanoparticles include nanoparticle properties, conditions of exposure to nanoparticles and pre-existing medical conditions of the subject.

Nanoparticles may be toxic simply due to their extremely small size, 1 billion times smaller than a meter, which allow them to enter organisms and cells. Nanoparticles can be internalized via different routes, such as inhalation, ingestion, injection, dermal exposure, or injection. Nanoparticles can be inhaled because their size is smaller than lung alveoli (200 µm) [2]. From the lung they are able to translocate to the circulatory system, being smaller than the blood capillaries (5 µm). They are even able to enter cells, having a smaller size than cells (red blood cells have a size of ~ 5 µm) and even cell organelles, like mitochondria (500 nm). When nanoparticles localize within organs and cells they might disrupt basic cellular processes, resulting in inflammation and disease [2,56–59].

2.3.1 Biodistribution of gold and silver nanoparticles

Both gold and silver nanoparticles will reach and locate in organs when their size is small enough to pass fenestrated cells, and circulatory routes. Fig. 2.10 shows a schematics of silver nanoparticle routes of exposure, biodistribution, and toxicity [60]. Nanoparticle exposure may occur via oral ingestion, inhalation, dermal or intravenous injection. Depending on their size, morphology and surface functionalization, the nanoparticles are able to pass gastric mucosal, lung epithelial, and skin barriers [60]. From there they can travel via the circulatory [61,62]

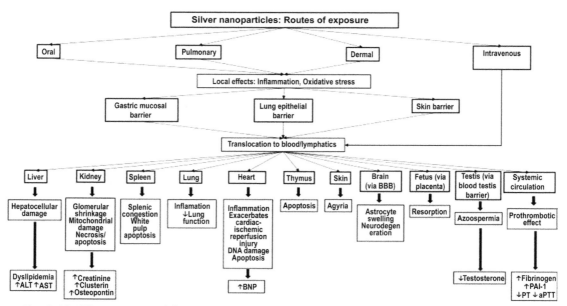

Fig. 2.10: Silver nanoparticles routes of exposure, biodistribution and adverse health effects.
Reproduced from Ferdous, Z. and A. Nemmar, Int. J. Mol. Sci. 2020 21(7) according to a CC.BY license 4.0
[60].

and lymphatic systems [63] to organs, such as: liver, kidney, spleen, heart, thymus, brain and to the fetus via the placenta [60]. A brief description of adverse health effects of silver nanoparticles in every organ is specified in Fig. 2.10, including but not limited to inflammation, oxidative stress, mitochondrial damage, necrosis, apoptosis, and DNA damage. While gold nanoparticles are able to translocate in a similar way to silver nanoparticles, their toxicity is much reduced, if any, as described in the following.

Experiments in mice show that injected gold nanoparticles with diameter of 15, 50, 100, and 200 nm travel and distribute to organs, mostly in liver, lung and spleen but can be found also in heart, brain, and pancreas [64]. Their accumulation depends on nanoparticle size, the smaller the nanoparticles, the higher their accumulation in organs [64]. If nanoparticles have a size smaller than 50 nm, they can cross the blood-brain barrier and localize in brain tissue [64]. Fast-cleared nanoparticles will have a lower toxicity due to less interaction time with biological systems compared to that of nanoparticles with a long-term residency. Intravenously injected gold nanoparticles with a 40 nm size were still found in the liver of mice after 6 months following nanoparticle exposure [65].

Ingested nanoparticles enter the gastro-intestinal tract and most of them are excreted [66]. However, a fraction is absorbed within the tract and will travel to organs [66]. Experiments on mice show that ingested gold nanoparticles with size between 4 and 28 nm travel through the circulatory system and can be found in the brain, lung, heart, kidney, spleen, liver [66]. Nanoparticles with size larger than 58 nm were usually not detected in most analyzed samples [66].

2.3.2 Cellular uptake of silver and gold nanoparticles

Size is a very important nanoparticle characteristic that decides whether or not it is going to be internalized by cells. The maximum uptake of gold nanoparticles by nonphagocytic cells occurs for nanoparticles with size in the 50 nm range [67,68]. Fig. 2.11 illustrates the amount of cellular uptake of spherical gold nanoparticle by HeLa cells as a function of size and time [67]. The highest amount of internalization by cells occurs within the first 2 hours, with the highest amount of internalization for nanoparticles with a size of 50 nm. This distribution is in agreement with findings related to gold nanoparticle cytotoxicity having a maximum value for sizes of 40–50 nm [69].

The uptake of nanoparticles by cells can occur in various locations within the cells [70–73]. This fact is illustrated in Fig. 2.12, where nanoparticles of silver and gold are located in (A) the cytoplasm, (B) vacuoles, (C) mitochondrion, and (D) nucleus.

Gold nanoparticles internalization amount and location depends on nanoparticle size, surface functionalization, and cell type, such as normal or cancel cell [72,73]. For example, gold nanoparticles are internalized in healthy and cancer lung cells within endosomes/lysosomes. Then, in healthy cells this process is followed by their subsequent elimination, while in lung cancer cells they subsequently suffer an uptake by mitochondria, which is then followed by cell death [73].

2.3.3 Gold nanoparticles toxicity

While bulk gold is chemically inert and biocompatible, one cannot say the same about gold nanoparticles. In occupational settings, where exposure to large amounts of gold occurs during a long period of time, or in animal experiments, where exposure to large amounts of gold occurs during shorter durations indicate that gold is toxic. Gold is considered a hepatotoxin and nephrotoxin in occupational setting and from experiments on animals [74]. An important toxic effect associated to gold toxicity is aplastic anemia [74]. Gold also induces skin and lung allergic reactions [74].

The assessment of gold nanoparticle toxicity usually shows negligible toxicity, or toxicity attributed to their surface functionalization [73,75]. However, in vitro studies demonstrate that gold nanoparticles are genotoxic in different cell lines [76]. It is suggested that exposure to gold nanoparticles with concentrations smaller than a threshold value does not induce cytotoxicity for nanoparticle with sizes down to 3 nm [77]. For smaller nanoparticles it is believed that, as a result of their irreversible binding to biomolecules, they will become cytotoxic [77].

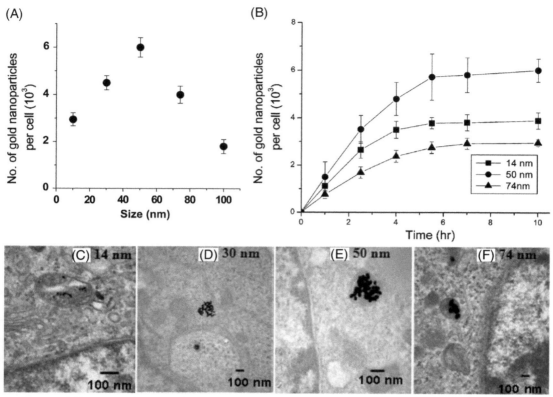

Fig. 2.11: (A) Internalization by HeLa cells of gold nanoparticles with several sizes. (B) Number of gold nanoparticle cell uptake versus the incubation time for gold nanoparticles with diameters of 14, 50, and 74 nm. (C –F) Transmission electron microscopy images of cells containing Au nanoparticles with different sizes: (C) 14 nm, (D) 30 nm, (E) 50 nm, and (F) 74 nm, respectively. *Images reprinted with permission from Chithrani B. D. et al., Nano letters, 6, 662-668. Copyright (2006) American Chemical Society [67].*

Other factors that might influence gold nanoparticle toxicity are the duration of exposure and nanoparticle retention time. Long term exposure to gold nanoparticles might induce cytotoxicity due to nanoparticles influence on cellular processes, such as cell metabolism and energy homeostasis [78]. Nanoparticles that become systemic will preferentially accumulate in the liver and their extended retention can result in hepatotoxicity [65]. Experiments in rats indicate that injected Au nanoparticles with size of 20 nm accumulate in the liver are result in modified gene expressions pertaining to detoxification, metabolism and cell cycle [79].

Nanoparticle charge is an important factor in their biological interaction. The charge of a nanoparticle can be modified from positive, neutral, to negative by means of surface functionalization. Electrostatic interactions with the negatively charged cell membranes will determine if a nanoparticle will be internalized by cells [80]. As expected, cationic nanoparticles

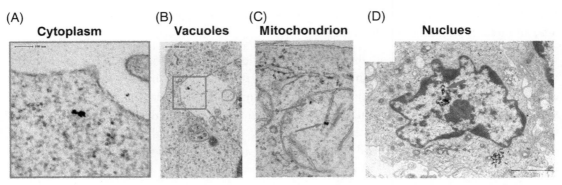

Fig. 2.12: Cell uptake of (A–C) Ag and (D) Au nanoparticles. Microscopy images show different localization of nanoparticles within the cells: (A) cytoplasm, (B) vacuoles, (C) mitochondrion, and (D) nucleus. Images (A–C), show 5–10 nm silver nanoparticles in TK6 cells [70] while (D) shows 4 nm sized gold nanoparticles in human fibroblast 1BR3G cells [72]. *Images (A–C) are adapted from Huk A. et al., Particle and Fiber Toxicology, 12, 2015 [70]. Image (D) is adapted from with permission from Ojea-Jimenez I. et al., ACS Nano, 6, 7692-7702, 2012. Copyright (2012) American Chemical Society [72].*

(functionalized with positively charged functional groups) will have a high cellular internalization as a result of the attractive electrostatic interactions between the positively charge nanoparticle surface and negatively charged cell membrane [68].

The amount of nanoparticles that are internalized by cells is correlated with the degree of nanoparticle toxicity. Cationic gold nanoparticles, which are internalized by cells, are moderately toxic, while the anionic gold nanoparticles (which are not internalized by cells) are nontoxic [72,81].

2.3.4 Silver nanoparticle toxicity

A large amount of data is available regarding the toxicity of silver. In this chapter we will outline only some important facts on nanosilver toxicity [82]. Due to its toxicity, silver nanoparticles are often used as antimicrobials [83].

Chronic inhalation of silver is associated to chronic bronchitis, and exposure to large amounts of silver leads to peripheral neuropathy, diminished mental status, and seizures [82]. Eighteen percent of orally ingested silver can be absorbed in the human tract [84]. The medical condition that describes the excessive absorption of silver is called argyria and manifests as a blue-gray discoloration of the skin [85]. Argyria occurs as a detoxification mechanism due to excessive amount of silver within the body, where silver is simply sequestered in the blood vessels and connective tissues and it is chemically bound to protein complexes or silver sulfide [84]. Intake of smaller amounts of silver over longer period of time results in liver and kidney damage. A variety of disorders are associated to exposure of humans and animals to materials containing silver, such as: autoimmune effects, blood disorders, chromosome break-age, and liver injury, among others [85].

Experiments in animals indicate that silver nanoparticles are toxic, their toxicity increasing in a dose dependent manner, having adverse health effects upon the immune system, blood chemistry and liver enzymes, manifestations including an enlarged heart, liver damage and weight loss [84,86–88]. In general, the adverse effects associated with exposure to silver nanoparticles encompass oxidative stress, genotoxicity, chromosomal aberration, breakage of the DNA, up-regulation of pro-inflammatory genes, neurodegeneration, cerebral microvascular damage and cell apoptosis [73,79,86,89–91].

Experimental evidence indicate that orally ingested Ag nanoparticles travel to all the organs, and can be found mostly in the intestinal tract and liver [87]. Silver aerosols that reach the lungs have a long residence time in rat lung tissue, with a third of the initial dose being found in lung tissue 56 days following exposure [92]. In general, the higher the exposure dose, the higher the Ag nanoparticles accumulation in every studied tissue [88].

Silver nanoparticle toxicity is due in part to the release of ions at the nanoparticle surface [93–95]. However, the main mechanism of silver nanoparticle toxicity seems to be the generation of reactive oxygen species and oxidative stress [96,97]. Smaller silver nanoparticles seem to have a higher toxicity compared to larger ones. Experiments in human and animal cells indicate that cell death occurs in a dose dependent manner as well as exposure duration manner [91].

Animal experiments employing radioactive labeling of Ag nanoparticles (35 nm) demonstrate that silver nanoparticles cross the placenta of rats, from where may be passed to fetuses [98].

While no direct causal relationship between the existence of nanoparticles in diseased tissue and the respective disease was proven yet, human studies clearly indicate the presence of Ag nanoparticles (together with other compositions) in human tissue where specific diseases are present [99,100]. Fig. 2.13 shows the presence of nanoparticles including silver, among other compositions, that are present at the site of (A) colon cancer [99], and (B) within an odontogenic tumor – ameloblastoma [100]. The microscopy images show the presence of the nanoparticles (lighter spots) in tissue while the energy dispersive spectroscopy images indicate their composition. Nanoparticles with various compositions, including silver, have been found in samples of patients with cancer and Crohn's disease.

2.3.5 Comparative toxicity of Ag nanoparticles

We have established that silver nanoparticles are toxic, however in order to comprehend the extent of their toxicity one must compare their toxic effects with those of other nanoparticles. Results of experiments in cells that show comparative toxicity of various nanoparticles are shown in Table 2.2 [54]. The base value "1" is being assigned to a known toxic substance – asbestos, which is a carcinogenic (mesothelioma). If a nanoparticle has a toxic value smaller

(A) **Ag nanoparticles in human biopsy of colon cancer**

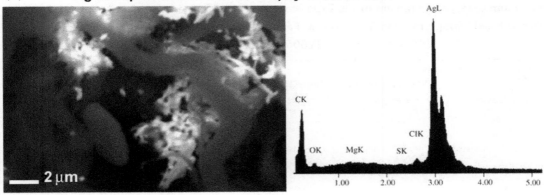

(B) **Ag nanoparticles in human tissue of ameloblastoma tumor**

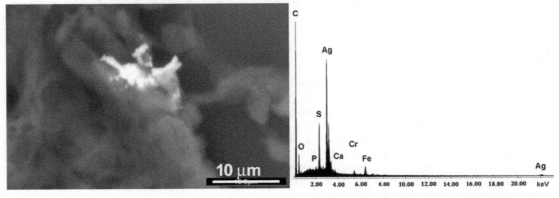

Fig. 2.13: (Left) Scanning Electron Microscopy and (right) energy dispersive spectroscopy images showing silver nanoparticles diseased tissue of (A) colon cancer and (B) ameloblastoma tumor. *Image (A) is adapted from Gatti A. M., Biomaterials, 25, 385-92, Copyright (2004), with permission from Elsevier [99]. Image (B) is adapted from Roncati L. et al., Ultrastruct Pathol, 39, 329-35, 2015, reprinted by permission of the publisher (Taylor & Francis Ltd, http://www.tandfonline.com) [100].*

than "1", they are less toxic than asbestos, while the opposite is true for values larger than "1". Silver is the only substance that in nanoparticle form is more cytotoxic than asbestos for concentrations of 5 µg/mL [54].

2.4 Conclusion

In this chapter we have presented selected properties of silver and gold nanoparticles that demonstrate their outstanding place at the confluence of quantum and solid-state-physics. While these properties make them attractive for a variety of new applications, one must mitigate their release into the environment and take into account their toxicity to organisms.

Table 2.2: Relative cytotoxicity of nanoparticles of Ag, asbestos, iron oxide, aluminum oxide, zirconium oxide, and titanium oxide. Experiments were performed on murine macrophage cells. Reproduced with permission from Soto K. F. et al., 2005, Journal of Nanoparticle Research, 7, 145-169.Copyright (2005) with permission of Springer.

Composition	Mean aggregate size (µm)	Mean nanoparticle size (nm)	Relative cytotoxicity index for 5 µg/mL	Relative cytotoxicity index for 10 µg/mL
Ag	0.4	30	1.8	0.1
Ag	1	30	1.5	0.8
Crysotile asbestos	0.5 – 15	15 – 40 diameter 0.5 – 15 µm length	1.0	1.0
Fe_2O_3	0.7	50	0.9	0.1
Al_2O_3	0.7	50	0.7	0.4
ZrO_2	0.7	20	0.7	0.6
Anatase TiO_2	1 – 2	20	0.4	0.1

References

[1] N.G. Khlebtsov, L.A. Dykman, Optical properties and biomedical applications of plasmonic nanoparticles, J. Quant. Spectrosc. Radiat. Transfer. 111 (1) (2010) 1–35.

[2] C. Buzea, I.I. Pacheco, K. Robbie, Nanomaterials and nanoparticles: Sources and toxicity, Biointerphases. 2 (4) (2007) MR17–MR71 and references therein.

[3] J.T. Lue, A review of characterization and physical property studies of metallic nanoparticles, J. Phys. Chem. Solids. 62 (9-10) (2001) 1599–1612.

[4] R. Krahne, et al., Physical properties of elongated inorganic nanoparticles, Phys. Rep.-Rev. Sec. Phys. Lett 501 (3-5) (2011) 75–221.

[5] K.J. Klabunde, et al., Nanocrystals as stoichiometric reagents with unique surface chemistry, J. Phys. Chem. 100 (30) (1996) 12142–12153.

[6] C. Buzea, I. Pacheco, Nanomaterials and their classification, in: A.K. Shukla (Ed.), EMR/ESR/EPR Spectroscopy for Characterization of Nanomaterials, Springer, New Delhi, India, 2017, pp. 3–45.

[7] R.L. Johnston, C.N.R. Rao, P.P. Edwards, The Development of metallic behavior in clusters [and discussion], Philos. Trans. A Math. Phys. Eng. Sci. 356 (1735) (1998) 211–230.

[8] A.S. Sharipov, B.I. Loukhovitski, Small atomic clusters: Quantum chemical research of isomeric composition and physical properties, Struct. Chem. 30 (6) (2019) 2057–2084.

[9] I. Pacheco, C. Buzea, Metal nanoparticles and their toxicity, in: S. Thota, D.C. Crans (Eds.), Metal Nanoparticles. Synthesis and Applications in Pharmaceutical Sciences, Wiley-VCH Verlag GmbH, Singapore, 2018.

[10] W.P. Halperin, Quantum size effects in metal particles, Rev. Mod. Phys. 58 (3) (1986) 533–606.

[11] O. Kamigaito, What can be improved by nanometer composites? J. Jpn. Soc. Powder Metal. 38 (1991) 315–321.

[12] R.S. Berry, B.M. Smirnov, Configurational transitions in processes involving metal clusters, Phys. Rep.-Rev. Sect. Phys. Lett. 527 (4) (2013) 205–250.

[13] S.K. Das, A.R. Das, A.K. Guha, Microbial synthesis of multishaped gold nanostructures, Small. 6 (9) (2010) 1012–1021.

[14] M.L. Personick, et al., Synthesis and isolation of {110}-faceted gold bipyramids and rhombic dodecahedra, J. Am. Chem. Soc. 133 (16) (2011) 6170–6173.

[15] B. Wiley, Y.G. Sun, Y.N. Xia, Polyol synthesis of silver nanostructures: Control of product morphology with Fe(II) or Fe(III) species, Langmuir. 21 (18) (2005) 8077–8080.

[16] N. Zhao, et al., Controlled synthesis of gold nanobelts and nanocombs in aqueous mixed surfactant solutions, Langmuir. 24 (3) (2008) 991–998.

[17] A.G. Mark, et al., Hybrid nanocolloids with programmed three-dimensional shape and material composition, Nat. Mater. 12 (9) (2013) 802–807.

[18] Y.S. Xia, Z.Y. Tang, Monodisperse hollow supraparticles via selective oxidation, Adv. Funct. Mater. 22 (12) (2012) 2585–2593.

[19] B. Cheng, Y. Le, J.G. Yu, Preparation and enhanced photocatalytic activity of $Ag@TiO_2$ core-shell nanocomposite nanowires, J. Hazard. Mater. 177 (1-3) (2010) 971–977.

[20] B. Issa, et al., Magnetic nanoparticles: Surface effects and properties related to biomedicine applications, Int. J. Mol. Sci. 14 (11) (2013) 21266–21305.

[21] H. Hori, et al., Anomalous magnetic polarization effect of Pd and Au nano-particles, Phys. Lett. A. 263 (4-6) (1999) 406–410.

[22] H. Hori, et al., Diameter dependence of ferromagnetic spin moment in Au nanocrystals, Phys. Rev. B. 69 (17) (2004) 174411, 5 pages.

[23] Y. Yamamoto, et al., Direct observation of ferromagnetic spin polarization in gold nanoparticles, Phys. Rev. Lett. 93 (11) (2004) 116801, 4 pages.

[24] U. Maitra, et al., Ferromagnetism exhibited by nanoparticles of noble metals, Chem. Phys. Chem. 12 (12) (2011) 2322–2327.

[25] C.Y. Li, et al., Intrinsic magnetic moments of gold nanoparticles, Phys. Rev. B. 83 (17) (2011).

[26] G.L. Nealon, et al., Magnetism in gold nanoparticles, Nanoscale. 4 (17) (2012) 5244–5258.

[27] S. Reich, G. Leitus, Y. Feldman, Observation of magnetism in Au thin films, Appl. Phys. Lett. 88 (22) (2006).

[28] K.S. Krishna, et al., Chemically induced magnetism in atomically precise gold clusters, Small. 10 (5) (2014) 907–911.

[29] T. Shinohara, T. Sato, T. Taniyama, Surface ferromagnetism of Pd fine particles, Phys. Rev. Lett. 91 (19) (2003) 197201, 4 pages.

[30] R. Greget, et al., Magnetic properties of gold nanoparticles: A room-temperature quantum effect, Chem. Phys. Chem. 13 (13) (2012) 3092–3097.

[31] R. Sato, et al., Magnetic order of Au nanoparticle with clean surface, J. Magn. Magn. Mater. 393 (2015) 209–212 and references therein.

[32] M. Pereiro, D. Baldomir, J.E. Arias, Unexpected magnetism of small silver clusters, Phys. Rev. A 75 (6) (2007) 063204, 6 pages.

[33] X.H. Huang, et al., Gold nanoparticles: Interesting optical properties and recent applications in cancer diagnostic and therapy, Nanomedicine. 2 (5) (2007) 681–693 and references therein.

[34] Y. Qiu, et al., Surface chemistry and aspect ratio mediated cellular uptake of Au nanorods, Biomaterials. 31 (30) (2010) 7606–7619.

[35] C. Ferrag, et al., Polyacrylamide hydrogels doped with different shapes of silver nanoparticles: Antibacterial and mechanical properties, Colloids Surf. B Biointerfaces. 197 (2021) 111397, 9 pages.

[36] R.B. King, Chirality and handedness - The Ruch "shoe-potato" dichotomy in the right-left classification problem. Ann. N. Y. Acad. Sci. 988, 2003 158–170.

[37] G. Schmid, B. Corain, Nanoparticulated gold: Syntheses, structures, electronics, and reactivities, Eur. J. Inorg. Chem. (17) (2003) 3081–3098.

[38] I. Shyjumon, et al., Structural deformation, melting point and lattice parameter studies of size selected silver clusters, Eur. Phys. J. D. 37 (3) (2006) 409–415.

[39] H.B. Liu, et al., Melting behavior of nanometer sized gold isomers, Surf. Sci. 491 (1-2) (2001) 88–98.

[40] M.A. Asoro, J. Damiano, P.J. Ferreira, Size effects on the melting temperature of silver nanoparticles: In-situ TEM observations, Microsc. Microanal. 15 (2009) 706–707.

[41] S.L. Lai, et al., Size-dependent melting properties of small tin particles: Nanocalorimetric measurements, Phys. Rev. Lett. 77 (1) (1996) 99–102.

[42] W.H. Qi, Size effect on melting temperature of nanosolids, Physica. B Condens. Matter. 368 (1-4) (2005) 46–50.

[43] W.H. Qi, Nanoscopic thermodynamics, Acc. Chem. Res. 49 (9) (2016) 1587–1595.

[44] P.R. Couchman, W.A. Jesser, Thermodynamic theory of size dependence of melting temperature in metals, Nature. 269 (5628) (1977) 481–483.

[45] A. Aguado, M.F. Jarrold, Melting and freezing of metal clusters, Annu. Rev. Phys. Chem. 62 (2011) 151–172 Vol 62and references therein.

[46] X. Yu, Z. Zhan, The effects of the size of nanocrystalline materials on their thermodynamic and mechanical properties, Nanoscale Res. Lett. 9 (1) (2014) 516.

[47] K. Koga, T. Ikeshoji, K. Sugawara, Size- and temperature-dependent structural transitions in gold nanoparticles, Phys. Rev. Lett. 92 (11) (2004) 115507, 4 pages.

[48] T. Castro, et al., Size-dependent melting temperature of individual nanometer-sized metallic clusters, Phys. Rev. B. 42 (13) (1990) 8548–8556.

[49] N. von Goetz, et al., Migration of silver from commercial plastic food containers and implications for consumer exposure assessment, Food Addit. Contam. Part A-Chem. Anal. Control Expo. Risk Assess. 30 (3) (2013) 612–620.

[50] T. Benn, et al., The release of nanosilver from consumer products used in the home, J. Environ. Qual. 39 (6) (2010) 1875–1882.

[51] T.M. Benn, P. Westerhoff, Nanoparticle silver released into water from commercially available sock fabrics, Environ. Sci. Technol. 42 (11) (2008) 4133–4139.

[52] Y.M. Huang, et al., Nanosilver migrated into food-simulating solutions from commercially available food fresh containers, Packag. Technol. Sci. 24 (5) (2011) 291–297.

[53] T.V. Duncan, K. Pillai, Release of engineered nanomaterials from polymer nanocomposites: Diffusion, dissolution, and desorption, ACS Appl. Mater. Interfaces 7 (1) (2015) 2–19.

[54] K.F. Soto, et al., Comparative in vitro cytotoxicity assessment of some manufactured nanoparticulate materials characterized by transmission electron microscopy, J. Nanopart. Res. 7 (2-3) (2005) 145–169.

[55] A. Hahn, et al., Metal ion release kinetics from nanoparticle silicone composites, J. Control Release. 154 (2) (2011) 164–170.

[56] A. Nel, et al., Toxic potential of materials at the nanolevel, Science. 311 (5761) (2006) 622–627.

[57] T.S. Wu, M. Tang, Review of the effects of manufactured nanoparticles on mammalian target organs, J. Appl. Toxicol. 38 (1) (2018) 25–40.

[58] A. Karmakar, Q. Zhang, Y. Zhang, Neurotoxicity of nanoscale materials, J. Food Drug Anal. 22 (1) (2014) 147–160.

[59] A.M. Schrand, et al., Metal-based nanoparticles and their toxicity assessment, Wiley Interdiscip. Rev. Nanomed. Nanobiotechnol. 2 (5) (2010) 544–568.

[60] Z. Ferdous, A. Nemmar, Health impact of silver nanoparticles: A review of the biodistribution and toxicity following various routes of exposure, Int. J. Mol. Sci. 21 (7) (2020) 2375, 31 pages.

[61] A.M. Gatti, S. Montanari, Retrieval analysis of clinical explanted vena cava filters, J. Biomed. Mater. Res. B Appl. Biomater. 77 (2) (2006) 307–314.

[62] A.M. Gatti, et al., Detection of micro- and nano-sized biocompatible particles in the blood, J. Mater. Sci. Mater. Med. 15 (4) (2004) 469–472.

[63] R. Landsiedel, et al., Toxico-/biokinetics of nanomaterials, Arch. Toxicol. 86 (7) (2012) 1021–1060.

[64] G. Sonavane, K. Tomoda, K. Makino, Biodistribution of colloidal gold nanoparticles after intravenous administration: Effect of particle size, Colloids Surf. B Biointerfaces. 66 (2) (2008) 274–280.

[65] B. Wang, et al., Metabolism of nanomaterials in vivo: Blood circulation and organ clearance, Acc. Chem. Res. 46 (3) (2013) 761–769.

[66] J.F. Hillyer, R.M. Albrecht, Gastrointestinal persorption and tissue distribution of differently sized colloidal gold nanoparticles, J. Pharm. Sci. 90 (12) (2001) 1927–1936.

[67] B.D. Chithrani, A.A. Ghazani, W.C.W. Chan, Determining the size and shape dependence of gold nanoparticle uptake into mammalian cells, Nano Lett. 6 (4) (2006) 662–668.

[68] K. Kettler, et al., Cellular uptake of nanoparticles as determined by particle properties, experimental conditions, and cell type, Environ. Toxicol. Chem. 33 (3) (2014) 481–492.

[69] W. Jiang, et al., Nanoparticle-mediated cellular response is size-dependent, Nat. Nanotechnol. 3 (3) (2008) 145–150.

[70] A. Huk, et al., Impact of nanosilver on various DNA lesions and HPRT gene mutations - effects of charge and surface coating, Part. Fibre Toxicol. 12 (2015) 25, 20 pages.

[71] R.R. Sathuluri, et al., Gold nanoparticle-based surface-enhanced raman scattering for noninvasive molecular probing of embryonic stem cell differentiation, PLoS One. 6 (8) (2011) e22802, 13 pages.

[72] I. Ojea-Jimenez, et al., Facile preparation of cationic gold nanoparticle-bioconjugates for cell penetration and nuclear targeting, ACS Nano. 6 (9) (2012) 7692–7702.

[73] L.C. Cheng, et al., Nano-bio effects: Interaction of nanomaterials with cells, Nanoscale 5 (9) (2013) 3547–3569.

[74] G. NCBI National center for biotechnology information (2022). PubChem Compound Summary for CID 23985, Gold. Retrieved June 3, 2022 from https://pubchem.ncbi.nlm.nih.gov/compound/Gold.

[75] A. Gerber, et al., Gold nanoparticles: Recent aspects for human toxicology, J. Occup. Med. Toxicol. 8 (1) (2013) 32.

[76] N. Hadrup, et al., Toxicological risk assessment of elemental gold following oral exposure to sheets and nanoparticles - A review, Regul. Toxicol. Pharmacol. 72 (2) (2015) 216–221.

[77] N. Khlebtsov, L. Dykman, Biodistribution and toxicity of engineered gold nanoparticles: A review of in vitro and in vivo studies, Chem. Soc. Rev. 40 (3) (2011) 1647–1671.

[78] N. Chen, et al., Long-term effects of nanoparticles on nutrition and metabolism, Small. 10 (18) (2014) 3603–3611.

[79] A. Kermanizadeh, et al., Toxicological effect of engineered nanomaterials on the liver, Br. J. Pharmacol. 171 (17) (2014) 3980–3987.

[80] S. Salatin, S.M. Dizaj, A.Y. Khosroushahi, Effect of the surface modification, size, and shape on cellular uptake of nanoparticles, Cell Biol. Int. 39 (8) (2015) 881–890.

[81] C.M. Goodman, et al., Toxicity of gold nanoparticles functionalized with cationic and anionic side chains, Bioconjugate. Chem. 15 (4) (2004) 897–900.

[82] S. NCBI National Center for Biotechnology Information (2022). PubChem Compound Summary for CID 23954, Silver. Retrieved June 3, 2022 from https://pubchem.ncbi.nlm.nih.gov/compound/Silver.

[83] B. Le Ouay, F. Stellacci, Antibacterial activity of silver nanoparticles: A surface science insight, Nano Today. 10 (3) (2015) 339–354.

[84] N. Hadrup, H.R. Lam, Oral toxicity of silver ions, silver nanoparticles and colloidal silver–A review, Regul. Toxicol. Pharmacol. 68 (1) (2014) 1–7.

[85] F.M. Christensen, et al., Nano-silver - feasibility and challenges for human health risk assessment based on open literature, Nanotoxicology. 4 (3) (2010) 284–295.

[86] S. Ahlberg, et al., PVP-coated, negatively charged silver nanoparticles: A multi-center study of their physicochemical characteristics, cell culture and in vivo experiments, Beilstein J. Nanotechnol. 5 (2014) 1944–1965 and references therein.

[87] S. Gaillet, J.M. Rouanet, Silver nanoparticles: Their potential toxic effects after oral exposure and underlying mechanisms–A review, Food Chem. Toxicol. 77 (2015) 58–63.

[88] Y.S. Kim, et al., Twenty-eight-day oral toxicity, genotoxicity, and gender-related tissue distribution of silver nanoparticles in Sprague-Dawley rats, Inhal Toxicol. 20 (6) (2008) 575–583.

[89] S. Hackenberg, et al., Silver nanoparticles: Evaluation of DNA damage, toxicity and functional impairment in human mesenchymal stem cells, Toxicol. Lett. 201 (1) (2011) 27–33.

[90] S. Kim, D.Y. Ryu, Silver nanoparticle-induced oxidative stress, genotoxicity and apoptosis in cultured cells and animal tissues, J. Appl. Toxicol. 33 (2) (2013) 78–89.

[91] F. Liu, et al., Effects of silver nanoparticles on human and rat embryonic neural stem cells, Front Neurosci. 9 (2015) 115.

[92] D.S. Anderson, et al., Influence of particle size on persistence and clearance of aerosolized silver nanoparticles in the rat lung, Toxicol. Sci. 144 (2) (2015) 366–381.

[93] Y. Arai, T. Miyayama, S. Hirano, Difference in the toxicity mechanism between ion and nanoparticle forms of silver in the mouse lung and in macrophages, Toxicology. 328 (2015) 84–92.

[94] C. Beer, et al., Toxicity of silver nanoparticles - nanoparticle or silver ion? Toxicol. Lett. 208 (3) (2012) 286–292.

[95] A.R. Gliga, et al., Size-dependent cytotoxity of silver nanoparticles in human lung cells: The role of cellular uptake, agglomeration and Ag release, Part. Fibre Toxicol. 11 (2014) 11.

[96] A. Avalos, et al., Cytotoxicity and ROS production of manufactured silver nanoparticles of different sizes in hepatoma and leukemia cells, J. Appl. Toxicol. 34 (4) (2014) 413–423.

[97] C. Carlson, et al., Unique cellular interaction of silver nanoparticles: Size-dependent generation of reactive oxygen species, J. Phys. Chem. B. 112 (43) (2008) 13608–13619.

[98] E.A. Melnik, et al., Transfer of silver nanoparticles through the placenta and breast milk during in vivo experiments on rats, Acta. Naturae. 5 (3) (2013) 107–115.

[99] A.M. Gatti, Biocompatibility of micro- and nano-particles in the colon. Part II, Biomaterials. 25 (3) (2004) 385–392.

[100] L. Roncati, et al., ESEM detection of foreign metallic particles inside ameloblastomatous cells, Ultrastruct. Pathol. 39 (5) (2015) 329–335.

Implementation of gold and silver nanoparticles in sensing and bioengineering

Geetika Bhardwaj[a], Navneet Kaur[a], Narinder Singh[b]

[a]*Department of Chemistry, Centre for Advanced Studies in Chemistry, Panjab University, Chandigarh, India* [b]*Department of Chemistry, Indian Institute of Technology Ropar (IIT Ropar), Rupnagar, Punjab, India*

3.1 Introduction

Nanotechnology is a very wide and fascinating field. The variety of revolutionary developments in the field of nanotechnology made the research more valuable and interesting. Nanotechnology includes nanomaterials like nanoparticles, quantum dots, nanorods, nanowires, etc. [1]. All these nanostructures can be widely used in the field of diagnostics, biomarkers, cell-labeling, antimicrobial agents, drug delivery systems, sensing, detection of harmful agents, and treatment of cancer [2]. The existence of nanoparticles (NPs) in nature has been reported since ancient times [3]. These NPs are in various forms like organic including proteins, polysaccharides, viruses, etc.; inorganic including iron oxyhydroxides, aluminum silicates, metals, and naturally produced in the atmosphere by weathering, volcano eruptions, wildfires, and microbial processes [4]. Metallic nanomaterials include metal nanoparticles (gold nanoparticles, silver nanoparticles, copper nanoparticles, titanium nanoparticles, platinum nanoparticles, zinc nanoparticles, magnesium nanoparticles, and many more), metal oxide nanoparticles (titanium oxide, silver oxide, zinc oxide, and more), metal sulfides (AgS, CuS, FeS) and metal-organic framework (MOFs) for example, Zn-based MOFs, Cu-based MOFs, Mn-based MOFs [5]. MNPs are generally nanosized metals with dimensions and sizes in the range of 1–100 nm. Their size, surface area, shape, doping, and interaction with the surrounding controls their optical properties. Gold and silver nanoparticles are great examples of MNPs exhibiting profuse properties. They display different colors due to specific interactions with incident light. The size and shape-to-volume ratio affect the wave-like properties of electrons inside the matter as well as their quantum effects [6]. Large interfacial area, increased surface tension, increased number of molecules on particle interface, and quantum electromagnetic inductions, all are due to the small size of NPs. They thus, exhibit spectacular effects and find a special place in the field of nanotechnology. Their inherent properties such

Gold and Silver Nanoparticles.
DOI: https://doi.org/10.1016/B978-0-323-99454-5.00003-2

as surface-to-volume ratio, Rayleigh scattering, surface-enhanced Raman scattering, strong plasma absorption, biological system imaging, Quantum confinement, increased number of kinks, and large surface energies allow them to interact with other particles and facilitates their use in various fields like sensing, drug delivery, recognition, etc. Metallic nanoparticles (MNPs) also show a variety of applications which includes conversion to nanowires, making of high-temperature superconductivity material, exhibition of magnetic properties, selective, highly active, and long lifetime catalyst, uses in electrical conductive pastes and battery material, prevention of environmental pollution from coal and gasoline, use in medical treatment of cancer and use in sunscreen to protect from UV light [7]. The size of these NPs depends on their method of preparation and can be detected on the basis of their color. Various chemical and biological methods are reported for the synthesis of gold and silver nanoparticles. Many researchers have done chemical modifications of gold and silver nanoparticles to develop sensing elements for metal ions and various biomolecules. This book chapter mainly focuses on the variety of methods for the synthesis of gold and silver nanoparticles, their modification, with different organic moieties and biomolecules like DNA, proteins, etc., and their application in the field of sensing and biosensing.

3.2 Synthesis of nanoparticles of gold and silver

Turkevich et al., in 1951 discovered a technique for the synthesis of nonfunctionalized gold nanoparticles (AuNPs) [8]. For which trisodium citrate dehydrate was added rapidly to the boiling solution of auric chloride ($HAuCl_4$) with continuous stirring as shown in Fig. 3.1A. This procedure gives AuNPs (**1**) of size 20 nm. This method is popularly known by his name as the Turkevich method. With time another method is the Brust-Schiffrin method discovered in 1994 in which $AuCl_4$ was transferred from the aqueous phase to the toluene phase using tetraoctyl ammonium bromide followed by reduction with $NaBH_4$ in the presence of dodecanethiol [9]. The same method was also applied for the synthesis of silver nanoparticles (AgNPs) with certain modifications by Oliveira et al., [10] Synthesis of spherical AgNPs (**2**) was also reported by Kim et al., by a modified precursor injection technique using polyol process (Fig. 3.1B). In this method, the injection rate and reaction temperature controlled the uniform size and high monodispersity of AgNPs of size 17±2 nm [11]. Methods such as microemulsion technique [12], photo-induced reduction [13], microwave-assisted technique [14], electrochemical technique [15] have also been reported well for the synthesis of nonfunctionalized gold and silver NPs. Nowadays abundant chemical modifications on the surface of the gold and silver NPs have been done to enhance their properties and thus to produce multi functionalized noble MNPs.

Huang et al. use quaternary ammonium-based room temperature ionic liquids [QAILS (**3**)] for the preparation of surface-modified AuNPs. Aqueous solutions of $HAuCl_4.4H_2O$ was added dropwise to QAILS and stirred vigorously with continuous heating at 135–145°C for

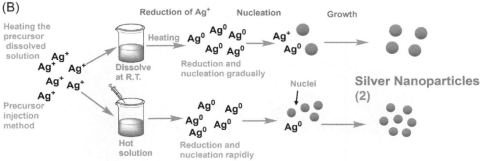

Fig. 3.1: (A) Schematic representation of synthesis of AuNPs (1) by Turkevich method. (B) Schematic illustration of two synthesis methods: heating the precursor dissolved solution and injection of the precursor solution.

15 minutes. The size of **4** depends on the reaction temperature and can be varied from 20 to 40 nm (Fig. 3.2A) [16]. While Shen et al. used the direct reduction capability of carbon dots to convert Ag^+ to Ag^0 (Fig. 3.2B). Changes in the signals of C-dots in FT-IR spectra confirmed the growth of AgNPs over C-dots (**5**) [17].

To promote green synthesis, gold and silver NPs have been prepared by using biological extracts of various plants. Chen et al., prepared the AuNPs using *Chenopodiumformasanum* (Djulis) shell extract and **6** as shown in Fig. 3.3A and evaluated their antibacterial properties [18]. Likewise, Oluwafemi et al. used water hyacinth leaf extract for the same purpose. FT-IR spectra showing shifts in O-H, C-H, and C-O stretching vibrations of water hyacinth leaves confirms the modifications of AgNPs (Fig. 3.3B) [19]. Numerous other methods for the biological synthesis were reported using *Aspergilum sp. WL-Au* [20], *Aspergillus niger NCIM* [21], *Plectonemaboryanum* (filamentous cyanobacteria) [22], *Klebsiellapneumoniae RCTC 2242* and *Escherichia coli DH5a* [23], *Saccharomyces cerevisiae* [24], *Eclipta* leaf [25],*Geranium* leaf extract [26].

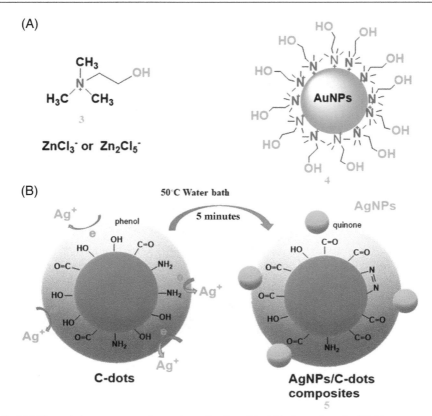

Fig. 3.2: (A) Pictorial representation of synthesis of AuNPs by using 3. (B) Representation of reduction of Ag to AgNPs over the surface of carbon dots.

These methods are mainly classified into two types:

1. Dispersion method
2. Condensation method

The dispersion method involves the disruption of the crystal lattice of the material and is commonly known as the top-down approach. These include laser ablation, cathode sputtering, and the electric arc dispersion method as in Fig. 3.4 [27]. On the other hand, condensation methods are based on condensation reactions which involve the reduction of metal salt in solution followed by nanoparticles formation, stabilization and precipitation. The condensation method mainly works on the phenomenon that a high reduction force favors a high reaction rate resulting in the formation of small-size nanoparticles. However, by changing the conditions of the same synthetic procedure different distributions of nanoparticles with different sizes and shapes can be obtained [27]. Some reducing agents which can be used for the preparation are sodium citrate, alcohols, Na_2S, sodium borohydride, hydrogen gas, sugars, etc. The size and shape can be controlled by controlling the amount of reducing agent and

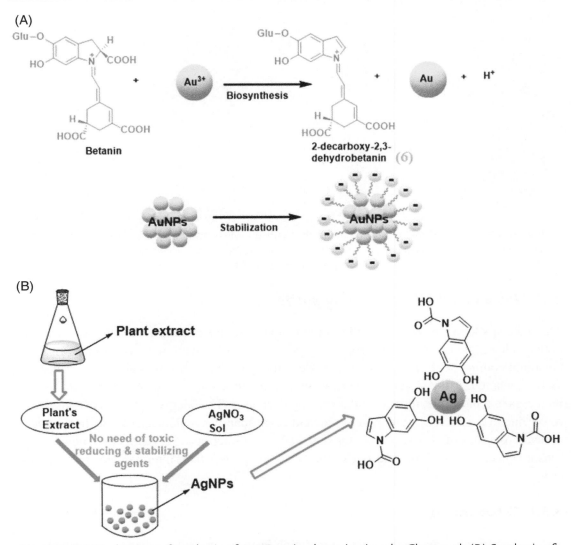

Fig. 3.3: (A) Mechanism of synthesis of AuNPs using betanin given by Chen et al. (B) Synthesis of AgNPs using water hyacinth extract by Oluwafemi et al.

stabilizer. In some cases, reducing agents also acts as a stabilizer. The size and shape of gold and silver nanoparticles, in turn, control their optical properties. These properties followed by their adaptability, low cost, versatility, and high sensitivity gold and silver nanoparticles in their original and modified form have been widely used for metal detection [28], proteins [29,30], DNA [31,32], small molecules [33,34] and enzymes [35]. Other analytical techniques like fluorescence spectroscopy, UV-Visible spectroscopy, voltammetry, high-performance liquid chromatography, circular dichroism, and electrophoresis have been used for the ultra-sensitive and fast detection of metal ions and biomolecules.

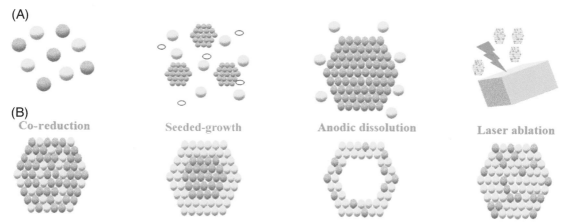

(A)

(B)

Co-reduction   Seeded-growth   Anodic dissolution   Laser ablation

Fig. 3.4: Synthesis of metal nanoparticles: Bottom-up synthesis by (A) Co-Reduction method. (B) Seeded-growth method (C) Anodic dissolution method. Top-down synthesis by (D) Laser ablation technique [36].

3.3 Noble metal NPs as sensing platform

The scale at which the size of the NPs complements with the characteristic length scale, the optical, electronic and magnetic properties become size dependent, the property known as Quantum confinement. In free electron model, energy of the electronic states is described as $1/L^2$, where L is the system dimensions in definite directions. As the number of atoms increases in a molecule, the molecular spacing decreases. The energy band gap become narrower due to the decrement in system size and consequently, the delocalized electronic and optical properties of NPs vary [37]. Gold and silver NPs acts as a very attractive moiety as a sensor because of the shape and size of the NPs which raises their innumerable optical properties.

3.3.1 Cation sensing

The change in color and mode of the aggregation forms a very simple, exact, and reliable colorimetric sensor. The diverse colors of the solutions are enough to tell their properties. Various metal ions show acceptable colorimetric and optical changes with gold and silver NPs. Obare et al., synthesized AuNPs of size 4 nm by following the standard Turkevich method of reducing the gold salt (HAuCl$_4$) with sodium citrate and sodium borohydride. The AuNPs were functionalized with 2,9-dibutyl-1,10-phenanthroline-5,6-aminoethanethiol (**7**). The 2.5×10^4 Msolution of both the ligands and AuNPs was prepared in ethanol and allowed to stir for 12 hours. After 12 hours, the solution was centrifuged and re-dispersed in water for the sensing of Li$^+$. The TEM image confirms the coating of AuNPs surfaced by ligands via the Au-S bond. The formation of the Au-S bond leaves the binding site for Li$^+$ fully exposed. So, when Li$^+$ was added to the solution the orange color of the solution darkens showing a

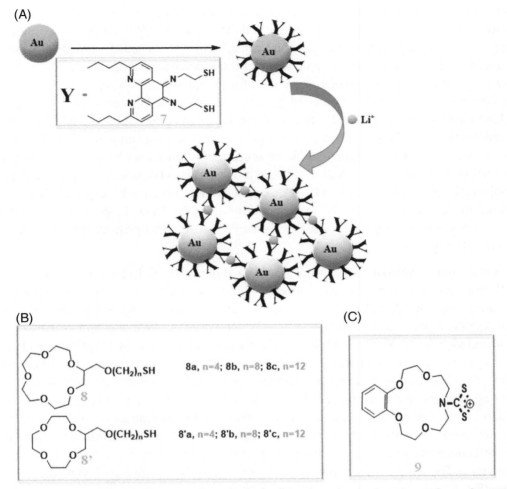

Fig. 3.5: (A) Detection scheme of Li$^+$ in which the ligand 7 adsorbed to the surface of AuNPs binds with Li$^+$ in a bidentate fashion resulting in aggregation. (B) Crown thiols (8,8') synthesized by Lin et al., for the sensing of K$^+$ ion. (C) Aza crown ether (9) used by Li et al., for the selective sensing of Ba^{2+} ion.

redshift in UV-spectrum. Nonfunctionalized AuNPs show aggregation with Li$^+$ resulting in precipitation with Li$^+$ as in Fig. 3.5A. During titration with the increased concentration of Li$^+$, the color of the solution changes from dark orange to grey but there were no precipitates in the solution. Selectivity of functionalized AuNPs was also checked by recording its UV spectrum with K$^+$ and Na$^+$. Out of both, Na$^+$ shows little interference while K$^+$ shows no interference in the selective sensing of Li$^+$ via functionalized AuNPs [38]. In alternative study, Lin et al., synthesized crown thiols **8,8'** (Fig. 3.5B) functionalized AuNPs and used them for the sensing of K$^+$ and Na$^+$. Citrate stabilized AuNPs were prepared by reducing

HAuCl$_4$ by sodium citrate. For the functionalization with thioctic amine, thioctic amine solution in methanol were mixed with citrate stabilized AuNPs. Upon addition the solution changes to purple, depicting aggregation and when 1 M HCl was added, immediately the solution turned red again. Further, to these thioctic acid/amine-functionalized AuNPs, crown thiols in ethanol were added. When crown thiol functionalized AuNPs were mixed with K$^+$, the solution instantly turns blue from red giving a redshift in the UV spectrum. These results are due to the formation of a sandwich complex between the crown thiol and K$^+$. Whereas on the introduction of thioctic acid/amine in the process and the bi-functionalized AuNPs with both thioctic acid/amine and crown thiol, the sensing of K$^+$ increases by 4 orders in magnitude and can also sense other alkali metal such as Na$^+$. This is because the negative charge of TA adjusts the orientation of crown thiol leaving the binding site for K$^+$ exposed. The method successfully detects the Na$^+$ and K$^+$ in human urine also [39]. Li et al., show the recognition of AgNPs towards Ba^{2+} by functionalizing the surface of AgNPs by aza-crown ether [ACE (**9**)]shown in Fig. 3.5C.

The surface-functionalized AgNPs were synthesized in two steps. In the first step, a solution of CS$_2$ in ethanol mixed dropwise to aqueous solution of ACE followed by 5 minutes sonication. On the other hand, AgNO$_3$ dissolved in DI water was reduced by NaBH$_4$ yielding colloidal AgNPs. In the second step, the prepared solution of CS$_2$-ACE and AgNPs was mixed and stirred vigorously. The formed ACE-AgNPs were characterized by UV, FT-IR, and TEM and analyzed for the recognition behavior towards Ba^{2+} via colorimetric sensing. Selectivity of Ba^{2+} was checked among alkali and alkaline earth metals, and it was seen that only Ba^{2+} causes aggregation in AgNPs by coordinating with N- from ACE and forming a sandwich structure. In the UV spectrum addition of Ba^{2+} results in broadening and redshift of the band and a new peak are formed at 600 nm. Interference was checked by functionalizing AgNPs with diethanolamine and aniline accordingly as if the benzene ring or ether ring affects the sensitivity of Ba^{2+}. The sensor shows good recognition of Ba^{2+} with a 10^{-8} M detection limit [40].

Wu et al. functionalized the surface of AgNPs with thioglycolic acid (TGA) and used them for the naked-eye detection of Mg^{2+}, Ca^{2+}, Sr^{2+}, and Ba^{2+}. The surface of AgNPs was functionalized by first reducing the aqueous solution of AgNO$_3$ with NaBH$_4$ under dark. Light yellow color of the solution confirmed the formation of dispersed AgNPs. The prepared AgNPs were then mixed with TGA for half-hour to assure the binding of TGA sulfur with Ag. The pH of the solution was adjusted to 12 and used for the sensing of alkali and alkaline earth metals. With the four metal ions (Mg^{2+}, Ca^{2+}, Ba^{2+}, and Sr^{2+}) the color of the TGA-AgNPs (**10**) changed from yellow to orange or red. In the UV spectrum, a new band at 542, 560, 568, and 563 nm respectively were formed. The titration experiment shows the gradual increase of these bands with a slow change in color from yellow to orange and red giving detection limits of 8.33, 2.08, 0.833, and 1.25 μM respectively. It was also checked that presence of other metal ions doesn't affect the sensing of these ions (Fig. 3.6A) [41].

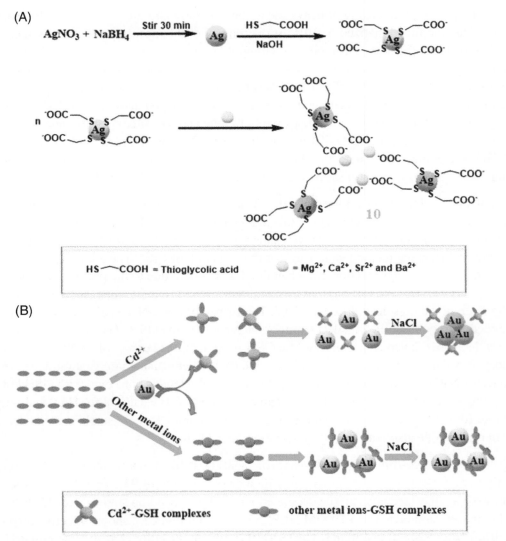

Fig. 3.6: (A) Schematic representation of the aggregation of TGA-AgNPs (10) induced by alkaline earth metals. (B) Diagrammatic representation of effect of GSH on aggregation induced by NaCl in AuNPs.

The major pollutants are not alkali and alkaline earth metals but heavy metals. Heavy metal pollutants like mercury, cadmium, nickel, copper, arsenic, lead, etc. affect both environment and health. These metals can affect the brain, heart, kidney, immune system, and lungs and can stay in our body for a very long time. The main sources of these heavy metals are industrial, agricultural, pharmaceutical and atmospheric waste which merge with the river water and enters our body through food chain. So, it can be said that sensing of heavy metal ions is more required as compared to alkali and alkaline earth metals. Many reports have been published regarding the sensing of heavy metal ions via modified gold and silver NPs. Varela et al. reported the functionalized AuNPs for the sensing of the most important heavy metal As^{3+}. The surface of AuNPs was modified with glutathione, DL-dithiothreitol (DTT), and cysteine respectively. The whole experiment was conducted in the dark to limit the photo reduction of gold. The functionalized AuNPs show stability over a month. Sensing of As^{3+} via these functionalized AuNPs was done on VIS spectroscopy. Upon addition, the As^{3+} causes aggregation resulting in a color change from red to blue and a change in absorbance in VIS spectroscopy. In particular, the DTT functionalized AuNPs show degradation from As^{5+} to As^{3+}. This method gives the detection limit of 2.5 µg/L and has the ability to sense the concentration of As (III) in a water sample with a 99% regression coefficient [42].

Guo et al. detected the concentration of Cd^{2+} in rice samples taking the concept that the addition of excess salt to AuNPs causes this aggregation but the addition of glutathione (GSH) prevents the aggregation of AuNPs shown in Fig. 3.6B. The bare AuNPs prepared by the standard Turkevich method were used for the detection of Cd^{2+}. The addition of excess NaCl to these causes' aggregation but when GSH was added; due to the presence of one sulfhydryl group, two free carboxyl groups, and one amino group, it stops the aggregation in AuNPs even in the presence of excess NaCl. To this AuNP-GSH-NaCl solution when Cd^{2+} was added, being thiophilic metal, it strongly interacts with the GSH. The sulfhydryl group of GSH decreases the concentration of free GSH present on the surface of AuNPs. These free surfaces of AuNPs, in turn, start interlinking with each other causing aggregation due to the presence of NaCl in solution. However, Hg^{2+} also have the tendency to bind with GSH but the binding energy from DFT calculations came out to be −420 Kcal/mol for Hg^{2+} and −621 Kcal/mol for Cd^{2+} which depicts that the GSH-Cd(II) complex is more stable and can detect Cd^{2+} in rice sample also with the 5 µM detection limit [43].

Dang et al., used 5-5'-dithiobis (2-nitro benzoic acid) (DTNBA) functionalized AuNPs for the detection of a Cr^{3+} in an aqueous medium with a detection limit of 93.6 ppb. AuNPs of size 13 nm were synthesized and surface decorated with DTNB. Binding studies with 15 different metal ions were conducted in a 10 mM HEPES buffer solution. Cr^{3+} shows selective response by redshift from 524 nm to 650 nm. The redshift results in the color change from pink to blue. The selective sensing of Cr^{3+} is due to the phenyl-linked nitryl and carboxyl group interaction with Cr^{3+}. These interactions were confirmed by recording the profile of unmodified DTNB with Cr^{3+} giving a minor blue shift. Further confirmations were done by preparing the TNBA analog DTNB-OMe and modifying it with AuNPs. Even 10 µM addition of Cr^{3+} could not induce aggregation in the solution which confirms the crucial role of the carboxyl group in selective sensing of Cr^{3+} ion [44]. Ali et al. modified the screen-printed electrode (SPE) based on 1,4-bis(8-mercapto octyl oxy)-benzene (**11**) (Fig. 3.7A) coated to the surface of AuNPs and used them for the electrochemical sensing of Ce^{3+} in a water sample. AuNPs prepared by the Turkevich method were mixed with the saturated solution of (**11**) and stirred vigorously till the color of the solution changed. The solution was centrifuged and washed

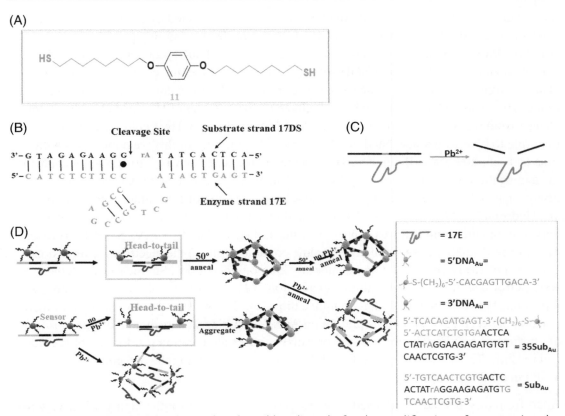

Fig. 3.7: (A) Ligand 11 synthesized and used by Ali et al., for the modification of screen-printed electrode and sensing of Ce^{3+}. (B) Secondary structure of "8-17 DNAzyme. (C) Cleavage of 17DS in the presence of Pb^{2+}. (D) Systematic diagram of previous and new Pb^{2+} colorimetric sensor design respectively.

with cold water. The resulting solid was mixed with homemade carbon ink and the working electrode was modified for the potentiometric sensing of Ce^{3+}. The modified electrode shows a linear range of $3.25 \times 10^{-10} - 1.0 \times 10^{-1}$ mol/L and Nernstian slopes 19.95 ± 0.97 mV/decade. It shows a 3.25×10^{-10} limit of detection value with a response time of 4 seconds and can be used for at least 7 months [45]. Liu et al. assembled the AuNPs with DNAzymes for the colorimetric sensing of Pb^{2+}. AuNPs were prepared by reducing the $HAuCl_4$ with sodium citrate with refluxing. The resulting AuNPs came out to be of size 42 nm. On the other side, thiol-modified DNA was prepared by incubating DNA with DTT for 1 hour and washing the mixture to remove excess DTT. After 1-day of incubation, buffer was added dropwise and incubated again for 2 days followed by centrifugation. The obtained solid was redispersed in tris acetate and were mixed with NaCl and respective metal ions for the sensing purpose. The functionalized AuNPs show aggregation within 8 minutes but on the addition of Pb^{2+}, the aggregation suppresses giving a color change from blue to red (Fig. 3.7B–D) [46]. The most abundant pollutant of the environment i.e., Hg^{2+} was detected by Si et al. by peptide-functionalized AuNPs. The peptide-AuNPs were synthesized by mixing peptide (NH_2-Leu-Aib-Tyr-OMe) in methanol with DI water and adding it to $HAuCl_4$ aqueous solution. pH was maintained at 11 by using NaOH solution. After some time, the color of the solution changes from yellow to red by the oxidation of tyrosine. To this solution, an aqueous solution of Hg^{2+} was added (4 ppm) which results in the broadening of UV spectra with a color change from red to purple. However, with the addition of 8 ppm Hg^{2+} color changes from red to purple to blue with a new band in UV spectra. The intensity of this band increases with time and saturates after 10 minutes. The change in DLS size from 11.7 to 105.4 nm confirms the aggregation of the particles [47]. Whereas Liu et al., prepared quaternary ammonium group capped AuNPs and used them for the selective sensing of Hg^{2+} among different metals. Hydrophilic (11-mercapto-undecyl)-trimethyl ammonium (MTA) was used as a quaternary ammonium group and was capped over the surface of AuNPs via Au-S bond formation. The prepared AuNPs were stabilized in acidic conditions due to the electrostatic repulsions between positively charged H^+ and quaternary ammonium cations. So, the sensor studies were performed in an acidic medium. On addition of Hg^{2+} to this solution, these electrostatic repulsions between quaternary ammonium cations and H^+ and Hg^{2+} hasten the displacement reaction between Hg^{2+} and thiols present on AuNPs resulting in detachment of ligand from the surface of AuNPs. The free surface of AuNPs causes aggregation resulting in blue color and redshift. The method shows a 30 nM detection limit satisfying the detection limit set by WHO in drinking water [48].

Apart from this, Farhadi, et al. used bare biosynthesized AgNPs for the highly selective sensing of Hg^{2+} with a detection limit of 2.2×10^{-6} mol /L. The AgNPs were synthesized using soap root plant extract. The soaproot plant extract acts as a stabilizer in the solution. Now to reduce the $AgNO_3$ solution, the manna of Hedysarum plant extract was added at 86°C. Mixture was centrifuged and extracted showing an appearance of AgNPs in EDS. The prepared solution was then diluted and used for the colorimetric response towards Hg^{2+} among various alkali metals, alkaline earth metals, and transition metals. Among all, Hg^{2+} shows a colorimetric response by changing the color of the solution from yellow-brown to light yellow. The color fades with the increasing concentration of Hg^{2+} giving colorless solution on the addition of 10^{-3} mol /L Hg^{2+} solution (Fig. 3.8A) The prepared solution was also evaluated for Hg^{2+} sensing in lake water showing a relative standard deviation of 2.2 % only [49]. Li et al. also use biosynthesized AgNPs but for the sensing of Ni^{2+} ions. GSH stabilized AgNPs were prepared by reducing the $AgNO_3$ solution with sodium borohydride solution dropwise. As the color of the solution turns yellow, solution of GSH was added dropwise and stirred for 2 hours. For the recognition of Ni^{2+} metal solutions were prepared and added to GSH-AgNPs. On addition of Ni^{2+}, $-NH_2$, and -COOH groups interact with it resulting in a color change from yellow to orange. With the successive increase in concentration, there was a redshift in wavelength with the broadening of UV spectra giving a linear correlation of 10^{-6} to 10^{-4} mol /L. It was seen that with the addition of ethylenediamine to this solution, the

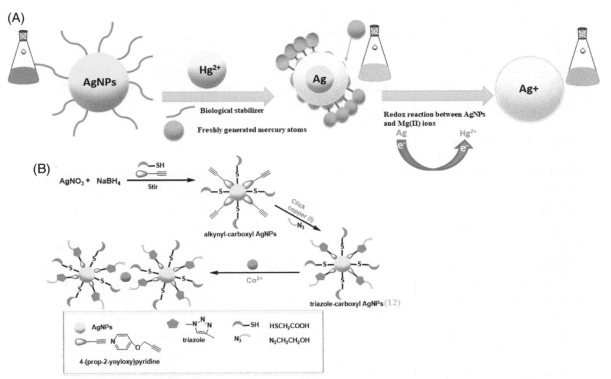

Fig. 3.8: (A) Plausible mechanism of Hg^{2+} sensing proposed by Farhadi et al. (B) Synthesis of triazole-carboxyl AgNPs (12) and their aggregation in the presence of Co^{2+}.

aggregation caused by Ni^{2+} diminishes causing redispersion of AgNPs. This is due to the fact that ethylenediamine interacts with Ni^{2+} more feasibly as compared to GSH dispersing the AgNPs again. So, the prepared solution can also act as a potential sensor for ethylenediamine also [50]. Kirubaharan et al., used biomediated AgNPs for the selective recognition of Cu^{2+}. Biomediated AgNPs-Rh$_6$G was synthesized using Azadirachta indica extract. Copper shows the selective response towards AgNPs-Rh$_6$G by a two-step mechanism. In the first step, Rh$_6$G dye gets adsorbed or fixed on the surface of AgNPs due to which the original purple color of the dye vanishes. In the second step, when Cu^{2+} was added, it replaces the Rh$_6$G from the surface of AgNPs as the color of the solution turns purple due to the presence of bare Rh$_6$G. This was also confirmed by quenching in UV absorbance and enhancement in fluorescence spectroscopy [51]. Yao et al., synthesized the biofunctionalized Triazole-carboxyl AgNPs (**12**) via click reaction in which the AgNO$_3$ solution was reduced by sodium borohydride and mixed with 4-(prop-2-ynyloxy) pyridine and 2-mercaptoacetic acid at r.t for 2 hours. To this solution, 2-azidoethanol, copper sulfate and sodium ascorbic acid was added and stirred at 60°C for 3 hours. The solution was centrifuged, redispersed in water, and used for metal ion sensing. Metal ion solution was added to the functionalized AgNPs and out of which Co^{2+} shows a colorimetric change from yellow to red. In UV absorbance, the intensity of the 405 nm band decreases with an increase in the intensity of the 550 nm band on adding Co^{2+}. The selectivity is due to the carboxyl group of mercaptoacetic acid and nitrogen of triazole which forms a six coordinated structure with Co^{2+}. All these outcomes show the selective sensing of Co^{2+} by **12** (Fig. 3.8B). The resulting **12** able to sense the concentration of Co^{2+} up to 1×10^{-5} M in drinking water [52].

3.3.2 Anion sensing

Other than cations, various anions are also toxic in nature, for example, fluoride in the form of SbF_6^- and PF_6^-. Kumar et al., developed polyacrylate modified AuNPs for the colorimetric sensing of both metal cation (Al^{3+}) and anion (F^-). The reduction of $HAuCl_4$ was done in the presence of sunlight having polyacrylic acid (PAA) as a reducing as well as a stabilizing agent. In the first 5 minutes of exposure to the sun the solution turns black in color but after 120 minutes, the color changes to the characteristic color of AuNPs that is, red showing band at 530 nm in UV absorbance. The particle size came out to be 7 nm determined by TEM. The solution was then diluted by DI water to which Al^{3+}/ F^- stock solutions was added for their detection. The solutions were left to equilibrate for 10 minutes and UV absorbance spectrum was recorded. In the case of Al^{3+}, the color of the solution changed to blue, giving a redshift from 530 to 630 nm. To this PAA-AuNPs-Al^{3+} solution, when F^- was added, the color of the solution turns red while there was no such change in the case of the other 13 anions. This could be that the addition of F^- stops the aggregation induced by Al^{3+} dispersing AuNPs in solution giving red color. So, a functionalized AuNPs was developed which can sense Al^{3+} and F^- simultaneously with a detection limit of 2 μM and 18 μM respectively [53]. Kubo et al., also functionalized the AuNPs by isothiouronium derived amphiphilic units **13** having an average size of 5.4 ± 3 nm (Fig. 3.9A) and use that for the sensing. To the solution of functionalized AuNPs, oxoanions (ACO^-, HPO_4^{2-}) were added for the colorimetric response. In the UV absorption spectrum, there is a redshift during which the color of the solution changes to reddish-violet from red. The change was due to the aggregation of the particles which was confirmed by Fe-SEM images [54]. Extending this work, even after 3 years, the modified AuNPs were also evaluated for the other anions. The solution was analyzed for the detection of F^- by adding PF_6^- and observing the color change to bluish violet from red showing a redshift of 26 nm in the UV spectrum. The changes correspond to the aggregation of AuNPs in solution. The response was also evaluated with the other anions and SCN^- and I^- shows a similar response of aggregation but with lesser intensity. It may be due to phase transfer from organic to aqueous resulting in aggregation of AuNPs in aqueous medium as there was no aggregation in the organic layer as seen by TEM [55]. Bartl et al., functionalized the surface of AuNPs by cyclopeptides and benzylamine (Fig. 3.9B). The modified method of Turkevich was used for the surface modification in which an aqueous solution of $HAuCl_4$ and sodium citrate was boiled at 100°C followed by the addition of **14** and **15** (carboxylic acid) collectively to yield AuNPs-COOH. The prepared AuNPs-COOH was then allowed to react

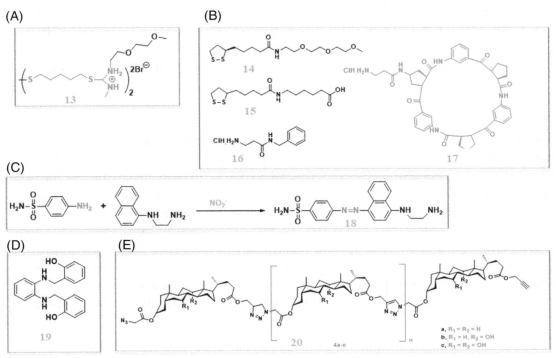

Fig. 3.9: (A) Isothiouronium derived amphiphilic units (13) synthesized by Kubo et al., for the sensing properties. (B) Cyclopeptides synthesized by Bartl et al. (C) Griess reaction performed by Daniel et al., for the sensing of nitrite ions. (D) Structure of probe (19) synthesized by Kumar et al., for the selective sensing of nitrite ions. E) Structure of bile acid-based polymer (20) for the selective sensing of iodide ions.

by two different pathways with **16** (benzylamine derivative) and **17** (cyclopeptide derivative). Out of these two, the cyclopeptide-derived AuNPs show a colorimetric change from red to purple with the sodium salt of sulfate while benzyl amine-functionalized AuNPs show no response with any anion. A titration experiment was conducted which shows that Na_2SO_4 can be detected with a minimum value of 2 mM [56]. Daniel et al., performed a Griess reaction with sulfanilamide and naphthyl ethylene diamine and prepared azo dye **(18)** shown in Fig. 3.9C, which was coated over the surface of AuNPs. At first, AuNPs were coated with 5-[1,2]dithiolan-3-yl pentanoic acid [2-(4-aminophenyl)ethyl]amide (DPAA) which is due to its hydrophobicity leads to precipitation of AuNPs in water. To increase the solubility of functionalized AuNPs in an aqueous medium, hydrophilic MTA was added to the surface of AuNPs. These are categorized as "aniline AuNPs". Following the same procedure, "naphthalene AuNPs" were prepared in which DPAA and MTA was used. Both the solutions were red in color when dispersed in water. To this solution nitrite ion was added, an amine group present on aniline gets converted to diazonium salt and couples with naphthalene AuNPs forming a crosslinking between the particles due to which it precipitates out and the solution turns colorless. The formation of diazonium salt was confirmed by Mass spectroscopy.

Selectivity towards nitrite ions was confirmed by performing the same experiment with F^-, SO_4^{2-}, Br^-, ClO_4^-, CH_3COO^-, $S_2O_3^{2-}$, $C_2O_4^{2-}$, N_3^-, HCO_3^- also. No color change is observed in other cases [57]. Kumar et al., also developed the probe(Fig. 3.9D) for the selective sensing of nitrite ions by modifying AgNPs by phenolic chelating ligand, N, N'-bis(2-hydroxy benzyl)-1,2-diaminobenzene (**19**). To the yellow solution of modified AgNPswhen NO_2^- and NO_3^- were added, the color of the solution disappears forming a band at 360 nm in the UV absorbance spectrum. The selectivity of NO_2^- and NO_3^- was due to the fact that on the addition of NO_2^-, AgNPs gets oxidized to Ag^0 which was confirmed by adding NaCl to the solution resulting in AgCl white-colored ppt. In the case of NO_3^-, the response of color change was slow which can be attributed to the conversion of NO_3^- to NO_2^- in the first step followed by oxidation of AgNPs to Ag^0. Titration experiment shows the gradual increase of 360 nm band giving a detection limit of 10^{-7} M and has the ability to sense in pond, tap, river, and groundwater [58]. Kumar et al., synthesize bile acid-based polymers **20** as in Fig. 3.9E, which can stabilize AgNPs and selectively sense iodide ions colorimetrically. The TEM image shows the formation of AgNPs of average size 4–4.2 nm. Tetra butyl ammonium salt solutions of various ions were used for the selective recognition of iodide ions. Out of all, only I^- shows a color change from dark yellow to colorless due to aggregation in AgNPs giving a detection limit of 250 µM [59].

Massue et al., show the sensing of phosphate ions by Eu (III)-cyclen-conjugated (**21**.Eu) AuNPs as shown in Fig. 3.10A. The functionalized AuNPs were prepared using the Brust method according to which DMAP-AuNPs were prepared which was then substituted by **21**. Eu by simple stirring at r.t for 12 hours. The shift in UV spectrum from 520 to 510 nm confirms the substitution of on the surface of AuNPs. The particles formed show the size of 5 nm. The prepared **21**.Eu AuNPs were then allowed to interact with **22** in the presence of water. The introduction of **22** shows the phenomenon of luminescence which was accomplished by the formation of the ternary complex. The formation of the ternary complex was evaluated by fluorescence at ex 336 nm which gives emission at 595, 616, 685, and 700 nm. This complex was then evaluated with creatine, coumaric acid, pantothenic acid, Na_2CO_3, and H_2PO_4. All these show a very less effect on luminescence of the complex while on the addition of Flavin mononucleotide (**23**), luminescence completely diminishes due to the substitution of **22** by **23** on the surface of **21**. Eu [60]. Dong et al., synthesize functionalized AgNPs (**24**) in the presence of $HAuCl_4.3H_2O$ as catalyst. The aqueous solutions of $AgNO_3$ and $HAuCl_4.3H_2O$ were mixed under vigorous stirring. To the stirring mixture, Tannic acid was added and stirred for 30 minutes, during which the color of the solution turns yellow. On adding $S_2O_3^{2-}$ to the solution, a redox reaction occurs between the phenolic hydroxyl end of tannic acid and $S_2O_3^{2-}$, due to which tannic acid adsorbed on the surface of AgNPs gets removed and the reduced form of $S_2O_3^{2-}$ that is S^{2-} combines with Ag forming Ag_2S which in results forms a layer over the surface of AgNPs. The redox reaction between tannic acid and $S_2O_3^{2-}$ results in a color change from yellow to brown accompanied by the decrease in intensity of UV visible absorption band at 419 nm (Fig. 3.10B). The selectivity of $S_2O_3^2$ was evaluated by performing the same

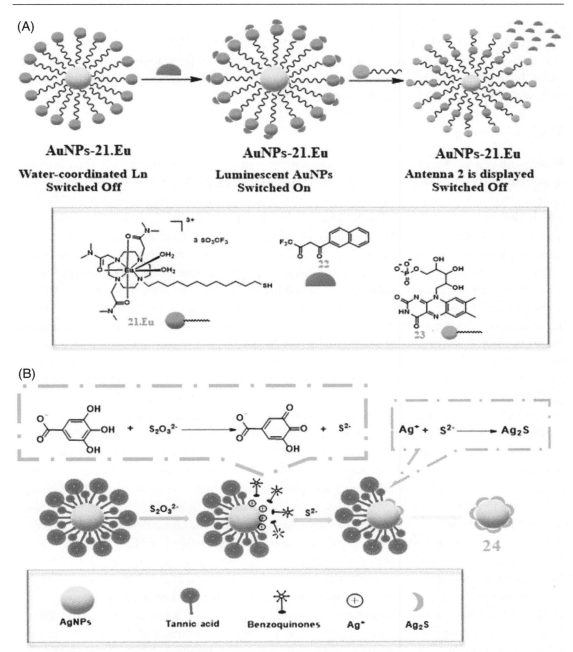

Fig. 3.10: (A) The self-assembly formation of nonluminescent AuNP-21.Eu and 22 to give AuNPs-21.Eu-22. Displacement of 22 by Flavin shown in second step. (B) Mechanism of AgNPs-based colorimetric sensor for $S_2O_3^{2-}$.

experiment with CO_3^{2-}, HCO_3^-, $C_2O_4^{2-}$, NO_3^-, NO_2^-, PO_4^{3-}, $P_2O_7^{4-}$, HPO_4^{2-}, $H_2PO_4^-$, S_2^-, SO_3^{2-}, SO_4^{2-}, $S_2O_7^{2-}$, F^-, Cl^-, Br^-, I^-, ClO_4^-, BrO_3^-, AC^-, $Cr_2O_7^{2-}$, and MnO_4^-. None of the ions shows color change or any change in UV visible spectrum. By calculating the RGB of the color change due to $S_2O_3^{2-}$, a smartphone-based device has been fabricated which can detect the level of $S_2O_3^{2-}$ in environmental samples [61].

3.3.3 Biosensing

Biosensing is a technique to detect the concentration of various biomolecules in the environment, agriculture as well as various animals and human beings. The usage of pesticides in agriculture causes various effects on human health due to their accumulation in the body which ultimately affects the nervous system, the digestive system of our body. The concentration of biomolecules already presents in our body also affects human health by causing diseases like diabetes. Due to these long-life diseases, the need for biosensors for these biomolecules is immense. In these, AuNPs and AgNPs plays important role in biosensing due to their sensing and selectivity. Aslan et al., coated the surface of gold with dextran. To coat the surface of AuNPs with dextran, it needs to be modified first with a long chain, 16-Mercaptohexadecanoic acid (16-MHDA). The terminal carboxyl group gets activated by N-3-(dimethyl aminopropyl)-N'-ethyl carbodiimide (EDC) and N-hydroxy-2,5-pyrrolidinedione (NHS) which forms NHS-esters bonds that are active in nature. In order to introduce the hydroxyl-terminated layer on the surface of AuNPs, the formed NHS esters were allowed to react with (2-(2-aminoethoxy) ethanol (AEE). These hydroxyl groups are activated with epichlorohydrin and allowed to be coupled covalently with dextran. The prepared dextran-coated AuNPs shows a complexation and decomplexation reaction with concanavalin A (Con A) and glucose respectively. The addition of Con A causes aggregation of AuNPs while addition of glucose results inredispersion accompanied by a blue shift in the UV spectrum. On varying the concentration of dextran-coated AuNPs and glucose, a ratiometric response was observed by an increase in the intensity of 560 nm and a negligible change in the 630 nm band. Dextran-coated AuNPs are able to sense glucose up to several millimolar concentrations [62]. Youk et al., functionalized the AuNPs by o-(trifluoroacetyl)carboxanilide [TFACA(**25**)] as in Fig. 3.11A, for the selective recognition of fumarate by considering the property of TFACA to form a covalent adduct with anions which is reversible in nature. The AuNPs of size 13 nm were prepared by reducing $HAuCl_4$ with sodium citrate and capped with TA derivative TFACA which was then stabilized by poly(vinyl alcohol) to form a full sensor system. The 2 mL of the resulting solution was mixed with 0.1 M aqueous solution of the sodium salt of fumarate, maleate, oxalate, malonate, succinate, glutarate, propanoate, and 4- pentanoate. Within 10 minutes the color of the solution containing fumarate changes to purple from red whereas in the UV visible spectrum it also shows a redshift from 520 to 550 nm. The selective recognition of fumarate over all the carboxylates is due to its geometry which favors the crosslinking of AuNPs resulting in aggregation. The TEM images were recorded to confirm

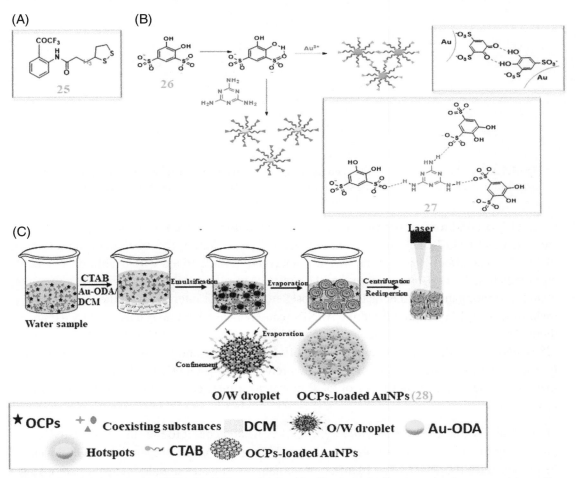

Fig. 3.11: (A) Structure of TFACA (25) synthesized by Youk et al., for the selective recognition of fumarate. (B) Schematic illustration of synthesis of PD capped AuNPs and involvement of melamine in generation of AuNPs. (C) Diagrammatic representation of in-situ extraction and self-assembly of AuNPs for detection of OCP.

the aggregation of AuNPs caused by fumarate over its cis over maleate [63]. Wu et al., prepare the AuNPs by using pyrocatechol-3,5-disodium sulfonate [PD(**26**)]on the basis of forming an intramolecular hydrogen bonding (Fig. 3.11B). In the process, PD acts as a reducing agent due to which its concentration in preparation of AuNPs was well monitored and fixed to be 1.6×10^{-4} M for the detection of melamine. The prepared AuNPs show green color in the absence of melamine but with the addition of 1.6×10^{-7} M melamine to it, the color changes to yellow with the disappearance of 600 nm band in the UV spectrum. It was also noticed that the addition of melamine to the solution decreases the reaction time of formation of AuNPs from 50 minutes. This can be due to the interruption of intramolecular H-bonds between PD

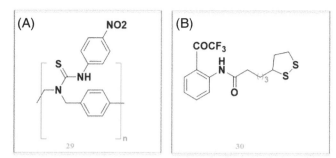

Fig. 3.12: (A) Polymer (29) synthesized by Singh et al., for the detection of quinalphos in water samples. (B) Naphthalimide based sensor (30) used by Singh et al., for the recognition of shellfish poison okadaic acid.

by an O-NH bond formed between melamine and PD due to which α-phenolic OH of PD was free and reactive (**27**). The free α-phenolic OH of PD accelerates the reaction of the formation of AuNPs. The effect of melamine was also analyzed in milk mixed with trichloroacetic acid and methanol. And it is also shown that the presence of other analytes (Vitamin A, glutamine, alanine, tryptophan, tyrosine, lactose, maltose, etc.) does not interfere with the detection of melamine [64]. Li et al., modified the AuNPs using octadecyl amine (ODA) and by using them trapped the organochlorine pesticides (OCP) in water (Fig. 3.11C). These OCP-loaded (**28**) AuNPs were then dispersed in an ethanol-water mixture and UV spectrum was recorded. It is seen that the loading of OCPs results in a redshift of Au-ODA from 525 to 555 nm. The method is able to detect four types of OCP in water samples collected from different sources [65].

Singh et al., synthesized the polymer **29** (Fig. 3.12A) and coat it to the surface of AuNPs by fabricating them in the form of organic nanoparticles (ONP). The prepared Au@ONP was further used for the selective recognition of quinalphos in water samples. The prepared Au@ONP was first used for the selective recognition of metals, in which Cu(II) shows different behavior in the fluorescence spectrum with a blue shift from 405 to 355 nm. On the basis of the results, the Au@ONP were complexed with Cu (II) and used for the selective recognition of quinalphos. In the presence of quinalphos, there was a red shift in fluorescence spectrum which was not observed in the case of other analytes, solid-state analysis was also done by coating Cu-Au@ONP over silica beads. The solid-state also, quinalphos shows recognition properties by a color change to greenish-yellow under UV light elimination with a detection limit of 2.4 nm [66]. In another study, Singh et al., used AuNPs for the detection of okadaic acid also. Naphthalimide based receptor (**30**) as shown in Fig. 3.12B, was prepared which was fabricated in the form of ONP and Au@ONP was prepared in this case also. The probe solution was used for the detection of diarrheic acid shellfish poisoning (DSP) which is caused by eating okadaic acid contaminated shellfish in seafood. For its detection, a composite was prepared by mixing ONP and Au@ONP in a 4:6 v/v ratio. In UV spectrum composite shows 2 bands at 450 and 530 nm, on adding okadaic acid to the solution, a ratiometric response was

(A)

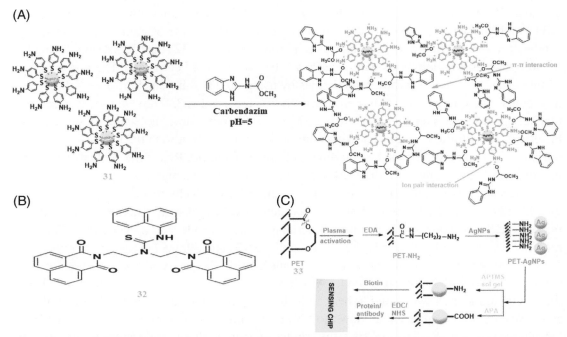

(B)   (C)

Fig. 3.13: (A) Schematic representation for the colorimetric detection of Carbendazim using ABT-AgNPs as colorimetric probe. (B) Structure of thiourea-based ligand synthesized by Singh et al., for the detection of ketoprofen. (C) Preparation of SNOPS strip for biosensing of streptavidin.

observed in which the 450 nm band was increased with a simultaneous decrease in 530 nm band while in fluorescence okadaic acid shows quenching in 535 nm band with a detection limit of 20 nM. Biological toxicity in *HeLa cells* and real sample analysis in water samples were carried out by using the prepared solution and it shows the recovery % above 95% [67].

Patel et al., synthesized 4-aminobenzenethizol functionalized AgNPs [ABT-AgNPs (**31**)] for the selective detection of fungicide *Carbendazim*. The prepared ABT-AgNPs were stable for even a month. The formed material was characterized using FT-IR and DLS showing a size of 3.6 nm. The ABT-AgNPs were then exposed to *Carbendazim* of various concentrations ranging from 1-100 μM. The addition of carbendazim causes the color to change to orange from yellow (Fig. 3.13A). The detection was done at various pH ranges from 2 to 10. And it was seen that the addition of *Carbendazim* shows the change in UV spectrum only up to pH 5 in which it shows a redshift from 395 to 510 nm. In pH range 6-10 there is no change in UV spectrum showing the recognition change in acidic conditions is better. The surface of ABT-AgNPs has a positive charge while *Carbendazim* has a negative surface charge. The mixing of both these causes ion-pair interactions which result in aggregation and color change from yellow to orange with a detection limit of 1.4 μM. The practical applicability was tested by water samples, fruit, and vegetables, and the results came out satisfactory with an RSD value of less than 5%. It was also shown that the same effect of *Carbendazim* cannot be obtained with ABT-AuNPs. There was no color change or aggregation in the case of gold [68]. Singh et al., synthesized three different ligands based on thiourea derivative **32** (Fig. 3.13B) and formed organic-inorganic nanohybrids with AgNPs for the detection of ketoprofen. The prepared nanohybrids were used for the sensor studies by using 20 μM solutions of different drugs molecules. The fluorescence emission spectrum was recorded at 338 nm and it was observed that ketoprofen shows enhancement in the 338 nm while other drug molecules show negligible effect. The titrations results showing continuous enhancement with the increasing concentration of ketoprofen confirmed its binding with the nanohybrid. Real-time analysis was done in tablets and gels available in the market giving a recovery percentage above 99% [69]. Fan et al., prepared silver nanoparticles on a plastic sensor (SNOPS) by modifying polyethylene terephthalate [PET (**33**)]. The PET surface was activated by a plasma oven and immersed in ethylenediamine due to which amino functionality was introduced on the surface. These modified strips were incubated in the presence of AgNPs and allowed to react with either 3-aminopropyltriethoxy silane or 3-aminopropanoic acid. The prepared SNOPS strips then came in contact with biotin-streptavidin which results in a redshift in the UV spectrum. The limit of quantification was found to be 9.5 nM for streptavidin (Fig. 3.13C) [70]. Lu et al., synthesized graphitic nitride nanodots (g-C_3N_4-dots) and used them as a reducing agent for the synthesis of AgNPs. The UV spectrum shows quenching that is, a decrease in intensity of prepared AgNPs in presence of biothiols like cysteine, homocysteine and, glutathione. This is due to the formation of the Ag-S bond between AgNPs and biothiols which stops the growth of AgNPs resulting in decreasing intensity. The scheme was applied for the detection of cysteine in human serum and it shows a recovery percentage above 95% which confirms the practical utility of the sensor [71].

3.4 Conclusion and future trends

To summarize, there are various routes synthesize gold and silver nanoparticles via chemical as well as biological pathways. The main step in the synthesis includes the reduction of metal salts $HAuCl_4$ and $AgNO_3$ respectively. Moreover, their optical properties and localized surface plasmon that is, UV visible spectrum proves to very useful in sensing applications. They are easily able to sense metal ions, anions, biomolecules and can also be used for the techniques like gene engineering, drug delivery, therapeutics and many more. Due to the color of the NPs, their colorimetric sensing proves to be easy, fast and naked eye detection and proves to be efficient optical sensors also. However, use of gold and silver nanoparticles in chips and device fabrication is now taking a new start in the field of sensing. Cooperative alliance between chemo sensing and electronics by using portable smart phones or fabricating lateral flow device are increasing the efficiency of the sensors day by day. These handmade devices or diagnostic kits are making it easy for homebodies to test the harmful analytes in their food, water and even body and building their life healthy and easy.

References

[1] C.N. Rao, A. Muller, A.K. Cheetham, The Chemistry of Nanomaterials: Synthesis, Properties and Applications, WILEY-VCH Verlag GmbH & Co, KGaA, Weinheim, 2004. ISBN: 3-527-30686-2.

[2] R. Singh, H.S. Nalwa, Medical applications of nanoparticles in biological imaging, cell labeling, antimicrobial agents, and anticancer nanodrugs, J. Biomed. Nanotechnol. 7 (2011) 489–503.

[3] D. Schaming, H. Remita, Nanotechnology: From the ancient time to nowadays, Found. Chem. 17 (2015) 187–205.

[4] F.J. Heiligtag, M. Niederberger, The fascinating world of nanoparticle research, Mater. Today. 16 (2013) 262–271.

[5] A.A. Yaqoob, H. Ahmad, T. Parveen, A. Ahmad, M. Oves, I.M.I. Ismail, H.A. Qari, K. Umar, M.N. Mohamad Ibrahim, Recent advances in metal decorated nanomaterials and their various biological applications: A review, Front Chem 8 (2020) 1–23.

[6] M.C. Roco, Nanoparticles and nanotechnology research, J. Nanoparticle Res. 1 (1999) 1–6.

[7] N. Venkatesh, Metallic nanoparticle: A review, Biomed. J. Sci. Tech. Res. 4 (2018) 3765–3775.

[8] J. Turkevich, P.C. Stevenson, J. Hillier, A study of the nucleation and growth processes in the synthesis of colloidal gold, Discuss. Faraday Soc. 11 (1951) 55–75.

[9] M. Brust, M. Walker, D. Bethell, D.J. Schiffrin, R. Whyman, Synthesis of Thiol-derivatised Gold Nanoparticles in, (2000) 801–802.

[10] M.M. Oliveira, D. Ugarte, D. Zanchet, A.J.G. Zarbin, Influence of synthetic parameters on the size, structure, and stability of dodecanethiol-stabilized silver nanoparticles, J. Colloid Interface Sci. 292 (2005) 429–435.

[11] D. Kim, S. Jeong, J. Moon, Synthesis of silver nanoparticles using the polyol process and the influence of precursor injection, Nanotechnology 17 (2006) 4019–4024.

[12] W. Zhang, X. Qiao, J. Chen, Synthesis of nanosilver colloidal particles in water/oil microemulsion, Colloids Surfaces A Physicochem. Eng. Asp. 299 (2007) 22–28.

[13] D.G. Shchukin, I.L. Radtchenko, G.B. Sukhorukov, Photoinduced reduction of silver inside microscale polyelectrolyte capsules, Chem. Phys. Chem 4 (2003) 1101–1103.

[14] M.N. Nadagouda, T.F. Speth, R.S. Varma, Microwave-assisted green synthesis of silver nanostructures, Acc. Chem. Res. 44 (2011) 469–478.

[15] M. Starowicz, B. Stypuła, J. Banaś, Electrochemical synthesis of silver nanoparticles, Electrochem. Commun. 8 (2006) 227–230.

[16] K.S. Mayya, F. Caruso, Phase transfer of surface-modified gold nanoparticles by hydrophobization with alkylamines, Langmuir 19 (2003) 6987–6993.

[17] L. Shen, M. Chen, L. Hu, X. Chen, J. Wang, Growth and stabilization of silver nanoparticles on carbon dots and sensing application, Langmuir 29 (2013) 16135–16140.

[18] M.N. Chen, C.F. Chan, S.L. Huang, Y.S. Lin, Green biosynthesis of gold nanoparticles using Chenopodium formosanum shell extract and analysis of the particles' antibacterial properties, J. Sci. Food Agric. 99 (2019) 3693–3702.

[19] O.S. Oluwafemi, J.L. Anyik, N.E. Zikalala, E.H.M. Sakho, Biosynthesis of silver nanoparticles from water hyacinth plant leaves extract for colourimetric sensing of heavy metals, Nano-Struct. Nano-Objects 20 (2019) 100387.

[20] Y. Qu, X. Pei, W. Shen, X. Zhang, J. Wang, Z. Zhang, S. Li, S. You, F. Ma, J. Zhou, Biosynthesis of gold nanoparticles by Aspergillum sp. WL-Au for degradation of aromatic pollutants, Phys. E Low-Dimensional Syst. Nanostructures. 88 (2017) 133–141.

[21] R. Bhambure, M. Bule, N. Shaligram, M. Kamat, R. Singhal, Extracellular biosynthesis of gold nanoparticles using Aspergillus niger - Its characterization and stability, Chem. Eng. Technol. 32 (2009) 1036–1041.

[22] M.F. Lengke, M.E. Fleet, G. Southam, Biosynthesis of silver nanoparticles by filamentous cyanobacteria from a silver(I) nitrate complex, Langmuir 23 (2007) 2694–2699.

[23] K. Luo, S. Jung, K.H. Park, Y.R. Kim, Microbial biosynthesis of silver nanoparticles in different culture media, J. Agric. Food Chem. 66 (2018) 957–962.

[24] H. Korbekandi, S. Mohseni, R.M. Jouneghani, M. Pourhossein, S. Iravani, Biosynthesis of silver nanoparticles using Saccharomyces cerevisiae, Artif. Cells, Nanomed. Biotechnol 44 (2016) 235–239.

[25] A.K. Jha, K. Prasad, V. Kumar, K. Prasad, Biosynthesis of silver nanoparticles using eclipta leaf, Biotechnol. Prog. 25 (2009) 1476–1479.

[26] S.S. Shankar, A. Ahmad, M. Sastry, Geranium leaf assisted biosynthesis of silver nanoparticles, Biotechnol. Prog. 19 (2003) 1627–1631.

[27] P. Slepicka, N.S. Kasalkova, J. Siegel, Z. Kolska, V. Svorcik, Methods of gold and silver nanoparticles preparation, Materials (Basel) 13 (2020) 1.

[28] D.L. Zhong, F.L. Yuan, L. Lian, Z.H. Cheng, A localized surface plasmon resonance light-scattering assay of mercury (II) on the basis of Hg^{2+}-DNA complex induced aggregation of gold nanoparticles, Environ. Sci. Technol. 43 (2009) 5022–5027.

[29] F. Xia, X. Zuo, R. Yang, Y. Xiao, D. Kang, A. Vallee-Belisle, X. Gong, J.D. Yuen, B.B.Y. Hsu, A.J. Heeger, K.W. Plaxco, Colorimetric detection of DNA, small molecules, proteins, and ions using unmodified gold nanoparticles and conjugated polyelectrolytes, Proc. Natl. Acad. Sci. U. S. A. 107 (2010) 10837–10841.

[30] I.E. Paul, A. Rajeshwari, T.C. Prathna, A.M. Raichur, N. Chandrasekaran, A. Mukherjee, Colorimetric detection of melamine based on the size effect of AuNPs, Anal. Methods. 7 (2015) 1453–1462.

[31] M. Lepoitevin, M. Lemouel, M. Bechelany, J.M. Janot, S. Balme, Gold nanoparticles for the bare-eye based and spectrophotometric detection of proteins, polynucleotides and DNA, Mikrochim. Acta 182 (2015) 1223–1229.

[32] R. Kanjanawarut, X. Su, Colorimetric detection of DNA using unmodified metallic nanoparticles and peptide nucleic acid probes, Anal. Chem. 81 (2009) 6122–6129.

[33] A. Ravindran, V. Mani, N. Chandrasekaran, A. Mukherjee, Selective colorimetric sensing of cysteine in aqueous solutions using silver nanoparticles in the presence of Cr3+, Talanta 85 (2011) 533–540.

[34] Z.S. Wu, S.B. Zhang, M.M. Guo, C.R. Chen, G.L. Shen, R.Q. Yu, Homogeneous, unmodified gold nanoparticle-based colorimetric assay of hydrogen peroxide, Anal. Chim. Acta. 584 (2007) 122–128.

[35] T. Serizawa, Y. Hirai, M. Aizawa, Detection of enzyme activities based on the synthesis of gold nanoparticles in HEPES buffer, Mol. Biosyst. 6 (2010) 1565–1568.

[36] K. Loza, M. Heggen, M. Epple, Synthesis, structure, properties, and applications of bimetallic nanoparticles of noble metals, Adv. Funct. Mater. 30 (2020) 1909260.

[37] J.A. Adekoya, K.O. Ogunniran, T.O. Siyanbola, E.O. Dare, N. Revaprasadu, Band structure, morphology, functionality, and size- dependent properties of metal nanoparticles, Noble Precious Met. - Prop. Nanoscale Eff. Appl. (2018).

[38] S.O. Obare, R.E. Hollowell, C.J. Murphy, Sensing strategy for lithium ion based on gold nanoparticles, Langmuir 18 (2002) 10407–10410.

[39] S.Y. Lin, C.H. Chen, M.C. Lin, H.F. Hsu, A cooperative effect of bifunctionalized nanoparticles on recognition: Sensing alkali ions by crown and carboxylate moieties in aqueous media, Anal. Chem. 77 (2005) 4821–4828.

[40] H. Li, L. Zhang, Y. Yao, C. Han, S. Jin, Synthesis of aza-crown ether-modified silver nanoparticles as colorimetric sensors for Ba^{2+}, Supramol. Chem. 22 (2010) 544–547.

[41] X. Wu, W. Tang, C. Hou, C. Zhang, N. Zhu, Colorimetric and bare-eye detection of alkaline earth metal ions based on the aggregation of silver nanoparticles functionalized with thioglycolic acid, Mikrochim. Acta 181 (2014) 991–998.

[42] R. Dominguez-González, L.G Varela, P. Bermejo-Barrera, Functionalized gold nanoparticles for the detection of arsenic in water, Talanta 118 (2014) 262–269.

[43] Y. Guo, Y. Zhang, H. Shao, Z. Wang, X. Wang, X. Jiang, Label-free colorimetric detection of cadmium ions in rice samples using gold nanoparticles, Anal. Chem. 86 (2014) 8530–8534.

[44] Y.Q. Dang, H.W. Li, B. Wang, L. Li, Y. Wu, Selective detection of trace Cr3+ in aqueous solution by using 5,5′-dithiobis (2-Nitrobenzoic acid)-modified gold nanoparticles, ACS Appl. Mater. Interfaces 1 (2009) 1533–1538.

[45] T.A. Ali, G.G. Mohamed, E.M.S. Azzam, A.A. Abd-Elaal, Thiol surfactant assembled on gold nanoparticles ion exchanger for screen-printed electrode fabrication. Potentiometric determination of Ce(III) in environmental polluted samples, Sens. Actuators B Chem 191 (2014) 192–203.

[46] J. Liu, Y. Lu, Accelerated color change of gold nanoparticles assembled by DNAzymes for simple and fast colorimetric Pb^{2+} detection, J. Am. Chem. Soc. 126 (2004) 12298–12305.

[47] S. Si, A. Kotal, T.K. Mandal, One-dimensional assembly of peptide-functionalized gold nanoparticles: An approach toward mercury ion sensing, J. Phys. Chem. 111 (2007) 1248–1255.

[48] D. Liu, W. Qu, W. Chen, W. Zhang, Z. Wang, X. Jiang, Highly sensitive, colorimetric detection of mercury(II) in aqueous media by quaternary ammonium group-capped gold nanoparticles at room temperature, Anal. Chem. 82 (2010) 9606–9610.

[49] K. Farhadi, M. Forough, R. Molaei, S. Hajizadeh, A. Rafipour, Highly selective Hg 2+ colorimetric sensor using green synthesized and unmodified silver nanoparticles, Sens Actuators B Chem 161 (2012) 880–885.

[50] H. Li, Z. Cui, C. Han, Glutathione-stabilized silver nanoparticles as colorimetric sensor for Ni2+ ion, Sens. Actuators. B Chem 143 (2009) 87–92.

[51] C.J. Kirubaharan, D. Kalpana, Y.S. Lee, A.R. Kim, D.J. Yoo, K.S. Nahm, G.G. Kumar, Biomediated silver nanoparticles for the highly selective copper(II) ion sensor applications, Ind. Eng. Chem. Res. 51 (2012) 7441–7446.

[52] Y. Yao, D. Tian, H. Li, Cooperative binding of bifunctionalized and click-synthesized silver nanoparticles for colorimetric Co^{2+} Sensing, ACS Appl. Mater. Interfaces 2 (2010) 684–690.

[53] A. Kumar, M. Bhatt, G. Vyas, S. Bhatt, P. Paul, Sunlight induced preparation of functionalized gold nanoparticles as recyclable colorimetric dual sensor for aluminum and fluoride in water, ACS Appl. Mater. Interfaces. 9 (2017) 17359–17368.

[54] Y. Kubo, S. Uchida, Y. Kemmochi, T. Okubo, Isothiouronium-modified gold nanoparticles capable of colorimetric sensing of oxoanions in aqueous MeOH solution, Tetrahedron Lett. 46 (2005) 4369–4372.

[55] T. Minami, K. Kaneko, T. Nagasaki, Y. Kubo, Isothiouronium-based amphiphilic gold nanoparticles with a colorimetric response to hydrophobic anions in water: a new strategy for fluoride ion detection in the presence of a phenylboronic acid, Tetrahedron Lett 49 (2008) 432–436.

[56] J. Bartl, L. Reinke, M. Koch, S. Kubik, Selective sensing of sulfate anions in water with cyclopeptide-decorated gold nanoparticles, Chem. 56 (2020) 10457–10460.

[57] W.L. Daniel, M.S. Han, J.S. Lee, C.A. Mirkin, Colorimetric nitrite and nitrate detection with gold nanoparticle probes and kinetic end points, J. Am. Chem. Soc. 131 (2009) 6362–6363.

[58] V.V. Kumar, S.P. Anthony, Highly selective silver nanoparticles based label free colorimetric sensor for nitrite anions, Anal. Chim. Acta 842 (2014) 57–62.

[59] A. Kumar, R.K. Chhatra, P.S. Pandey, Synthesis of click bile acid polymers and their application in stabilization of silver nanoparticles showing iodide sensing property, Org. Lett. 12 (2010) 24–27.

[60] J. Massue, S.J. Quinn, T. Gunnlaugsson, Lanthanide luminescent displacement assays: The sensing of phosphate anions using Eu(III)-cyclen-conjugated gold nanoparticles in aqueous solution, J. Am. Chem. Soc. 130 (2008) 6900–6901.

[61] C. Dong, Z. Wang, Y. Zhang, X. Ma, M.Z. Iqbal, L. Miao, Z. Zhou, Z. Shen, A. Wu, High-performance colorimetric detection of thiosulfate by using silver nanoparticles for smartphone-based analysis, ACS Sens 2 (2017) 1152–1159.

[62] K. Aslan, J.R. Lakowicz, C.D. Geddes, Nanogold plasmon resonance-based glucose absorption bands due to electron oscillations induced by, Anal. Chem. 77 (2005) 2007–2014.

[63] K.S. Youk, K.M. Kim, A. Chatterjee, K.H. Ahn, Selective recognition of fumarate from maleate with a gold nanoparticle-based colorimetric sensing system, Tetrahedron Lett. 49 (2008) 3652–3655.

[64] Z. Wu, H. Zhao, Y. Xue, Q. Cao, J. Yang, Y. He, X. Li, Z. Yuan, Colorimetric detection of melamine during the formation of gold nanoparticles, Biosens. Bioelectron. 26 (2011) 2574–2578.

[65] N. Fahimi-Kashani, M.R. Hormozi-Nezhad, Gold-nanoparticle-based colorimetric sensor array for discrimination of organophosphate pesticides, Anal. Chem. 88 (2016) 8099–8106.

[66] R. Rani, P.T Mayank, N. Singh, Fine tuning of polymer-coated gold nanohybrids: Sensor for the selective detection of quinalphos and device fabrication for water purification, ACS Appl. Nano Mater 2 (2019) 1–5.

[67] M. Verma, M. Chaudhary, A. Singh, N. Kaur, N. Singh, Naphthalimide-gold-based nanocomposite for the ratiometric detection of okadaic acid in shellfish, J. Mater. Chem. B. 8 (2020) 8405–8413.

[68] G.M. Patel, J.V. Rohit, R.K. Singhal, S.K. Kailasa, Recognition of carbendazim fungicide in environmental samples by using 4-aminobenzenethiol functionalized silver nanoparticles as a colorimetric sensor, Sens. Actuators B. Chem 206 (2015) 684–691.

[69] A. Saini, M. Kaur, A.K Mayank, N. Kaur, N. Singh, Hybrid nanoparticle based fluorescence switch for recognition of ketoprofen in aqueous media, Mol. Syst. Des. Eng. 5 (2020) 1428–1436.

[70] M. Fan, M. Thompson, M.L. Andrade, A.G. Brolo, Silver nanoparticles on a plastic platform for localized surface plasmon resonance biosensing, Anal. Chem. 82 (2010) 6350–6352.

[71] Q. Lu, H. Wang, Y. Liu, Y. Hou, H. Li, Y. Zhang, Graphitic carbon nitride nanodots: As reductant for the synthesis of silver nanoparticles and its biothiols biosensing application, Biosens. Bioelectron. 89 (2017) 411–416.

CHAPTER 4

Glucose biosensing with gold and silver nanoparticles for real-time applications

R. Balamurugan[a], **S. Siva Shalini**[a], **M.P. Harikrishnan**[a], **S. Velmathi**[b], **A. Chandra Bose**[a]

[a]*Nanomaterials Laboratory, Department of Physics, National Institute of Technology, Tiruchirappalli, Tamil Nadu, India* [b]*Organic and Polymer Synthesis Laboratory, Department of Chemistry, National Institute of Technology, Tiruchirappalli, Tamil Nadu, India*

4.1 Introduction to glucose sensing

This chapter describes the fundamentals and working of the biosensor with its parameters, electrochemical glucose sensing, colorimetric glucose sensing, and the role of gold and silver nanoparticles in real-time glucose sensing applications. Before that, let's see the historical developments of glucose sensors and the importance of glucose detection.

Diabetes mellitus has become the first endocrine disease that affects approximately 425 million people worldwide. Most people in the age group 20–79 years are probably affected [1]. According to the world health organization (WHO), persons with diabetes were approximately (6.4%) of 285 million adults in 2010, and it's expected to increase (7.7%) to 439 million adults by 2030 [2–4]. Type 1 (T1D) and type 2 (T2D) diabetes are the most common form of existing diabetes. T1D is usually identified in children and young adults as a genetic disorder that is often revealed in early life. Type 2 diabetes is more common in US people. It is typically caused by lifestyle [5,6].

Various attempts have been carried out to measure the glucose in urine from the mid of 18th century. After several measures, Benedict evolved a copper reagent for monitoring glucose in urine in 1908, which was utilized with simple modifications for more than 50 years [7]. In 1945 cumbersome methodology of heating became more familiar with modified copper reagent as a tablet in which glucose is oxidized, and the quantity of glycosuria was proportionate to the color of the heated solution. In 1962, the first enzymatic glucose sensors were introduced by Clark and Lyons, in which consumption of oxygen was observed based on catalytic oxidation of glucose [8,9]. In 1965, Ames developed dextrostix using glucose oxidase enzyme is the first glucose testing strip and filed patent. It requires a large amount of blood sample to be placed on the strip. After 60s the strip was cleaned away, and the generated color was then compared with the standard chart. In 1973, the first amperometric enzyme glucose

Gold and Silver Nanoparticles.
DOI: https://doi.org/10.1016/B978-0-323-99454-5.00002-0

sensor was evolved in which hydrogen peroxide production was characterized instead of oxygen reduction current [10]. Due to lack of accuracy, in 1980 the dextrometer was commercialized at a lower price and required only fewer amount of blood samples. This meter consists of dextrostix with a digital display, which patients can measure from their homes [11,12].

In-depth treatment for various types of diabetes was carried out by the introduction of new methods of monitoring glycemic control and new ways of insulin delivery were introduced in the late 1970s and 1980s [13]. Self-monitoring of blood glucose (SMBG) in diabetic patients plays a major role and helps diabetic patient to better understand their glycemic status to undertake appropriate remedies with hyper and hypoglycemia [12,14–16]. In contrast, multiple daily injections or insulin inject therapy which, helps to reduce the risks of severe hypoglycemia and control postprandial hyperglycemia. Self-monitoring blood glucose sensor uses test strips containing either hexokinase or glucose oxidase. A little drop of blood is placed, and it undergoes numerous chemical reactions, and glucose ranges identified [17–20]. SMBG was commercially used in numerous recent benchmark clinical trials, which mainly focus on studies involving cardiovascular disease, which leads to cause morbidity and mortality in diabetic patients [21]. Some drawbacks of SMBG are, it will increase the anxiety about blood sugar level and state of health, the physical pain of fingerpicking [22].

Continuous glucose monitoring (CGM) is an excellent path to manage the normal blood glucose range in the human body. A diverse research test has been focused on the usage of commercial CGM sensors to anticipate and protect hypoglycemic and hyperglycemic levels [23,24]. CGM is used to measure the glucose level continuously in the interstitial fluid every 10 seconds and gives out mean values every 5 minutes. In this method, the device is just placed under the skin, likely in the patient's thigh, abdomen, or upper arm. CGM is more effective than SMBG in type 2 diabetes. CGM also suffers from certain drawbacks such as expensive, prescription only, limited food, and administration agency approval, and sometimes it shows inaccurate results at low glucose ranges. Some of the CGM sensors required finger stick glucose testing to calibrate the device frequently [25–27]. Typically, two different CGM systems are available they are isCGM (intermittently scanned CGM) and rtCGM (real-time CGM). IsCGM provides glucose values and direct arrows when scanned by the user? rtCGM continuously provides data and notify at extremely low and high glucose levels. Both CGMs are costly. When compared to isCGM, the rtCGM system is more expensive [28–30].

4.1.1 Importance of glucose sensing

Diabetes is a well-known widespread disease caused due to the rise in blood glucose when the insulin produced by the pancreas is not utilized properly or due to insufficient production of insulin in the human body [31]. It is a metabolic syndrome that affects over 420 million people throughout the world and causes more than 3.2 million death every year [32]. Detection of glucose is very important in present days because cells present in the human body need the energy to develop, reproduce, and function. Biomolecules such as carbohydrates, including glucose, are one of the predominant energy source for human beings [33]. The glucose level in blood should be maintained in the range of (3800–6900 µM) to meet individual health requirements. Therefore, monitoring of glucose levels in blood is very important to maintain a balanced healthy state, where the high or low level of glucose in blood cause incurable health problems such as cancer and cardiovascular disease [34]. Also, monitoring glucose concentration in blood has various applications in the food industry. Therefore, effective blood glucose monitoring is important for diagnosing and treating diabetes. Various high sensitivity, low cost, and excellent glucose-sensing methods have been developed over the past few decades [35].

4.2 Fundamentals and working of biosensor

A biosensor is a device that is used to measure the concentration of the desired bioanalyte by generating a signal corresponding to the induced biological or chemical, or physical reactions on it. The schematic diagram of the biosensors is shown in Fig. 4.1. In the biosensor, the bioreceptor adsorbs the biosample and reacts with the desired bioanalyte to produce a signal. The transducer receives the signal produced by the bioreceptor and converts it to the corresponding electrical signal. Finally, the signal is processed and displays the concentration of the bioanalyte [36,37].

A substance of interest in a sample that needs to be detected is called an analyte. If the analyte is in biosample, then the analyte is called bioanalyte. For example, in blood glucose level sensor needed bio-sample is blood, and the bioanalyte is glucose.

A substance or a molecule that specifically recognizes the desired bioanalyte is known as a bioreceptor. Some of the examples of bioreceptors are antibodies, enzymes, nucleic acid, cells, and aptamers. The bioreceptors interact with the bioanalyte to produce the signal corresponding to the concentration of the bioanalyte. [36,37]

4.2.1 Transducers

A transducer is a device or an element, which converts one form of energy into another readable signal. In most places, a transducer converts one form of signal into electrical signals.

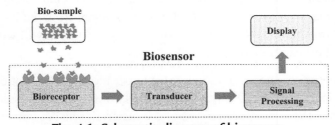

Fig. 4.1: Schematic diagram of biosensor.

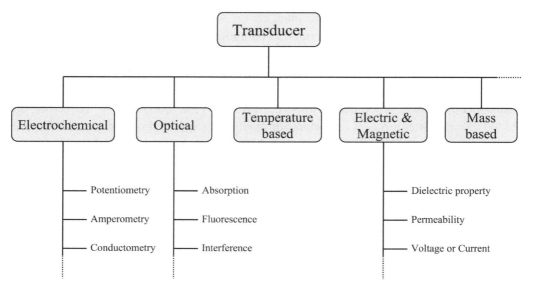

Fig. 4.2: Categories of transducers and its technologies.

Transducers are categorized based on the converting signal. A few categories of transducers and their technologies are shown in Fig. 4.2.

In electrochemical biosensors, the transducers receive the chemical signal from the bioreceptors, and it converts the signal into an electrical signal. The electrochemical transducers convert the signal by three major technologies as potentiometry, amperometry, and conductometry. The optical transducers convert the optical signal to an electrical signal corresponding to the absorption or fluorescence or interference of the light signal by the sample. The temperature-based and mass-based transducers produce the electrical signal corresponding to the temperature and mass of the analyte, respectively. The electric and magnetic transducers convert the variation of dielectric property or permeability or other electromagnetic properties of the bioanalyte into a readable electrical signal [36–39].

4.2.2 Parameters of a biosensor

Selectivity: A biosensor's selectivity depends on the bioreceptor's ability to detect the desired bioanalyte in the biosample containing other interference species and contaminants. The major interference species for electrochemical glucose sensing from the blood are ascorbic acid, uric acid, and dopamine. The interaction of an antigen (bioanalyte) with the antibody (bioreceptor) is the best example of selectivity.

Sensitivity: A biosensor's response to the increase in per molar concentration of the bioanalyte is termed the biosensor's sensitivity. The sensitivity of the biosensors depends on many factors, which include the active surface area of the material, sensing temperature, etc.

Limit of detection (LOD): A biosensor's limit of detection is the minimum amount of analyte that can be detected.

Linearity: The value of sensitivity of a biosensor within the detection range should be constant, and it is defined as the linearity of the biosensor. The linearity of a biosensor is associated with the resolution power or accuracy of the biosensor. If the sensitivity of the sensor is more linear, then the measured concentration of the analyte is more accurate.

Reproducibility: The reproducibility of the biosensor is the ability of the sensor to give identical results to the same sample for duplication of the experiment setup. Reproducibility is an important parameter for the commercialization of the product.

Stability: A biosensor's sustainability over the ambient condition for a greater number of days is called stability of the biosensor. The biosensors sensitivity and all other properties should not be affected by the ambient disturbance. The stability of the biosensor is the most crucial feature in an application; otherwise, long incubations are needed. [36,40]

4.3 Electrochemical glucose sensors

The electrochemical sensors are majorly classified into two categories as enzymatic sensors and nonenzymatic sensors. In the enzymatic electrochemical sensors, enzymes act as a bioreceptor to produce the chemical signals corresponding to the concentration of the desired bioanalyte. The enzymatic sensors are more selective and sensitive to the desired analyte. But the enzymes are not stable; it is affected by pH value, humidity, temperature, etc. Due to the unstable nature of the enzymes, researchers moved toward the nonenzymatic sensors. In the nonenzymatic sensors, instead of enzymes, carbon, metals, alloys, and metal oxide nanoparticles are used as a bioreceptor. [41–45]

The electrochemical glucose sensors developments are categorized into four generations. The first three generations of the glucose sensors are enzymatic glucose sensors, and the fourth-generation glucose sensors are nonenzymatic glucose sensors. In the enzymatic glucose sensors, the glucose 1-oxidase (GOx) is utilized as the best catalytic enzyme. The GOx exhibits relatively high stability, sensitivity, and selectivity, compared to other enzymes. It is a typical Flavin enzyme with Flavin adenine dinucleotide (FAD) as a redox prosthetic group. In the enzymatic glucose sensor, the redox prosthetic group is reduced during the interaction with glucose and produces the redox product of gluconolactone [10,41]. The redox reaction is given as,

$$\mathrm{GOx}\left(\mathrm{FAD}\right) + \text{glucose} \rightarrow \mathrm{GOx}\left(\mathrm{FADH_2}\right) + \text{gluconolactone} \qquad (4.1)$$

The first-generation enzyme-based glucose sensor was reported by Clark in 1962 [43]. In the first-generation glucose sensor, regenerate GOx (FAD) from GOx(FADH$_2$) and transfer the chemical signal to the electrode (transducer); the dissolved oxygen is used as an

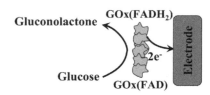

Fig. 4.3: Schematic representation of third-generation glucose sensors.

intermediator. But it is very tedious work to maintain the concentration of oxygen. So, researchers replace the oxygen intermediator with artificial electron-accepting mediators (M), such as ferrocene derivates, quinones, ferrocyanide, and organic salts. The glucose sensors with artificial mediators are categorized into second-generation glucose sensors [41–46].

4.3.1 Third generation glucose sensor

In the third-generation glucose sensors, electrons are directly transferred from the enzyme to the electrode to avoid the usage of artificial and natural mediators. The schematic representation of the third-generation glucose sensor is shown in Fig. 4.3. In this method, the enzymes are immobilized at the electrode, and the enzyme is reacted with the glucose by the same way of the first- and second-generation glucose sensors. The electron transfer between the enzyme and the electrode is given the reaction (2). The direct transfer of electrons reduces the response time of the sensors [41–47].

$$GOx\left(FADH_2\right) \rightarrow GOx\left(FAD\right) + 2H^+ + 2e^- \tag{4.2}$$

4.3.2 Nonenzymatic glucose sensor

The enzymes used in the first three-generation glucose sensors are needed to incubate for long-term usage. The enzymes are mainly affected by the pH value, humidity, temperature, etc. To overcome these, researchers started working on enzyme-free glucose sensors, and it is classified as fourth-generation glucose sensors and also called nonenzymatic glucose sensors. The schematic diagram of the fourth-generation glucose sensor is shown in Fig. 4.4.

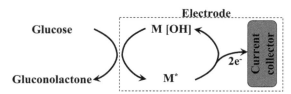

Fig. 4.4: Schematic diagram of nonenzymatic glucose sensor.

In the nonenzymatic glucose sensor, the electrode materials, carbon (carbon nanofiber, carbon nanotube, activated carbon, etc.), metals (Cu, Au, Pt, etc.), metal alloys (PtRu, PtPb, etc.), and metal oxides (CuO, CoO, etc.) act as a catalyst to oxidize glucose and produce the electrical signal corresponding to the concentration of glucose. Here the electrode material itself acts as a bioreceptor and transducer [48–50].

4.4 Colorimetric glucose sensors

Colorimetry, also known as visible spectrophotometry, is a technique for decisiving the concentration of colored analyte in a solution based on Beer- Lambert's equation, which states that at a given wavelength of maximum absorption, there is a direct relationship between absorbance and concentration. Colorimetry is a device that measures the absorbance of a given wavelength of light to find the concentration of analyte in a solution. The schematic representation of colorimetric sensing is shown in Fig. 4.5. For colorimetric analysis, the selectivity and specificity of the reagent have to be considered. Since in colorimetric technique, the selectivity depends on the nature of the reagent, temperature, pH of the medium, the oxidation state of the metal ion, the order in which the reagents are mixed, aging of reagents, and the careful assessment of the absorbance properties and stability of the chromophore (an atom or group whose presence is responsible for the color of a compound) generated [51,52].

4.5 The role of gold and silver nanoparticles in real-time application

Noble metal nanoparticles with highly controllable localized surface plasmon resonance characteristics provide a promising platform for developing sensitive glucose sensor applications [52–55]. Commonly, gold nanoparticles (Au NPs) are typically used to make nonenzymatic glucose sensors. Small-size Au NPs hold excellent glucose oxidase (GOx) like activity nowadays, suggesting that they could be employed as GOx substitutes in glucose detection systems [56–58]. Meanwhile, biocompatibility, significant shape-dependent optical characteristics, and antibacterial capabilities have aroused interest in silver nanoparticles in recent years [53,57,58]. In this section, we briefly summarized the recent advancement of gold and silver nanoparticles in different glucose detection technology from the research articles published in the year ranging from 2017 to 2021.

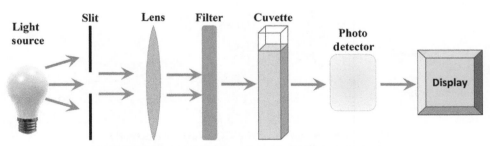

Fig. 4.5: Schematic illustration of Colorimetric Sensor.

4.5.1 Recent advancements on Au- and Ag-based electrochemical glucose sensors

Yan Long et al. synthesized the 3D coral-like gold/carbon paper (Au/CP) electrode by simple one-step electrodeposition process aid by hydrogen evolution without adding any additives, and it's utilized as a support of the enzyme (GOx), as shown in Fig. 4.6. Amid these high specific surface area materials, porous gold has been dominated more owing to its excellent conductivity, good biocompatibility, and high specific surface area. To prepare Au/CP electrode the carbon paper was cleaned using ethanol, deionized water, and after drying, the cleaned CP was immersed in $HAuCl_4$. To activate the carboxyl groups in the prepared electrode various chemical processes has carried out. GOx-Au/CP electrode is fabricated by placing the activated electrode into GOx, which is dissolved in phosphate buffer saline. Many efforts have been made to design a high surface area electrode substrate and the fabrication procedure of the modified GOx electrode to elevate the performance of the enzymatic glucose biosensors. The fabricated unique biosensor electrode material exhibits the sensitivity of 96.27 $\mu A\ mM^{-1}\ cm^{-2}$ to glucose within the linear range of 0.002–21.97 mM with the LOD of 0.6 μM [59].

Yu Mengke et al. successfully fabricated a sensitive and flexible electrochemical biosensor using the electrodeposition method to grow gold nanopine needles (AuNNs) on the flexible electrode substrate. Polyethylene naphthalate (PEN) substrate was used as a base substrate on that Au electrode pattern produced by photolithography, and the AuNNs were grown on it. The grown AuNNs structure gives a large surface area to carry more enzymes, and it raised

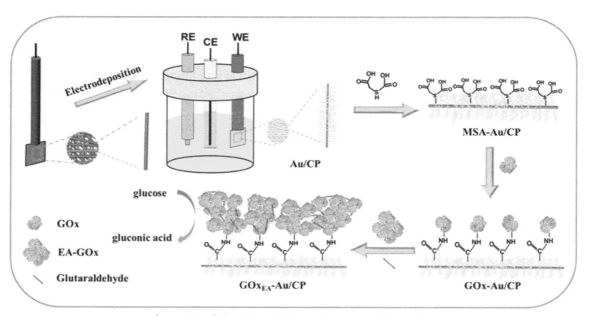

Fig. 4.6: Fabrication of the GOx-Au/CP electrode.
"Reprinted (adapted) with permission from ref. 59 Copyright (2020) Elsevier".

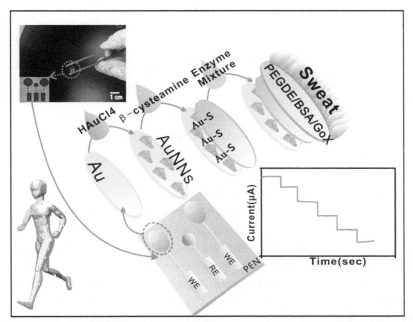

Fig. 4.7: Schematic structure of the flexible and wearable AuNNs biosensor and the photography of the fabricated device.

"Reprinted (adapted) with permission from ref. 60 Copyright (2021) Elsevier".

the sensitivity. This flexible biosensor detects glucose levels from the sweat, and its structure is as shown in Fig. 4.7. This sweat-biosensor achieved 7 μM L^{-1} LOD for glucose detection, and it exhibited more selectivity, stability, and reproducibility [60].

Dong Shuang et al. synthesized noble metal (Ag, Au, and Pd) nanoparticles loaded on ordered microporous carbon using a polydopamine (PDA) based approach for biosensing application. The PDA-based approach has the dominance of versatility, high efficiency, and simplicity. However, decorating active metal nanoparticles on porous carbon substrate have been treated as an efficient method for the fabrication of huge performance electrochemical biosensors. Especially for electrochemical sensing applications of noble metals such as Ag, Au, Pd are used as alternatives to enzymes due to direct redox reactions with small biomolecules. The uniform distribution of Au nanoparticles on the pore surface exhibited better efficiency and versatility. Due to the harmonious effect of scaffold macroporous carbon and the active metal nanoparticles loading on OMC/N-C (ordered macroporous carbon/ N doped carbon thin layer), nonenzymatic biosensor exhibited better electrochemical performance. Au@OMC/N-C nanomaterial exhibited high activity and faster electron exchange towards the direct redox reaction of glucose. It exhibits within the linear range for glucose concentration from 40 μM to 19.56 mM within the LOD of 4 μM. Au@OMC/N-C shows better electrocatalytic performance for the detection of glucose [61].

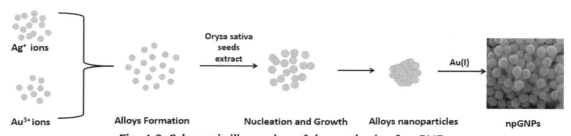

Fig. 4.8: Schematic illustration of the synthesis of npGNPs.
"Reprinted (adapted) with permission from ref. 62 Copyright (2018) Elsevier".

Verma Nishant synthesized nanoporous gold nanoparticles (npGNPs) using alloying and dealloying process as shown in Fig. 4.8. Recently npGNPs has attracted the research community due to its higher specific area, high electrical conductivity, and longer stability, which are assets for sensing application. Nanomaterials are synthesized based on three parameters, environment-friendly solvent, nontoxic capping agents, and less hazardous reducing agents. In many reports, starch is used as the reducing agent for the synthesis of other metallic nanoparticles. Oryza sativa (rice) extract is used as a reducing agent for the synthesis of core Au-Ag alloys nanostructure. Sensing performance for modified Au-Ag alloy nanoparticles electrode and Au electrode was compared and analyzed. By adjusting the dilution of Ag content in alloys, the size of the npGNPs can be tuned, which makes the easy transfer of ions between the electrode material and analyte. The sensitivity of the prepared porous nanoparticles was measured to be ~ 6.67 $\mu A \, \mu M^{-1} \, cm^2$ with the LOD of 0.1 μM. While compared to other metals, synthesized npGNPs exhibited high efficiency for glucose sensing due to their 3D porous framework [62].

Zhu Bicheng et al. synthesized gold nanoparticles (AuNPs) improved laser-scribed graphene electrode (LSGE) for disposable and portable nonenzymatic glucose sensors. Due to their low cost and mass production, screen-printed electrodes (SPEs) are commonly used as electrode strips for monitoring glucose level. The first graphene electrode strip was constructed using a rapid, simple, one-step laser scribing process and schematic representation of the fabrication, as shown in Fig. 4.9. The wavelength of the laser was 10.6 μm. Gold nanoparticles behave as an electrocatalyst for glucose oxidation, which were electrodeposited on the LSGEs. The

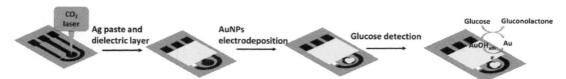

Fig. 4.9: Schematic representation on fabrication procedure of AuNPs-improved LSGE and the mechanism for glucose detection.
"Reprinted (adapted) with permission from ref. 63 Copyright (2021) Elsevier".

unique grass-like morphology and optimized AgNPs on the LSG strips show better selectivity and sensitivity. Fabricated AuNPs modified-LSGEs nonenzymatic glucose sensor delivers linear detection range for glucose between 0.01 mM and 10 mM with LOD of 6.3 μM [63].

Shamsabadi Amir Shahin et al. developed a nonenzymatic glucose sensor by decorating glassy carbon (GC). A glassy carbon electrode is the most widely used electrode for nonenzymatic glucose sensors due to their better performance. In order to enhance the sensing performance of the GC, it is incorporated with noble metal nanoparticles. GC is coated by incorporating gold nanoparticles into graphene nanofibers (GNP/GNF). The layered flakes-like structure withstands the aggregated gold nanoparticles, which helps the core of nanofibers to ease the movement of ions for glucose oxidation. Mostly, a higher surface area of graphene leads to better performance of graphene decorated GC sensor for glucose. The schematic representation of the glucose oxidation at the electrode surface due to GNP/GNF is shown in Fig. 4.10. The GNP/GNF shows a linear sensing response to glucose in the concentration range of 0.5–0.9 mM with a LOD of 55 μM. The fabricated sensor exhibits a sensitivity of 1.1437 μA mM^{-1} cm^{-2} and the response time of the electrode is 1.7 seconds. Interference species including ascorbic acid, NaCl, methanol also tested, and it showed negligible effects [64].

Awais Azka et al. synthesized perpendicularly aligned Au-ZnO nanorods of FTO surface using a hydrothermal method for efficient nonenzymatic glucose sensing. First, ZnO nanorods were directly grown on the FTO electrode at low temperature (80°C for 3 hours) using a hydrothermal setup. After deposition of ZnO nanorods on the FTO electrode, gold nanoparticles were deposited on the nanorods by hydrothermal method, and its schematic illustration is shown in Fig. 4.11. Gold nanoparticles deposited ZnO nanorods having high surface area exhibited better electrochemical behavior for glucose sensing with excellent selectivity, stability, reproducibility, and repeatability. The fabricated nonenzymatic glucose

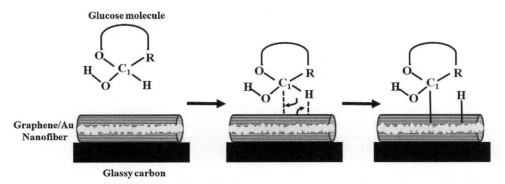

Fig. 4.10: Schematic representation of the glucose oxidation at the electrode surface due to GNP/GNF (activated chemisorption model).

"Reprinted (adapted) with permission from ref. 64 Copyright (2020) Elsevier".

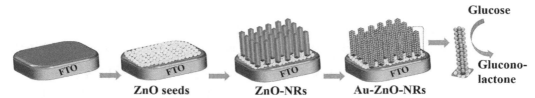

Fig. 4.11: Schematic representation of the construction of ZnO nanorods (NRs) on the FTO electrode, followed by Au nanoparticle's deposition for glucose sensing.

"Reprinted (adapted) with permission from ref. 65 Copyright (2021) Elsevier".

sensor has a wide linear range (0.001–15 mM), LOD (0.12 µM), and good sensitivity (4.416 mA mM^{-1} cm^{-2}). The fabricated Au-ZnO nanorods showed dominant electrochemical performance compared to ZnO nanorods, Au nanoparticles, and other electrodes due to their synergistic effects [65].

Wang Ruru et al. synthesized 3D Au@Pt core-shell nanoparticles (Au@Pt NPs) through the seed-mediated growth method. The GO/MWCNT was prepared by graphene oxide (GO) dispersed on multiwalled carbon nanotubes (MWCNTs). The GO/MWCNT assembled in GCE is then decorated by Au@Pt NPs for enzyme-free electrochemical detection of glucose. The schematic illustration of the preparation of the sensor electrode is shown in Fig. 4.12. The Au@Pt NPs core-shell structure and its electrocatalytic activity exhibit synergistic performance for glucose sensing. The 3D Au@Pt NPs electrochemical glucose sensor demonstrates a two wide linear range from 0.05 µM to 0.1 mM and 0.1 mM to 2.5 mM, a LOD as 0.042 µM at a remarkable negative potential of -0.013 V versus Ag/AgCl [66].

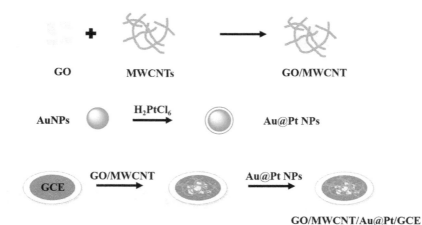

Fig. 4.12: Schematic illustration of synthesis of GO/MWCNT, Au@Pt NPs and assembly of the electrochemical sensor.

"Reprinted (adapted) with permission from ref. 66 Copyright (2022) Elsevier".

Chaandini, J.P and her research group fabricated a nonenzymatic electrochemical glucose sensor for sensing glucose in a neutral medium. Au/rGO nanocomposite is used as an electro-catalyst for the oxidation of glucose synthesized by the one-pot method using chemical reduction of graphene, chloroauric acid, and sodium borohydride. Au/rGO coated screen-printed electrode shows high catalytic activity, which is due to a large raise in the surface area offered by graphene and the augmented electron transfer property of Au nanoparticles. Glucose detection on disposable screen-printed electrodes predicts its pertinency for point of concern of detecting glucose. The developed sensor exhibited the linear range of glucose up to 60 mM and sensitivity of 0.245 and 0.069 mA mM^{-1} cm^{-2} for oxidation and reduction peak, respectively [67].

Zhao Zhenting et al. have fabricated a portable electrochemical glucose sensor by combining it with application specific integrated circuit (ASIC), smartphones, and AuNPs@CuO NWs/Cu$_2$O/CF-based sensing elements, as shown in Fig. 4.13. The deposition of gold nanoparticles (AuNPs) is optimized to improve the performance of the sensor. Au$_{100}$NPs@CuO NWs/Cu$_2$O/CF electrode exhibited better electrocatalytic activity towards the analyte, which is due to the synergistic effect obtained from the exclusive properties of AuNPs and CuO NWs/Cu$_2$O/CF nanostructure. The relation between ΔI and electrodeposition times concludes that enormous Au alteration will suppress the catalytic behavior of the sensor toward analyte oxidation and decreases the catalytic current. The constructed sensor exhibited a sensitivity of 1.619 µA µM^{-1} cm^{-2} with the linear range from 2.8 to 2000 µM [68].

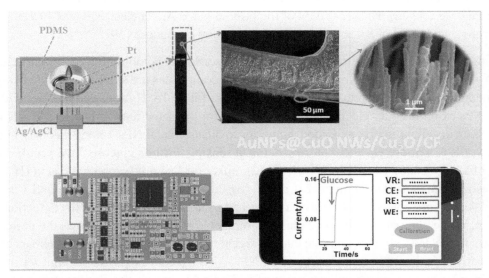

Fig. 4.13: Schematic representation of the AuNPs@CuO NWs/Cu$_2$O/CF nanostructure based compact glucose detection system.
"Reprinted (adapted) with permission from ref. 68 Copyright (2021) Elsevier".

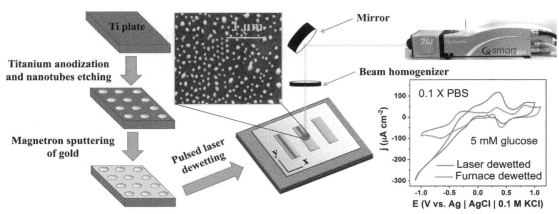

Fig. 4.14: Schematic illustration of fabrication of Au-Ti based glucose sensing electrode and its response to the glucose.

"Reprinted (adapted) with permission from ref. 69 Copyright (2021) Elsevier".

Olejnik Adrian et al. developed Au-Ti based glucose-sensing electrodes by the laser-assisted method. Au was coated on the titanium template by magnetron sputtering, and its surface was modified by two different methods, pulsed laser dewetting and furnace dewetting. The schematic illustration of the fabrication of Au-Ti based electrode is shown in Fig. 4.14. The laser-treated electrode gives a better response compared to the furnace dewetted electrode. On laser treatment, the sudden change in temperature has improved the electroactive surface area of the Au nanoparticle, resulting in a markedly better response for electrochemical nonenzymatic glucose sensing. The sensitivity of the laser-assisted Au-Ti electrode was 49.1 ± 1.3 µA cm^{-2} mM^{-1} and LOD of 8.4 ± 0.8 µM for the detection of glucose [69].

Qu Lingli et al. fabricated a nonenzymatic glucose sensor by combining the nanoparticles (NPs) with coordination polymers (CPs). Ag@CPs composites are synthesized by solvothermal method for electrochemical sensing, as shown in Fig. 4.15. Hybridizing the CPs with metal NPs exhibits better electrochemical behavior for the detection of glucose. CPs also serve as an excellent matrix and good supporter to diffuse metal NPs homogeneously in the frameworks to achieve excellent electrochemical sensing towards glucose. Two peculiar CPs [Mn (PIAD)]$_n$ (1) and [Pb(HPIAD)(NO$_3$)]$_n$ (2), linker 6-(3-pyridyl)isophthalic acid (H$_2$PIAD) synthesized and named as compound 1 and 2, respectively. Ag NPs with compound 1 is named Ag@1, are coated on the glassy carbon and used as an electrode for the detection of glucose. While comparing with bare glassy carbon electrode, the Ag@1 shows the dramatic increase in the peak with appropriate current signal in cyclic voltammetry. This enhancement in the activity is due to conductive Ag metal NPs. Ag@1 electrode possesses high sensitivity of 0.156 µA µM^{-1} cm^{-2} in the linear range of 5–3500 µM towards the glucose. From these results, it was concluded that CPs with Ag NPs make the modified glassy carbon electrode an encouraging electrocatalyst with a satisfactory response and great accuracy [70].

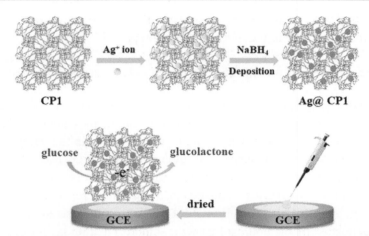

Fig. 4.15: Schematic illustration of synthesis of Ag@1 and glucose detection.
"Reprinted (adapted) with permission from ref. 70 Copyright (2021) Elsevier".

Yang Ziyin and Xiao Bai synthesized Au core flower surrounding with sulfur-doped thin Co_3O_4 shell ($Au@Co_3O_4$-S) by the hydrothermal process for electrochemical sensing of glucose. Au flower core improves the conductivity of Co_3O_4 without occupying its active sites, which helps to possess excellent electrocatalytic activity towards the oxidation of glucose. The synergistic effect of cobalt oxide and Au made the sensor exhibit better performance for glucose detection. $Au@Co_3O_4$-S coated on a glassy carbon electrode (GCE) exhibits excellent electrocatalytic activity towards glucose oxidation, where the modification of Au may speed up the conductivity of Co_3O_4. Both Au modification and sulfur doping are employed to boost the catalytic activity of Co_3O_4. Due to this advantage, $Au@Co_3O_4$-S/GCE is chosen as the working electrode to detect glucose. The practical application of $Au@Co_3O_4$-S/GCE was accomplished by determining the glucose in sports drinks, and the exhibited results are well agreed with the reported values. The fabricated electrode material exhibits a high sensitivity of 1127.3 $\mu A\ mM^{-1}\ cm^{-2}$ with a wide linear range of 0.2 μM–3.1 mM [71].

Gowthaman, N.S.K. and his research group synthesized gold-platinum bimetallic nanoparticles (BMNPs) on SWCNTs by electroless deposition method and tested the electrocatalytic activity in relation to the oxidation of glucose and reduction of hydrogen peroxide (HP). Au-Pt BMNPs have earned many considerations in various applications due to their large surface to volume ratio and exclusive electronic properties. The galvanic displacement mechanism was done for deposition of Au-Pt NPs on SWCNTs. The reduction of Au^{3+} ions by SWCNTs followed by the reduction of Pt^{4+} ions by AuNPs in which the deposition takes place with no support of any reducing agent and the bonding electrons obtained from the substrate lattice functioned as reducing electrons. Morphology of the SWCNTs on the gold substrate looks like a jungle-gym structure and when the deposition time is raised the size of

Au-Pt NPs was found to be increased. While raising the glucose concentration from 50 to 750 µM, there were linear increase in the glucose oxidation peak [72].

4.5.2 Recent advancements on Au- and Ag-based colorimetric glucose sensors

Liu Ao et al. developed Ag@Au core/shell triangular nanoplates (TNPs) with dual enzyme-like properties for the colorimetric sensing of glucose. Nanozymes can be stored stably and are less affected by the environment, which can be broadly used in the biological and medical fields. Highly tunable localized surface plasmon resonance properties of noble metal nanoparticles offer better convenience to fabricate sensitive sensors for colorimetric sensing applications. Ag TNPs show superior localized surface plasmon resonance due to the high degree of anisotropy in their structures. First Ag TNPs were synthesized above that thin layer of Au shell was deposited by the epitaxial growth method. The thin layer of the Au shell holds several advantages, like it enhances the stability of Ag TNPs, enabling the subsequent glucose oxidation. The horseradish peroxidase-like activity of Au can etch the Ag TNPs and alter the localized surface plasmon resonance properties. Also, Ag TNPs can be used for visual readout as a signal transducer. These optimistic characters made Ag@Au core/shell TNPs as suitable nanoprobes for the calorimetric detection of glucose and its schematic illustration, as shown in Fig. 4.16. The fabricated sensor is free from the use of enzymes and chromogenic agents. The significant change in UV-vis absorption peak is due to the change of localized surface

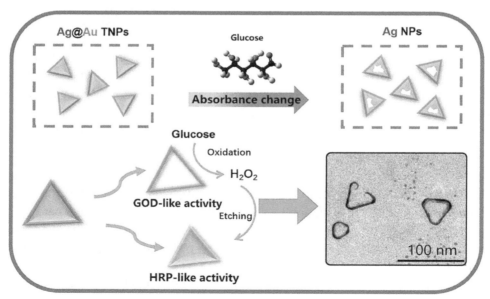

Fig. 4.16: Schematic representation of the colorimetric detection of glucose based on Ag@Au core/shell TNPs.
"Reprinted (adapted) with permission from ref. 57 Copyright (2021) Elsevier".

plasmon resonance characteristics, thus recognizing the colorimetric sensing of glucose without the presence of glucose oxidase enzyme [57].

Huang Pin-Hsuan et al. have demonstrated that judicious model of Ag@Au nanoprisms covered with phosphorescent MOFs can significantly improve the luminescence strength of phosphorescent MOFs. This model provides increased surface area as reaction sites for colliding with oxygen, as well as visual intensification via energy transfer. The luminosity of the Ag@Au nanoprism structures was higher than that of pure Ag nanoprisms. By conjugating rough Ag@Au heterometallic nanoprisms to act as both an enlarged platform for greater oxygen adsorption and a reaction site. Simultaneously, Ag@Au nanoprisms placed next to a luminophore considerably increase the phosphorescence radiation strength and hence enhance the glucose-sensing detection limit. To ensure that the glucose solution penetrates through the specified area, a fixed zone of wax is embossed onto a paper substrate. The oxidation of glucose at GOx is followed by oxygen utilization, which boosts phosphorescence emission intensity. Their approach expands the linear sensing range of ordinary paper-based glucose optical sensors and can be used for clinical assessment and illness diagnosis [58].

Gao Yan et al. present a simple colorimetric approach for glucose detection based on small Au NPs and silver nanoparticles (Ag NPs). The proposed approach was also shown to be capable of detecting serum glucose in clinic samples. Fig. 4.17 depicts the sensing method. Au NPs catalyze the oxidation of glucose to gluconic acid and H_2O_2 in the presence of oxygen. The product of H_2O_2 then etches the Ag NPs in the solution, causing a shift in the localized surface plasmon resonance (LSPR) band intensity. The hue of the AuNPs-AgNPs solution was originally yellow and it was due to the color of AgNPs. With AgNP etching, the yellow of the AgNPs faded gradually. The red color of AuNPs was detected when AgNPs were entirely dissolved by H_2O_2 generated by the oxidation of glucose. As a result, this method uses a couple of Au NPs and Ag NPs to colorimetrically detect glucose without the use of enzymes or chemical chromogenic agents [73].

Jabariyan Shaghayegh et al. synthesized lanthanum-incorporated MCM-41, and its peroxidase-like activity was demonstrated in Fig. 4.18. In the presence of hydrogen peroxide, La-MCM-41 was utilized to oxidize the peroxidase substrate 3,3 ′,5,5′-tetramethylbenzidine (TMB) to create a blue hue in aqueous media, which can lead to a straightforward technique

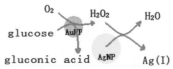

Fig. 4.17: Schematic representation of the responses for colorimetric detection of glucose based on Au NPs coupled with Ag NPs.

"Reprinted (adapted) with permission from ref. 73 Copyright (2017) Elsevier".

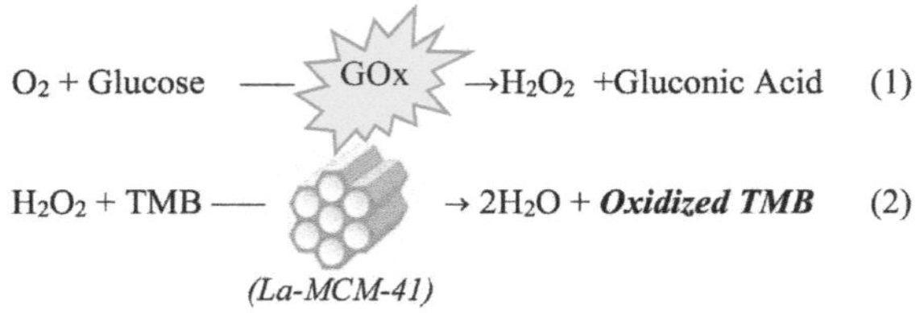

$$O_2 + Glucose \quad \underline{\quad\quad} \quad \text{GOx} \quad \rightarrow H_2O_2 + Gluconic\ Acid \quad (1)$$

$$H_2O_2 + TMB \quad \underline{\quad\quad} \quad \rightarrow 2H_2O + \textit{Oxidized TMB} \quad (2)$$

(La-MCM-41)

Fig. 4.18: Schematic representation of colorimetric detection of glucose by utilizing glucose oxidase (GOx) and La-MCM-41-catalyzed reactions.
"Reprinted (adapted) with permission from ref. 74 Copyright (2018) Elsevier".

for colorimetric detection of H_2O_2, as hydrogen peroxide is one of the principal products of the glucose oxidation reaction catalyzed by glucose oxidase, this colorimetric method established a foundation for glucose detection via a hydrogen peroxide mediated mechanism. This finding could help to advance research into the use of stabilized metal-substituted mesoporous materials as peroxidase mimics in medical diagnostics [74].

Nguyen Nghia Duc et al. developed a directed colorimetric approach for the direct detection of hydrogen peroxide using AgNPs/GQDs (H_2O_2). Based on combination with glucose oxidase, a colorimetric glucose sensor has been devised and produced (GOx), and its schematics is as shown in Fig. 4.19. For hydrogen peroxide and glucose detecting, the constructed sensors have outstanding sensitivity and selectivity. Furthermore, the purported sensor has been used to detect glucose dilutions in human urine samples with a great performance LOD value of 30 μM. Based on its impressive results, the suggested colorimetric glucose sensor appears to be a great prospective nominee for glucose level monitoring without the requirement for blood and without the use of needles [75].

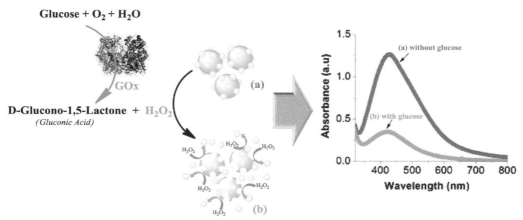

Fig. 4.19: Schematic representation of detection mechanism of proposed label-free and reagentless colorimetric sensor for hydrogen peroxide and glucose using AgNPs/GQDs.
"Reprinted (adapted) with permission from ref. 75 Copyright (2018) Elsevier".

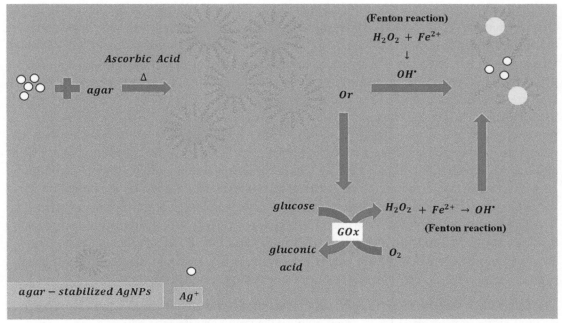

Fig. 4.20: Schematic illustration of synthesis of AgNPs and the detection of glucose.
"Reprinted (adapted) with permission from ref. 76 Copyright (2018) Elsevier".

Basiri Sedigheh and his group developed, a multifunctional platform for very sensitive colorimetric detection of Fe^{2+}, H_2O_2 and glucose, was developed using a new technique based on Fenton reaction combined with silver nanoparticles (AgNPs). AgNPs were produced in a green approach utilizing agar and ascorbic acid for this purpose and then used as a colorimetric sensor (Fig. 4.20). Since glucose oxidase generates H_2O_2 from glucose, this technique was also used to determine glucose levels. The LOD of glucose detection was calculated as 0.29 µM using the above method. The constructed sensing system was used to measure glucose in human blood serum sample, due to H_2O_2 production during the oxidation of glucose by GOx. In a two-step approach, Fenton reaction-mediated discoloration of AgNPs solution was employed to determine glucose. The colorimetric test based on the AgNPs–Fe^{2+}–H_2O_2 combination has a high sensitivity for determining glucose. The selectivity of the Fenton reaction-based approach to glucose detection (0.04 mM), various probable interferences including lactose (0.2 mM), maltose (0.2 mM), and fructose (0.2 mM) was tested at identical conditions. Due to GOx's great substrate selectivity, these species had no effect on glucose determination. As a result, a new colorimetric sensor for glucose detection was developed, based on the discoloration of AgNPs solution via Fenton reaction and glucose oxidation via GOx [76].

Maruthupandy Muthuchamy et al. proposed a study about a simple method for the rapid colorimetric and visual detection of glucose molecules in water medium. A prominent surface plasmon resonance band at 429 nm was seen in the silver nanoparticles distributed on the chitosan surface (CS/Ag NCs), which vanished when accelerative amounts of glucose molecules were added and were accompanied by a color change from yellow to colorless. Glucose molecules were detected by UV-vis spectrophotometer in the concentration range of 0–100 μM, with a LOD of 5 μM, and was observed by naked eyes (from yellow to purple-grey). Glucose molecules can be adsorbed on the surface of CS/Ag NCs in this system, resulting in -OH radicals. The created -OH radicals may be maintained by CS/Ag NCs by limited electron exchange contact, which may promote to CS/Ag NCs' detecting ability, causing them to react with glucose and generate a strong color reaction. Because of the interaction of Ag with chitosan, the positioning of nanoparticles can be affected, reducing the plasmonic coupling between the two nanomaterials. Large SPR shifts and increased sensitivity are typically the outcomes. The colorimetric detection of glucose molecules is controlled by a process in which glucose controls the colorimetric reactivity of CS/Ag NCs via a possible physical contact between Ag and chitosan. The porous nanostructures of Ag NPs and the wide surface areas of chitosan increased the interacting sites between CS/Ag NCs and glucose molecules, resulting in faster Ag electron transfer and excellent biosensor sensitivity [77].

Liu Wei et al., based on Au@Ag NPs and C-dots, developed a fluorometric and colorimetric dual-signal sensor for H_2O_2-assisted glucose detection and its schematics, as shown in Fig. 4.21. The Au@Ag NPs were mixed into a solution of C-dots made by the hydrothermal

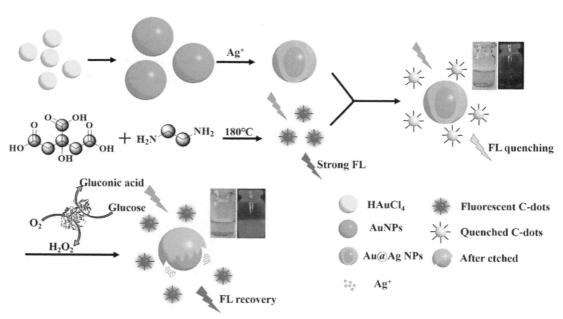

Fig. 4.21: Representation of the fluorometric and colorimetric assay for glucose detection based on Au@Ag NPs and C-dots.

"Reprinted (adapted) with permission from ref. 78 Copyright (2018) Elsevier".

technique. C-dots were utilized as efficient energy donors, and Au@Ag NPs dramatically reduced fluorescence (silver shell as acceptor). Enzymatic-triggered oxidation of glucose produces H_2O_2 and gluconic acid, which etches silver shell (Ag^0) into Ag^+ and exposes AuNPs. The absorbance of the silver shell decreases, resulting in a marked color change from orange-yellow to pink and C-dot fluorescence recovery. As a result, a dual-signal glucose sensing system based on the recording of fluorescence production and UV–vis absorption spectra are assembled. Due to the high molar extinction coefficient of AgNPs, this procedure had a higher sensitivity than previously used methods, and no extra chromogenic agent was required. Its other benefits, such as direct readout with the naked eye, ease of operation, and reproducibility in complex samples, were critical for practical implementation. Fresh serum samples were evaluated using a commercial Beckman Coulter AU5821 chemical analyzer at Chengdu First People's Hospital as an assessment criterion to examine the performance of their approach [78].

Tran, Hoang V., et al. used silver nanoparticles (AgNPs@rGO) to adorn reduced graphene oxide sheets and showed that they have peroxidase-like activity. For the oxidation of 3,3',5,5', - tetramethylbenzidine (TMB) in the presence of hydrogen peroxide, AgNPs@rGO has demonstrated outstanding catalytic activity. They used this catalyst in a low-cost colorimetric assay for glucose monitoring as an example of application, and they showed that it is as effective as genuine conventional (but expensive) hospital procedures for determining glucose in human blood serum. The working principle is demonstrated in the schematic diagram shown in Fig. 4.22. When glucose, GOx, and AgNPs@rGO are combined only cause a color shift, which means any of the components, such as GOx, glucose, or AgNPs@rGO, were missing cause no change in absorbance, indicating that no side reactions occurred. The UV-Vis spectra are used to determine this variation in absorbance rate. Control studies using lactose, fructose, saccharose, galactose, and ascorbic acid, all at a concentration of 0.8 mM, were used to assess the selectivity of the colorimetric glucose test. As a result of the aforesaid findings,

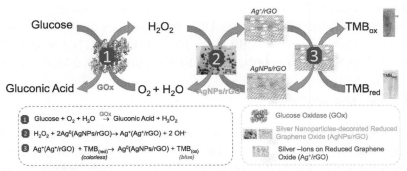

Fig. 4.22: Illustration of working principle of glucose biosensor based on AgNPs/rGO –a novel peroxidase-like activity nanomaterial.

"Reprinted (adapted) with permission from ref. 79 Copyright (2020) Elsevier".

AgNPs@rGO can be used as an artificial peroxidase enzyme to construct a selective colorimetric glucose biosensor with a LOD of 40 μM [79].

The glucose concentration in urine or blood samples is crucial in clinical diagnosis when assessing a patient's health or physical condition. H_2O_2 sensing can aid in the colorimetric detection of glucose because it is an intermediary in the glucose oxidation pathway catalyzed by GOx. Ma Lian et al. in their study, found that increasing the glucose concentration gradually results in an increased absorption peak gradually and shows a color change from colorless to blue color. They reported the mechanism of detection as in equation form. GOx may quantitatively catalyze the oxidation of glucose to create H_2O_2; the resulting H_2O_2 can then react with TMB via the catalysis of PB-Au@MoS$_2$ NFs, leading to a color shift. As a result, when the glucose concentration rises, the solution's absorbance rises as well. They have explained the steady-state kinetic research, which revealed that PB-Au@MoS$_2$ NFs had a greater affinity for the two substrates of H_2O_2 and TMB than the normal HRP enzyme. Colorimetric sensing techniques for H_2O_2 and glucose were developed based on these findings. The sensor demonstrated good sensitivity and selectivity, suggesting that it could be used in biological and clinical diagnosis in the future [80].

Xu Shenghao et al. present an ultrasensitive multicolor colorimetric method for blood glucose detection based on the enzyme-catalyzed reaction-induced etching of Au NBPs. In this, H_2O_2 (produced by glucose oxidation) is broken down into hydroxyl radicals (•OH) with a high oxidizability beneath the catalysis of horseradish peroxidase (HRP), which can speed up the etching of Au NBPs, resulting in vivid color changes and a blue shift in the localized surface plasmon resonance (LSPR). The estimate glucose levels can be determined instinctively and conveniently by the color changes with the naked eye based on this. Most notably, this suggested multicolor colorimetric sensor can reliably visually discriminate healthy persons from diabetes patients with the naked eye, proving its great promise in clinical blood glucose detection. The suggested colorimetric glucose test achieves a dynamic range of 0.05–90 μM with a detection limit of 0.02 μM, which is nearly three orders of magnitude lower than the gold nanorod-based assay (10 μM). Fig. 4.23 schematically illustrates the sensing mechanism of the multicolor colorimetric glucose assay. Different quantities of glucose cause varying degrees of etching, resulting in a blue shift in the LSPR band of Au NBPs, as well as various color changes (such as light brown, blue, and pink) that can be seen with the naked eye [81].

4.6 Conclusion and scope

In this chapter, we summarized the role of Au- and Ag-based materials role in real-time glucose sensing applications and the basics behind the two broad classifications of glucose sensing technology (electrochemical and colorimetric glucose-sensing). In both glucose sensing technologies, sensitivity, selectivity, linear detection range, and LOD were depended on the reactivity of the materials with the glucose, surface morphology, and surface to volume ratio.

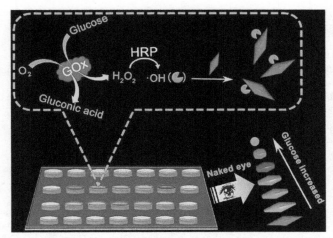

Fig. 4.23: Schematic illustration of the sensing mechanism of the multicolor colorimetric glucose assay.
"Reprinted (adapted) with permission from ref. 81 Copyright (2019) Elsevier".

To make for real-time application, the materials must be biologically compatible and with high physical and chemical stability [82]. Au- and Ag-based materials have been prepared with a wide variety of morphologies by different synthesis techniques, such as nanowires [83–86], cubes [87–89], triangles [90–92], rods [93–96], and spheres [97–99]. Au- and Ag-based materials have biological compatibility and high physical and chemical stability [82]. Nowadays, we are all moving towards flexible electronic devices. A lot of smart bands and smart devices are coming into our life, that all had lots of biosensors. So, recently scientists have been focusing on fabricating flexible sensors with high sensitivity, selectivity, and a wide linear range of detection [100–102]. Au- and Ag-based materials have a lot of advantages towards sensing applications. At the same time, we need to use these advantages of Au and Ag nanomaterial to fabricate flexible nonenzymatic glucose sensing devices.

References

[1] M. Klimek, J. Knap, M. Reda, M. Masternak, History of glucose monitoring: Past, present, future, J. Phys. Educ. Sport 9 (9) (2019) 222–227.

[2] E.H. Yoo, S.Y. Lee, Glucose biosensors: An overview of use in clinical practice, Sensors 10 (5) (2010) 4558–4576.

[3] J.E. Shaw, R.A. Sicree, P.Z. Zimmet, Global estimates of the prevalence of diabetes for 2010 and 2030, Diabetes Res. Clin. Pract. 87 (1) (2010) 4–14.

[4] S. Wild, G. Roglic, A. Green, R. Sicree, H. King, Global prevalence of diabetes: Estimates for the year 2000 and projections for 2030, Diabetes Care. 27 (5) (2004) 1047–1053.

[5] G. Xu, B. Liu, Y. Sun, Y. Du, L.G. Snetselaar, F.B. Hu, W. Bao, Prevalence of diagnosed type 1 and type 2 diabetes among US adults in 2016 and 2017: Population based study, BMJ. 362 (2018) 1497–1503.

[6] L. Chen, D.J. Magliano, P.Z. Zimmet, The worldwide epidemiology of type 2 diabetes mellitus—present and future perspectives, Nat. Rev. Endocrinol. 8 (4) (2012) 228–236.

[7] S.F. Clarke, J.R. Foster, A history of blood glucose meters and their role in self-monitoring of diabetes mellitus, Br. J. Biomed. Sci. 69 (2) (2012) 83–93.

[8] L.C. Clark Jr, C. Lyons, Electrode systems for continuous monitoring in cardiovascular surgery, Ann. N Y Acad. Sci. 102 (1) (1962) 29–45.

[9] B. Singh, V. Bhatia, V.K. Jain, Electrostatically functionalized multi-walled carbon nanotubes based flexible and non-enzymatic biosensor for glucose detection, Sens. Transducers 146 (11) (2012) 69–77.

[10] K.E. Toghill, R.G. Compton, Electrochemical non-enzymatic glucose sensors: A perspective and an evaluation, Int. J. Electrochem. Sci. 5 (9) (2010) 1246–1301.

[11] I.B. Hirsch, Introduction: History of glucose monitoring. In: Role of Continuous Glucose Monitoring in Diabetes Treatment, American Diabetes Association, Arlington (VA), 2018.

[12] Diabetes Control and Complications Trial Research GroupD.M. Nathan, S. Genuth, J. Lachin, P. Cleary, O. Crofford, M. Davis, L. Rand, C. Siebert, The effect of intensive treatment of diabetes on the development and progression of long-term complications in insulin-dependent diabetes mellitus, N. Engl. J. Med. 329 (14) (1993) 977–986.

[13] E. Boland, T. Monsod, T. Delucia, C.A. Brandt, S. Fernando, W.V. Tamborlane, Limitations of conventional methods of self-monitoring of blood glucose: Lessons learned from 3 days of continuous glucose sensing in pediatric patients with type 1 diabetes, Diabetes Care. 24 (11) (2001) 1858–1862.

[14] H.K. Sia, C.T. Kor, S.T. Tu, P.Y. Liao, J.Y. Wang, Self-monitoring of blood glucose in association with glycemic control in newly diagnosed non-insulin-treated diabetes patients: A retrospective cohort study, Sci. Rep. 11 (1) (2021) 1–9.

[15] N.M S.Tomah, A. Mottalib, D.M. Pober, M.W. Tasabehji, S. Ashrafzadeh, O. Hamdy, Frequency of self-monitoring of blood glucose in relation to weight loss and A1C during intensive multidisciplinary weight management in patients with type 2 diabetes and obesity, BMJ. Open Diabetes Res. Care. 7 (1) (2019) 659–665.

[16] U.L. Malanda, L.M. Welschen, I.I. Riphagen, J.M. Dekker, G. Nijpels, S.D. Bot, Self-monitoring of blood glucose in patients with type 2 diabetes mellitus who are not using insulin, Cochrane Database Syst. Rev. (1) (2012) 1–92, https://doi.org/10.1002/14651858.CD005060.pub3.

[17] S.R. Patton, M.A. Clements, Continuous glucose monitoring versus self-monitoring of blood glucose in children with type 1 diabetes-are there pros and cons for both? US Endocrinol 8 (1) (2012) 27–29.

[18] J. Silverstein, G. Klingensmith, K. Copeland, L. Plotnick, F. Kaufman, L. Laffel, L. Deeb, M. Grey, B. Anderson, L.A. Holzmeister, N. Clark, Care of children and adolescents with type 1 diabetes: A statement of the American Diabetes Association, Diabetes Care. 28 (1) (2005) 186–212.

[19] E.M. Benjamin, Self-monitoring of blood glucose: The basics, Clin. Diabetes. 20 (1) (2002) 45–47.

[20] D.E. Goldstein, R.R. Little, R.A. Lorenz, J.I. Malone, D.M. Nathan, C.M. Peterson, Tests of glycemia in diabetes, Diabetes Care. 26 (2003) 106–108.

[21] O. Schnell, M. Hanefeld, L. Monnier, Self-monitoring of blood glucose: A prerequisite for diabetes management in outcome trials, J. Diabetes Sci. Technol 8 (3) (2014) 609–614.

[22] P.A. Porter, B. Keating, G. Byrne, T.W. Jones, Incidence and predictive criteria of nocturnal hypoglycemia in young children with insulin-dependent diabetes mellitus, J. Pediatr. 130 (3) (1997) 366–372.

[23] H. Yoon, X. Xuan, S. Jeong, J.Y. Park, Wearable, robust, non-enzymatic continuous glucose monitoring system and its in vivo investigation, Biosens. Bioelectron. 117 (2018) 267–275.

[24] D. Deiss, J. Bolinder, J.P. Riveline, T. Battelino, E. Bosi, N. Tubiana-Rufi, D. Kerr, M. Phillip, Improved glycemic control in poorly controlled patients with type 1 diabetes using real-time continuous glucose monitoring, Diabetes Care. 29 (12) (2006) 2730–2732.

[25] R.N. Janapala, J.S. Jayaraj, N. Fathima, T. Kashif, N. Usman, A. Dasari, N. Jahan, I. Sachmechi, Continuous glucose monitoring versus self-monitoring of blood glucose in type 2 diabetes mellitus: a systematic review with meta-analysis, Cureus. 11 (9) (2019) 5634–5642.

[26] D.C. Klonoff, Personalized medicine for diabetes, J Diabetes Sci. Technol 2 (3) (2008) 335–341.

[27] J.E. Fradkin, M.C. Hanlon, G.P. Rodgers, NIH precision medicine initiative: Implications for diabetes research, Diabetes Care. 39 (7) (2016) 1080–1084.

[28] K. Nørgaard, S. Schmidt, Is reimbursement for alerts and real-time continuous glucose monitoring needed? Lancet. 397 (10291) (2021) 2230–2232.

[29] S.V. Edelman, N.B. Argento, J. Pettus, I.B. Hirsch, Clinical implications of real-time and intermittently scanned continuous glucose monitoring, Diabetes Care. 41 (11) (2018) 2265–2274.

[30] M.M. Visser, S. Charleer, S. Fieuws, C. De Block, R. Hilbrands, L. Van Huffel, T. Maes, G. Vanhaverbeke, E. Dirinck, N. Myngheer, C. Vercammen, Comparing real-time and intermittently scanned continuous glucose monitoring in adults with type 1 diabetes (ALERTT1): A 6-month, prospective, multicentre, randomised controlled trial, Lancet. 397 (10291) (2021) 2275–2283.

[31] P. Chakraborty, N. Deka, D.C. Patra, K. Debnath, S.P. Mondal, Salivary glucose sensing using highly sensitive and selective non-enzymatic porous NiO nanostructured electrodes, Surf. Interfaces 26 (2021) 101324–101332.

[32] X. Sun, Glucose detection through surface-enhanced Raman spectroscopy: a review, Anal. Chim. Acta 1206 (2021) 339226–339238.

[33] M. Palmer, M. Masikini, L.W. Jiang, J.J. Wang, F. Cummings, J. Chamier, O. Inyang, M. Chowdhury, Enhanced electrochemical glucose sensing performance of CuO: NiO mixed oxides thin film by plasma assisted nitrogen doping, J. Alloys Compd. 853 (2021) 156900–156910.

[34] A. Swaidan, A. Addad, J.F. Tahon, A. Barras, J. Toufaily, T. Hamieh, S. Szunerits, R. Boukherroub, Ultrasmall CuS-BSA-Cu$_3$(PO$_4$)$_2$ nanozyme for highly efficient colorimetric sensing of H$_2$O$_2$ and glucose in contact lens care solutions and human serum, Anal. Chim. Acta 1109 (2020) 78–89.

[35] S. eun Kim, A. Muthurasu, Metal-organic framework–assisted bimetallic Ni@ Cu microsphere for enzyme-free electrochemical sensing of glucose, J. Electroanal. Chem. 873 (2020) 114356–114363.

[36] B. Nikhil, J. Pawan, F. Nello, E. Pedro, Introduction to biosensors, Essays Biochem. 60 (1) (2016) 1–8.

[37] Y. Wu, S. Hu, Biosensors based on direct electron transfer in redox proteins, Mikrochim. Acta. 159 (1-2) (2007) 1–17.

[38] J.X.J. Zhang, K. Hoshino, Chapter 4 electrical transducers electrochemical sensors and semiconductor molecular sensors. In: Molecular Sensors and Nanodevices, Elsevier, Oxford, 2014, pp. 169–232.

[39] P. Martinkova, A. Kostelnik, T. Válek, M. Pohanka, Main streams in the construction of biosensors and their applications, Int. J. Electrochem. Sci. 12 (8) (2017) 7386–7403.

[40] S.S. Shalini, R. Balamurugan, A.J.C. Mary, A.C. Bose, Hierarchical porous carbon nanoparticles derived from Helianthus annuus for glucose-sensing application, Emergent. Mater 4 (3) (2021) 755–760.

[41] H. Teymourian, A. Barfidokht, J. Wang, Electrochemical glucose sensors in diabetes management: An updated review (2010–2020), Chem. Soc. Rev. 49 (21) (2020) 7671–7709.

[42] V.B. Juska, M.E. Pemble, A critical review of electrochemical glucose sensing: Evolution of biosensor platforms based on advanced nanosystems, Sensors. 20 (21) (2020) 6013–6040.

[43] P. Avari, M. Reddy, N. Oliver, Is it possible to constantly and accurately monitor blood sugar levels, in people with Type 1 diabetes, with a discrete device (non-invasive or invasive)? Diabet. Med. 37 (4) (2020) 532–544.

[44] K.J. Cash, H.A. Clark, Nanosensors and nanomaterials for monitoring glucose in diabetes, Trends Mol. Med. 16 (12) (2010) 584–593.

[45] D. Bruen, C. Delaney, L. Florea, D. Diamond, Glucose sensing for diabetes monitoring: Recent developments, Sensors. 17 (8) (2017) 1866–1886.

[46] M.H. Hassan, C. Vyas, B. Grieve, P. Bartolo, Recent advances in enzymatic and non-enzymatic electrochemical glucose sensing, Sensors. 21 (14) (2021) 4672–4697.

[47] C. Chen, Q. Xie, D. Yang, H. Xiao, Y. Fu, Y. Tan, S. Yao, Recent advances in electrochemical glucose biosensors: a review, RSC Adv 3 (14) (2013) 4473–4491.

[48] R. Ahmad, N. Tripathy, M.S. Ahn, K.S. Bhat, T. Mahmoudi, Y. Wang, J.Y. Yoo, D.W. Kwon, H.Y. Yang, Y.B. Hahn, Highly efficient non-enzymatic glucose sensor based on CuO modified vertically-grown ZnO nanorods on electrode, Sci. Rep. 7 (1) (2017) 1–10.

[49] H. Xu, C. Xia, S. Wang, F. Han, M.K. Akbari, Z. Hai, S. Zhuiykov, Electrochemical non-enzymatic glucose sensor based on hierarchical 3D Co$_3$O$_4$/Ni heterostructure electrode for pushing sensitivity boundary to a new limit, Sens. Actuators B Chem. 267 (2018) 93–103.

[50] N.I. Chandrasekaran, M. Manickam, A sensitive and selective non-enzymatic glucose sensor with hollow Ni-Al-Mn layered triple hydroxide nanocomposites modified Ni foam, Sens. Actuators B Chem. 288 (2019) 188–194.

[51] G. Sowjanya, K. Mohana, Colorimetric approaches to drug analysis and applications–A review, Am. J. PharmTech Res. 9 (1) (2019) 14–37.

[52] S. Xu, L. Jiang, Y. Liu, P. Liu, W. Wang, X. Luo, A morphology-based ultrasensitive multicolor colorimetric assay for detection of blood glucose by enzymatic etching of plasmonic gold nanobipyramids, Anal. Chim. Acta 1071 (2019) 53–58.

[53] V. Naresh, N. Lee, A review on biosensors and recent development of nanostructured materials-enabled biosensors, Sensors. 21 (4) (2021) 1109–1143.

[54] O. Adeniyi, S. Sicwetsha, P. Mashazi, Nanomagnet-silica nanoparticles decorated with Au@ Pd for enhanced peroxidase-like activity and colorimetric glucose sensing, ACS Appl. Mater. Interfaces 12 (2) (2019) 1973–1987.

[55] H. Wang, H. Rao, M. Luo, X. Xue, Z. Xue, X. Lu, Noble metal nanoparticles growth-based colorimetric strategies: From monocolorimetric to multicolorimetric sensors, Coord. Chem. Rev. 398 (2019) 113003–113025.

[56] Y. Xia, J. Ye, K. Tan, J. Wang, G. Yang, Colorimetric visualization of glucose at the submicromole level in serum by a homogenous silver nanoprism–glucose oxidase system, Anal. Chem. 85 (13) (2013) 6241–6247.

[57] A. Liu, M. Li, J. Wang, F. Feng, Y. Zhang, Z. Qiu, Y. Chen, B.E. Meteku, C. Wen, Z. Yan, J. Zeng, Ag@ Au core/shell triangular nanoplates with dual enzyme-like properties for the colorimetric sensing of glucose, Chin. Chem. Lett. 31 (5) (2020) 1133–1136.

[58] P.H. Huang, C.P. Hong, J.F. Zhu, T.T. Chen, C.T. Chan, Y.C. Ko, T.L. Lin, Z.B. Pan, N.K. Sun, Y.C. Wang, J.J. Luo, Ag@ Au nanoprism-metal organic framework-based paper for extending the glucose sensing range in human serum and urine, Dalton Trans. 46 (21) (2017) 6985–6993.

[59] L. Yan, P. Ma, Y. Liu, X. Ma, F. Chen, M. Li, 3D coral-like gold/carbon paper electrode modified with covalent and cross-linked enzyme aggregates for electrochemical sensing of glucose, Microchem. J. 159 (2020) 105347–105354.

[60] M. Yu, Y.T. Li, Y. Hu, L. Tang, F. Yang, W.L. Lv, Z.Y. Zhang, G.J. Zhang, Gold nanostructure-programmed flexible electrochemical biosensor for detection of glucose and lactate in sweat, J. Electroanal. Chem. 882 (2021) 115029–115036.

[61] S. Dong, Z. Yang, B. Liu, J. Zhang, P. Xu, M. Xiang, T. Lu, (Pd, Au, Ag) nanoparticles decorated well-ordered macroporous carbon for electrochemical sensing applications, J. Electroanal. Chem. 897 (2021) 115562–115569.

[62] N. Verma, A green synthetic approach for size tunable nanoporous gold nanoparticles and its glucose sensing application, Appl. Surf. Sci. 462 (2018) 753–759.

[63] B. Zhu, L. Yu, S. Beikzadeh, S. Zhang, P. Zhang, L. Wang, J. Travas-Sejdic, Disposable and portable gold nanoparticles modified-laser-scribed graphene sensing strips for electrochemical, non-enzymatic detection of glucose, Electrochim. Acta 378 (2021) 138132–138139.

[64] A.S. Shamsabadi, H. Tavanai, M. Ranjbar, A. Farnood, M. Bazarganipour, Electrochemical non-enzymatic sensing of glucose by gold nanoparticles incorporated graphene nanofibers, Mater Today Commun 24 (2020) 100963–100970.

[65] A. Awais, M. Arsalan, X. Qiao, W. Yahui, Q. Sheng, T. Yue, Y. He, Facial synthesis of highly efficient non-enzymatic glucose sensor based on vertically aligned Au-ZnO NRs, J. Electroanal. Chem. 895 (2021) 115424–115432.

[66] R. Wang, X. Liu, Y. Zhao, J. Qin, H. Xu, L. Dong, S. Gao, L. Zhong, Novel electrochemical non-enzymatic glucose sensor based on 3D Au@ Pt core-shell nanoparticles decorated graphene oxide/multi-walled carbon nanotubes composite, Microchem. J. 174 (2022) 107061–107068.

[67] J.P. Chaandini, P.V. Suneesh, T.S. Babu, Gold nanoparticle decorated reduced graphene oxide for the nonenzymatic electrochemical sensing of glucose in neutral medium, Mater Today Proc 33 (2020) 2414–2420.

[68] Z. Zhao, Q. Li, Y. Sun, C. Zhao, Z. Guo, W. Gong, J. Hu, Y. Chen, Highly sensitive and portable electrochemical detection system based on AuNPs@ CuO NWs/Cu$_2$O/CF hierarchical nanostructures for enzymeless glucose sensing, Sens. Actuators B Chem 345 (2021) 130379–130390.

[69] A. Olejnik, G. Śliwiński, J. Karczewski, J. Ryl, K. Siuzdak, K. Grochowska, Laser-assisted approach for improved performance of Au-Ti based glucose sensing electrodes, Appl. Surf. Sci. 543 (2021) 148788–148798.

[70] L. Qu, L. Zhao, T. Chen, J. Li, X. Nie, R. Li, C. Sun, Two novel coordination polymers and their hybrid materials with Ag nanoparticles for non-enzymatic detection of glucose, J. Solid State Chem. 297 (2021) 122086–122092.

[71] Z. Yang, X. Bai, Synthesis of Au core flower surrounding with sulphur-doped thin Co$_3$O$_4$ shell for enhanced non-enzymatic detection of glucose, Microchem. J. 160 (2021) 105601–105607.

[72] N.S.K. Gowthaman, S.A. John, M. Tominaga, Fast growth of Au-Pt bimetallic nanoparticles on SWCNTs: Composition dependent electrocatalytic activity towards glucose and hydrogen peroxide, J. Electroanal. Chem. 798 (2017) 24–33.

[73] Y. Gao, Y. Wu, J. Di, Colorimetric detection of glucose based on gold nanoparticles coupled with silver nanoparticles, Spectrochim. Acta A Mol. Biomol. Spectrosc. 173 (2017) 207–212.

[74] S. Jabariyan, M.A. Zanjanchi, M. Arvand, S. Sohrabnezhad, Colorimetric detection of glucose using lanthanum-incorporated MCM-41, Spectrochim. Acta A Mol. Biomol. Spectrosc. 203 (2018) 294–300.

[75] N.D. Nguyen, T. Van Nguyen, A.D. Chu, H.V. Tran, L.T. Tran, C.D. Huynh, A label-free colorimetric sensor based on silver nanoparticles directed to hydrogen peroxide and glucose, Arab. J. Chem. 11 (7) (2018) 1134–1143.

[76] S. Basiri, A. Mehdinia, A. Jabbari, A sensitive triple colorimetric sensor based on plasmonic response quenching of green synthesized silver nanoparticles for determination of Fe^{2+}, hydrogen peroxide, and glucose, Colloids Surf. A Physicochem. Eng. Asp 545 (2018) 138–146.

[77] M. Maruthupandy, G. Rajivgandhi, T. Muneeswaran, T. Vennila, F. Quero, J.M. Song, Chitosan/silver nanocomposites for colorimetric detection of glucose molecules, Int. J. Biol. Macromol. 121 (2019) 822–828.

[78] W. Liu, F. Ding, Y. Wang, L. Mao, R. Liang, P. Zou, X. Wang, Q. Zhao, H. Rao, Fluorometric and colorimetric sensor array for discrimination of glucose using enzymatic-triggered dual-signal system consisting of Au@ Ag nanoparticles and carbon nanodots, Sens. Actuators B Chem. 265 (2018) 310–317.

[79] H.V. Tran, N.D. Nguyen, C.T. Tran, L.T. Tran, T.D. Le, H.T. Tran, B. Piro, C.D. Huynh, T.N. Nguyen, N.T. Nguyen, H.T. Dang, Silver nanoparticles-decorated reduced graphene oxide: A novel peroxidase-like activity nanomaterial for development of a colorimetric glucose biosensor, Arab. J. Chem. 13 (7) (2020) 6084–6091.

[80] L. Ma, J. Zhu, C. Wu, D. Li, X. Tang, Y. Zhang, C. An, Three-dimensional MoS$_2$ nanoflowers supported Prussian blue and Au nanoparticles: A peroxidase-mimicking catalyst for the colorimetric detection of hydrogen peroxide and glucose, Spectrochim. Acta A Mol. Biomol. Spectrosc. 259 (2021) 119886–119893.

[81] S. Xu, L. Jiang, Y. Liu, P. Liu, W. Wang, X. Luo, A morphology-based ultrasensitive multicolor colorimetric assay for detection of blood glucose by enzymatic etching of plasmonic gold nanobipyramids, Anal. Chim. Acta 1071 (2019) 53–58.

[82] T. Xiao, J. Huang, D. Wang, T. Meng, X. Yang, Au and Au-based nanomaterials: Synthesis and recent progress in electrochemical sensor applications, Talanta 206 (2020) 120210–120228.

[83] Y. Yu, F. Cui, J. Sun, P. Yang, Atomic structure of ultrathin gold nanowires, Nano Lett 16 (5) (2016) 3078–3084.

[84] A. Loubat, L.M. Lacroix, A. Robert, M. Imperor-Clerc, R. Poteau, L. Maron, R. Arenal, B. Pansu, G. Viau, Ultrathin gold nanowires: Soft-templating versus liquid phase synthesis, a quantitative study, J. Phys. Chem. 119 (8) (2015) 4422–4430.

[85] S. Gong, W. Schwalb, Y. Wang, Y. Chen, Y. Tang, J. Si, B. Shirinzadeh, W. Cheng, A wearable and highly sensitive pressure sensor with ultrathin gold nanowires, Nat. Commun. 5 (1) (2014) 1–8.

[86] A.G. Ricciardulli, S. Yang, X.F G.J.A.Wetzelaer, P.W. Blom, Hybrid silver nanowire and graphene-based solution-processed transparent electrode for organic optoelectronics, Adv. Funct. Mater. 28 (14) (2018) 1706010–1706015.

[87] R. Omar, A.E. Naciri, S. Jradi, Y. Battie, J. Toufaily, H. Mortada, S. Akil, One-step synthesis of a monolayer of monodisperse gold nanocubes for SERS substrates, J. Mater. Chem. 5 (41) (2017) 10813–10821.

[88] M. Thiele, J.Z.E. Soh, A. Knauer, D. Malsch, R.M O.Stranik, A. Csaki, T. Henkel, J.M. Koehler, W. Fritzsche, Gold nanocubes–Direct comparison of synthesis approaches reveals the need for a microfluidic synthesis setup for a high reproducibility, Chem. Eng. J. 288 (2016) 432–440.

[89] X. Wang, X. Bai, Z. Pang, H. Yang, Y. Qi, Investigation of surface plasmons in Kretschmann structure loaded with a silver nano-cube, Results Phys 12 (2019) 1866–1870.

[90] L. Scarabelli, M. Coronado-Puchau, J.J. Giner-Casares, J. Langer, L.M. Liz-Marzán, Monodisperse gold nanotriangles: Size control, large-scale self-assembly, and performance in surface-enhanced Raman scattering, ACS Nano. 8 (6) (2014) 5833–5842.

[91] S.R. Bhattarai, P.J. Derry, K. Aziz, P.K. Singh, A.M. Khoo, A.S. Chadha, A. Liopo, E.R. Zubarev, S. Krishnan, Gold nanotriangles: Scale up and X-ray radiosensitization effects in mice, Nanoscale. 9 (16) (2017) 5085–5093.

[92] G. Hu, W. Zhang, Y. Zhong, G. Liang, Q. Chen, W. Zhang, The morphology control on the preparation of silver nanotriangles, Curr. Appl. Phys. 19 (11) (2019) 1187–1194.

[93] M. Azimzadeh, M. Rahaie, N. Nasirizadeh, K. Ashtari, H. Naderi-Manesh, An electrochemical nanobiosensor for plasma miRNA-155, based on graphene oxide and gold nanorod, for early detection of breast cancer, Biosens. Bioelectron 77 (2016) 99–106.

[94] S.C. Nguyen, Q. Zhang, K. Manthiram, X. Ye, J.P. Lomont, C.B. Harris, H. Weller, A.P. Alivisatos, Study of heat transfer dynamics from gold nanorods to the environment via time-resolved infrared spectroscopy, ACS Nano. 10 (2) (2016) 2144–2151.

[95] L. Zhang, K. Xia, Z. Lu, G. Li, J. Chen, Y. Deng, S. Li, F. Zhou, N. He, Efficient and facile synthesis of gold nanorods with finely tunable plasmonic peaks from visible to near-IR range, Chem. Mater. 26 (5) (2014) 1794–1798.

[96] X. Zhuo, H.K. Yip, X. Cui, J. Wang, H.Q. Lin, Colour routing with single silver nanorods, Light Sci. Appl. 8 (1) (2019) 1–11.

[97] E. Tan, B. Erwin, S. Dames, K. Voelkerding, A. Niemz, Isothermal DNA amplification with gold nanosphere-based visual colorimetric readout for herpes simplex virus detection, Clin. Chem. 53 (11) (2007) 2017–2020.

[98] Z.H. Kim, S.R. Leone, High-resolution apertureless near-field optical imaging using gold nanosphere probes, J. Phys. Chem. B 110 (40) (2006) 19804–19809.

[99] S. Zhang, L. Zhang, K. Liu, M. Liu, Y. Yin, C. Gao, Digestive ripening in the formation of monodisperse silver nanospheres, Mater. Chem. Front. 2 (7) (2018) 1328–1333.

[100] K. Grochowska, J. Ryl, J. Karczewski, G. Śliwiński, A. Cenian, K. Siuzdak, Non-enzymatic flexible glucose sensing platform based on nanostructured TiO_2–Au composite, J. Electroanal. Chem. 837 (2019) 230–239.

[101] M. Jiang, P. Sun, J. Zhao, L. Huo, G. Cui, A flexible portable glucose sensor based on hierarchical arrays of Au@ Cu $(OH)_2$ nanograss, Sensors. 19 (22) (2019) 5055–5068.

[102] M. Xu, Y. Song, Y. Ye, C. Gong, Y. Shen, L. Wang, L. Wang, A novel flexible electrochemical glucose sensor based on gold nanoparticles/polyaniline arrays/carbon cloth electrode, Sens. Actuators B Chem. 252 (2017) 1187–1193.

Atomically precise gold and silver nanoclusters: Synthesis and applications

Rajanee Nakum[a], Raj Kumar Joshi[b], Suban K. Sahoo[a]

[a]Department of Chemistry, Sardar Vallabhbhai National Institute of Technology, Surat, Gujarat, India
[b]Department of Chemistry, Malaviya National Institute of Technology (MNIT), Jaipur, India

5.1 Introduction

There is an expedited growth in the development of fluorescent nanomaterials for the real-time monitoring of multiple targets in the field of disease diagnosis, forensic analysis, food safety, pesticides, environmental monitoring, explosives detection, etc. [1]. The fluorescent nanomaterials based on metal-based quantum dots (QDs), carbon dots (CDs), graphene quantum dots (GQDs), polymeric dots (Pdots), metal nanoclusters (NCs) etc. have been proven to be excellent nanoprobes with high sensitivity and selectivity for detecting target analytes because of their strong photoluminescence (PL), long fluorescence lifetime, large Stokes shift, photobleaching resistance, controllable synthesis and easy surface modification with other molecules [2]. Different synthetic methods produce nanoprobes with varying fluorescence properties, allowing them to recognize various targets. Also, the surface of the fluorescent nanomaterials can be modified with different molecules for probing different target analytes selectively [2].

Recently, there has been bourgeoning interest in the synthesis of luminescent metal nanoclusters (NCs) with intriguing properties like ultra-small size, large Stokes shifts (> 100 nm), high chemical and photostability, good biocompatibility, and low cytotoxicity [3]. The metal NCs consist of several to tens of metal atoms (typically Au, Ag, Cu, and Pt) with diameters smaller than 2 nm [4]. Because of the ultra-small size, the diameters of NCs are approaching the Fermi wavelength of electrons that breaks up the continuous density of states into discrete energy levels due to the strong quantum confinement effect, and the NCs with a well-defined structure behave like a molecule. The metal NCs are expected to be the missing link between the metal atoms and the larger plasmonic nanoparticles (NPs) (Fig. 5.1). Therefore, the non-conducting metal NCs do not exhibit surface plasmon resonance (SPR) absorbance observed in the larger nanoparticles but showed excellent fluorescence from visible to the near-infrared (NIR) region with other advantageous properties of high photostability, tunable fluorescence, large Stokes shift, etc. (Fig. 5.2) [5]. Among the various metal NCs, the fluorescent noble

Gold and Silver Nanoparticles.
DOI: https://doi.org/10.1016/B978-0-323-99454-5.00010-X

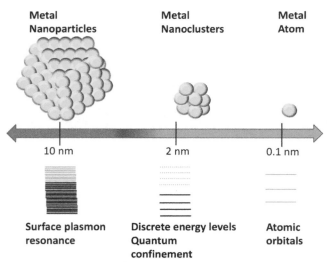

Fig. 5.1: Electronic structures of a single metal atom, metal nanoclusters and nanoparticles.

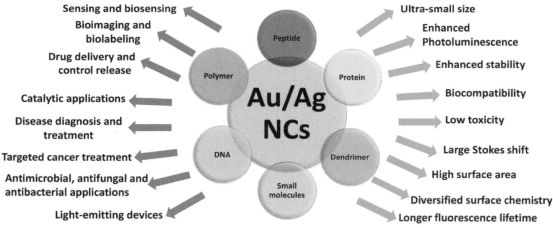

Fig. 5.2: Some advantageous properties and applications of functionalized Au/Ag NCs.

metal NCs of Au and Ag are extensively employed in biosensing, bioimaging, catalysis, solid-state lighting, drug and gene delivery, immunostimulatory agents, biomedical applications etc., owing to the high chemical inertness of Au and Ag atoms, biocompatibility, low cytotoxicity, aqueous solubility, size-dependent fluorescence behavior, high quantum yield, high surface reactivity and high adsorption capacity [6–8]. Also, because of the synthetic simplicity and low toxicity, the scientific community gives much interest to Au and Ag NCs as compared to semiconductor QDs and organic dyad [10,11]. This chapter was narrated to discuss the important synthetic approaches, properties, and applications of atomically precise fluorescent AuNCs, AgNCs, and Au-Ag alloy NCs.

5.2 Synthesis of Au and Ag NCs

The synthesis of ultra-small metal NCs has attracted increasing attention due to their fascinating physicochemical properties and growing acceptance in biomedical applications [10,11]. Over the last two decades, significant progress has been made in developing new synthetic processes, understanding characteristics, and achieving NCs biocompatibility [12]. The fluorescent Au and Ag NCs can be prepared commonly through different methods such as template method, seed growth method, ultrasonic synthesis, microwave-assisted synthesis, phase transfer synthesis, template-assisted synthesis, sol-gel method, microemulsion method, etching method, etc. can be broadly classified into two types: (1) bottom-up and (2) top-down (Fig. 5.3) [13]. The controlled reduction of metal ions precursors of Au^{3+} and Ag^+ to their zerovalent metal atoms in the presence of suitable protecting ligands formed luminescent Au/Ag NCs upon adopting a bottom-up approach [14]. The presence of protecting ligands stabilized the NCs from aggregation to form larger NPs, and also played a key role in tuning the fluorescence behavior and photostability [15,16]. The most common protecting ligands are small thiolated molecules, DNA, polymers, and dendrimers, peptides and proteins,

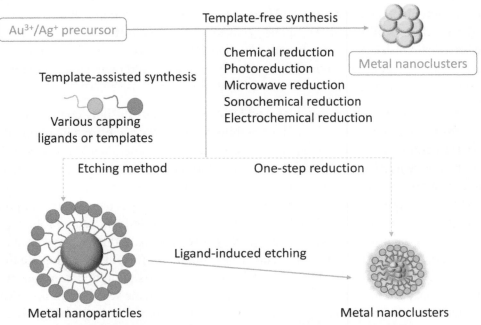

Fig. 5.3: Schematic route for the common synthesis of metal NCs.

oligonucleotides, etc. In top-down methods, the larger nanoparticles can be etched to ultra-small NCs by adopting a ligand-induced and/or solvent-induced etching approach. Roughly, Au/Ag NCs with the desire size, shape, and functionalities can be obtained in three ways: direct chemical reduction of Au and Ag precursors in the presence of suitable capping ligand, etching of larger NPs and by the postsynthetic ligand exchange process. In the post-synthetic ligand exchange process, the surface ligands of the synthesized NCs, either by bottom-up or top-down approach are exchanged partially or fully with other suitable ligands to improve the stability and properties of the NCs [17].

5.2.1 *"From atoms to nanoclusters" or bottom-up approach*

The bottom-up procedures are the most common and simplest method to obtain stable and monodispersed NCs by the controlled reduction of metal ions precursors in the presence of a suitable capping or stabilizing agent [17]. The reducing agents, such as sodium hydroxide (NaOH), sodium borohydride ($NaBH_4$), hydrazine hydrate, citrate, tetrakis (hydroxymethyl) phosphonium chloride (THPC), and ascorbic acid are commonly used to reduce the Au^{3+} and Ag^+ ions into their respective zerovalent atoms. The bottom-up approaches include the Brust-Schiffrin method, microwave-assisted synthesis, ultrasound-assisted synthesis, template-based synthesis, sonochemical, electrochemical, photoreduction synthesis, bioreduction (by plants, fungi, and bacteria) synthesis, laser pyrolysis, chemical vapor deposition (CVD), plasma or flame spraying synthesis, and bio-assisted synthesis, etc. (Fig. 5.3) [18]. The optical properties of NCs during synthesis can be tuned by altering the metal to ligand ratio, type of surface ligand, reaction time, reducing agents, pH, and temperature. For example, blue to NIR-emitting nanoclusters of Au and Ag can be synthesized by using different biomolecules, like bovine serum albumin (BSA), human serum albumin (HSA), DNA, peptides, amino acids, etc., as protecting ligands and sodium hydroxide/$NaBH_4$ as a mild reducing agent at varied conditions of medium pH, temperature and composition (Fig. 5.4) [19,20].

Biomolecules, such as proteins, peptides, and DNA etc. could exquisitely control or facilitate the formation of metal NCs through a combination of concerted efforts involving the complex formation, metal ion reduction, NCs protection, and mineralization processes in the bottom-up approach [21]. Biomolecules enable the formation of metal NCs at room temperature, most often in an aqueous solution, without toxic reagents, resulting in a smaller environmental footprint, which could be loosely classified as a "green chemistry" approach for the synthesis of metal NCs. Biomolecules can also serve as reducing cum protecting agents for the production of a variety of metal NCs. The proteins and DNA have been used as templates to direct the formation of metal NCs as the most efficient NCs template to date [22,23], where a simple yet effective method for direct synthesis of highly luminescent Au and Ag NCs was developed by using a commercial protein, bovine serum albumin (BSA). In general, some external reducing agents, most often sodium borohydride ($NaBH_4$), are needed

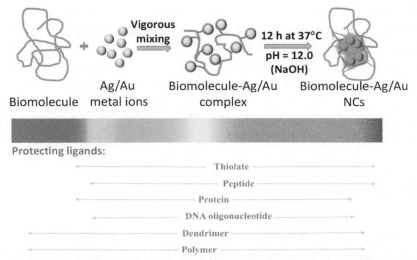

Protecting ligands:

Thiolate

Peptide

Protein

DNA oligonucleotide

Dendrimer

Polymer

Fig. 5.4: Ligands or biomolecules (proteins, DNA, peptides) protected template-based synthesis of Au and Ag nanoclusters emitting from visible to NIR region.

in the reaction solution to reduce Au^{3+}/Ag^+ to Au/Ag NCs in a protein matrix. In addition to BSA, others have adopted this efficient method to synthesize NCs by using proteins, such as HSA, ovalbumin, lysozyme, pepsin, trypsin and insulin. Biomacromolecule is always relative to a specific biomolecular recognition function, multifunctional groups (-SH, -COOH, -OH, and -NH$_2$), and excellent biocompatibility, showing promising potential in the development of various biofunctional metal NCs. Proteins with biological activities, such as enzymes can also be used as template for the synthesis of metal NCs [24]. Metal NCs with different emission colors can also be synthesized by using the protein-directed synthesis approach. Although there are several successful attempts at the protein-directed synthesis of metal NCs, in-depth and clear understandings of the formation process of protein-protected Au and Ag NCs are presently lacking. Some energy inputs, such as sonochemical and microwave, could further improve the synthesis efficiency of metal NCs by using proteins as templates. In addition to large biomolecules, stable NCs can also be synthesized by using small molecules as capping ligands like amino acids (for example, cysteine, histidine, proline, tryptophan, proline, etc.), glutathione (GSH), dithiothreotol (DTT), 11-mercaptoundecanoic acid (11-MUA), thiosalicylic acid, mercaptopropionic acid (MPA), lipoic acid etc. In some synthetic pathways of NCs, the protecting ligands like GSH can simultaneously act as a stabilizing and reducing agent in the synthesis of fluorescent Au and Ag NCs [25,26].

The chemical reduction of Au^{3+} and Ag^+ in the presence of a suitable protecting ligand is the most commonly used synthetic method for obtaining Au and Ag NCs. The template-based synthesis and monolayer-protected nanoclusters (MPCs) synthesis are the two approaches to add the precursors for the stabilization of NCs. The template-assisted synthetic method is facile, and the reduction of Au^{3+} and Ag^+ precursors takes place in the presence of a template. The reduction occurs in the presence of a mild or no reductant, the clustering rate is controlled by the template, and also the NCs are stabilized and segregated by the template. In addition to the role of templates in controlling the size, shape and morphology of NCs, the template provides multiple reactive functional groups for post-functionalization with other molecules for diversified applications. The biomolecules (e.g., proteins, DNA and peptides), dendrimers (e.g., poly(amidoamine) (PAMAM)), polymers (e.g., poly(vinylpyrrolidone), poly(methacrylic acid) (PMAA), polyethyleneimine (PEI)) etc. possessing multiple functional groups (e.g., -OH, -COOH, -SH and $-NH_2$) are commonly used as templates to synthesize ultra-small NCs with good water solubility, excellent optical properties and diversified surface chemistry.

On the other hand, the Brust-Schiffrin method gain popularity in the synthesis of nanoparticles and nanoclusters after the first two-phase synthetic of organic soluble alkanethiol-stabilized AuNPs reported in 1994 by using the phase transfer reagent tetra-octylammonium bromide (TOAB) and $NaBH_4$ as reducing agent [27]. This method used thiol (-SH) containing ligands, such as dodecanethiol, mercaptoundecanoic acid, thiolate-α-cyclodextrin, D-penicillamine etc. as capping agents. The ligand formed strong Au-S/Ag-S bonds on the surface to stabilize the nanoclusters. This method is useful in forming ultra-fine-sized and monodispersed metal NCs with excellent stability. In addition to thiolated ligands, molecules with amines and phosphine functional groups are also used as capping agents.

5.2.2 "Nanoparticles to nanoclusters" by a top-down approach

The top-down strategy involves the production of larger Au and Ag NPs starting with bulk material and cracking it into nanoparticles using various processes [28–30]. From the larger NPs, a ligand (such as thiol compounds) is used as an etching agent to prepare ultra-small nanoclusters by size reduction or core etching. There are two types of etching methods: ligand etching and solvent etching. In general, the large plasmonic Au/Ag nanoparticles can be converted to highly fluorescent AuNCs and AgNCs by the reduction of NPs size due to the etching of Ag/Au atoms by the excess etching ligands, or by the removal of Ag/Au atoms from the surface of NPs followed by the formation of NCs by the excess etching ligands. The ligands such as alkylated thiols, thiols, phosphines, amines, polymers, etc., are commonly used to etch the core of metallic NPs. During chemical etching or ligand-induced etching, strong ones typically replace weak nucleophilic ligands on the surface to form NCs, and the process is also referred as ligand exchange process. In solvent etching or phase transfer, controlled addition of etching ligands and metal ion precursor results in fairly monodisperse clusters of definite nuclearity. This interfacial etching process occurs in a water-organic biphasic system.

5.2.3 Postfunctionalization of nanoclusters

The fluorescent Au/Ag NCs can be further postfunctionalized with other biomolecules, antibodies, fluorescent dyad and nanomaterials to improve their water solubility, stability, optical properties, biocompatibility, toxicity, therapeutic and targeting ability and to achieve the desired surface for catalytic and biomedical applications [31,32]. The methods adopted for the postfunctionalization of NCs can be classified broadly into (1) bioconjugation, (2) postsynthesis ligand exchange, and (3) electrostatic or noncovalent interactions (Fig. 5.5). The surface of ligand protected NCs can be decorated with additional functionalities or biomolecules by using the covalent bioconjugation. The protecting ligand with reactive and terminal functional groups, such as primary amine, alcohol, carboxylic acid or thiols can be bioconjugated to NCs to add extra functionalities to improve the structural stability, biocompatibility and targeting ability. For example, the EDC/NHS (N-ethyl-N'-(3-(dimethylamino)propyl) carbodiimide/N-hydroxysuccinimide) can activate coupling between carboxylic group and amine [33], whereas glutaraldehyde can be used to connect two different molecules having terminal primary amine group. The conjugation of NCs can be done with small molecules,

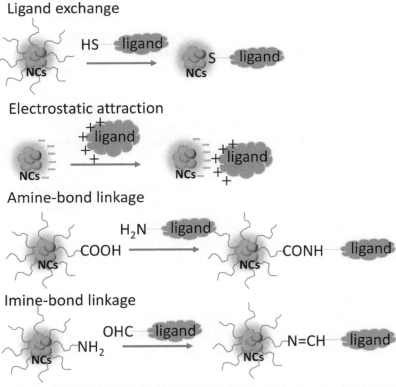

Fig. 5.5: Schematic illustration of different postfunctionalization approaches on as-synthesized NCs.

oligonucleotides, peptides, proteins, viruses, etc. to achieve the desired properties for bio-medical applications. In the ligand exchange method, the original ligand coated on the as-synthesized NCs can be partially or completely exchanged with another suitable ligand that changes the surface properties of NCs with a change of structure, functionalities, and composition. The noncovalent interactions, such as electrostatic attraction is used to postfunctionalized charged biomolecules on the surface of an as-synthesized NCs with the coated ligands that possess opposite charge.

5.3 Characterization of Au and Ag NCs

The size, composition, bonding between the metal core and the capping ligand, and the synthetic pathway greatly influenced the optical, electronic, and structural features of AuNCs and AgNCs. Therefore, different analytical techniques are used for the characterization of NCs, such as UV-Vis absorption, fluorescence and Fourier transform infrared (FTIR) spectroscopy, fluorescence life-time measurements, high resolution transmission electron microscopy (HRTEM), X-ray photoelectron spectroscopy (XPS), dynamic light scattering (DLS), mass spectrometry (MS), single crystal X-ray crystallography and so on.

The larger plasmonic NPs of Au and Ag with sizes 5–100 nm showed intense SPR peak at about 525 and 420 nm, respectively, while the absence of SPR peak supported the successful formation of subnanometer sized metal nanoclusters. In many synthetic pathways of NCs, the ligand plays the role of both reductant as well as a stabilizing agent. Therefore, the reaction and bonding between Au/Ag atoms and the capped ligand can be examined by recording their absorption spectrum separately for the capping ligand and ligand-capped NCs. In addition to surface ligands, the cluster size can also affect the absorption behavior of NCs. For example, the smallest cluster $Au_{10-12}GSH_{10-12}$ has an absorption onset around 400 nm, and shifted to a longer wavelength with the increase in cluster size from $Au_{15}GSH_{13}$, $Au_{18}GSH_{14}$ to $Au_{25}GSH_{18}$ [34].

The fluorescence-based spectroscopic and imaging techniques are very useful in characterizing the photoluminescence properties of NCs. The fluorescence spectra were recorded by exciting NCs at a fixed or different excitation wavelength. The absolute and/or relative quantum yield of NCs is determined to examine the efficiency of photon emission from NCs. A higher fluorescence quantum yield value indicates the brighter the NCs with a stronger fluorescence signal intensity. The main factors that can be altered to enhance the fluorescence intensity and quantum yield of NCs are the size of the central metal core, oxidation state of the metal, the valence electron count, cluster configuration, and ligand type [35]. The doping or fabricating of alloy NCs is an important approach to improve the brightness of the NCs fluorescence. The most common dopants for bimetallic Au/Ag NCs are Ag/Au, Pt, Cu, Zn, Ni, Pd, Cd, etc. The Au/Ag NCs doped with foreign atoms is helpful in tuning the electronic and geometrical structure of the cluster core that influences the optical properties, catalytic activity, and thermal stability. The luminescence of pure gold and silver nanoclusters ($Au_{25}(SH)_{18}^-$ and $Ag_{25}(SH)_{18}^-$), as well as systems in which one dopant atom was incorporated ($Au_{24}Ag(SH)_{18}^-$ and $Ag_{24}Au(SH)_{18}^-$) was investigated to examine the factors involved in the luminescence of doped nanocluster systems [36,37]. The aggregation of Au/Ag NCs can also enhance fluorescence [12]. The aggregation of NCs can be achieved by changing the solvents, temperature, pH or by adding appropriate ions. The synthetic procedure and shape of NCs also greatly influence the photoluminescence properties. For example, Kawasaki et al. demonstrated the pH-dependent synthesis of pepsin mediated Au clusters with blue, green and red fluorescence, which corresponded to $Au_5(Au_8)$, Au_{13}, and Au_{25} nanoclusters, respectively [38]. In another study, the Ag_{62}NCs were synthesized by adopting two different methods, that is, by one-pot method $[Ag_{62}S_{12}(SBu^t)_{32}]^{2+}$ (denoted as Ag_{62}-I) and a two-phase ligand exchange method $[Ag_{62}S_{13}(SBu^t)_{32}]^{4+}$ (denoted as Ag_{62}-II). Even though both NCs had similar morphology, the Ag_{62}-II had complete fluorescence quenching due to the presence of free valence electrons, whereas Ag_{62}-I showed strong red fluorescence [39].

The Au and Ag NCs showed a longer fluorescence life-time with large Stokes shifts useful for fluorescence sensing and bioimaging applications. The mono-exponential nanosecond fluorescence lifetimes mostly observed in Au and Ag NCs, however some NCs decorated with polymer and peptide are reported to show biexponential lifetime decays [40]. The NCs with submicrosecond to microsecond lifetimes are ideally considered for fluorescence sensing and bioimaging owing to their ability to eliminate autofluorescence from the biological tissues. Because of the longer life-time (>100 ns) with emission wavelengths ranging from visible to NIR regions, the Au and AgNCs are also applied for fluorescence lifetime imaging study of live cells.

The FT-IR is used to characterize the participation of ligand functional groups in the formation of ligand-protected NCs. For this, the free ligand FT-IR was compared with the ligand-protected NCs. For example, in thiol-stabilized Au/Ag NCs, the formation of Au/Ag-S bond can be confirmed from the disappearance of –SH stretching vibrational mode of thiol observed around 2525 cm^{-1}. In addition to FT-IR, ^{1}H and ^{13}C NMR can also be recorded for the free ligand and compared with the ligand-protected NCs to examine the purity of the NCs and the functional groups of ligand participate in the interaction with the metal core.

XPS is a useful surface analysis technique for obtaining quantitative elemental composition, oxidation states of metal and the binding states of the elements present in the NCs. The oxidation state of the core can influence the optical properties of metal NCs because of the enhancement of electropositivity of the metal core that promotes the interactions between the surface metals and ligands. Therefore, the determination of oxidation states of metal core in NCs is very important. Sahoo et al. characterized the pyridoxal conjugated BSA-AuNCs with XPS and found two peaks at 84.1 and ~ 88.0 eV respectively for the Au $4f_{7/2}$ and Au $4f_{5/2}$. The binding energy >84.0 eV supported the presence of both Au(0) and Au(I) in the core and surface of AuNCs [41]. The deconvulated peaks of C1s, N1s and O1s supported the different linkages present on the surface of the NCs. The same group analyzed the polyethyleneimine (PEI) coated AgNCs by XPS [42], and observed two peaks at 368.1 eV and 373.8 eV for the Ag 3d5/2 and Ag 3d3/2, respectively. The binding energy 368.1 eV for Ag 3d5/2 is correspond to Ag(0) that support the reduction of Ag(I) into Ag(0).

The size and shape of NCs are the important factors to decide the optical, electronic and catalytic properties. Because of the ultra-small size, the morphology and size of the NCs is determined by HRTEM with the highest resolution of ~0.1 nm. In this technique, the diluted solution of NCs is carefully taken in a TEM specimen to avoid aggregation. Except HRTEM, the scanning transmission electron microscopy (STEM), atomic force microscopy (AFM) and scanning tunneling microscopy (STM) are also employed for the morphological studies of NCs. In addition to, the number of metal atoms presence in the cluster core also affects the optoelectronic properties of NCs. The mass spectrometry based on electrospray ionization mass spectrometry (ESI-MS), fast atom bombardment ionization mass spectrometry (FAB-MS), laser desorption/ionization time-of-flight, laser desorption/ionization mass spectrometry and matrix-assisted laser desorption ionization mass spectrometry (MALDI-MS) are used to provide accurate results on the exact core diameter of NCs including the number of metal atoms and capping ligands.

5.4 Applications of Au and Ag NCs

Because of the unique and fascinating surface chemistry, high surface area, chemical and thermal stability, optoelectronic properties, the ultra-small AuNCs, Ag NCs, and alloy NCs have been extensively applied for sensing and biosensing, biolabeling, bioimaging, multi-modal imaging, drug delivery and control release, gene and siRNA delivery, antimicrobial applications, catalytic applications, disease diagnosis and treatment, antifungal and antibacterial applications, photosensitizers, light-emitting devices, targeted cancer treatment, radiation therapy, photodynamic therapy etc. [35,43–55]. This section covers the important sensing and biosensing applications of Au/Ag NCs.

In the field of sensing and biosensing, the ultra-small as-synthesized fluorescent Au/Ag NCs and/or postfunctionalized with other suitable molecules/dyad/fluorescent nanomaterials are applied for the detection of various analytes, such as metal ions, anions, small biomolecules, protein and enzymatic activity, drug molecules, pesticides and explosives, pH, temperature etc.

5.4.1 Detection of metal ions and anions

The fluorescent Au/Ag NCs and their alloy NCs are extensively applied for the detection of metal ions and anions. The heavy metal ions, such as Cu^{2+}, Hg^{2+}, and Pb^{2+} can cause serious toxicity to human health, and therefore the permissible limit in drinking water is respectively limit to 21, 10, and 72 μM by the US Environmental Protection Agency. The fluorescence sensing by Au/Ag NCs is possible due to the analyte directed changes in the Au/Ag cluster core and/or the surface properties. The change in the valence state of the Au/Ag core upon interaction, cluster aggregation, complex formation, electron/energy transfer can perturb the fluorescence emission of NCs [56]. For example, the AuNCs with Au^+ ($5d^{10}$) on the surface of the cluster core can interact with Hg^{2+} ($5d^{10}$) because of the metallophilic $Au^+\cdots Hg^{2+}$ interaction leads to the quenching of fluorescence. Similar fluorescence quenching was also observed with Ag^+-rich Ag NCs. Thus, the valence state of the cluster core also plays a key role in the detection of the analyte [57]. In addition, the as-synthesized and post-functionalized Au/Ag NCs possessing surface ligands with multiple functionalities can apply for the sensing of bioactive and toxic metal ions and anions. Some recent fluorescent Au/Ag NCs based sensors for metal ions and anions are discussed here.

The fluorescence of keratin-templated AuNCs was enhanced at 725 nm after modifying with Ag^+ due to the "silver effect". The modified AuNCs showed fluorescence turn-off response for Hg^{2+} with the detection limit down to 2.31 nM. It was proposed that after adsorption of Hg^{2+} on the surface of cluster core, Hg^{2+} interacted with Au^+ and also oxidized Ag^0 to Ag^+ [58]. Sahoo et al. synthesized BSA-AuNCs followed by conjugated with pyridoxal for the fluorescent turn-off sensing of Hg^{2+} [41]. The quenching occurred due to the metallophilic bonding between $5d^{10}$ Hg^{2+} and $5d^{10}$ Au^+ upon high specificity of $Au^+\cdots Hg^{2+}$ interactions. The dual ligand 11-mercaptoundecanoic acid (MUA) and threonine (Thr) co-functionalized AuNCs emitting at 606 nm was applied for the fluorescence turn-off detection of Hg^{2+} with a detection limit of 8.6 nM. Addition of Hg^{2+} caused an aggregation-induced fluorescence quenching [59]. The blue-emitting tryptophan (Trp)-coated AgNCs was applied for the fluorescence turn-off detection of Cu^{2+} with the detection limit of 0.029 μM. The fluorescence of Trp-AgNCs was quenched due to the charge transfer occurred upon complexation of Cu^{2+}. The nanoprobe was applied for the bioimaging in cells [60]. In another work, the Cu^{2+} interacted with the Au core or the BSA-template lead to the energy loss of the excited AuNCs, which allowed to detect Cu^{2+} down to 0.83 μM [61].

The turn-on and ratiometric fluorescence sensors are widely employed in biochemical sensing and bioimaging compared to turn-off sensors because of their visibility, sensitivity and accuracy. The AuAg alloy NCs emitting at 637 nm was synthesized by interfacial etching of mercaptosuccinic acid (MSA)-capped AgNPs followed by galvanic exchange reaction. The surface of MSA-AgAuNCs was modified further with m-PEG-NH_2 by using EDC. The PEGylated MSA-AgAu NCs showed significant fluorescence enhancement in the presence of Al^{3+} with a sensitivity limit of 0.8 μM. The enhancement occurred due to the deposition of Al^{3+} on the surface of the cluster core, but not due to the complexation-induced aggregation of NCs [62]. Sahoo et al. synthesized the blue-emitting PEI-AgNCs followed by postfunctionalized with pyridoxal 5'phosphate (PLP) by forming imine-linkage between the $-NH_2$

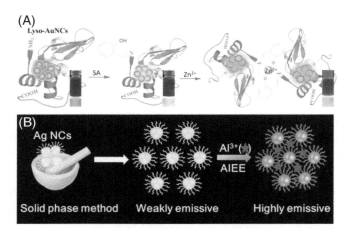

Fig. 5.6: (A) The schematic representation for the change in fluorescence of Lyso-AuNCs for the relay detection of SA and Zn^{2+}. *(Adapted with permission from ref. [64]).* (B) Solid-state synthesis of GSH-AgNCs and its application in Al^{3+} detection. *(Adapted with permission from ref. [65]).*

group of PEI and –CHO group of PLP. The PLP conjugated PEI-AgNCs showed selective fluorescence enhancement in the presence of Zn^{2+} and Cd^{2+} ions due to the complexation-induced aggregation of NCs with the detection limit of 50.5×10^{-8} M and 59.0×10^{-8} M, respectively. The developed nanoprobe was chemically decorated over cellulose strips and applied for the fluorescent visual detection of Zn^{2+} and Cd^{2+} [42]. Similarly, the PLP and salicylaldehyde (SA, Fig. 5.6A) are conjugated over lysozyme (Lyso)-templated AuNCs, and applied for the fluorescence turn-on sensing of Zn^{2+} [63,64]. The solid-state synthetic strategy was adopted to prepare GSH-AgNCs. Upon addition of Al^{3+}, the fluorescence of AgNCs was enhanced because of the aggregation-induced emission enhancement with a detection limit of $0.1\,\mu M$ (Fig. 5.6B) [65].

The Ag^+ selective ratiometric turn-on fluorescent probe CSA was designed by coupling carbon dots-silicon spheres composites (CSs) and AuNCs. The CSA showed two fluorescence emissions at 448 and 610 nm, respectively due to the CSs and AuNCs. The addition of Ag^+ enhanced the fluorescence intensity of AuNCs, which led to the fluorescent color change from blue to purple and then to pink (Fig. 5.7). The composite CSA showed a detection limit of 1.6 nM for Ag^+, and the agarose hydrogels of CSA was developed for the visual detection of Ag^+ [66]. In another approach, the composite of carbon quantum dots (CQDs)/AuNCs was developed as a dual-signal ratiometric fluorescent probe for the detection of L-cysteine (Cys) and Ag^+. The emission at 472 nm of CQDs act as a reference signal, whereas the emission at 605 nm of AuNCs functioned as a reporter. The gray-orange color emitting CQDs/AuNCs upon interaction with Ag^+ enhanced the orange fluorescence, whereas the sequential addition of Cys quenched the enhanced fluorescence due to the complexation occurred between the

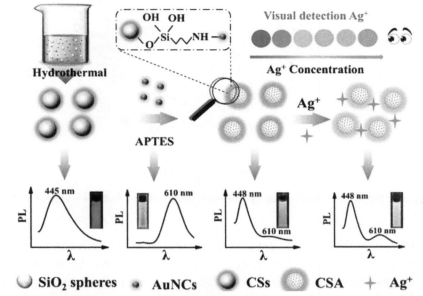

Fig. 5.7: CSA composite synthesis and Ag⁺ detection process flow diagram.
(Adapted with permission from ref. [66]).

Ag⁺ and Cys. During the sensing of Ag⁺ and Cys, the fluorescence of CQDs remains unaltered. The paper strips coated with CQDs/AuNCs were developed for the visual detection of Ag⁺ and Cys [67]. The dipeptide L-cysteinyl-L-cysteine was used as a template to prepare AuNCs and applied for the fluorescence turn-on sensing of As^{3+} with a detection limit of 53.7 nM. The fluorescence of AuNCs is enhanced due to the electron transfer process along with the increase in the radiative decay rate of AuNCs in the presence of As^{3+} [68].

Similar to metal ions, anions are ubiquitous and play a crucial role in many biological processes. Some anions, such as S^{2-}, CN^- and NO^{2-} can also cause toxic effects to human health. Therefore, the fluorescent Au/Ag NCs are applied for the detection of anions. The BSA-AuNCs was applied for the selective detection of CN^- based on the mechanism of cyanide etching-induced fluorescence quenching of AuNCs. This nanoprobe can detect CN^- down to 200 nM, which is much lower than the permissible limit of CN^- (2.7 μM) in drinking water by the World Health Organization (WHO) [69]. The toxic sulfide ions (S^{2-}) was detected by papain-AuNCs. The fluorescence of AuNCs was quenched due to the reaction occurred between S^{2-} and the Au atoms/ions, which led to the formation of Au_2S and the aggregation of the AuNCs (Fig. 5.8). This nanoprobe was applied to detect S^{2-} down to 0.38 μM [70]. In another approach, the red-emitting Lyso-AgNCs was applied for the fluorescent turn-off sensing of S^{2-} with the detection limit of 1.1 μM. The a\Addition of S^{2-} formed Ag_2S that quenched the fluorescence of Lyso-AgNCs [71].

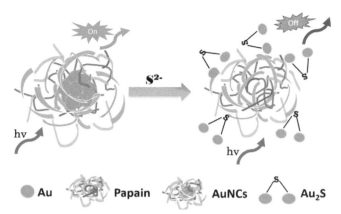

Fig. 5.8: Schematic illustration of the quenching of the AuNCs induced by S^{2-}.

The mixture of dimer DNA AgNCs and carbon dots (CDs) in 5:1 ratio was applied for the detection of I^- down to 19.8 nM. The orange fluorescence of dDNA-AgNCs at 577 nm was quenched in the presence of I^- due to the I^--induced oxidative etching and aggregation of AgNCs. The blue fluorescent CDs at 446 nm was remains unaltered. The test paper of the dual-emissive probe was developed for the visual detection of I^- [72]. In another study, the NO_2^- was detected by using the co-functionalized with BSA and 3-mercaptopropionic acid AuNCs (BSA/MPA-AuNCs). Upon addition of NO_2^- under acidic conditions, the oxidation of Fe^{2+} to Fe^{3+} quenched the fluorescence of BSA/MPA-AuNCs through nonradiative electron-transfer mechanism with the sensitivity limit of 0.7 μM [73].

5.4.2 Detection of small molecules

The strong fluorescence and attractive surface properties of Au/Ag NCs are employed for the facile detection of small bioactive molecules, such as biothiols, glucose, dopamine (DA), urea, uric acid, bilirubin, cholesterol, creatinine, vitamin B12, 5-hydroxytryptamine etc. Some important examples are discussed here.

Glucose is an important energy source for human beings. The imbalance of glucose in the blood leads to serious metabolic disorders, such as hypoglycemia and diabetes mellitus. The functionalized Au and Ag NCs are employed for the accurate fluorescence detection of glucose. For example, the mixture of AIE active molecule sodium 1,2-bis [4-(3-sulfonatopropoxyl) phenyl]-1,2-diphenylethene (BSPOTPE) and BSA-AuNCs is used for the detection of glucose (Scheme 5.1). In the presence of H_2O_2, the fluorescence intensity of NCs at 680 nm was successively decreased while the AIE band remain constant at 490 nm. It is well-known that glucose oxidase (GOx) can catalyze glucose to produce H_2O_2 that allowed the ratiometric fluorescence detection of glucose in the range 1–8 mM [74]. In another approach, the nanocomposite of bromelain-AuNCs and MnO_2 nanosheets in the presence of GOx was applied

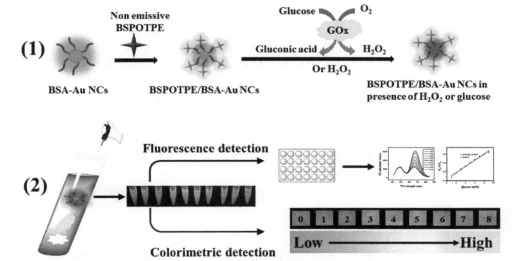

Scheme 5.1: Schematic illustration of the BSPOTPE/BSA-AuNCs ratiometric fluorescence probe for sensing of glucose.
(Adapted with permission from ref. [74]).

for the detection of glucose with a sensitivity limit of 6.7 μM. The fluorescence of AuNCs was quenched in the presence of MnO_2 nanosheets due to the inner filter effect (IFE). The formation of H_2O_2 upon the enzymatic reaction of GOx with glucose reduced the MnO_2 nanosheets to Mn^{2+} ions, and restored the fluorescence emission of AuNCs [75]. In addition to AuNCs, the BSA-AgNCs was also employed for the detection of H_2O_2 and glucose [76]. In this approach, the fluorescence intensity of BSA-AgNCs at 635 nm was enhanced in the presence of H_2O_2 due to the etching and the formation of smaller AgNCs with the sensitivity limit of 1.1 μM. Using the enzymatic reaction of GOx, the BSA-AgNCs can be used to detect glucose concentration down to 3.4 μM. These developed methods are applied for the determination of glucose in human blood samples.

The neurotransmitter dopamine (DA) play an important role in behavior responses of feeling and the normal functioning of the brain, but its facile detection in human bodies is very challenging. The red-emitting GSH-AgNCs upon the hydrogen bonding interaction between the amino group of DA and the carboxyl groups of GSH-AgNCs resulted an increase in the fluorescence intensity at 645 nm. The fluorescence turn-on response of GSH-AgNCs is linear between 20 nM to 220 nM with the sensitivity limit of 0.35 nM. The developed nanoprobe was applied for the detection of DA in spiked human urine samples [77]. In another approach, the BSA stabilized Au-AgNCs was applied for the fluorescence turn-on detection of DA. The fluorescence intensity of Au-AgNCs was enhanced and red-shifted due to the reduction of nanoclusters in the presence of DA. With the detection of 6.9 nM, the nanoprobe Au-AgNCs was successfully applied for the quantification of DA in serum samples [78]. In addition to turn-on probes, the fluorescence resonance energy transfer (FRET) mechanism was employed for the ratiometric detection of DA by coupling BSA-AuNCs (λ_{em} = 670 nm) and CQDs (λ_{em} = 446 nm). The nanoprobe CQDs-AuNCs was obtained by coupling amidated modified CQDs with BSA-AuNCs. In the presence of DA, the electron transfer from CQDs to DA quenched the purple fluorescence of the CQDs-AuNCs and the FRET process from CQDs and BSA-AuNCs is inhibited. This nanoprobe is selective for DA with a limit of detection of 2.66 nM, and applied for the measurement of DA in serum samples [79]. The AuNCs decorated with GSH and 11-mercaptoundecanoic acid (11-MUA) was further post-function with amino pillar[5]arene (AP5) through EDC/NHS condensation reaction. The fluorescence of the nanocomposite AP5-AuNCs is quenched due to the electron transfer from the oxidized DA with a sensitivity limit of 1.5 nM. The nanoprobe was applied for the quantification of DA in human urine [80].

Biological biothiols, such as GSH, cysteine (Cys) and homocysteine (Hcy) play an important role in maintaining human health. The fluorescent intensity of BSA-AgNCs was quenched upon addition of biothiols due to the strong affinity between the -SH group of biothiols and the AgNCs. With this nanoprobe, the biothiols Cys, Hcy and GSH can be facilely detected down to 0.81, 1.0 and 1.1 µM, respectively, and applied to detect biothiols in human plasma [81]. In compared to turn-off probes, the ratiometric response improve the performance of a probe by eliminating the influence of external factors. In order to design a ratiometric probe, the nitrogen-doped carbon dots (NCDs, λ_{em} = 450 nm) was mixed with red-emitting AgNCs (λ_{em} = 650 nm). The IFE quenched the fluorescence of NCDs in the presence of AgNCs. Upon addition of biothiols, the fluorescence intensity of AgNCs was decreased that restored the fluorescence of NCDs at 450 nm. Using Cys as the model, the ratiometric response (I_{450}/I_{650}) from this nanoprobe can be applied to detect down to 0.14 µM Cys and real samples analyses were performed successfully in urine and serum samples [82]. In a similar study, the DNA-AgNCs was applied for the fluorescence turn-off detection of Cys down to 0.05 nM. Because of the strong affinity between -SH group of Cys and Ag, the addition of Cys quenched the fluorescence intensity of DNA-AgNCs [83].

5.4.3 Detection of enzymatic activity

Enzymes are protein known for their catalytic reactions in the biological processes. The irregularity of enzymatic activity causes abnormalities in cellular functions, which leads to serious health problems. Therefore, the detection of enzymatic activity is used as a biomarker for the diagnosis of different diseases.

Sahoo and his coworkers developed several fluorescence bioassays using AuNCs for the selective detection of the enzyme alkaline phosphatase (ALP) [84,85]. ALP in biological processes known to catalyze the dephosphorylation of a variety of monophosphate ester substrates, such as phenyl phosphate, pyridoxal 5′phosphate (PLP), *p*-nitrophenyl phosphate,

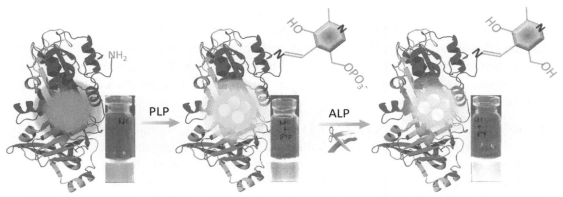

Scheme 5.2: The diagram depicts the sensing mechanism of ALP activity with the probe PLP_ OVA@AuNCs.

(Adapted with permission from ref. [84]).

adenosine triphosphate (ATP), ascorbic acid-2-phosphate, 5-bromo-4-chloro-3-indolyl-phosphate, amifostine, etc. The red-emitting ovalbumin (OVA)-stabilized AuNCs and the PLP as the monophosphate ester substrate was applied for the selective sensing of ALP. The fluorescence emission at 655 nm of OVA@AuNCs was quenched, and a new emission band formed at 532 nm due to the formation of imine linkage between the PLP and the free –NH$_2$ group of OVA. The yellow emitting PLP_OVA@AuNCs turned to pale-white emitting nanoclusters upon addition of ALP due to the catalytic dephosphorylation of PLP into pyridoxal (Scheme 5.2). The nanoprobe PLP_OVA@AuNCs can detect ALP down to 3.9 U/L [84]. In another approach, the red-emitting Lyso-AuNCs was applied for the selective detection ALP using the PLP as monophosphate ester substrate with a sensitivity limit of 0.002 U/L [85]. The developed nanoprobes were applied for the quantification of ALP in various biological samples, such as blood plasma and serum. The OVA/Lyso-AuNCs can be decorated with optimized amounts of PLP and pyridoxal to generate pure white-light emission. In addition to PLP, the *p*-nitrophenyl phosphate (PNPP) was used as a monophosphate ester substrate that suppressed the fluorescence emission of MUA-AuNCs due to the IFE. The excitation spectrum of MUA-AuNCs and the absorption spectrum of PNPP overlap well to promote the IEF, which caused fluorescence quenching at 600 nm. The catalytic conversion of PNPP into *p*-nitrophenol restored the fluorescence emission of AuNCs with the sensitivity limit of 0.002 U/L [86]. In a similar approach, the histidine and β-cyclodextrin functionalized AgNCs was applied for the detection of ALP in the presence of PNPP. The IFE effect due to PNPP quenched the fluorescence emission of AgNCs was restored upon addition of ALP. This nanoprobe showed a detection limit of 0.0046 U/L for ALP [87].

The phosphorylation of proteins is catalyzed by the enzyme protein kinase A (PKA). In order to detect PKA, a dual emission nanohybrids QDs@SiO$_2$@peptide-AuNCs was fabricated by combining blue-emitting peptide-AuNCs (λ_{em} = 415 nm) and red-emitting CdS/ZnS@

SiO$_2$ (λ_{em} = 630 nm). Upon addition of carboxypeptidase Y (CPY), the blue-fluorescence of peptide-AuNCs is quenched due to the digestion of peptide by CPY without altering the emission intensity at 630 nm. However, the phosphorylation of the peptide in the presence of PKA suppressed the fluorescence quenching of peptide-AuNCs. The change in fluorescence intensities ratio (I$_{415}$/I$_{630}$) of the nanohybrids can be employed to detect PKA down to 0.004 U/mL [88].

The enzyme acetylcholinesterase (AChE) in the process of nerve impulse transmission can catalyze the hydrolysis of acetylcholine (ACh) to choline and acetic acid. The DNA-templated Cu/AgNCs was applied for the facile detection of AChE. In this approach, hydrolysis of acetylthiocholine (ATCh) by AChE formed choline that quenched the fluorescence of DNA-Cu/AgNCs. With a linearity range from 0.05 to 2.0 mU/mL, this nanoprobe can be applied to detect AChE down to 0.05 mU/mL. In addition, the organophosphorus pesticides (OPs) and tacrine were used to eliminate the fluorescence quenching of DNA-Cu/AgNCs by suppressing the hydrolysis of ATCh by AChE [89]. Compared to AChE, the enzyme butyrylcholinesterase (BChE) is abundantly found in plasma and catalyzes the hydrolysis of choline biomolecules. For detecting BChE activity, the fluorescence of AuNCs is quenched by the spindle-like FeOOH nanorods. In the presence of acetylthiocholine (ATCh) and BChE, FeOOH nanorods is decomposed by the enzymatic hydrolysate (thiocholine) that restored the fluorescence of AuNCs. This nanoprobe can detect BChE activity down to 4 ng/mL, and applied to quantify BChE in tiny finger blood [90].

The digestive enzyme pepsin present actively in the animal stomachs is used by the food industry for the preparation of bioactive peptides. The Lyso@AuNCs was used for the detection of pepsin activity. The enzymatic digestion of Lyso@AuNCs in acidic medium (pH 3.0) by pepsin decrease the fluorescence intensity at 420 nm with a sensitivity limit of 0.256 µg/mL. The nanoprobe Lyso@AuNCs was applied for the monitoring of pepsin in biological and food samples [91]. In a similar study, the peptidase activity was determined by using fluorescent peptide-AuNCs and β-site amyloid precursor protein-cleaving enzyme 1 (BACE1) as a model peptidase. It was observed that the fluorescence intensity of AuNCs is inversely related with the peptide length coated on the AuNCs surface. The cleaving of peptide substrates on AuNCs by BACE1 caused a fluorescence enhancement at 405 nm with the detection limit of 1 U/mL. The bioassay was applied for the detection of BACE1 activity in spiked cell lysates [92]. The S1 nuclease play an important role in DNA replication, transcription, repair and recombination. The polycytosine oligonucleotide (dC$_{12}$)-templated Ag NCs was employed for the label-free fluorescence detection of S1 nuclease activity. The fluorescence intensity at 611 nm of dC$_{12}$-AgNCs was quenched in the presence of S1 nuclease due to the degradation of dC$_{12}$ to mono- or oligonucleotide fragments (Fig. 5.9). With a linearity from 5×10^{-7} to 1×10^{-3} U/µL, the nanoprobe dC$_{12}$-AgNCs can detect S1 nuclease activity down to 5×10^{-8} U/µL [93].

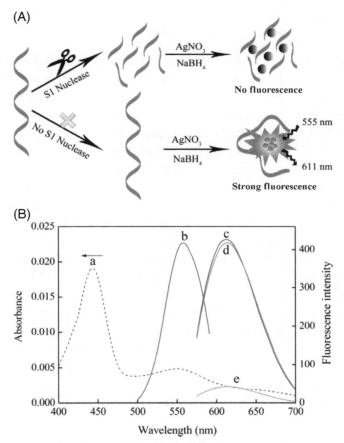

Fig. 5.9: (A) Schematic diagram of S1 nuclease activity assay based on dC$_{12}$-Ag NCs. (B) UV-Vis (a) and fluorescence spectra of dC$_{12}$-Ag NC probe: excitation curve (b), emission curve (c), incubation with inactivated S1 nuclease (d), and incubation with activated S1 nuclease (e). Maximum wavelengths of excitation and emission are 555 and 611 nm, respectively. *(Adapted with permission from ref. [93]).*

5.4.4 Detection of pesticides and explosives

Pesticides are extensively used to protect plants and crops from harmful organisms, weeds, unwanted pests, infections etc. The pesticides, such as OPs, triazines, carbamates, pyrethroids, chloroacetanilides, etc. are used for pest control. Because of the high effectiveness in pest control and less accumulation in the human body, the OPs and carbamates are mostly used as pesticides. However, the excessive use of pesticides can cause negative impacts to human health and also destroy soil fertility. The OPs, such as azinphos-methyl, paraoxon, diazinon, dichlorvos, malathion, chlorpyrifos (CP), parathion, methyl parathion, tetrachlorvinphos, phosmet, oxydemeton, triazophos etc. can restrict AChE activity to cause neurotoxicity. Therefore, facile fluorescent-based analytical methods including the fluorescent Au/Ag NCs are employed for the detection of OPs.

The chicken egg white (CEW) protected AuNCs was employed for the selective detection of OPs by restricting the enzymatic activity of tyrosinase (TYR). The fluorescence emission at 630 nm of AuNCs was dynamically quenched due to the catalytic oxidation of dopamine to dopaminechrome by TYR. The presence of OPs can suppress the enzymatic activity of TYR that restored the fluorescence of AuNCs. Using paraoxon as OPs model, the AuNCs and TYR can achieve the sensitivity limit of 0.1 ng/mL, and the paper-based test strips were developed for the cost-effective visual detection of OPs [94]. In another approach, the OPs was detected by using BSA-AuNCs and AChE enzyme. In this approach, AChE hydrolytically catalyzes the acetylthiocholine to produce thiocholine. The thiocholine is absorbed over the surface of AuNCs by forming Au-S bond and electrostatic interaction, which caused aggregation and AuNCs and quenching of fluorescence at 630 nm. When paraoxon as OPs model was added, the AChE activity was inhibited and the fluorescence quenching of AuNCs was reduced. This bioassay allowed to detect OPs with a detection limit of 3.3×10^{-14} M [95]. The nonthiolate DNA was used as template to synthesize red-emitting DNA-AuNCs and applied for the selective detection of organophosphorothioate pesticides by taking chlorpyrifos (CP) as model analyte. As depicted in Scheme 5.3, the hydrolysis of chlorpyrifos (CP) formed the diethyposphorthioate (DEP). The DEP formed immediately replaced the nonthiolate DNA from ultra-small AuNCs to form plasmonic AuNPs, which caused fluorescence quenching at 630 nm. This assay showed a sensitivity limit for CP down to 0.50 µM, and the practical utility was validated by quantifying CP in real samples [96].

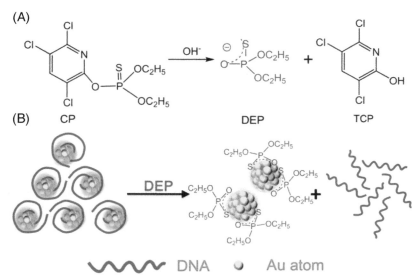

Scheme 5.3: (A) The hydrolysis process of CP in an alkaline solution. (B) Schematic illustration of the DNA-AuNCs for the detection of CP. *(Adapted with permission from ref. [96]).*

The selective and sensitive detection of nitroaromatic compounds, such as picric acid (PA), 2,4,6-trinitrotoluene (TNT), and 4-nitrophenol (4-NP), etc. is very important due to their explosive nature and threat to living society. The PEI-AgNCs emitting at 450 nm were applied for the fluorescence turn-off sensing of PA with a sensitivity limit of 0.1 nM. The absorption spectrum of PA overlaps with the emission of PEI-AgNCs, which caused efficient energy transfer from AgNCs to PA. In addition, the fluorescence quenching was occurred due to the electron-transfer from electron-rich PEI-AgNCs to electron-deficient PA. This nanoprobe was applied for the detection of PA in real samples [97]. The red-emitting BSA-AuNCs was applied for the simultaneous fluorescence turn-off detection of TNT and 4-NP with the detection limit of 10 nM and 1 nM, respectively. The quenching was occurred due to the transfer of electron from electron-rich BSA-AuNCs to electron-deficient TNT/4-NP. The test paper coated with BSA-AuNCs was applied to detect vapour of TNT and 4-NP with a sensitivity limit of 10 pM and 1 pM, respectively [98]. The Tb-doped BSA-AuNCs was applied to detect TNT in an aqueous medium. The doping of Tb improved the complexation between TNT and AuNCs. The paper strip of Tb-doped BSA-AuNCs was also developed for the visual detection of TNT [99]. In another approach, the GSH-AuNCs was modified with zeolitic imidazolate framework (ZIF-8) showed enhanced aggregation-induced fluorescence emission was quenched in the presence of TNT due to the energy transfer process with the detection limit of 5 nM [100].

5.5 Conclusions and future perspectives

This chapter discussed the sensing and biosensing applications of the ultra-small AuNCs, AgNCs and their alloy NCs. The synthesis and important properties of the Au/Ag NCs are also summarized and discussed. The unique properties of Au/Ag NCs are ultrasmall size, excellent photoluminescence, photostability, large Stokes shift, longer fluorescence lifetime, chemical inertness of Au and Ag atoms, biocompatibility, low toxicity, synthetic simplicity and post-functionalized with suitable other biomolecule for targeted applications. Because of the unique chemical and optical properties, the Au/Ag NCs are extensively applied for sensing and biosensing, bioimaging, biolabeling, drug delivery and control release, gene and siRNA delivery, catalytic applications, disease diagnosis and treatment, antimicrobial applications, antifungal and antibacterial applications, photosensitizers, light-emitting devices, targeted cancer treatment, radiation therapy, photodynamic therapy etc.

In the field of sensing and biosensing, the advantageous features of fluorescence emission of Au/Ag NCs from visible to NIR region, water solubility and surface with rich functionalities are used for designing nanoprobes for the detection of various analytes, such as metal ions, anions, small biomolecules, drug, pesticides, explosives, protein, enzymatic activity, DNA, nucleic acid etc. In designing nanoprobes with NCs, the majority of the NCs are synthesized by adopting the "bottom-up" approaches of reduction of Au^{3+}/Ag^+ precursor in the presence of a suitable template, compared to the chemical etching methods.

There are multiple options for choosing appropriate templates, such as thiolated small molecules, peptides, DNA, proteins and enzymes, polymer, dendrimer etc. The Au/Ag NCs properties can be tuned by doping other metal atoms, including capping multiple different ligands. Size is ultra-small, but there is numerous chemistry at cluster core, cluster surface and at the interface of cluster core and surface that can be focused to tune the properties of NCs for desired applications. However, there are still some challenges remain that need to be resolved. For example, the challenge still remains in the synthesis of monodisperse NCs with improved stability. Despite multiple claims on the biocompatibility of Au/Ag NCs, toxicity remains a big concern when the applications are based on biomedical. The majority of the available ligands are used in the synthesis of NCs. Future research is required on the designing and synthesizing new surface protecting ligands for preparing NCs with advantageous features. The biogenic sources may also be explored for the synthesis of NCs with improved biocompatibility for biomedical applications. This chapter will definitely be useful in opening new directions on the synthesis and applications of Au/Ag NCs and their alloy NCs.

References

[1] A. Han, S. Hao, Y. Yang, X. Li, X. Luo, G. Fang, J. Liu, S. Wang, Perspective on recent developments of nanomaterial based fluorescent sensors: Applications in safety and quality control of food and beverages, J. Food Drug Anal. 28 (4) (2020) Article 2.

[2] S.K. Sahoo, Sensing and Biosensing With Optically Active Nanomaterials: A note, sensing and biosensing with optically active nanomaterials, Elsevier, Philadelphia, 2022, pp. 1–7.

[3] Y.-S. Lin, Y.-F. Lin, A. Nain, Y.-F. Huang, H.-T. Chang, A critical review of copper nanoclusters for monitoring of water quality, Sens. Actuators Rep. 3 (2021) 100026.

[4] L. Shang, S. Dong, G.U. Nienhaus, Ultra-small fluorescent metal nanoclusters: Synthesis and biological applications, Nano Today. 6 (4) (2011) 401–418.

[5] J. Xu, X. Zhu, X. Zhou, F.Y. Khusbu, C. Ma, Recent advances in the bioanalytical and biomedical applications of DNA-templated silver nanoclusters, TrAC Trends Anal. Chem. 124 (2020) 115786.

[6] Y. Zheng, L. Lai, W. Liu, H. Jiang, X. Wang, Recent advances in biomedical applications of fluorescent gold nanoclusters, Adv. Colloid Interface Sci. 242 (2017) 1–16.

[7] Y. Zhang, C. Zhang, C. Xu, X. Wang, C. Liu, G.I. Waterhouse, Y. Wang, H. Yin, Ultrasmall Au nanoclusters for biomedical and biosensing applications: A mini-review, Talanta. 200 (2019) 432–442.

[8] Z. Wang, B. Chen, A.L. Rogach, Synthesis, optical properties and applications of light-emitting copper nanoclusters, Nanoscale Horiz. 2 (3) (2017) 135–146.

[9] L. Liu, A. Corma, Metal catalysts for heterogeneous catalysis: From single atoms to nanoclusters and nanoparticles, Chem. Rev. 118 (10) (2018) 4981–5079.

[10] X. Kang, Y. Li, M. Zhu, R. Jin, Atomically precise alloy nanoclusters: Syntheses, structures, and properties, Chem. Soc. Rev. 49 (17) (2020) 6443–6514.

[11] Y. Tao, M. Li, J. Ren, X. Qu, Metal nanoclusters: Novel probes for diagnostic and therapeutic applications, Chem. Soc. Rev. 44 (23) (2015) 8636–8663.

[12] D. Li, Z. Chen, X. Mei, Fluorescence enhancement for noble metal nanoclusters, Adv. Colloid Interface Sci. 250 (2017) 25–39.

[13] N. Goswami, K. Zheng, J. Xie, Bio-NCs–the marriage of ultrasmall metal nanoclusters with biomolecules, Nanoscale. 6 (22) (2014) 13328–13347.

[14] L. Zhu, M. Gharib, C. Becker, Y. Zeng, A.R. Ziefuß, L. Chen, A.M. Alkilany, C. Rehbock, S. Barcikowski, W.J. Parak, Synthesis of fluorescent silver nanoclusters: Introducing bottom-up and top-down approaches to nanochemistry in a single laboratory class, J. Chem. Educ. 97 (1) (2019) 239–243.

[15] X. Huang, Z. Li, Z. Yu, X. Deng, Y. Xin, Recent advances in the synthesis, properties, and biological applications of platinum nanoclusters, J. Nanomat. 2019 (2019) Article ID 6248725.

[16] Q. Yao, Z. Wu, Z. Liu, Y. Lin, X. Yuan, J. Xie, Molecular reactivity of thiolate-protected noble metal nanoclusters: Synthesis, self-assembly, and applications, Chem. Sci. 12 (1) (2021) 99–127.

[17] L. Farzin, M. Shamsipur, L. Samandari, S. Sadjadi, S. Sheibani, Biosensing strategies based on organic-scaffolded metal nanoclusters for ultrasensitive detection of tumor markers, Talanta. 214 (2020) 120886.

[18] R.L. Pérez, M. Cong, S.R. Vaughan, C.E. Ayala, W.I.S. Galpothdeniya, J.K. Mathaga, I.M. Warner, Protein discrimination using a fluorescence-based sensor array of thiacarbocyanine-GUMBOS, ACS Sens. 5 (8) (2020) 2422–2429.

[19] J.M. Obliosca, C. Liu, H.-C. Yeh, Fluorescent silver nanoclusters as DNA probes, Nanoscale. 5 (18) (2013) 8443–8461.

[20] L.-Y. Chen, C.-W. Wang, Z. Yuan, H.-T. Chang, Fluorescent gold nanoclusters: Recent advances in sensing and imaging, Anal. Chem. 87 (1) (2015) 216–229.

[21] D. Bain, S. Maity, A. Patra, Opportunities and challenges in energy and electron transfer of nanocluster based hybrid materials and their sensing applications, Phys. Chem. Chem. Phys. 21 (11) (2019) 5863–5881.

[22] A. Latorre, A. Somoza, DNA-mediated silver nanoclusters: Synthesis, properties and applications, ChemBioChem 13 (7) (2012) 951–958.

[23] H. Li, W. Zhu, A. Wan, L. Liu, The mechanism and application of the protein-stabilized gold nanocluster sensing system, Analyst. 142 (4) (2017) 567–581.

[24] F. Meng, H. Yin, Y. Li, S. Zheng, F. Gan, G. Ye, One-step synthesis of enzyme-stabilized gold nanoclusters for fluorescent ratiometric detection of hydrogen peroxide, glucose and uric acid, Microchem. J. 141 (2018) 431–437.

[25] F. Mo, Z. Ma, T. Wu, M. Liu, Y. Zhang, H. Li, S. Yao, Holey reduced graphene oxide inducing sensitivity enhanced detection nanoplatform for cadmium ions based on glutathione-gold nanocluster, Sens. Actuators B 281 (2019) 486–492.

[26] T. Tanziela, S. Shaikh, F. ur Rehman, F. Semcheddine, H. Jiang, Z. Lu, X. Wang, Cancer-exocytosed exosomes loaded with bio-assembled AgNCs as smart drug carriers for targeted chemotherapy, Chem. Eng. J. 440 (2022) 135980.

[27] J.F. Hicks, A.C. Templeton, S. Chen, K.M. Sheran, R. Jasti, R.W. Murray, J. Debord, T.G. Schaaff, R.L. Whetten, The monolayer thickness dependence of quantized double-layer capacitances of monolayer-protected gold clusters, Anal. Chem. 71 (17) (1999) 3703–3711.

[28] T.-H. Chen, C.-C. Nieh, Y.-C. Shih, C.-Y. Ke, W.-L. Tseng, Hydroxyl radical-induced etching of glutathione-capped gold nanoparticles to oligomeric Au I–thiolate complexes, RSC Adv. 5 (56) (2015) 45158–45164.

[29] T.-H. Chen, C.-J. Yu, W.-L. Tseng, Sinapinic acid-directed synthesis of gold nanoclusters and their application to quantitative matrix-assisted laser desorption/ionization mass spectrometry, Nanoscale. 6 (3) (2014) 1347–1353.

[30] X. Jiang, B. Du, Y. Huang, J. Zheng, Ultrasmall noble metal nanoparticles: Breakthroughs and biomedical implications, Nano Today. 21 (2018) 106–125.

[31] C.-A.J. Lin, C.-H. Lee, J.-T. Hsieh, H.-H. Wang, J.K. Li, J.-L. Shen, W.-H. Chan, H.-I. Yeh, W.H. Chang, Synthesis of fluorescent metallic nanoclusters toward biomedical application: recent progress and present challenges, J. Med. Biol. Eng. 29 (6) (2009) 276–283.

[32] S. Zhang, X. Zhang, Z. Su, Biomolecule conjugated metal nanoclusters: Bio-inspiration strategies, targeted therapeutics, and diagnostics, J. Mater. Chem. B 8 (19) (2020) 4176–4194.

[33] C.-A.J. Lin, T.-Y. Yang, C.-H. Lee, S.H. Huang, R.A. Sperling, M. Zanella, J.K. Li, J.-L. Shen, H.-H. Wang, H.-I. Yeh, Synthesis, characterization, and bioconjugation of fluorescent gold nanoclusters toward biological labeling applications, ACS Nano. 3 (2) (2009) 395–401.

[34] K.G. Stamplecoskie, P.V. Kamat, Size-dependent excited state behavior of glutathione-capped gold clusters and their light-harvesting capacity, J. Am. Chem. Soc. 136 (31) (2014) 11093–11099.

[35] P. Gao, X. Chang, D. Zhang, Y. Cai, G. Chen, H. Wang, T. Wang, Synergistic integration of metal nanoclusters and biomolecules as hybrid systems for therapeutic applications, Acta. Pharm. Sin B 11 (5) (2021) 1175–1199.

[36] C.M. Aikens, Electronic and geometric structure, optical properties, and excited state behavior in atomically precise thiolate-stabilized noble metal nanoclusters, Acc. Chem. Res. 51 (12) (2018) 3065–3073.

[37] R.D. Senanayake, A.V. Akimov, C.M. Aikens, Theoretical investigation of electron and nuclear dynamics in the $[Au_{25}(SH)_{18}]^{-1}$ thiolate-protected gold nanocluster, J. Phys. Chem. C 121 (20) (2017) 10653–10662.

[38] H. Kawasaki, K. Hamaguchi, I. Osaka, R. Arakawa, ph-Dependent synthesis of pepsin-mediated gold nanoclusters with blue green and red fluorescent emission, Adv. Funct. Mater. 21 (18) (2011) 3508–3515.

[39] S. Jin, S. Wang, Y. Song, M. Zhou, J. Zhong, J. Zhang, A. Xia, Y. Pei, M. Chen, P. Li, Crystal structure and optical properties of the [Ag62S12 (SBut) 32] 2+ nanocluster with a complete face-centered cubic kernel, J. Am. Chem. Soc. 136 (44) (2014) 15559–15565.

[40] S. Choi, R.M. Dickson, J. Yu, Developing luminescent silver nanodots for biological applications, Chem. Soc. Rev. 41 (5) (2012) 1867–1891.

[41] S. Bothra, Y. Upadhyay, R. Kumar, S.A. Kumar, S.K. Sahoo, Chemically modified cellulose strips with pyridoxal conjugated red fluorescent gold nanoclusters for nanomolar detection of mercuric ions, Biosens. Bioelectron. 90 (2017) 329–335.

[42] S. Bothra, P. Paira, A. Kumar SK, R. Kumar, S.K. Sahoo, Vitamin B6 cofactor-conjugated polyethyleneimine-passivated silver nanoclusters for fluorescent sensing of Zn2+ and Cd2+ using chemically modified cellulose strips, ChemistrySelect. 2 (21) (2017) 6023–6029.

[43] R. Jin, C. Zeng, M. Zhou, Y. Chen, Atomically precise colloidal metal nanoclusters and nanoparticles: Fundamentals and opportunities, Chem. Rev. 116 (18) (2016) 10346–10413.

[44] J. Yang, R. Pang, D. Song, M.-B. Li, Tailoring silver nanoclusters via doping: Advances and opportunities, Nanoscale Adv. 3 (9) (2021) 2411–2422.

[45] Y. Sun, X. Cheng, Y. Zhang, A. Tang, X. Cai, X. Liu, Y. Zhu, Precisely modulating the surface sites on atomically monodispersed gold-based nanoclusters for controlling their catalytic performances, Nanoscale. 12 (35) (2020) 18004–18012.

[46] Q. Zhu, X. Huang, Y. Zeng, K. Sun, L. Zhou, Y. Liu, L. Luo, S. Tian, X. Sun, Controllable syntheses and electrocatalytic applications of atomically precise gold nanoclusters, Nanoscale Adv. (2021).

[47] Y.-C. Shiang, C.-C. Huang, W.-Y. Chen, P.-C. Chen, H.-T. Chang, Fluorescent gold and silver nanoclusters for the analysis of biopolymers and cell imaging, J. Mater. Chem. 22 (26) (2012) 12972–12982.

[48] Y. Zheng, J. Wu, H. Jiang, X. Wang, Gold nanoclusters for theranostic applications, Coord. Chem. Rev. 431 (2021) 213689.

[49] Z. Liu, Z. Wu, Q. Yao, Y. Cao, O.J.H. Chai, J. Xie, Correlations between the fundamentals and applications of ultrasmall metal nanoclusters: Recent advances in catalysis and biomedical applications, Nano Today. 36 (2021) 101053.

[50] Z. Qiao, J. Zhang, X. Hai, Y. Yan, W. Song, S. Bi, Recent advances in templated synthesis of metal nanoclusters and their applications in biosensing, bioimaging and theranostics, Biosens. Bioelectron. 176 (2021) 112898.

[51] N. Amaly, P. Pandey, A.Y. El-Moghazy, G. Sun, P.K. Pandey, Cationic microcrystalline cellulose–Montmorillonite composite aerogel for preconcentration of inorganic anions from dairy wastewater, Talanta 242 (2022) 123281.

[52] B. Sreenivasulu, B. Ramji, M. Nagaral, A review on graphene reinforced polymer matrix composites, Mater. Today: Proc. 5 (1) (2018) 2419–2428.

[53] J. Yang, R. Jin, New advances in atomically precise silver nanoclusters, ACS Mater. Lett. 1 (4) (2019) 482–489.

[54] Y. Du, H. Sheng, D. Astruc, M. Zhu, Atomically precise noble metal nanoclusters as efficient catalysts: A bridge between structure and properties, Chem. Rev. 120 (2) (2019) 526–622.

[55] X. Hu, Y. Zheng, J. Zhou, D. Fang, H. Jiang, X. Wang, Silver-assisted thiolate ligand exchange induced photoluminescent boost of gold nanoclusters for selective imaging of intracellular glutathione, Chem. Mater. 30 (6) (2018) 1947–1955.

[56] S. Chatterjee, X.-Y. Lou, F. Liang, Y.-W. Yang, Surface-functionalized gold and silver nanoparticles for colorimetric and fluorescent sensing of metal ions and biomolecules, Coord. Chem. Rev. 459 (2022) 214461.

[57] S. Qian, Z. Wang, Z. Zuo, X. Wang, Q. Wang, X. Yuan, Engineering luminescent metal nanoclusters for sensing applications, Coord. Chem. Rev. 451 (2022) 214268.

[58] J. Wang, S. Ma, J. Ren, J. Yang, Y. Qu, D. Ding, M. Zhang, G. Yang, Fluorescence enhancement of cysteine-rich protein-templated gold nanoclusters using silver (I) ions and its sensing application for mercury (II), Sens. Actuators B 267 (2018) 342–350.

[59] S. Xu, X. Li, Y. Mao, T. Gao, X. Feng, X. Luo, Novel dual ligand co-functionalized fluorescent gold nanoclusters as a versatile probe for sensitive analysis of Hg2+ and oxytetracycline, Anal. Bioanal.Chem. 408 (11) (2016) 2955–2962.

[60] S. Li, G. Li, H. Shi, M. Yang, W. Tan, H. Wang, W. Yang, A fluorescent probe based on tryptophan-coated silver nanoclusters for copper (II) ions detection and bioimaging in cells, Microchem. J. 175 (2022) 107222.

[61] A.-M. Hada, M. Zetes, M. Focsan, T. Nagy-Simon, A.-M. Craciun, Novel paper-based sensing platform using photoluminescent gold nanoclusters for easy, sensitive and selective naked-eye detection of Cu2+, J. Mol. Struct. 1244 (2021) 130990.

[62] T.-Y. Zhou, L.-P. Lin, M.-C. Rong, Y.-Q. Jiang, X. Chen, Silver–gold alloy nanoclusters as a fluorescence-enhanced probe for aluminum ion sensing, Anal. Chem. 85 (20) (2013) 9839–9844.

[63] S. Bothra, L.T. Babu, P. Paira, S. Ashok Kumar, R. Kumar, S.K. Sahoo, A biomimetic approach to conjugate vitamin B6 cofactor with the lysozyme cocooned fluorescent AuNCs and its application in turn-on sensing of zinc (II) in environmental and biological samples, Anal. Bioanal.Chem. 410 (1) (2018) 201–210.

[64] V. Bhardwaj, T. Anand, H.-J. Choi, S.K. Sahoo, Sensing of Zn (II) and nitroaromatics using salicyclaldehyde conjugated lysozyme-stabilized fluorescent gold nanoclusters, Microchem. J. 151 (2019) 104227.

[65] X. Liu, C. Shao, T. Chen, Z. He, G. Du, Stable silver nanoclusters with aggregation-induced emission enhancement for detection of aluminum ion, Sens. Actuators B 278 (2019) 181–189.

[66] J. An, R. Chen, M. Chen, Y. Hu, Y. Lyu, Y. Liu, An ultrasensitive turn-on ratiometric fluorescent probes for detection of Ag+ based on carbon dots/SiO2 and gold nanoclusters, Sens. Actuators B 329 (2021) 129097.

[67] B. Han, Y. Li, X. Hu, Q. Yan, J. Jiang, M. Yu, T. Peng, G. He, based visual detection of silver ions and l-cysteine with a dual-emissive nanosystem of carbon quantum dots and gold nanoclusters, Anal. Methods 10 (32) (2018) 3945–3950.

[68] S. Roy, G. Palui, A. Banerjee, The as-prepared gold cluster-based fluorescent sensor for the selective detection of As III ions in aqueous solution, Nanoscale 4 (8) (2012) 2734–2740.

[69] Y. Liu, K. Ai, X. Cheng, L. Huo, L. Lu, Gold-nanocluster-based fluorescent sensors for highly sensitive and selective detection of cyanide in water, Adv. Funct. Mater. 20 (6) (2010) 951–956.

[70] L. Wang, G. Chen, G. Zeng, J. Liang, H. Dong, M. Yan, Z. Li, Z. Guo, W. Tao, L. Peng, Fluorescent sensing of sulfide ions based on papain-directed gold nanoclusters, New J. Chem. 39 (12) (2015) 9306–9312.

[71] H. Sun, D. Lu, M. Xian, C. Dong, S. Shuang, A lysozyme-stabilized silver nanocluster fluorescent probe for the detection of sulfide ions, Anal. Methods. 8 (22) (2016) 4328–4333.

[72] P. Chen, X. Xu, J. Ji, J. Wu, T. Lu, Y. Xia, L. Wang, J. Fan, Y. Jin, L. Zhang, Specific and visual assay of iodide ion in human urine via redox pretreatment using ratiometric fluorescent test paper printed with dimer DNA silver nanoclusters and carbon dots, Anal. Chim. Acta 1138 (2020) 99–107.

[73] H.-H. Deng, K.-Y. Huang, M.-J. Zhang, Z.-Y. Zou, Y.-Y. Xu, H.-P. Peng, W. Chen, G.-L. Hong, Sensitive and selective nitrite assay based on fluorescent gold nanoclusters and Fe2+/Fe3+ redox reaction, Food Chem. 317 (2020) 126456.

[74] X. Wu, P. Wu, M. Gu, J. Xue, Ratiometric fluorescent probe based on AuNCs induced AIE for quantification and visual sensing of glucose, Anal. Chim. Acta 1104 (2020) 140–146.

[75] S. Liu, C. Lv, R. Liu, G. Yang, S. Li, L. Zuo, P. Xue, A label-free "turn-on" fluorescence platform for glucose based on AuNCs@ MnO 2 nanocomposites, New J. Chem. 43 (33) (2019) 13143–13151.

[76] Y. Chen, T. Feng, L. Chen, Y. Gao, J. Di, Bovine serum albumin stabilized silver nanoclusters as "signal-on" fluorescent probe for detection of hydrogen peroxide and glucose, Opt. Mater. 114 (2021) 111012.

[77] R. Rajamanikandan, M. Ilanchelian, Highly selective and sensitive biosensing of dopamine based on glutathione coated silver nanoclusters enhanced fluorescence, New J. Chem. 41 (24) (2017) 15244–15250.

[78] T. Zhou, Z. Su, Y. Tu, J. Yan, Determination of dopamine based on its enhancement of gold-silver nanocluster fluorescence, Spectrochim. Acta Part A 252 (2021) 119519.

[79] Y. Zhang, H. Xu, Y. Yang, F. Zhu, Y. Pu, X. You, X. Liao, Efficient fluorescence resonance energy transfer-based ratiometric fluorescent probe for detection of dopamine using a dual-emission carbon dot-gold nanocluster nanohybrid, J. Photochem. Photobiol. A 411 (2021) 113195.

[80] X. Liu, X. Hou, Z. Li, J. Li, X. Ran, L. Yang, Water-soluble amino pillar [5]arene functionalized gold nanoclusters as fluorescence probes for the sensitive determination of dopamine, Microchem. J. 150 (2019) 104084.

[81] Z. Chen, D. Lu, Z. Cai, C. Dong, S. Shuang, Bovine serum albumin-confined silver nanoclusters as fluorometric probe for detection of biothiols, Luminescence. 29 (7) (2014) 722–727.

[82] S. Zhang, B. Lin, Y. Yu, Y. Cao, M. Guo, L. Shui, A ratiometric nanoprobe based on silver nanoclusters and carbon dots for the fluorescent detection of biothiols, Spectrochim. Acta Part A 195 (2018) 230–235.

[83] W. Wang, J. Li, J. Fan, W. Ning, B. Liu, C. Tong, Ultrasensitive and non-labeling fluorescence assay for biothiols using enhanced silver nanoclusters, Sens. Actuators B 267 (2018) 174–180.

[84] Y. Upadhyay, R. Kumar, A.K. SK, S.K. Sahoo, Vitamin B6 cofactors conjugated ovalbumin-stabilized gold nanoclusters: Application in alkaline phosphatase activity detection and generating white-light emission, Microchem. J. 156 (2020) 104859.

[85] Y. Upadhyay, R. Kumar, S.K. Sahoo, Developing a cost-effective bioassay to detect alkaline phosphatase activity and generating white light emission from a single nano-assembly by conjugating vitamin B6 cofactors with lysozyme-stabilized fluorescent gold nanoclusters, ACS Sustain. Chem. Eng. 8 (10) (2020) 4107–4113.

[86] H. Liu, M. Li, Y. Xia, X. Ren, A turn-on fluorescent sensor for selective and sensitive detection of alkaline phosphatase activity with gold nanoclusters based on inner filter effect, ACS Appl. Mater. Interfaces 9 (1) (2017) 120–126.

[87] M. Wang, S. Wang, L. Li, G. Wang, X. Su, β-Cyclodextrin modified silver nanoclusters for highly sensitive fluorescence sensing and bioimaging of intracellular alkaline phosphatase, Talanta. 207 (2020) 120315.

[88] W. Song, R.-P. Liang, Y. Wang, L. Zhang, J.-D. Qiu, Gold nanoclusters-based dual-emission ratiometric fluorescence probe for monitoring protein kinase, Sens. Actuators. B 226 (2016) 144–150.

[89] W. Li, W. Li, Y. Hu, Y. Xia, Q. Shen, Z. Nie, Y. Huang, S. Yao, A fluorometric assay for acetylcholinesterase activity and inhibitor detection based on DNA-templated copper/silver nanoclusters, Biosens. Bioelectron. 47 (2013) 345–349.

[90] X.-P. Zhang, C.-X. Zhao, Y. Shu, J.-H. Wang, Gold nanoclusters/iron oxyhydroxide platform for ultrasensitive detection of butyrylcholinesterase, Anal. Chem. 91 (24) (2019) 15866–15872.

[91] W. Li, Z. Gao, R. Su, W. Qi, L. Wang, Z. He, Scissor-based fluorescent detection of pepsin using lysozyme-stabilized Au nanoclusters, Anal. Methods. 6 (17) (2014) 6789–6795.

[92] J. Luo, A. Rasooly, L. Wang, K. Zeng, C. Shen, P. Qian, M. Yang, F. Qu, Fluorescent turn-on determination of the activity of peptidases using peptide templated gold nanoclusters, Microchim. Acta 183 (2) (2016) 605–610.

[93] L. Wang, K. Ma, Y. Zhang, Label-free fluorometric detection of S1 nuclease activity by using polycytosine oligonucleotide-templated silver nanoclusters, Anal. Biochem. 468 (2015) 34–38.

[94] X. Yan, H. Li, T. Hu, X. Su, A novel fluorimetric sensing platform for highly sensitive detection of organophosphorus pesticides by using egg white-encapsulated gold nanoclusters, Biosens. Bioelectron. 91 (2017) 232–237.

[95] B. Liang, L. Han, Displaying of acetylcholinesterase mutants on surface of yeast for ultra-trace fluorescence detection of organophosphate pesticides with gold nanoclusters, Biosens. Bioelectron. 148 (2020) 111825.

[96] Q. Lu, T. Zhou, Y. Wang, L. Gong, J. Liu, Transformation from gold nanoclusters to plasmonic nanoparticles: A general strategy towards selective detection of organophosphorothioate pesticides, Biosens. Bioelectron. 99 (2018) 274–280.

[97] J.R. Zhang, Y.Y. Yue, H.Q. Luo, N.B. Li, Supersensitive and selective detection of picric acid explosive by fluorescent Ag nanoclusters, Analyst 141 (3) (2016) 1091–1097.

[98] X. Yang, J. Wang, D. Su, Q. Xia, F. Chai, C. Wang, F. Qu, Fluorescent detection of TNT and 4-nitrophenol by BSA Au nanoclusters, Dalton Trans. 43 (26) (2014) 10057–10063.

[99] S.M. Anju, R.K. Anjana, N. Vijila, A. Aswathy, J. Jayakrishna, B. Anjitha, J. Anjalidevi, S. Adhya, S. George, Tb-doped BSA–gold nanoclusters as a bimodal probe for the selective detection of TNT, Anal. Bioanal.Chem. 412 (17) (2020) 4165–4172.

[100] Y. Zhao, M. Pan, F. Liu, Y. Liu, P. Dong, J. Feng, T. Shi, X. Liu, Highly selective and sensitive detection of trinitrotoluene by framework-enhanced fluorescence of gold nanoclusters, Anal. Chim. Acta. 1106 (2020) 133–138.

CHAPTER 6

Array-based sensing using gold and silver nanoparticles

Forough Ghasemi[a], **Samira Abbasi-Moayed**[b], **Zahra Jafar-Nezhad Ivrigh**[b], **M. Reza Hormozi-Nezhad**[b]

[a]*Department of Nanotechnology, Agricultural Biotechnology Research Institute of Iran (ABRII), Agricultural Research, Education, and Extension Organization (AREEO), Karaj, Iran* [b]*Department of Chemistry, Sharif University of Technology, Tehran, Iran*

6.1 Array sensing

6.1.1 Design

Sensors can be classified into two categories in terms of recognition elements: (1) the ones with specific recognition elements and (2) the ones with cross-reactive recognition elements. In the first category (Fig. 6.1A), a highly selective probe is designed to respond to one particular analyte based on the strategy so-called "lock-and-key". Though sensitive and specific, these ideal sensors usually require expensive sensing units such as enzymes, antibodies, aptamers, etc. Moreover, time-consuming synthesis/design processes might be needed to achieve such specific sensors [1]. In the second category (Fig. 6.1B), more than one analyte can be detected and identified by using a group of cross-reactive recognition elements arranged in an array format which is called a "sensor array". The sensor array mimics the mammalian gustatory/olfactory principle in which all responses collected from nonspecific receptors (cross-reactive or semiselective recognition elements) represent a fingerprint-like pattern for each test/odor [2].

Generally, three steps must be taken to develop a sensor array: (1) selection and design of sensor elements (recognition elements) which is the principal step in developing a sensor array. Sensor elements should be as cross-reactive as possible to produce an expanded response space. This sensor array can respond to a large number of similar analytes even in a complex matrix, which is the aim of utilizing a sensor array instead of a specific sensor. However, partial selectivity toward some particular analytes is needed to avoid highly correlated data and low verification and to ameliorate the accuracy of discrimination. In addition to the cross-reactivity issue, the number of sensor elements must also be given special attention. The number of sensor elements should be chosen in an optimum number to avoid overfitting (in case of a high number of sensor elements) or unacceptable discrimination (in case of a low number of sensor elements) [4]. (2) Selection of a data acquisition method to record signals

Gold and Silver Nanoparticles.
DOI: https://doi.org/10.1016/B978-0-323-99454-5.00008-1

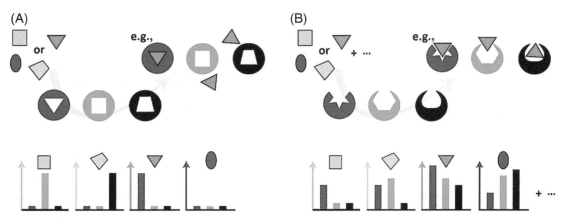

Fig. 6.1: Schematic illustration of (A) three lock-and-key sensors in which each sensor specifically interacts with only one analyte and (B) a sensor array in which three cross-reactive sensors interact with several analytes (could be more than three) and produce a distinctive pattern for each analyte [3].

from the interaction of sensor elements with the analytes. We will describe these methods in Section 6.1.2. (3) Selection of a method/a few methods to analyze the collected multivariate data, illustrate the discrimination among analytes, and identify the unknown analyte(s). This topic will be covered in Section 6.1.3.

6.1.2 Signal readout

In optical sensor arrays based on silver and gold nanoparticles (AgNPs and AuNPs), the interaction of electromagnetic radiation with the sensor elements in the presence of target analytes is surveyed in the visible/ultraviolet region, depending on the type of nanomaterials selected as sensor elements. Absorption or fluorescence, and in some cases phosphorescence, chemiluminescence, scattering, reflection, or refraction are the signals that are recorded as sensor array responses. For example, in colorimetric sensors, the electric field of an electromagnetic beam interacts with the conduction band of electrons of NPs and creates a specific oscillation frequency called localized surface plasmon resonance (LSPR) which leads to the appearance of a particular color for the NPs [5]. The spectral position of the LSPR peak is highly dependent on the size, shape, type, and composition of the metal NPs, the density state (distance between the particles), and the refractive index of the dielectric medium around the NPs [6]. By altering any of the predetermined factors, the analyte causes changes in the LSPR peak and the corresponding color of metallic NPs. These changes can be considered colorimetric sensor array responses. The LSPR band of AuNPs and AgNPs is highly strong compared to other plasmonic NPs, introducing them as prominent components in the design of calorimetric sensors.

In fluorimetric sensors, the emission intensity of the fluorescence sensor elements in the presence of the analytes decreases or increases due to the quenching effect or removal of the quenching effect, respectively. In addition, the fluorescence peak shift to shorter or longer wavelengths can also be considered as readout signals for these sensors. Gold and silver nanoclusters (AuNCs and AgNCs) with size-dependent properties, high biocompatibility, the possibility of excitation by visible light, and emission with high Stokes shifts are potential candidates to design fluorescence sensor elements [7].

As mentioned, in optical sensor arrays, one method to collect response signals is recording alters in the absorption or emission profiles of sensor elements before and after interacting with the analyte using a spectrophotometer or spectrofluorimeter. Pixel-by-pixel digital subtraction of red/green/blue (RGB) values, which are collected from images taken by a smartphone from the sensor elements before and after the addition of the analyte, is another way to display analytical signals. This method is mainly employed in portable sensors in which the sensor elements are immobilized on a substrate. In this case, a color difference map is created for each analyte, which is a unique response pattern like a fingerprint for each analyte. It is important to note that the brightness of the environment during shooting should be constant so as not to affect the final results. Finally, using statistical data analysis methods, these response patterns are attributed to an analyte, and the target analytes can be distinguished from each other [8]. The data analysis methods will be explained in the next section.

6.1.3 Signal analysis

There are usually some cross-reactive sensing elements in the array-based sensors that produce multiple response signals for each analyte. Especially in plasmonic NPs based sensor arrays, the variation of many wavelengths for each sensing element is chosen which produces a large-scale multivariate data matrix. Therefore, data visualization and data analysis methods are necessary to extract the most representative data to determine and discriminate target analytes. In the following, data visualization approaches and some common pattern recognition methods will be described briefly.

6.1.3.1 Data visualization

Simple and user-friendly plots according to significant differences in colors or signals can provide unique graphical patterns for each analyte to discriminate them visually. Bar plots, heatmaps, radar plots, and color difference maps are the most useful approaches that are generally employed to follow the profiles of signals collected in the designed sensor array (Fig. 6.2). These visual results can help the evaluation of the arrays' response and their ability to discriminate the class or concentration of analytes [4]. There is a brief introduction of each data visualization method in the following.

6.1.3.1.1 Bar Plots

As illustrated in Fig. 6.2A and B, bar plots provide a district pattern based on two subsets: the response profiles of each sensor element grouped by analytes or analytes clustered by each sensor element.

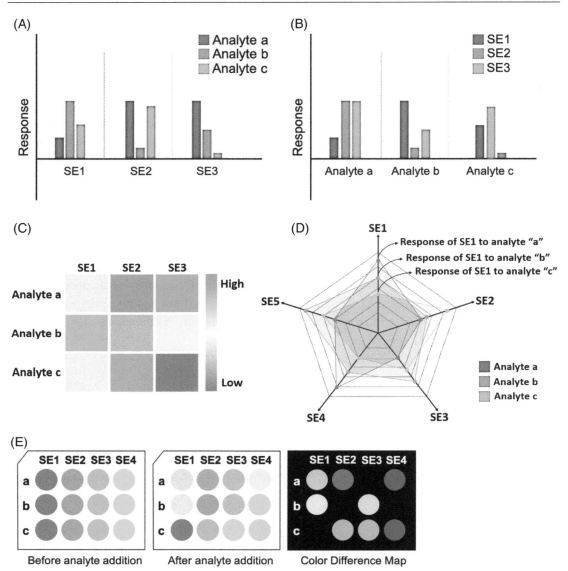

Fig. 6.2: Bar plots illustrating (A) cross-reactivity in sensor elements and (B) unique response profile for each analyte. (C) A heatmap illustrating the unique response profiles for each analyte via thermometer color pallets. (D) Schematic illustration of a radar plot and (E) a color difference map (CDM) [4].

6.1.3.1.2 Heat maps

The heat map is also a simple way for the visual illustration of response profiles of sensor elements toward each analyte. In this approach, raw data are utilized to create color pallets. This is much easier and more convenient to follow colors rather than a set of numerical values, especially when the variations are not enough significant in the responses profiles. Therefore, the response signals are magnitude by color in a heat map. Fig. 6.2C illustrates a heat map for three different analytes using three sensor elements. In this heat map, each column represents the response of each sensor element and each row displays a distinctive color pattern corresponding to each analyte; thus, each analyte is visually distinguished from the others [4]. The red color shows a high signal value while decreasing in signal value, the color changes to blue.

6.1.3.1.3 Cluster heat maps

A heat map can be merged with a cluster analysis dendrogram to produce a cluster heat map. A cluster heat map contains a colorful pallet like a heat map in which cluster analysis is performed on both sample and sensor element direction at the same time. This cluster analysis in the rows and columns direction is achieved based on the proximity of the similar rows and columns [4]. It should be mentioned that it is not mandatory to apply clustering analysis in both rows and columns in cluster heatmaps, but it depends on the clustering aim.

6.1.3.1.4 Radar plots

A radar plot also called a polar plot has been more recently utilized to display the raw data of the response profile of the sensor array. Fig. 6.2D demonstrates a radar plot for discriminating three analytes with five sensor elements. Each vector of the radar plot represents the responses of each sensor element. The signal of each sensor element is different in the presence of each analyte, thus the overall shape of the radar plot is varied [4].

6.1.3.1.5 Color difference maps

The signal difference of sensor elements before and after the addition of analytes can also be visually shown by colorful spots called color difference maps (CDM) [4,9]. The CDM provides a visual representation of large-scale data in sensor array by evaluation of red/green/blue (RGB) values, before and after the addition of analytes. If there is no color variation in the colorimetric response of sensing elements before and after the addition of analytes, the corresponding color spot of CDM shows a black spot. However, if obvious color alternation happens during the interaction of an analyte with a sensor element, a colorful spot appears in the CDM (Fig. 6.2E). There are two scenarios to create these CDMs. The first one is used when the data is collected with photographs of sensor elements solution before and after the addition of analyte. Therefore, ΔRGB values are extracted directly by subtracting the difference in the RGB of taken images to establish the CDM.

The second strategy is based on using spectral information upon the interaction of the sensor elements and the analytes. In this approach, RGB values can be produced by mathematical transformation of absorbance spectra information at different wavelengths into corresponding RGB values. These ΔRGB values belong to the differences in absorbance spectrum information in the absence and presence of the analyte, which create colorful spots in the CDM. From both first and second strategies, significant or slight interaction between analytes and sensing elements would lead to produce bright, colorful, or black spots in CDM. It is worth mentioning that the colorful spots provided by two different strategies may have RGB values. However, both of them can provide a unique colorful pattern for each analyte.

6.1.3.2 Pattern recognition methods

The raw data collected by the previously mentioned readout system usually show a unique fingerprint pattern that can be applied to discriminate among various analytes with the help of pattern recognition methods. These pattern recognition methods classify the samples into distinct groups according to their similarity. The pattern recognition methods are generally categorized into two types, which are called, supervised and unsupervised approaches. In the supervised method, the model is built to classify the sample by knowing their class membership, and later it is used to identify the unknown sample. However, there is no expectation about the class of samples in unsupervised methods. Thus, the unsupervised methods can just classify the known sample and do not have a prediction ability to recognize the unknown sample [2,4,10,11]. In the following, some important supervised or unsupervised pattern recognition methods, which are often used in the array-based sensors, are briefly introduced.

Principle component analysis (PCA). PCA is employed as the simplest unsupervised pattern recognition method in the data analysis of sensor arrays. This method reduces the dimension of data by transforming the initial variables which might be partially correlated to each other, into a set of linearly uncorrelated variables called principal components (PCs). In the new orthogonal space that is defied by PCs, the patterns in the data are provided with fewer components and finally, samples with similar behavior are displayed as data points that can be clustered near each other in the PCA graph. Moreover, the loading values indicate the degree of the contributions of the sensor elements in the PCs, thus determine the significance of sensor elements [12,13].

Hierarchical cluster analysis (HCA). HCA is an unsupervised pattern recognition method in which similar samples are clustered based on the distances (the Euclidean distances) of their data points. Analytes with shorter distances between their data points have shown more similarity and will be grouped together. Then, clusters are made by merging the next analyte or cluster with high similarity to an existing cluster. The result of HCA is a hierarchical diagram named a dendrogram in which a visual illustration of clustering is achieved. However, when a sample is added to the data set the whole process of clustering should be done again which is a major drawback in HCA [13,14].

Linear discriminant analysis (LDA). LDA is one of the most commonly used supervised pattern recognition methods in which known samples with predefined classes' identifiers are considered as training matrices. In the training step, the linear combinations of the initial variables (sensor elements) are employed to maximize the between-class variance and minimize the within-class variance. In the next step, an unknown sample can be appertained to a predefined class according to its similar response to a specific class in training data [15,16].

6.1.4 Application area

Sensor arrays have attracted considerable interest for various applications in the fields of beverage assessment, analysis of agricultural and food products [17], environmental monitoring [18], detection of biological and chemical hazards [19,20], point-of-care assays [21], pharmaceutical purposes [22], and disease diagnostics [3,23]. Optical sensor arrays can be designed by different kinds of chemoresponsive dyes and nanomaterials which have found extensive applications, however, they are beyond the topic of this chapter. For more details on these types of sensor arrays, the readers are referred to the publications by pioneers in the field [2–4,24]. In this chapter, we will particularly focus on the application of gold and silver NPs-based sensor arrays (mainly gold and silver NPs and NCs).

Gold and silver NPs and NCs (i.e., gold NPs (AuNPs), silver NPs (AgNPs), gold NCs (AuNCs), and silver NCs (AgNCs)) are promising sensing elements in sensor array design owing to their physicochemical properties, easy synthesis, chemical functionalization, and color visualization. A variety of analytes including metal ions [25,26], bacteria [27], catecholamine neurotransmitters [28], amino acids [29], peptides [30], proteins [31,32], pesticides [33,34], opioids [35], biogenic amines [36], cancer cells [37], nitrophenol isomers [38], antioxidant [39] etc. have been efficiently detected and discriminated by gold and silver NPs/NCs-based sensor arrays. In the following sections, we will cover colorimetric, luminescent (fluorescence and chemiluminescence), and multi-channel sensor arrays, in which AuNPs, AgNPs, AuNCs, and AgNCs are employed as fundamental materials to design cross-reactive recognition elements. We will highlight the sensing mechanisms while discussing the diverse applications of gold and silver NPs/NCs-based sensor arrays.

6.2 Colorimetric array sensing

Colorimetric array sensing using plasmonic NPs mainly relies on the surface plasmon resonance (SPR) behavior of NPs. The use of intrinsic plasmonic absorption of AuNPs and AgNPs in the visible region can help to monitor the colorimetric responses of the probe with the naked eye or available instruments such as smartphones and scanners. Exploitation of AuNPs/AgNPs as sensor elements in the sensor array, pay the way toward the development of sensitive colorimetric sensor arrays with wide color variation.

6.2.1 Gold nanoparticles

It is a well-known phenomenon that aggregation of AuNPs leads to a decline in the SPR band of AuNPs along with appearing a new peak in a longer wavelength. Therefore, the visible color of dispersed AuNPs changes from red to purple-blue due to the aggregation of NPs [40,41]. This causes aggregation as a desirable mechanism for designing AuNPs-based colorimetric sensor array [42]. In this regard, eight antidepressants (ADs) were discriminated by using the aggregation behavior of unmodified AuNPs [43]. As illustrated in Fig. 6.3, the cross-reactivity of sensor elements toward the target analytes has been achieved by changing pH (the composition of buffer) and ionic strength (buffer concentration). Moreover, color difference maps, LDA, and partial least-squares regression (PLS-R) provided fingerprint patterns, qualification, and quantification discrimination.

Moa and coworkers proposed an AuNPs-based colorimetric sensor array for protein discrimination [44]. In this study, DNA (A21) protected AuNPs were utilized as sensing elements. These AuNPs at high concentrations of salt exhibited different degrees of aggregation and consequently color changes after the addition of proteins. Moreover, A21-AuNPs showed different color variations after the direct addition of proteins. The growth of the A21-AuNPs due to the reduction of $HAuCl_4$ by NH_2OH as reducing agents was also different in the presence of target proteins. Therefore, using three responses of conjugated AuNPs diverse response patterns were generated for each protein that can discriminate them.

The antiaggregation mechanism is another promising approach for developing AuNPs-based colorimetric sensor arrays. In this mechanism, the target analyte has a higher affinity than the

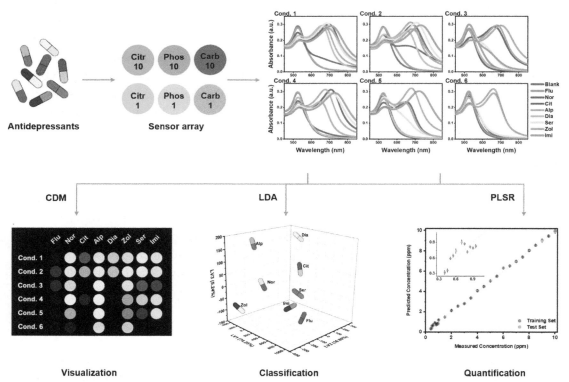

Fig. 6.3: Schematic illustration of the sensor array designed by changing pH (the composition of buffer) and ionic strength (buffer concentration), UV-Vis spectra of the sensor elements in the presence of analytes, the resulting color difference map, linear discriminant analysis (LDA), and partial least square regression (PLSR) [43].

aggregation reagent, so prevents the aggregation process [45]. For example, Najafzadeh and coworkers described an AuNPs-based antiaggregation colorimetric sensor array by using cysteine, arginine, and melamine as aggregation reagents [25]. In the presence of target analytes (Hg^{2+}, Fe^{3+}, Ag^+, and Pb^{2+}), aggregation reagents were bound to them rather than AuNPs (Fig. 6.4A). Thus, this aggregation inhibition caused the color change from blue to red by increasing the analyte concentration. The bar plots of the absorbance responses provided a fingerprint pattern for each ion and the ions were discriminated accurately by LDA (Fig. 6.4B and C).

In another study, the excellent artificial enzymatic activity, mainly, the peroxidase-like activity of AuNPs has been applied to build an efficient AuNPs-based colorimetric sensor array. Guan and coworkers have utilized the different enzymatic activity of AuNPs in the presence of Ce^{3+}, Fe^{3+}, and Cr^{3+} as three sensing elements for protein discrimination [25]. According to the dissimilar interaction of these three metal ions and phosphates ions, the peroxidase activity of AuNPs for tetramethylbenzidine (TMB) oxidation has been different. Thus, diverse color changes have occurred for the different phosphate ions. Using radar plots and heat maps, the developed AuNPs-based colorimetric sensor array discriminated among phosphate ions at a concentration of 25 μmol L^{-1}.

The other reports on using diverse AuNPs with different sizes or capping agents as colorimetric sensor elements are listed in Table 6.1.

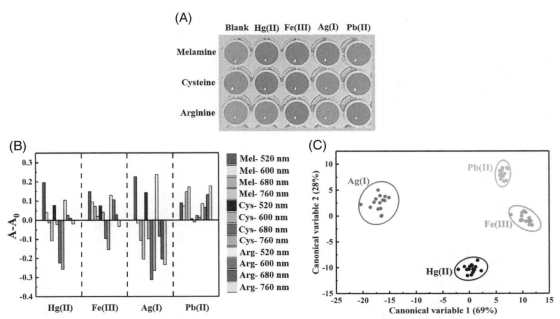

Fig. 6.4: (A) The color of sensor elements in the absence and presence of Hg^{2+}, Ag^+, Fe^{3+}, and Pb^{2+}. (B) A bar plot of sensor elements responses at 4 wavelengths against the ions at a concentration of 20 μmol L^{-1}. (C) Two-dimensional score plot illustrating the discrimination of the ions at a wide range of concentrations [25].

Table 6.1: Summary of colorimetric array sensing based on AuNPs.

Analytes	Mechanism	Sensor element	Data analysis method	Real sample	Refs.
Proteins (trypsine, human serum albumin, lysozyme, pepsin, egg white albumin, horseradish peroxidase, transferring, cytochrome c, bovine serum albumin, immunoglobulin G, hemoglobin, and concanavalin)	Aggregation	AuNPs in the presence of four concentration of NaCl	LDA and HCA	Urine	[46]
Proteins (human serum albumin, papain, hemoglobin, pepsin, lysozyme, egg white albumin, and myoglobin)	Aggregation	Three-aptamer protected AuNPs	LDA	Human cell	[47]
Metal ions (Fe^{3+}, Al^{3+}, Cd^{2+}, Cu^{2+}, Pb^{2+}, Hg^{2+}, and Cr^{3+})	Aggregation	11-Mercaptoundecanoic acid-capped AuNPs and five amino acids (cysteine, tyrosine, histidine, lysine, and 4-arginine)	HCA	Water	[48]
Adenosine, estradiol, riboflavin, and cholic acid	Aggregation	Four aptamer functionalized AuNPs	PCA	-	[49]
Cysteine, glutathione disulfide, and glutathione	Aggregation	AuNPs with various coatings (citrate, cetyl trimethylammonium bromide, and borohydride)	PCA and HCA	Blood and plasma	[50]
Pesticides (pirimiphos-methyl, chlorpyrifos, fenamiphos, phosalone, and azinphos-methyl)	Aggregation	AuNPs at various pHs and ionic strengths	LDA and HCA	Rice sample	[51]
Proteins (transferring, trypsin, hemoglobin, pepsine, bovine serum albumin, egg white albumin, cytochrome c, human serum albumin, myoglobin, and immunoglobulin G)	Aggregation and growth	Two different DNA functionalized AuNPs	LDA and HCA	Urine	[52]
Proteins (hemoglobin, human serum albumin, immunoglobulin G, bovine serum albumin, egg white albumin, and myoglobin)	Aggregation and growth	DNA protected AuNPs	LDA	Urine	[44]

Analytes	Mechanism	Sensor element	Data analysis method	Real sample	Refs.
Proteins (human serum albumin, trypsin, hemoglobin, bovine serum albumin, fibrinogen, transferring, pepsin, papain, horseradish peroxidase, cytochrome c, and concanavalin)	Catalytic activity of AuNPs for the reaction of 4-nitrophenol and $NaBH_4$	Three nonspecific DNA strands protected AuNPs	LDA and PCA	Human serum	[53]
Opioids (methadone, morphine, tramadol, codeine, oxycodone, noroxycodone, and thebaine)	Aggregation	AuNPs in four particle sizes	PCA and HCA	Urine	[35]
Proteins (trypsin, hemoglobin, lysozyme, bovine albumin, human serum albumin, pepsin, egg albumin, papain, myoglobin, horseradish peroxidase, transferrin, and immunoglobulin G)	Aggregation	AuNPs/CDs synthesized with different amino acids (lysine, glycine, and serine)	LDA	Urine	[54]
Proteins (urate oxidase, lysozyme, concanavalin, bovine serum albumin, hemoglobin, lipase, horseradish peroxidase, egg white albumin, pepsin, and transferrin)	Aggregation	DNA strands (15A, 15C, and 15T)/AuNPs	LDA	Serum	[55]
Aminoglycoside antibiotics (ribostamycin, vtobramycin, gentamicin, streptomycin, and kanamycin)	Aggregation	Two ssDNA-conjugated AuNPs	LDA	Milk	[56]
Ions (Cu^{2+}, Cd^{2+}, Cr^{3+}, Ba^{2+}, Mn^{2+}, Fe^{3+}, $Cr_2O_7^{2-}$, Sn^{4+}, Ni^{2+}, and Pb^{2+})	Aggregation	AuNPs/amino acids (L-lysine, L-histidine, L-methionine, and D-penicillium)	PLS-DA	River water	[57]
Proteins (trypsin, papain, hemoglobin, transferring, and horseradish peroxidase)	Catalytic activity on the TMB oxidation	Three DNA protected AuNPs	LDA	Urine	[58]

(Continued)

Table 6.1: (Cont'd).

Analytes	Mechanism	Sensor element	Data analysis method	Real sample	Refs.
Microorganism (staphylococcus aureus, shigella flexneri, staphylococcus epidermidis, pseudomonas aeruginosa, listeria monocytogenes, bacillus aceticus, Escherichia coli, bacillus subtilis, salmonella paratyphi, enterobacter sakazaki, vibrio parahemolyticus, candida albicans, clostridium putrefaciens, aspergillus flavus, and penicillium)	Aggregation	Mercaptopropionic acid capped AuNPs, mercaptosuccinic acid capped AuNPs, cysteamine capped AuNPs, and cetyltrimethylammonium bromide capped AuNPs	LDA	-	[59]
Proteins (bovine serum albumin, hemoglobin, egg white albumin, cytochrome c, pepsin, myoglobin, trypsin, and concanavalin)	Anti-etching	Three DNA functionalization of Cu–AuNPs	LDA	Serum	[60]
Metal ions (Fe^{3+}, Hg^{2+}, Ag^+, and Pb^{2+})	Anti-aggregation	AuNPs in the presence of three aggregation reagents (cysteine, melamine, and arginine)	LDA	Water and fish sample	[25]
Binary peptides (glycylglycine and alanylglutamine)	Anti-aggregation	Unmodified AuNPs and DNA functionalized AuNPs	PLS	Saliva sample	[61]
Proteins (trypsin, thrombin, myoglobin, urate oxidase, transferring, lysozyme, pepsin, horseradish peroxidase, egg white albumin, concanavalin, bovine serum albumin, lipase, cytochrome c, and fibrinogen)	Aggregation	Three-aptamer protected AuNPs	LDA	Human serum	[62]
Biothiols (mercaptosuccinic acid, dithiothreitol, L-cysteine, mercaptoethanol, mercaptoacetic acid, and glutathione)	Anti-aggregation	Polydopamine capped AuNPs in the presence of five metal ions (Ca^{2+}, Fe^{3+}, Ag^+, and Hg^{2+}) as five sensor elements)	LDA and HCA	Serum sample	[63]

Analytes	Mechanism	Sensor element	Data analysis method	Real sample	Refs.
Horseradish peroxidase, cytochrome C, lysozyme, myoglobin, transferrin, thrombin, hemoglobin, concanavalin, trypsin, and bovine serum albumin	Competition reaction between Zr-MOFs or proteins with AuNPs	Six DNA functionalized AuNPs with Zr-MOFs	LDA	Human serum	[64]
Metal ions (Pb^{2+}, Ti^{4+}, Fe^{3+}, Mn^{2+}, Cr^{3+}, and Sn^{4+})	Aggregation	AuNPs in the presence of three receptor unit	LDA and HCA	River water	[65]
Foodborne (acrylamide, n-tert-butylacrylamide, acetamide, nicotinamide, methacrylamide, diacetone acrylamide, and n-(hydroxymethyl) acrylamide)	Aggregation	Three functionalized (BSA, OVA, and BLG) AuNPs	PCA and HCA	Coffee sample	[66]
Foodborne pathogenic bacteria (vibrio parahaemolyticusz, staphylococcus aureus, bacillus subtilis, pseudomonas aeruginosa, shigella flexneri, escherichia coli, and salmonella typhimurium)	Aggregation	Unmodified AuNPs at low pH	-	-	[67]
Pesticides (bifenazate, azinphos-methyl, paraquat, thiometon, and parathion-methyl)	Aggregation	Two different AuNPs	LDA	Limon juice	[33]
Alkaloids berberine chloride, eserine, harmane, berberine, jatrorrhizine, palmatine, and galanthamine	Etching	Au@MnO$_2$ and MnO$_2$ nanostar	LDA	Chinese herbal medicine	[68]
Glucose, tannins, melatonin, glutathione, ascorbic acid, naringenin, tea polyphenols, and dopamine	Etching	Au@Ag NRs and Au@Ag NBPs	LDA	Human serum	[69]
Phosphates (pyrophosphate, adenosine triphosphate, adenosine monophosphate, adenosine diphosphate, and phosphate)	Catalytic activity on the TMB oxidized	AuNPs in the presence of Ce^{3+}, Fe^{2+}, and Cr^{3+}	LDA	Bovine serum	[70]
Amines (histamine, 1,6-hexadiamine, 1,7-heptanediamine, paraquat, 1,3-diaminopropane, putrescine, ethylenediamine, cadaverine, spermidine, and spermine)	Aggregation	4-Mercaptobenzoic acid, 6mercaptohexanoic acid, and 11-mercaptoundecanoic acid functionalized AuNPs	LDA	Raw fish	[71]

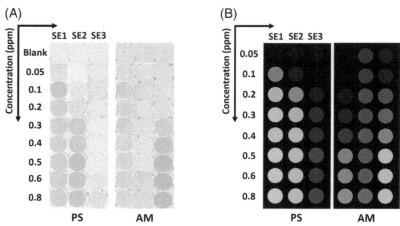

Fig. 6.5: The color of solutions and the color difference maps of sensor elements in the absence and presence of azinphos-methyl (AM) and phosalone (PS) at different concentrations [74].

6.2.2 Silver nanoparticles

As discussed earlier, most reports are based on using AuNPs as sensing elements in plasmonic NPs-based colorimetric sensor arrays and less attention has focused on using AgNPs. However, AgNPs illustrated superior advantages compared to the AuNPs with the same size because of their higher extinction coefficient. On the other hand, the simple and low-cost availability of AgNPs make them excellent candidates for designing sensing elements. During the aggregation process of AgNPs, the visible color of dispersed AgNPs alters from yellow to orange, red, and finally gray which can produce wide color variation [72–74]. Incorporating the utility of AgNPs as colorimetric sensor elements, Orouji and coworkers have developed an AgNPs-based colorimetric sensor array for identification of azinphos-methyl (AM) and phosalone (PS) pesticides [74]. As illustrated in Fig. 6.5, citrate capped AgNPs at three different pHs (4.5, 6.5, and 9.5) produced diverse color variations in the presence of two target pesticides, and the color difference map provided a distinct fingerprint for each pesticide. There are some other reports on using AgNPs to develop colorimetric sensor arrays which are summarized in Table 6.2.

6.2.3 Combination of gold and silver nanoparticles

In the design of plasmonic NPs-based colorimetric sensor arrays, the reports have shown that utilization of one kind of metal NPs has an excellent ability to discriminate diverse analytes. However, some reports employed the combination of AuNPs and AgNPs to design the

Table 6.2: Summary of colorimetric array sensing based on AgNPs.

Analytes	Mechanism	Sensor element	Data analysis method	Real Sample	Refs.
Pb^{2+} and Hg^{2+}	Aggregation for Pb^{2+} and amalgamation for Hg^{2+}	-	-	-	[75]
Catechol, hydroquinone, and resorcinol	Catalytic ability of N-GQDs/AgNPs for oxidation of analytes by Ag^+	N-GQDs/AgNPs in the presence of Ag^+	-	-	[76]
Gas pollutants (SO_2, H_2S, HCHO, HCO_2H, CH_3CO_2H, and O_3)	Loss of surface capping ligands and aggregation or physical adsorption of analytes on AgNPs	AgNPs capped with poly-vinylpyrrolidone (PVP), cysteine, and dode-canethiol	LDA and HCA	-	[77]
Pesticides (azinphosmethyl and phosalone)	Aggregation	Unmodified AgNPs at three different pHs	LDA	Apple	[74]
Carbonyl flavor compounds (hexanoic acid, heptanoic acid, isovaleric acid, octanoic acid, isobutyric acid, ethyl lactate, butyric acid, valeric acid, acetaldehyde, propionic acid, acetal, ethyl hexanoate, furfural, ethyl acetate, ethyl butyrate, ethyl valerate, 2,3-butanedione, acetoin, isobutyraldehyde, and ethyl isovalerate)	Aggregation	Four AgNPs were produced by the reduction of Ag^+ by four o-phenylenediamine derivatives	LDA	Chinese Baijiu	[78]

cross-reactive NPs based sensing elements. The different compositions of plasmonic NPs can demonstrate outstanding varied behavior and colors after the addition of the target analytes. In this regard, Ghasemi et al. proposed a colorimetric sensor array by using both AuNPs and AgNPs for producing a distinct pattern to distinguish 40- and 42-residue amyloid β peptides (i.e., Aβ40 and Aβ42) and human serum albumin (HSA) [30]. The target proteins were bound to the surface of AuNPs or AgNPs and the aggregation occurred after the addition of copper ions due to the different coordination capabilities of Aβ42 and Aβ40 toward copper ions. Barplot and heat map were produced fingerprint patterns for each analyte (Fig. 6.6).

In a relatively similar strategy, four pesticides including amitrole, prothioconazole, propiconazole, and tebuconazole were discriminated by aggregation behavior of AgNPs, AuNPs, and gold-silver bimetallic NPs (Au-AgNPs) [79]. Unique fingerprints were produced for each pesticide by the aggregation response of each sensor element. LDA provided the discrimination of target pesticides in a wide range of concentrations. More investigations on using two or more types of plasmonic NPs as sensing elements are reported in Table 6.3.

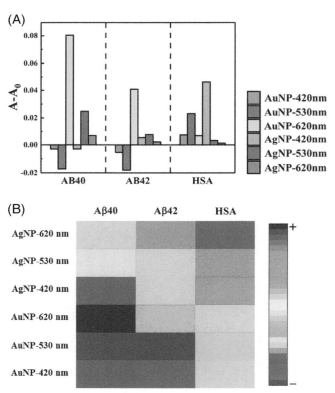

Fig. 6.6: (A) Bar plot and (B) heat map of sensor elements' responses against Aβ40, Aβ42, and HSA (300 nmol L^{-1}) [30].

Table 6.3: Summary of colorimetric array sensing based on combination of AuNPs and AgNPs.

Analytes	Mechanism	Sensor element	Data analysis method	Real Sample	Refs.
Proteins (Aβ40, Aβ42, and human serum albumin)	Aggregation	Unmodified AuNPs and AgNPs	LDA	Plasma	[30]
Antioxidant (gallic acid, p-coumaric acid, caffeic acid, 4-hydroxybenzoic acid, dopamine, butylated hydroxy toluene, flavanone, cynnamic acid, rutin, quercetin, cysteine, fisetin, narangin, catechin, ascorbic acid, glutathione, pyrogallol, citric acid, tannic acid, and catechol)	Aggregation	NaBH$_4$, CTAB, BSA, PVP, and glucose capped AuNPs/AgNPs	PCA and HCA	Lemon and tea	[80]
Mycotoxins (aflatoxin G1, aflatoxin B1, aflatoxin M1, ochratoxin A, and zearalenone)	Aggregation	Caffeic acid, dopamine, and PVP capped AuNPs or AgNPs	LDA and HCA	Pistachio, wheat and coffee milk	[81]
Pesticides (thiometon, phosalone, and prothioconazole)	Aggregation	Unmodified AuNPs and AgNPs	LDA	Water and cucumber	[34]
Antifreezes (ethylene glycol, ethanol, formamide, glycerin, and polyethylene glycol)	Aggregation	AuNPs, AgNPs and PBNPs (Prussian blue NPs)	PCA	Tap water	[82]
Proteins (lysozyme, cytochrome c, trypsin, myoglobin, papain, pepsin, and hemoglobin) Bacteria (CRPA, bacillus natto, staphylococcus, acetobacter aceti, rhodopseudomonas, E. coli, and bacillus)	Aggregation	Five AuNPs and two AgNPs with different sizes	LDA	Cancer cell	[83]
Anions (CO_3^{2-}, F^-, HCO_3^-, PO_4^{3-}, SO_4^{2-}, HPO_4^{2-}, IO_3^-, S^{2-}, $H_2PO_4^-$, $C_2O_4^{2-}$, N^{3-}, OH^-, CH_3COO^-, NO_3^-, ClO_4^-, Cl^-, SO_3^{2-}, I^-, SCN^-, CN^-, CrO_4^{2-}, Br^-, $Cr_2O_7^{2-}$, NO_2^-, arsenate, borate, benzoate, citrate, and bromate)	Catalytic activity of diverse AuNPs/AgNPs on the oxidization process of TMB	Chitosan-AgNPs, BSA-AgNPs, CTAB-AgNPs, CTAB-AuNPs, BSA-AuNPs, and glucose-AuNPs	PCA and HCA	-	[84]
Monofloral honey	Aggregation	Unmodified AuNPs/AgNPs, chitosan-AuNPs/AgNPs, CTAB-AuNPs/AgNPs, BSA-AuNPs/AgNPs, glucose-AuNPs/Ag-NPs, GSH-AuNPs/AgNPs, and cysteine-AuNPs/AgNPs	PCA	Honey	[85]
Pesticides (amitrole, propiconazole, prothioconazole, and tebuconazole)	Aggregation	Unmodified AuNPs, AgNPs, and Au-Ag NPs	LDA	Wheat flour and cucumber	[79]

6.2.4 Core-shell nanoparticles

In sensor array design, Au@Ag core-shell NPs and Ag@Au core-shell NPs are produced via the analyte-mediated growth of silver and gold shells on the surface of AuNPs and AgNPs, respectively. For example, AuNPs and various concentrations of Ag$^+$ ions were employed to develop a sensor array for the differentiation of antioxidants. Antioxidants (i.e., epicatechin 3-gallate, catechin, epigallocatechin 3-gallate, gallocatechin epigallocatechin, and epicatechin) reduced Ag$^+$ ions differently, causing various quantities of silver metallization on the surface of AuNPs. Six antioxidants were distinguished at concentrations of 20 nmol L^{-1}, 0.2 µmol L^{-1}, 2.0 µmol L^{-1}, 20 µmol L^{-1}, and 200 µmol L^{-1} employing LDA analysis [86].

In another study, AgNPs and various concentrations of Au^{3+} ions were employed to develop a sensor array for the distinction of antioxidants including dopamine, citric acid, glucose, glycine, and melatonin. Different antioxidants have various reducing strengths, causing gold deposition on the surface of AgNPs by different degrees. Using linear discriminant analysis (LDA) analysis, five different antioxidants were accurately identified at concentrations of 5, 10, 100, and 300 nmol L^{-1} [39]. There are some other reports of using spherical NPs to design core-shell-based sensor elements. However, silver/gold coating is usually accompanied by only changes in the intensity of SPR peak of core leading to monocolorimetric detection.

In contrast, it has been reported that silver deposition on the surface of gold nanorods (AuNRs) allows for multicorometric detection. AuNRs have two characteristic SPR peaks, that is, transversal and longitudinal peaks. The longitudinal peak can be tuned in a wide range of visible-near IR spectrum resulting in rainbowlike and distinctive color changes. For instance, the silver deposition on the surface of both spherical (AuNSs) and rod-shaped (AuNRs) NPs was employed to develop a multicolorimetric sensor array for the distinction of biogenic amines. As Fig. 6.7 displays, in the presence of biogenic amines (tyramine, spermine, tryptamine, spermidine, ethylenediamine, and histamine) growth of silver shell happened with various kinetic leading to diverse spectral and rainbow-like color variations. Fig. 6.8 illustrates the silver shells on the surface of AuNSs and AuNRs at a concentration of 60 µmol L^{-1} of spermine. Because of silver deposition, the aspect ratio of AuNRs (4.2) reduced to 2.8 and the diameter of AuNSs increased from 25 nm to 30 nm. Therefore, silver metallization resulted in a blue shift in the longitudinal peak along with increased intensity. Table 6.4 lists the reported colorimetric sensor arrays based on core-shell NPs.

6.3 Luminescent array sensing

6.3.1 Fluorescence array sensing

The fluorescence sensor arrays have attracted remarkable attention due to their low cost, high sensitivity, strong differentiation capability, and high changes in their emission profiles in the presence of the analytes. The size-dependent emission, good stability, and biocompatibility

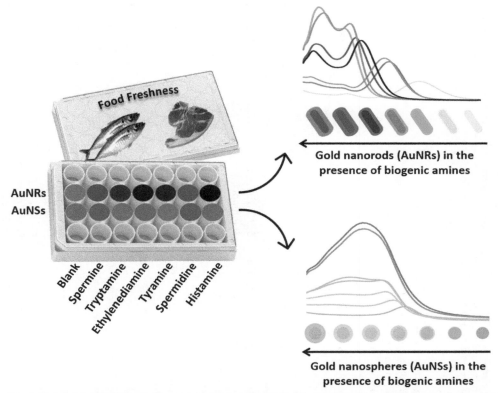

Fig. 6.7: Schematic illustration of the developed multicolorimetric sensor array based on anisotropic growth of silver shells on the surface of AuNPs and AuNRs [36].

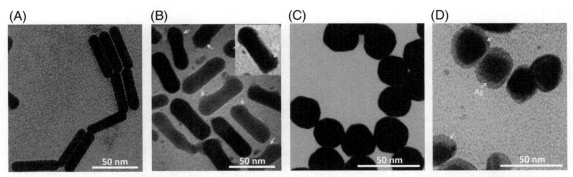

Fig. 6.8: TEM images of AuNRs and AuNSs in the absent (A and C) and presence (B and D) of spermine at a concentration of 60 μmol L⁻¹[36].

of metallic NCs especially, AuNCs and AgNCs have made them excellent candidate nanomaterials for developing fluorescence sensor arrays. Over the past decade, various fluorescence sensor arrays have been designed utilizing AuNCs and AgNCs to distinguish different analytes. For instance, Zhang et al. [92] introduced a sensor array based on AuNCs to identify the heavy

Table 6.4: Summary of colorimetric sensor arrays based on core-shell NPs.

NPs	Analytes	Sensing elements	Distinguished Concentrations	Data analysis method	Refs.
AuNRs and AuNSs	Biogenic amines (spermine, tryptamine, spermidine, ethylenediamine, tyramine, and histamine)	AuNRs/silver ions and AuNSs/silver ions	10 – 80 μmol L^{-1} 120-200 μmol L^{-1}	LDA and PLS	[36]
AuNRs	Reducing sugars (glyceraldehyde, fructose, glucose, maltose, and ribose)	Two negatively charged polydielectrics (sodium polymethacrylate (NaPMAA) and sodium polystyrenesulfonate (NaPSS))/AuNRs with positive charges	10 nmol L^{-1}, 100 nmol L^{-1}, 1.0 μmol L^{-1}, 10 μmol L^{-1}, 100 μmol L^{-1}	LDA and HCA	[87]
AuNRs	Chinese white spirits (Luzhou Laojiao, Guojiao 1573, Luzhou Laojiao, Shuijinagfang, Luzhou Laojiao, Luzhou Laojiao, Guojiao 1573, Yanghe, Jiangnanchun, Wuliangye, Gujingong, and Yanghe)	4 types of AuNRs with different aspect ratios (i.e., longitudinal LSPR peak at 659, 695, 743, and 763 nm)/silver ions/3 different reductants (i.e., catechol, o-phenylenediamine, and o-aminophenol)	0.05 g L^{-1}	PCA, HCA, and LDA	[88]
AgNPs	Antioxidants (glycine, glucose, citric acid, dopamine, and melatonin)	AgNPs/three concentrations of chloroauric acid (HAuCl$_4$)	5, 10, 100, and 300 nmol L^{-1}	LDA	[39]
AuNPs	Catechins (catechin, epicatechin gallate, gallocatechin, epicatechin gallate, epicatechin, and epigallocatechin)	AuNPs/silver ions/different reaction time intervals	200, 20, 2.0, 0.2, and 0.02 μmol L^{-1}	LDA	[89]
AuNPs	Antioxidants (catechins, epigallocatechin, gallocatechin, epicatechin, epigallocatechin 3-gallate, and epicatechin 3-gallate)	AuNPs/different concentrations of silver ions	20 nmol L^{-1}, 0.2 μmol L^{-1}, 2.0 μmol L^{-1}, 20 μmol L^{-1}, and 200 μmol L^{-1}	LDA and HCA	[86]
AuNRs	Dihydroxybenzene structural isomers (catechol, resorcinol, and hydroquinone)	AuNRs/different concentrations of silver ions/different time intervals	10-800 μmol L^{-1}	HCA and LDA	[90]
AuNRs	Catecholamine neurotransmitters (norepinephrine, epinephrine, and dopamine)	Different concentrations of AuNRs/different concentrations of silver ions	Epinephrine: 10-30 μg mL^{-1} Dopamine: 1-30 μg mL^{-1} Norepinephrine: 10-30 μg mL^{-1}	HCA and LDA	[28]
AuNPs	Cell lines (RAW264.7, 786-O, Hela, and L929)	DNA–AuNPs nanoconjugates/reductant (NH$_2$OH)/HAuCl$_4$	6,000 cells per well	LDA	[91]

metal ions (Fe^{2+}, Hg^{2+}, Pb^{2+}, Sn^{2+}, Mn^{2+}, Fe^{3+}, and Ag^+). Fluorescent AuNCs were synthesized via reducing Au^{3+} by 2-mercapto-1-methylimidazole (MMI) in the presence of polyvinylpyr-rolidone (PVP). Since the emission of the as-prepared AuNCs changed by the pH, PVP/MMI-AuNCs in two different pHs with yellow and red fluorescence were chosen as sensor elements (Fig. 6.9). LDA canonical score plots demonstrated that all heavy metal ions in tap and surface water samples are clustered into distinct categories without any misclassifications.

In another research, Shojaeifard et al. [93] applied two different types of receptors to construct a paper-based fluorescence sensor array for the distinction of twelve biogenic amines (BAs). Sensing elements comprised of four Pt(II) complexes ([PtMe(ppy)(SMe$_2$)], [Pt(p-tollyl)(ppy)(POPh)$_3$], [PtMe(ppy)(PPh$_3$)], and [PtMe(ppy)(POPh)$_3$]) and eight metallic NCs (AuNCs, CuNCs, and CdNCs with different capping agents). Intramolecular charge transfer, resonance energy transfer, electronic energy transfer, molecular rearrangement, photo-induced electron transfer, and inner filter effect can be responsible for fluorescence quenching of sensor elements in the presence of BAs. For example, amino-functional groups assemble on the surface of NCs and act as fluorescence moieties, which are liable to quench fluorescence, and the collision of BAs and NCs together can be the reason for the fluorescence quenching of NCs. Moreover, analyte uptake into the crystal lattice and direct ligation of an analyte to a metal center followed by the change in intermolecular interactions can lead to fluorescence quenching of Pt(II) complexes. The BAs were well classified based on the number and type of amine groups in their structures. The practicability of the developed sensor array has been demonstrated by detecting some BAs in five urine samples spiked with four various concentrations of each BA.

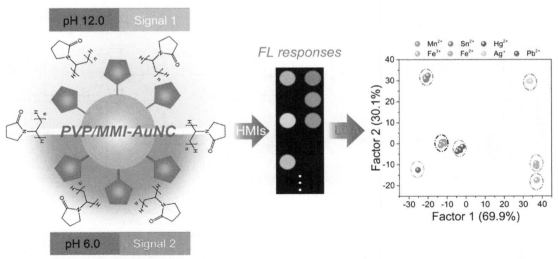

Fig. 6.9: PVP/MMI-AuNCs-based fluorescent sensor array for discrimination of heavy metal ions (Sn^{2+}, Hg^{2+}, Mn^{2+}, Fe^{2+}, Pb^{2+}, Fe^{3+}, and Ag^+) [92].

In another report, quenchers (Cu^{2+}/Fe^{3+}) and masking agents (EDTA/F^-) were added to three fluorophores (AuNCs, MoS_2-QDs, and WS_2-QDs) for discrimination among nitro-aromatic compounds (NACs) [93]. As displayed in Fig. 6.10, the quenching effect of NACs on three fluorophores (AuNCs, MoS_2-QDs, and WS_2-QDs), caused further quenching and followed by recovery of the fluorescence by adding a quencher and a masking agent, respectively, which created a fingerprint response pattern for NACs. The photo-excited electron transfer (PET) and the inner-filter effect were the possible quenching mechanisms of the receptors. The difference in fluorescence intensity of various channels in the absence and presence of the NACs (I_0-I) produced a data matrix (nine NACs × nine channels × nine replicates) analyzed by LDA to cluster NACs.

Cao et al. [95] proposed the use of four kinds of polymers templated AgNCs to discriminate seven heavy metal ions (Cd^{2+}, Pb^{2+}, Fe^{3+}, Co^{2+}, Cu^{2+}, Ni^{2+}, and Cr^{3+}) using a sensor array system. Metal ions interacted with the free amino or carboxyl groups of polymers and quenched the fluorescence of AuNCs by disturbing the ligand-to-metal charge transfer (LMCT) or ligand-to-metal–metal charge transfer (LMMCT) which were the basis of their emissions. On the other hand, the interaction of some metal ions and polymers decreased the nonradiative deactivations of AuNCs and increased the fluorescence of AuNCs through an aggregation-induced emission mechanism. They used the fluorescence intensity ratios of AuNCs in the presence and absence of each ion (I/I_0) as output signals and employed PCA for data analysis and distinguishing among metal ions. The proposed sensor array was successfully applied to discriminate metal ions in tap water samples.

Table 6.5 summarizes the reported fluorescence sensor arrays based on AuNCs and AgNCs.

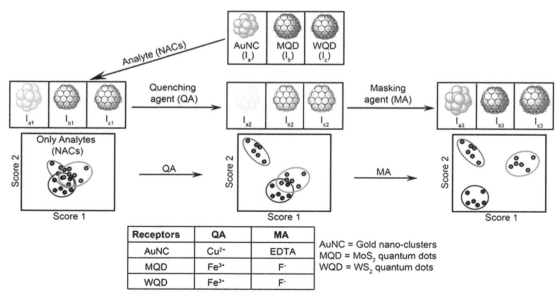

Receptors	QA	MA
AuNC	Cu^{2+}	EDTA
MQD	Fe^{3+}	F^-
WQD	Fe^{3+}	F^-

AuNC = Gold nano-clusters
MQD = MoS_2 quantum dots
WQD = WS_2 quantum dots

Fig. 6.10: Schematic description of the principle of operation of a nine-channel sensor array based on sequential on−off response [94].

Table 6.5: Summary of fluorescence sensor arrays based on AuNCs and AgNCs.

NPs	Analytes	Sensing elements	Discrimination mechanism	Signal	Data analysis method	Refs.
AuNCs	Seven kinds of heavy metal ions (Fe^{2+}, Hg^{2+}, Pb^{2+}, Sn^{2+}, Fe^{3+}, Mn^{2+}, and Ag^+)	Dual functionalization of AuNCs by 2-mercapto-1-methylimidazole and polyvinylpyrrolidone (PVP/MMI-AuNCs) at two different pHs	Different interactions between AuNCs with metal ions produced distinct fluorescence responses for analytes.	Fluorescence quenching/enhancement	HCA and LDA	[92]
AuNCs	16 biogenic amines (butylamine, cyclohexylamine, pentylamine, pyrrolidine, hexylamine, urea, creatinine, diisopropylamine, triethylamine, putrescine, ethylenediamine, ammonia, pyridine, dipropylamine, cadaverine, and piperidine)	Four kinds of Pt(II) complexes ([Pt(p-tollyl)(ppy)(POPh)$_3$], [PtMe(ppy)(SMe$_2$)], [PtMe(ppy)(PPh$_3$)] and [PtMe(ppy)(POPh)$_3$)]) and eight metallic nanoclusters with different capping agent (BSA-AuNCs, BSA-CdNCs, CEW-AuNCs, CEW-CuNCs, GSH-AuNCs, Lys-AuNCs, Try-AuNCs, and Cys-AuNCs)	Due to different structures of sensing elements and the number and nature of the BAs' amine groups, the responses collected from the sensor array were different for each BA. BAs can quench the fluorescence intensity of sensing elements through various mechanisms such as intramolecular charge transfer, photo-induced electron transfer, resonance energy transfer, electronic energy transfer, molecular rearrangement, and inner filter effect.	Fluorescence quenching	HCA, PCA, and PCA-LDA	[93]
AuNCs	Different nitroaromatic compounds (NACs) (4-nitrophenol, 2,4-dinitrophenol, 2-nitrophenol, 2-hydroxy-3,5-dinitrobenzoic acid, 2-nitrobenzene-1,3,5-triol, 2,6-dinitroaniline, 2-amino-6-nitrophenol, 2,4-dinitrobenzaldehyde, 4-nitroaniline, 2-amino-3-nitrophenol, 3-methyl-4-nitrophenol, 3,5-dinitrobenzoic acid, 3-nitrobenzoic acid, 3-chloro-2-nitrophenol, and 3-dinitrobenzene	An AuNCs and two kinds of QDs with a quencher and masker pair (AuNCs/Cu^{2+}/EDTA, MoS$_2$-QDs/Fe^{3+}/F$^-$ and WS$_2$-QDs/Fe^{3+}/F$^-$)	NACs quenched the fluorescence of receptors through photo-excited electron transfer (PET) and the inner-filter effect mechanisms. After adding a quencher, the fluorescence was further quenched and recovered again with a masking agent. The overall responses from all nine channels produced a unique fluorescence pattern for NACs.	Fluorescence quenching and enhancement	LDA	[94]

(Continued)

Table 6.5: (Cont'd).

NPs	Analytes	Sensing elements	Discrimination mechanism	Signal	Data analysis method	Refs.
AuNCs	Artificial bovine serum samples with three different levels of CA125 antigen	Five kinds of aptamer-AuNCs	The interaction between aptamers and AuNCs led to the formation of soft aggregations, followed by the change in AuNCs morphology and emission. In the presence of CA125, the competition between AuNCs and CA125 over the interaction with aptamer creates different degrees of aggregation and anti-aggregation. As a result, the emission changes again.	Fluorescence quenching/enhancement	LDA	[96]
AuNCs	Phosphate species (PPi, ADP, and ATP)	AuNCs-rare earth ions assembly (AuNCs-Ce^{3+}, AuNCs-Tm^{3+}, AuNCs-Yb^{3+}, and AuNCs-Gd^{3+})	Metal ions increased the emission of AuNCs through aggregation-induced emission (AIE). Due to the higher affinity of metal ions to phosphate species than AuNCs, phosphate species redispersed the assemblies of AuNCs and metal ions and reduced the enhanced fluorescence.	Fluorescence quenching	LDA and PCA	[97]
AuNCs	Ten proteins(pepsin, hemoglobin, myoglobin, cytochrome c, human serum albumin, transferrin, catalse, trypsin, lysozyme, and bovine serum albumin	A peptide sequence (Cys-Met-Met-Met-Met-Met) coated AuNCs (CMMMMM-AuNCs) at five different pH	Through electrostatic and partly hydrophobic interactions, proteins were adsorbed on the surface of CMMMMM-AuNCs. Different changes in the binding energy of Au 4f, fluorescence lifetime, and energy transfer between proteins and CM-MMMM-AuNCs in the presence of various proteins changed the emission of CMMMMM-AuNCs.	Fluorescence quenching/enhancement	LDA	[98]

NPs	Analytes	Sensing elements	Discrimination mechanism	Signal	Data analysis method	Refs.
AuNCs	Three nitrophenol isomers (m-nitrophenol, o-nitrophenol, and p-nitrophenol)	Three kinds of dual-ligands cofunctionalized Au NCs (β-CD/MUA, His/β-CD, GSH/β-CD cofunctionalized AuNCs)	Different host-guest interaction affinities between β-CDs and nitrophenols caused nitrophenols to be adsorbed on the surface of AuNCs to a different extent. After that, nitrophenols induced the various levels of fluorescence quenching of AuNCs via the inner filter effect.	Fluorescence quenching	LDA and HCA	[99]
AuNCs	Six bacteria (B. subtilis, S. aureus, MRSA, E. coli, KREC, and A. faecalis)	Four kinds of protein-encapsulated AuNCs (HSA-AuNCs, Lf-AuNCs, Lys-AuNCs, Vancomycin functionalized BSA-AuNCs)	AuNCs bound to bacterial surfaces with different interactions depending on the type of bacteria and protein scaffolds of AuNCs. After centrifugation, the various extent of this interaction caused the fluorescence of the supernatant to decrease differently.	Fluorescence quenching	LDA and HCA	[100]
AuNCs	Nine proteins (trypsin, pepsin, hemoglobin, casein, lysozyme, lipase, papain, and myohemoglobin) Six bacteria (pseudomonas aeruginosa, escherichia coli, bacillus natto, and acetobacter aceti)	Six kinds of metal ion-AuNCs (Cd²⁺-Lys-AuNCs, Zn²⁺-HSA-AuNCs, Zn²⁺-Lys-AuNCs, Zn²⁺-BSA-AuNCs, Cd²⁺-BSA-AuNCs, and Cd²⁺-HAS-AuNCs)	Protein-stabilized AuNCs purified with Cd²⁺ and Zn²⁺ interacted with proteins or bacteria differently. Some of the bacteria or proteins increased the fluorescence, while some of them caused fluorescence quenching.	Fluorescence quenching/ enhancement	LDA	[101]
AuNCs	Ten proteins (hemoglobin, human serum albumin, transferrin, lysozyme, pepsin, cytochrome C, CA 125, trypsin, catalase, and CA 153)	Six kinds of near infrared fluorescent dual ligand functionalized AuNCs (Iso/ MUA-AuNCs, Thr/MUA-AuNCs, Ser/MUA-AuNCs, Pro/ MUA-AuNCs, Met/ MUA-AuNCs and Ala/MUA-AuNCs)	After adsorption on the surface of AuNCs, mainly through hydrogen bonding and van der Waals forces, proteins changed the AuNCs' photophysical properties. Some of them caused direct electron transfer reactions at the surface of nano-clusters. Various proteins generated a unique response pattern due to the different changing in the fluorescence intensity of AuNCs.	Fluorescence quenching/ enhancement	LDA	[102]

(Continued)

Table 6.5: (Cont'd).

NPs	Analytes	Sensing elements	Discrimination mechanism	Signal	Data analysis method	Refs.
AuNCs	Five highly metastatic breast cancer cell lines (HCC1569, MDA-MB-231, MDA-MB-157, HCC1806 and Hs578T), four non or low metastatic cancer cell lines (MCF7, SKBR3, MDA-MB-436, and MDA-MB-468), and one non-neoplastic, fibrocystic breast cell line (MCF10A)	Seven dual ligand functionalized AuNCs (4-mercaptobenzoic acid-MUA-AuNCs, 6-mercapto-1- hexanol-MUA-AuNCs, (11-mercaptoundecyl)- N,N,N-trimethylammonium bromide-MUA-AuNCs, 3-mercaptopropionic acid-MUA-AuNCs, (2-mercaptoethyl)amine-MUA-AuNCs, glutathione-MUA-AuNCs, and mercaptosuccinic acid-MUA-AuNCs)	Interaction with AuNCs depended on the concentration and composition of the cell's surface components, such as proteins, lipids, and carbohydrates, and the charge state and chemistry of AuNCs. The energy/electron transfer between AuNCs and cells or the aggregation of AuNCs in the presence of cells caused the fluorescence of AuNCs is decreasing. Conversely, the cells can protect AuNCs from the quencher of the environment, such as O_2, and lead to enhancement of the Fluorescence intensity of AuNCs.	Fluorescence quenching/enhancement	LDA	[103]
AuNCs	10 proteins (trypsin, human serum albumin, catalase, pepsin, hemoglobin, cytochrome C, myoglobin, lysozyme, transferrin, and bovine serum albumin)	Six kinds of AuNCs (MUA-AuNCs, THPC/GSH-AuNCs, BSA-Au NCs, Pro-Au NCs, GSH-AuNCs and Lys-AuNCs)	Proteins bound to the surface of AuNCs through different forces such as hydrogen bonding and van der Waals, and alter their photophysical properties such as lifetime and binding energy of Au4f. Change in photophysical properties and surface charge of AuNCs affected their interaction with proteins and their emission changes in the presence of proteins.	Fluorescence quenching/enhancement	LDA	[31]

NPs	Analytes	Sensing elements	Discrimination mechanism	Signal	Data analysis method	Refs.
AuNCs	One non-neoplastic, fibrocystic breast cell line (MCF10A), four non- or low metastatic breast cancer cell lines (MDA-MB-436, SKBR3, MCF7, and MDA-MB-468) and five highly metastatic cancer cell lines (HCC1806, MDA-MB-231, MDA-MB-157, MDA-MB-175VII, and Hs578T)	Two kinds of AuNCs conjugated with graphene oxide (BSA-AuNCs-GO and Lys-AuNCs-GO), two kinds of graphene quantum dots conjugated with graphene oxide (GQDs-COOH-GO and GQDs-NH2-GO), polyethyleneimine functionalized carbon dots conjugated with graphene oxide (PEI-CDs-GO) and CdTe quantum dots conjugated with graphene oxide (QDs-GO)	Through hydrophobic, π–π stacking, or electrostatic interactions, GO formed complexes with nanodots and quenched their fluorescence. In the presence of cells, competition between cells and luminescent nanodots to react with GO led to the release of the nanodots. Therefore, the quenched fluorescence of nanodots was recovered. Different cells generated a unique pattern through different nanodots' fluorescence recoveries due to their various types and amounts of membrane proteins, carbohydrates and lipids.	Fluorescence enhancement	LDA	[104]
AuNCs	Nine proteins (trypsin, bovine serum albumin, histone, human serum albumin, myoglobin, cytochrome C, lysozyme, and streptavidin)	Eight kinds of dual-ligand cofunctionalized AuNCs (MEN/MUA, GSH/MUA, TSH/MUA, MSA/MUA, MH/ MUA, MBA/MUA, MPA/MUA, and NSH/MUA cofunctionalized AuNCs)	The electron or energy transfer between the surface ligands of AuNCs and adsorbed proteins or/ and changing surface charge or dispersion state of AuNCs in the presence of various proteins led to fluorescence intensity variations of AuNCs.	Fluorescence quenching/ enhancement	LDA and HCA	[105]
AuNCs	Eight proteins (lysozyme, human serum albumin, pepsin, egg white albumin, hemoglobin, catalse, trypsin, trandferrin)	Collagen modified AuNCs (Col-AuNCs) and Macerozyme R-10 modified AuNCs (Mac-AuNCs)	Hydrophobic interactions by adsorbing proteins on the surface of AuNCs caused the formation of the protein–AuNCs complexes with different photophysical properties compared with AuNCs. Aggregation of AuNCs could be another reason for the fluorescence wavelength shifts of AuNCs in the presence of various proteins.	Changing fluorescence wavelength or intensity	LDA	[106]

(Continued)

Table 6.5: (Cont'd).

NPs	Analytes	Sensing elements	Discrimination mechanism	Signal	Data analysis method	Refs.
AuNCs	12 amino acids (glutamine, threonine, arginine, valine, histidine, leucine, alanine, proline, tryptophan, serine, cysteine, and aspartic acid)	Metal ion-modulated BSA-AuNCs (BSA-AuNCs-Pb^{2+}, BSA-AuNCs-Ni^{2+}, BSA-AuNCs-Zn^{2+}, and BSA-AuNCs-Cd^{2+})	Metal ions bound to the BSA-AuNCs and through ISC, dynamic and static quenching, repairing the lattice defect of AuNCs, etc., affected their emission at both 425 and 635 nm wavelengths associated with oxides of BSA and AuNCs, respectively. After the addition of amino acids, the stronger interaction between metal ions and amino acids than BSA led to the dissociation of metal ions from BSA-AuNCs or aggregation of BSA-AuNCs, so produce different ratiometric responses.	Ratiometric responses (Fluorescence quenching/ enhancement)	–	[29]
AuNCs and AgNCs	Heavy metal ions (Mn^{2+}, Ag^{+}, Hg^{2+}, Cu^{2+}, Pb^{2+}, Cr^{3+}, and Cd^{2+})	Two semiconductor quantum dots (glutathione modified cadmium telluride and 3-mercaptopropionic acid modified cadmium telluride), two noble metal nanoclusters (poly (methacrylic acid) modified silver nanoclusters (AgNCs@ PMAA), and bovine serum albumin modified gold nanoclusters (AuNCs@BSA)) and two organic dyes (rhodamine derivative (RHD) and calcein blue (CB))	Due to different functional groups (carboxyl, amine, and thiol groups), both luminescent nanoprobes and organic dyes reacted differently with the heavy metal ions and caused producing a unique fluorescence response pattern for each ion.	Fluorescence quenching/ enhancement	LDA	[26]

NPs	Analytes	Sensing elements	Discrimination mechanism	Signal	Data analysis method	Refs.
AuNCs and AgNCs	10 proteins (glucose oxidase, β-galactosidase, bovine serum albumin, cytochrome C, proteinase K, horseradish peroxidase, lysozyme, urease, hemoglobin, and esterase)	Seven luminescent nanodots conjugated with graphene oxide including graphene quantum dots, CdTe quantum dots, carboxyl-carbon dots, lysozyme-templated gold nanoclusters, polyethyleneimine functionalized carbon dots, BSA-templated gold nanoclusters, and DNA-templated silver nanoclusters	Luminescent nanodots bound to GO through π–π attraction, hydrophobic, and electrostatic interactions and their fluorescence was quenched. Due to the proteins' higher affinity to GO, the presence of protein led to separate nanodots from GO and recovery of nanodots emission. If the affinity of the protein to nanodots was more than GO, proteins caused the aggregation of nanodots and decreased their emission.	Fluorescence quenching/enhancement	LDA	[107]
AgNCs	Seven heavy metal ions (Cu^{2+}, pb^{2+}, Fe^{3+}, Cr^{3+}, Co^{2+}, Cd^{2+}, and Ni^{2+})	Four kinds of polymers templated AgNCs (poly (methyl vinyl ether-alt-maleic acid)-AgNCs, poly (methacrylic acid)-AgNCs, polyethyleneimine-AgNCs, and poly (acrylic acid)-AgNCs)	Some metal ions interacted with some modified polymers and enhanced the fluorescence of the AgNCs through aggregation-induced emission. Some of these interactions decreased the emission of AgNCs by disturbing the origin of their emissions, including the ligand-to-metal charge transfer (LMCT) or ligand-to-metal-metal charge transfer (LMMCT).	Fluorescence quenching/enhancement	PCA	[95]

6.3.2 Chemiluminescence array sensing

The chemiluminescence (CL) methods have a high sensitivity and signal-to-noise ratio among spectroscopic methods because they do not need an external light source and thus source-induced interferences are eliminated. The use of NPs to enhance emission by catalyzing the reactions has recently been considered in the design of CL-based sensor arrays. Several CL-based sensor arrays catalyzed by AuNPs and AgNPs have been developed to distinguish different analytes. Table 6.6 summarizes the reported chemiluminescence sensor arrays based on AuNPs and AgNPs.

As an example, Shahrajabian et al. [108] developed a CL sensor array based on two different reagent pairs (AgNO$_3$-luminol and H$_2$O$_2$-luminol), which their CL enhanced in the presence of four various types of NPs. Citrate and NaBH$_4$ capped AuNPs, citrate-coated AgNPs, and thioglycolic acid-modified cadmium telluride quantum dots were chosen to enhance the CL of the AgNO$_3$-luminol and H$_2$O$_2$-luminol reagent pairs, respectively. Due to having various structures, biothiols (glutathione, cysteine, and glutathione disulfide) reacted with NPs and changed CL signals depending on the degree of the interactions. The factors such as the changes in the surface charge density of the NPs and the competitive reaction of the SH-reducing groups with luminol for oxygen-containing intermediate affected the CL responses. The difference in CL intensity before and after adding biothiols to sensor elements (I-I$_0$) was employed to confirm the potential of the sensor array to produce a fingerprint pattern for biothiols. Biothiols were classified into three expected classes employing LDA and HCA. Furthermore, the developed sensor array successfully classified biothiols in human urine samples.

6.4 Multichannel array sensing

In the design of sensor arrays, the focus is usually on using a single transduction channel such as fluorescence, colorimetry, light scattering, etc. To increase the discrimination ability and sensitivity of a sensor array, more information needs to be added to the data set. In the case of the single-channel sensor array, adding more information requires an increase in the number of sensing elements which is costly and time-consuming. An effective alternative strategy could be extracting more than one kind of transduction signal from one sensing element which is named multichannel sensor array. In a multi-channel sensor array, each sensor element plays as a virtual sensor array and provides different information. Subsequently, more sensitive and high-throughput discrimination of target analytes is resulted by a multichannel sensor array. However, it is complicated and challenging to collect more than one fundamentally different transduction signal from one sensor element. Therefore, the reported multichannel sensor arrays are rare [111–113].

As an example of multichannel array sensing, Gui et al. [114] have synthesized AuNCs by employing luminol as both the reducing and the capping agent. The functionalized AuNCs

Table 6.6: Summary of chemiluminescence sensor arrays based on AuNPs and AgNPs.

NPs	Analytes	Sensing elements	Discrimination mechanism	Signal	Data analysis method	Refs.
AuNPs and AgNPs	Biothiols (glutathione, cysteine, and glutathione disulfide)	$AgNO_3$-luminol enhanced CL with citrate and $NaBH_4$ capped AuNPs, H_2O_2-luminol enhanced CL with citrate capped AgNPs, and thioglycolic acid-modified cadmium telluride quantum dots	After the addition of biothiols, the CL intensity of the sensing elements changed to various degrees due to the different extent of interaction between biothiols and NPs, thus resulted in the distinct response pattern for biothiols.	Decreasing or increasing enhanced CL	HCA and LDA	[108]
AgNPs	Amino acids (L-threonine, L-cysteine, L-proline, L-arginine, L-phenylalanine, L-glutamic acid, and L-tyrosine)	Luminol functionalized AgNPs -H_2O_2	Amino acids bound to the surface of AgNPs, and the possible reaction of the adsorbed amino acids with active oxygen intermediates and direct hydrogen bonding between luminol and amino acids affected the CL responses. Three-channel properties of CL (time required for the CL to appear (Ta), the time required to reach the highest CL (Tp), and CL intensity) caused distinct response patterns due to various properties of amino acids.	Changing Ta, Tp, and CL intensity	PCA	[109]
AgNPs	Five pesticides (dimethoate, chlorpyrifos, dipterex, carbaryl, and carbofuran)	Luminol functionalized AgNPs-H_2O_2	Different structures of pesticides caused some of them could react with luminol and produce an unstable intermediate, so the time required for the CL to appear (Ta) and CL intensity are increased, and the time needed to reach the highest value of CL (Tp) was decreased. Unlike them, the existence of a reductive hydroxyl group (–OH) and amino group (–NH–) in some pesticide structures can consume the oxidants. In these cases, CL intensity decreased, and the time needed for the CL to appear (Ta) and the time required to reach the highest value of CL (Tp) were increased.	Changes in Ta, Tp, and CL intensity	PCA	[110]
AgNPs	Seven proteins (papain, γ-myoglobin, chymotrypsin, bovine serum albumin, superoxide dismutase, lysozyme, and hemoglobin)	Luminol functionalized AgNPs-H_2O_2	Different proteins can bind to the surface of the NPs through various interactions, such as electrostatic attraction, hydrogen bonding, and covalent bonding (Ag–S). The adsorbed proteins by influencing the accessibility of H_2O_2 to the surface of NPs, and changing the surface-to-volume ratio of the luminol–AgNPs led to different CL responses. On the other hand, the charge of proteins and the existence of metal in their structures, which affected the diffusion of H_2O_2 and the catalytic effect on the reaction of luminol and H_2O_2, respectively, also assisted in the creation of unique patterns by proteins.	Changing Ta, Tp, and CL intensity	PCA	[32]

with multioptical signal channels (i.e., chemiluminescence and fluorescence) were applied for heavy metal ion discrimination. Metal ions caused the aggregation of AuNCs and consequently changed the fluorescence intensity. Furthermore, metal ions could accelerate the chemical reactions of the chemiluminescence process accompanied by variation in the chemiluminescence signal. Principal component analysis (PCA) was applied to discriminate heavy metal ions (Pb^{2+}, Al^{3+}, Co^{2+}, Hg^{2+}, Zn^{2+}, Cu^{2+}, and Cr^{3+}) at a concentration of 1.0 μg mL^{-1}.

In another study, a multichannel sensor array has been developed based on designing a composite of positively charged cetyltrimethylammonium bromide (CTAB)-capped AuNPs and negatively charged vancomycin (Van)-templated AuNCs [115]. The negatively charged bacteria with various net charges and hydrophobicity disassembled the composites differently, leading to fluorescence recovery, a decrease in UV-Vis absorption, and a decrease in light scattering signal (Fig. 6.11). Employing linear discriminant analysis (LDA), ten kinds of gram-negative bacteria were discriminated accurately at a low concentration of OD_{600}= 0.015.

Table 6.7 summarizes the reported multichannel array sensing based on gold and silver NPs and NCs.

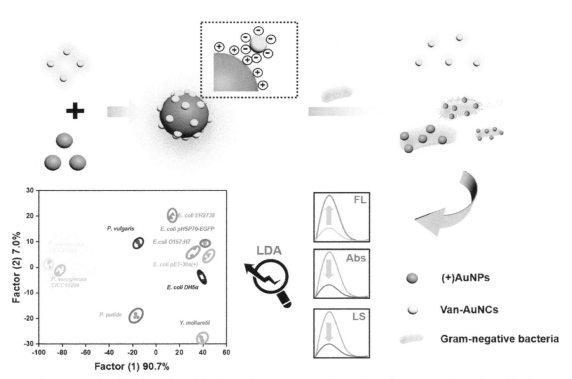

Fig. 6.11: The developed multi-channel sensor array based on the composite of positively charged CTAB-capped AuNPs and negatively charged Van-templated AuNCs to distinguish gram-negative bacteria [115].

Table 6.7: Multichannel array sensing based on gold and silver NPs and NCs.

NPs/NCs	Analytes	Channels	Sensing element	Signals	Distinguished Concentrations	Refs.
AuNCs	Heavy metal ions (Al^{3+}, Hg^{2+}, Co^{2+}, Zn^{2+}, Cr^{3+}, Cu^{2+}, and Pb^{2+})	Chemiluminescence and fluorescence	Luminol functionalized AuNCs	CL integrated intensity in 60 seconds, CL peak intensity, CL intensity at 60 sseconds, fluorescence intensity at 598 nm and fluorescence intensity at 424 nm	1.0 μg mL^{-1}	[114]
AuNPs and AuNCs	Bacteria (P. aeruginosa, P. vulgaris, E. coli O157:H7, S. aureus, and B. subtilis)	UV-Vis absorption, fluorescence, and light scattering	A composite of cetyltrimethylammonium bromide (CTAB)-capped AuNPs and vancomycin-templated AuNCs	Fluorescence intensity at 400 nm (λex 290 nm), absorption intensity at 525 nm and light scattering intensity at 390 nm	$OD_{600} = 0.015$	[115]
AuAg-NCs and AuNPs	Sulfur species (oxidized glutathione, cystine, methionine, disulfite, sulfite, n-acetyl-l-cysteine, sulfide, thiosulfate, peroxydisulfate, tetrathionate, reduced glutathione, homocysteine, and cysteine) and bacteria (thiobacimonas profunda, escherichia coli, citreicella thiooxidans, staphylococcus aureus, and acidithiobacillus caldus)	UV-Vis absorption, fluorescence, light scattering	Au-Ag alloy NCs on the surface of AuNPs	Fluorescence intensity at 440 nm (λex 280 nm) and absorption intensity at 535 nm	Sulfur species: 0.5 μmol L^{-1} Bacteria: $OD_{600} = 0.005$	[116]
AuNPs	Metal ions (Cd^{2+}, Hg^{2+}, Cr^{3+}, Sn^{4+}, Cu^{2+}, Ag^+, Pb^{2+}, Zn^{2+}, and Mn^{2+})	UV-Vis absorption and fluorescence	FAM-labeled DNA adsorbed onto AuNPs	Fluorescence intensity at 511 nm (λex 491 nm) and absorption ratio 620 nm/520 nm	500, 100, and 50 nmol L^{-1}	[117]
AuNPs	Proteins (trypsin, lysozyme, bovine serum albumin, myoglobin, hemoglobin, human serum albumin, pepsin, horseradish peroxidase, transferrin, papain, and egg white albumin)	UV-Vis absorption and fluorescence	Dye-labeled DNA sequences absorbed onto AuNPs	Fluorescence intensity at 520 nm (λex 470 nm) and absorption ratio 620 nm/520 nm	1.0 μmol L^{-1}	[112]

6.5 Conclusion

Taking advantage of the outstanding optical properties, the easy synthesis, surface functionalization process, and other unprecedented characteristics, there is no doubt that gold and silver nanostructures have emerged with great potential in the development of colorimetric and fluorimetric sensor arrays. In this chapter, at first we focused on the significant progress of AuNPs /AgNPs-based colorimetric sensor arrays in a variety of applications such as environmental monitoring, food quality control, and biomedical and clinical investigation. The recent successes in the design of these kinds of colorimetric sensor arrays, their portability, and their low price make them outstanding sensors providing signal visualization by the naked eye, which can enter to human daily life. The discrimination performance of the plasmonic NPs-based sensor array can be improved by using different sizes or different surface modifications of NPs to create more cross-reactive responses in the presence of the analytes with similar structures. Moreover, the combination of both AuNPs and AgNPs can be a promising approach for increasing the color tuneability of sensor arrays. Although most colorimetric sensor arrays designed by plasmonic NPs are based on aggregation mechanisms, one of the other directions in this field is to develop multicolorimetric sensor arrays with nonaggregation mechanisms, which are mainly focused on the formation of core-shell NPs. On the other hand, the size-dependent emission, good stability, and biocompatibility of metallic NCs such as AuNCs and AgNCs have made them as interesting fluorescence nanomaterials for developing fluorimetric sensor arrays. The application of portable test strips and digital imaging progress can move the developed sensors out of the lab. Thus, following the combination of the metallic nanostructures-based optical sensor array with the test strip/digital imaging technology promise a successful outlook in near future.

References

[1] Z.-H. Chen, Q.-X. Fan, X.-Y. Han, G. Shi, M. Zhang, Design of smart chemical 'tongue' sensor arrays for pattern-recognition-based biochemical sensing applications, TrAC Trends Anal. Chem. 124 (2020) 115794.
[2] A. Bigdeli, F. Ghasemi, H. Golmohammadi, S. Abbasi-Moayed, M.A.F. Nejad, N. Fahimi-Kashani, et al., Nanoparticle-based optical sensor arrays, Nanoscale 9 (2017) 16546–16563.
[3] Y. Geng, W.J. Peveler, V.M. Rotello, Array-based "chemical nose" sensing in diagnostics and drug discovery, Angew. Chem. Int. Ed. 58 (2019) 5190–5200.
[4] N. Fahimi-Kashani, F. Ghasemi, A. Bigdeli, S. Abbasi-Moayed, M.R. Hormozi-Nezhad, Nanostructure-based optical sensor arrays: Principles and applications, Sensing and Biosensing with Optically Active Nanomaterials. Elsevier, Amsterdam (2022) 523–565.
[5] M.A. García, Surface plasmons in metallic nanoparticles: Fundamentals and applications, J. Phys. D Appl. Phys. 44 (2011) 283001.
[6] E. Petryayeva, U.J. Krull, Localized surface plasmon resonance: Nanostructures, bioassays and biosensing—A review, Anal. Chim. Acta 706 (2011) 8–24.
[7] L.-Y. Chen, C.-W. Wang, Z. Yuan, H.-T. Chang, Fluorescent gold nanoclusters: Recent advances in sensing and imaging, Anal. Chem. 87 (2015) 216–229.
[8] M.C. Janzen, J.B. Ponder, D.P. Bailey, C.K. Ingison, K.S. Suslick, Colorimetric sensor arrays for volatile organic compounds, Anal. Chem. 78 (2006) 3591–3600.

[9] N.A. Rakow, K.S. Suslick, A colorimetric sensor array for odour visualization, Nature. 406 (2000) 710–713.

[10] E.R. Ziegel, Handbook of Chemometrics and Qualimetrics, part B, Taylor & Francis, Abingdon, 2000, pp. 218–219.

[11] J. Mller, J. Miller, Statistics and Chemometrics For Analytical Chemistry Harlow, Pearson Prentice Hall, 2005, pp. 263.

[12] R.G. Brereton, Applied Chemometrics for Scientists, John Wiley & Sons, Hoboken, 2007.

[13] K. Odziomek, A. Rybińska, T. Puzyn, Unsupervised learning methods and similarity analysis in chemoinformatics, Springer, Dordrecht, 2017.

[14] R.G. Brereton, Chemometrics: Data Analysis for the Laboratory and Chemical Plant, John Wiley & Sons, Hoboken, 2003.

[15] R.G. Brereton, Multivariate Pattern Recognition in Chemometrics: Illustrated by Case Studies, Elsevier, Amsterdam, 1992.

[16] J. Miller, J.C. Miller, Statistics and Chemometrics For Analytical Chemistry, Pearson education, London, 2018.

[17] M.M. Ali, N. Hashim, S. Abd Aziz, O. Lasekan, Principles and recent advances in electronic nose for quality inspection of agricultural and food products, Trends Food Sci. Technol. 99 (2020) 1–10.

[18] A.T. John, K. Murugappan, D.R. Nisbet, A. Tricoli, An outlook of recent advances in chemiresistive sensor-based electronic nose systems for food quality and environmental monitoring, Sensors. 21 (2021) 2271.

[19] K.L. Diehl, E.V. Anslyn, Array sensing using optical methods for detection of chemical and biological hazards, Chem. Soc. Rev. 42 (2013) 8596–8611.

[20] J.R. Askim, M. Mahmoudi, K.S. Suslick, Optical sensor arrays for chemical sensing: The optoelectronic nose, Chem. Soc. Rev. 42 (2013) 8649–8682.

[21] A.D. Wilson, Developing electronic-nose technologies for clinical practice, J. Med. Surg. Pathol. 3 (2018) 169–171.

[22] E.A. Baldwin, J. Bai, A. Plotto, S. Dea, Electronic noses and tongues: Applications for the food and pharmaceutical industries, Sensors. 11 (2011) 4744–4766.

[23] A.D. Wilson, Noninvasive Early Disease Diagnosis By Electronic-Nose and Related VOC-Detection Devices, Multidisciplinary Digital Publishing Institute, Basel, 2020, pp. 73.

[24] Z. Li, J.R. Askim, K.S. Suslick, The optoelectronic nose: Colorimetric and fluorometric sensor arrays, Chem. Rev. 119 (2018) 231–292.

[25] F. Najafzadeh, F. Ghasemi, M.R. Hormozi-Nezhad, Anti-aggregation of gold nanoparticles for metal ion discrimination: A promising strategy to design colorimetric sensor arrays, Sens. Actuators B 270 (2018) 545–551.

[26] H. Kang, L. Lin, M. Rong, X. Chen, A cross-reactive sensor array for the fluorescence qualitative analysis of heavy metal ions, Talanta. 129 (2014) 296–302.

[27] H. Ji, L. Wu, F. Pu, J. Ren, X. Qu, Point-of-care identification of bacteria using protein-encapsulated gold nanoclusters, Adv. Healthc. Mater. 7 (2018) 1701370.

[28] S. Jafarinejad, M. Ghazi-Khansari, F. Ghasemi, P. Sasanpour, M.R. Hormozi-Nezhad, Colorimetric fingerprints of gold nanorods for discriminating catecholamine neurotransmitters in urine samples, Sci. Rep. 7 (2017) 1–8.

[29] M. Wang, Q. Mei, K. Zhang, Z. Zhang, Protein-gold nanoclusters for identification of amino acids by metal ions modulated ratiometric fluorescence, Analyst. 137 (2012) 1618–1623.

[30] F. Ghasemi, M.R. Hormozi-Nezhad, M. Mahmoudi, Label-free detection of β-amyloid peptides (Aβ40 and Aβ42): A colorimetric sensor array for plasma monitoring of Alzheimer's disease, Nanoscale. 10 (2018) 6361–6368.

[31] S. Xu, Y. Wu, X. Sun, Z. Wang, X. Luo, A multicoloured Au NCs based cross-reactive sensor array for discrimination of multiple proteins, J. Mater. Chem. B 5 (2017) 4207–4213.

[32] Y. He, X. He, X. Liu, L. Gao, H. Cui, Dynamically tunable chemiluminescence of luminol-functionalized silver nanoparticles and its application to protein sensing arrays, Anal. Chem. 86 (2014) 12166–12171.

[33] M.R. Mirghafouri, S. Abbasi-Moayed, F. Ghasemi, M.R. Hormozi-Nezhad, Nanoplasmonic sensor array for the detection and discrimination of pesticide residues in citrus fruits, Anal. Methods. 12 (2020) 5877–5884.

[34] M. Koushkestani, S. Abbasi-Moayed, F. Ghasemi, V. Mahdavi, M.R. Hormozi-Nezhad, Simultaneous detection and identification of thiometon, phosalone, and prothioconazole pesticides using a nanoplasmonic sensor array, Food Chem. Toxicol. 151 (2021) 112109.

[35] N. Mohseni, M. Bahram, T. Baheri, Chemical nose for discrimination of opioids based on unmodified gold nanoparticles, Sens. Actuators B. 250 (2017) 509–517.

[36] A. Orouji, F. Ghasemi, A. Bigdeli, M.R. Hormozi-Nezhad, Providing multicolor plasmonic patterns with Au@ Ag core–shell nanostructures for visual discrimination of biogenic amines, ACS Appl. Mater. Interfaces 13 (2021) 20865–20874.

[37] Y. Tao, D.T. Auguste, Array-based identification of triple-negative breast cancer cells using fluorescent nanodot-graphene oxide complexes, Biosens. Bioelectron. 81 (2016) 431–437.

[38] H. Yang, F. Lu, Y. Sun, Z. Yuan, C. Lu, Fluorescent gold nanocluster-based sensor array for nitrophenol isomer discrimination via an integration of host–guest interaction and inner filter effect, Anal. Chem. 90 (2018) 12846–12853.

[39] X. Zhang, Z. Chen, X. Zuo, Chloroauric acid/silver nanoparticle colorimetric sensors for antioxidant discrimination based on a honeycomb Ag-Au nanostructure, ACS Sustain. Chem. Eng. 8 (2020) 3922–3928.

[40] J. Sun, Y. Lu, L. He, J. Pang, F. Yang, Y. Liu, Colorimetric sensor array based on gold nanoparticles: Design principles and recent advances, TrAC Trends Anal. Chem. 122 (2020) 115754.

[41] K. Saha, S.S. Agasti, C. Kim, X. Li, V.M. Rotello, Gold nanoparticles in chemical and biological sensing, Chem. Rev. 112 (2012) 2739–2779.

[42] W. Zhao, M.A. Brook, Y. Li, Design of gold nanoparticle-based colorimetric biosensing assays, ChemBioChem 9 (2008) 2363–2371.

[43] Z.J.-N. Ivrigh, A. Bigdeli, S. Jafarinejad, M.R. Hormozi-Nezhad, Multiplex detection of antidepressants with a single component condition-based colorimetric sensor array, Sens. Actuators. B 363 (2022) 131855.

[44] J. Mao, Y. Lu, N. Chang, J. Yang, J. Yang, S. Zhang, et al., A nanoplasmonic probe as a triple channel colorimetric sensor array for protein discrimination, Analyst. 141 (2016) 4014–4017.

[45] M.R. Hormozi-Nezhad, S. Abbasi-Moayed, A sensitive and selective colorimetric method for detection of copper ions based on anti-aggregation of unmodified gold nanoparticles, Talanta. 129 (2014) 227–232.

[46] F. Wang, Y. Lu, J. Yang, Y. Chen, W. Jing, L. He, et al., A smartphone readable colorimetric sensing platform for rapid multiple protein detection, Analyst. 142 (2017) 3177–3182.

[47] Y. Lu, Y. Liu, S. Zhang, S. Wang, S. Zhang, X. Zhang, Aptamer-based plasmonic sensor array for discrimination of proteins and cells with the naked eye, Anal. Chem. 85 (2013) 6571–6574.

[48] G. Sener, L. Uzun, A. Denizli, Colorimetric sensor array based on gold nanoparticles and amino acids for identification of toxic metal ions in water, ACS Appl. Mater. Interfaces 6 (2014) 18395–18400.

[49] J.L. Chávez, J.K. Leny, S. Witt, G.M. Slusher, J.A. Hagen, N. Kelley-Loughnane, Plasmonic aptamer–gold nanoparticle sensors for small molecule fingerprint identification, Analyst. 139 (2014) 6214–6222.

[50] F. Ghasemi, M.R. Hormozi-Nezhad, M. Mahmoudi, A colorimetric sensor array for detection and discrimination of biothiols based on aggregation of gold nanoparticles, Anal. Chim. Acta 882 (2015) 58–67.

[51] N. Fahimi-Kashani, M.R. Hormozi-Nezhad, Gold-nanoparticle-based colorimetric sensor array for discrimination of organophosphate pesticides, Anal. Chem. 88 (2016) 8099–8106.

[52] J. Mao, Y. Lu, N. Chang, J. Yang, S. Zhang, Y. Liu, Multidimensional colorimetric sensor array for discrimination of proteins, Biosens. Bioelectron. 86 (2016) 56–61.

[53] X. Wei, Z. Chen, L. Tan, T. Lou, Y. Zhao, DNA-catalytically active gold nanoparticle conjugates-based colorimetric multidimensional sensor array for protein discrimination, Anal. Chem. 89 (2017) 556–559.

[54] J. Sun, Y. Lu, L. He, J. Pang, F. Yang, Y. Liu, A colorimetric sensor array for protein discrimination based on carbon nanodots-induced reversible aggregation of AuNP with GSH as a regulator, Sens. Actuators B 296 (2019) 126677.

[55] H. Xi, W. He, Q. Liu, Z. Chen, Protein discrimination using a colorimetric sensor array based on gold nanoparticle aggregation induced by cationic polymer, ACS Sustain. Chem. Eng. 6 (2018) 10751–10757.

[56]   S. Yan, X. Lai, G. Du, Y. Xiang, Identification of aminoglycoside antibiotics in milk matrix with a colorimetric sensor array and pattern recognition methods, Anal. Chim. Acta 1034 (2018) 153–160.

[57]   L. Huang, X. Zhang, Z. Zhang, Sensor array for qualitative and quantitative analysis of metal ions and metal oxyanion based on colorimetric and chemometric methods, Anal. Chim. Acta 1044 (2018) 119–130.

[58]   J. Yang, Y. Lu, L. Ao, F. Wang, W. Jing, S. Zhang, et al., Colorimetric sensor array for proteins discrimination based on the tunable peroxidase-like activity of AuNPs-DNA conjugates, Sens. Actuators B 245 (2017) 66–73.

[59]   B. Li, X. Li, Y. Dong, B. Wang, D. Li, Y. Shi, et al., Colorimetric sensor array based on gold nanoparticles with diverse surface charges for microorganisms identification, Anal. Chem. 89 (2017) 10639–10643.

[60]   H. Qiang, X. Wei, Q. Liu, Z. Chen, Iodide-responsive Cu–Au nanoparticle-based colorimetric sensor array for protein discrimination, ACS Sustain. Chem. Eng. 6 (2018) 15720–15726.

[61]   L. Huang, X. Zhang, Z. Zhang, UV–vis sensor array combining with chemometric methods for quantitative analysis of binary dipeptide mixture (Gly-Gly/Ala-Gln), Spectrochim. Acta. Part A 221 (2019) 117205.

[62]   F. Jia, Q. Liu, W. Wei, Z. Chen, Colorimetric sensor assay for discrimination of proteins based on exonuclease I-triggered aggregation of DNA-functionalized gold nanoparticles, Analyst 144 (2019) 4865–4870.

[63]   X. Li, S. Li, Q. Liu, Z. Cui, Z. Chen, A triple-channel colorimetric sensor array for identification of biothiols based on color RGB (Red/Green/Blue) as signal readout, ACS Sustain. Chem. Eng. 7 (2019) 17482–17490.

[64]   Z. Sun, S. Wu, J. Ma, H. Shi, L. Wang, A. Sheng, et al., Colorimetric sensor array for human semen identification designed by coupling zirconium metal–organic frameworks with DNA-modified gold nanoparticles, ACS Appl. Mater. Interfaces 11 (2019) 36316–36323.

[65]   X. Li, S. Li, Q. Liu, Z. Chen, Electronic-tongue colorimetric-sensor array for discrimination and quantitation of metal ions based on gold-nanoparticle aggregation, Anal. Chem. 91 (2019) 6315–6320.

[66]   S.F. Wong, S.M. Khor, Differential colorimetric nanobiosensor array as bioelectronic tongue for discrimination and quantitation of multiple foodborne carcinogens, Food Chem. 357 (2021) 129801.

[67]   J. Du, Z. Yu, Z. Hu, J. Chen, J. Zhao, Y. Bai, A low pH-based rapid and direct colorimetric sensing of bacteria using unmodified gold nanoparticles, J. Microbiol. Methods 180 (2021) 106110.

[68]   Y. Li, Z. Qian, C. Shen, Z. Gao, K. Tang, Z. Liu, et al., Colorimetric sensors for alkaloids based on the etching of Au@ MnO2 nanoparticles and MnO2 nanostars, ACS Appl Nano Mater 4 (2021) 8465–8472.

[69]   R. Wu, K. Li, C. Shen, J. Huang, Z. Gao, K. Tang, et al., Antioxidant recognition by colorimetric sensor array based on differential etching of gold nanorods and gold nanobypyramids, ACS Appl Nano Mater 4 (2021) 8482–8490.

[70]   Y. Guan, Y. Lu, J. Sun, J. Zhao, W. Huang, X. Zhang, et al., Redox recycling-activated signal amplification of peroxidase-like catalytic activity based on bare gold nanoparticle–metal ion ensembles as colorimetric sensor array for ultrasensitive discrimination of phosphates, ACS Sustain. Chem. Eng. 9 (2021) 9802–9812.

[71]   L. Du, Y. Lao, Y. Sasaki, X. Lyu, P. Gao, S. Wu, et al., Freshness monitoring of raw fish by detecting biogenic amines using a gold nanoparticle-based colorimetric sensor array, RSC Adv. 12 (2022) 6803–6810.

[72]   Y. Lin, C. Chen, C. Wang, F. Pu, J. Ren, X. Qu, Silver nanoprobe for sensitive and selective colorimetric detection of dopamine via robust Ag–catechol interaction, Chem. Commun. 47 (2011) 1181–1183.

[73]   S. Hajizadeh, K. Farhadi, M. Forough, R.E. Sabzi, Silver nanoparticles as a cyanide colorimetric sensor in aqueous media, Anal. Methods 3 (2011) 2599–2603.

[74]   A. Orouji, S. Abbasi-Moayed, M.R. Hormozi-Nezhad, ThThnated Development of a pH assisted AgNP-based colorimetric sensor array for simultaneous identification of phosalone and azinphosmethyl pesticides, Spectrochim. Acta Part A 219 (2019) 496–503.

[75]   N. Emmanuel, R. Haridas, S. Chelakkara, R.B. Nair, A. Gopi, M. Sajitha, et al., Smartphone assisted colourimetric detection and quantification of Pb 2+ and Hg 2+ ions using Ag nanoparticles from aqueous medium, IEEE Sensors J. 20 (2020) 8512–8519.

[76]   B. Shi, Y. Su, J. Zhao, R. Liu, Y. Zhao, S. Zhao, Visual discrimination of dihydroxybenzene isomers based on a nitrogen-doped graphene quantum dot-silver nanoparticle hybrid, Nanoscale. 7 (2015) 17350–17358.

[77] Z. Li, Z. Wang, J. Khan, M.K. LaGasse, K.S. Suslick, Ultrasensitive monitoring of museum airborne pollutants using a silver nanoparticle sensor array, ACS Sens. 5 (2020) 2783–2791.

[78] M. Wu, H. Chen, Y. Fan, S. Wang, Y. Hu, J. Liu, et al., Carbonyl flavor compound-targeted colorimetric sensor array based on silver nitrate and o-phenylenediamine derivatives for the discrimination of Chinese Baijiu, Food Chem. 372 (2022) 131216.

[79] K. Kalantari, N. Fahimi-Kashani, M.R. Hormozi-Nezhada, Development of a colorimetric sensor array based on monometallic and bimetallic nanoparticles for discrimination of triazole fungicides, Anal. Bioanal. Chem. (2021) 1–12.

[80] M.M. Bordbar, B. Hemmateenejad, J. Tashkhourian, S. Nami-Ana, An optoelectronic tongue based on an array of gold and silver nanoparticles for analysis of natural, synthetic and biological antioxidants, Microchim. Acta 185 (2018) 1–12.

[81] A. Sheini, Colorimetric aggregation assay based on array of gold and silver nanoparticles for simultaneous analysis of aflatoxins, ochratoxin and zearalenone by using chemometric analysis and paper based analytical devices, Microchim. Acta 187 (2020) 1–11.

[82] H. Yu, D. Long, W. Huang, Organic antifreeze discrimination by pattern recognition using nanoparticle array, Sens. Actuators B 264 (2018) 164–168.

[83] D. Li, Y. Dong, B. Li, Y. Wu, K. Wang, S. Zhang, Colorimetric sensor array with unmodified noble metal nanoparticles for naked-eye detection of proteins and bacteria, Analyst. 140 (2015) 7672–7677.

[84] H. Sharifi, J. Tashkhourian, B. Hemmateenejad, An array of metallic nanozymes can discriminate and detect a large number of anions, Sens. Actuators B 339 (2021) 129911.

[85] M. Chaharlangi, J. Tashkhourian, M.M. Bordbar, R. Brendel, P. Weller, B. Hemmateenejad, A paper-based colorimetric sensor array for discrimination of monofloral European honeys based on gold nanoparticles and chemometrics data analysis, Spectrochim. Acta. Part A 247 (2021) 119076.

[86] Y. Li, Q. Liu, Z. Chen, A colorimetric sensor array for detection and discrimination of antioxidants based on Ag nanoshell deposition on gold nanoparticle surfaces, Analyst. 144 (2019) 6276–6282.

[87] X. Zhang, Z. Wang, Z. Liu, B. Liu, R. Wu, Z. Chen, et al., New application of a traditional method: colorimetric sensor array for reducing sugars based on the in-situ formation of core-shell gold nanorod-coated silver nanoparticles by the traditional Tollens reaction, Microchim. Acta 188 (2021) 1–11.

[88] J. Jia, M. Wu, S. Wang, X. Wang, Y. Hu, H. Chen, et al., Colorimetric sensor array based on silver deposition of gold nanorods for discrimination of Chinese white spirits, Sens. Actuators B 320 (2020) 128256.

[89] Y. Li, S. Wang, Q. Liu, Z. Chen, A chrono-colorimetric sensor array for differentiation of catechins based on silver nitrate-induced metallization of gold nanoparticles at different reaction time intervals, ACS Sustain. Chem. Eng. 7 (2019) 17306–17312.

[90] N. Fahimi-Kashani, M.R. Hormozi-Nezhad, Gold nanorod-based chrono-colorimetric sensor arrays: A promising platform for chemical discrimination applications, ACS Omega. 3 (2018) 1386–1394.

[91] X. Yang, J. Li, H. Pei, Y. Zhao, X. Zuo, C. Fan, et al., DNA–gold nanoparticle conjugates-based nanoplasmonic probe for specific differentiation of cell types, Anal. Chem. 86 (2014) 3227–3231.

[92] X.-P. Zhang, K.-Y. Huang, S.-B. He, H.-P. Peng, X.-H. Xia, W. Chen, et al., Single gold nanocluster probe-based fluorescent sensor array for heavy metal ion discrimination, J. Hazard. Mater. 405 (2021) 124259.

[93] Z. Shojaeifard, M.M. Bordbar, M.D. Aseman, S.M. Nabavizadeh, B. Hemmateenejad, Collaboration of cyclometalated platinum complexes and metallic nanoclusters for rapid discrimination and detection of biogenic amines through a fluorometric paper-based sensor array, Sens. Actuators B 334 (2021) 129582.

[94] P. Behera, A. Mohanty, M. De, Functionalized fluorescent nanodots for discrimination of nitroaromatic compounds, ACS Appl Nano Mater 3 (2020) 2846–2856.

[95] N. Cao, J. Xu, H. Zhou, Y. Zhao, J. Xu, J. Li, et al., A fluorescent sensor array based on silver nanoclusters for identifying heavy metal ions, Microchem. J. 159 (2020) 105406.

[96] Y. Shi, X. Gao, W. Wei, Y. Chen, Development of array-based gold nanoclusters for discrimination of CA125 overexpressed serum samples, Adv. Mater. Sci. Eng. 2021 (2021) 7326552.

[97] W. Miao, L. Wang, Q. Liu, S. Guo, L. Zhao, J. Peng, Rare earth ions-enhanced gold nanoclusters as fluorescent sensor array for the detection and discrimination of phosphate anions, Chem. Asian J., 16 (2021) 247–51.

[98]   S. Xu, W. Li, X. Zhao, T. Wu, Y. Cui, X. Fan, et al., Ultrahighly efficient and stable fluorescent gold nanoclusters coated with screened peptides of unique sequences for effective protein and serum discrimination, Anal. Chem. 91 (2019) 13947–13952.

[99]   H. Yang, F. Lu, Y. Sun, Z. Yuan, C. Lu, Fluorescent gold nanocluster-based sensor array for nitrophenol isomer discrimination via an integration of host-guest interaction and inner filter effect, Anal. Chem. 90 (2018) 12846–12853.

[100] H. Ji, L. Wu, F. Pu, J. Ren, X. Qu, Point-of-care identification of bacteria using protein-encapsulated gold nanoclusters, Adv Healthc Mater 7 (2018) e1701370.

[101] Y. Wu, B. Wang, K. Wang, P. Yan, Identification of proteins and bacteria based on a metal ion–gold nanocluster sensor array, Anal. Methods 10 (2018) 3939–3944.

[102] S. Xu, T. Gao, X. Feng, X. Fan, G. Liu, Y. Mao, et al., Near infrared fluorescent dual ligand functionalized Au NCs based multidimensional sensor array for pattern recognition of multiple proteins and serum discrimination, Biosens. Bioelectron. 97 (2017) 203–207.

[103] Y. Tao, M. Li, D.T. Auguste, Pattern-based sensing of triple negative breast cancer cells with dual-ligand cofunctionalized gold nanoclusters, Biomaterials. 116 (2017) 21–33.

[104] Y. Tao, D.T. Auguste, Array-based identification of triple-negative breast cancer cells using fluorescent nanodot-graphene oxide complexes, Biosens. Bioelectron. 81 (2016) 431–437.

[105] Z. Yuan, Y. Du, Y.-T. Tseng, M. Peng, N. Cai, Y. He, et al., Fluorescent gold nanodots based sensor array for proteins discrimination, Anal. Chem. 87 (2015) 4253–4259.

[106] S. Xu, X. Lu, C. Yao, F. Huang, H. Jiang, W. Hua, et al., A visual sensor array for pattern recognition analysis of proteins using novel blue-emitting fluorescent gold nanoclusters, Anal. Chem. 86 (2014) 11634–11639.

[107] Y. Tao, X. Ran, J. Ren, X. Qu, Array-based sensing of proteins and bacteria by using multiple luminescent nanodots as fluorescent probes, Small. 10 (2014) 3667–3671.

[108] M. Shahrajabian, M.R. Hormozi-Nezhad, Design a new strategy based on nanoparticle-enhanced chemiluminescence sensor array for biothiols discrimination, Sci. Rep. 6 (2016) 32160.

[109] Y. He, Y. Liang, H. Yu, Simple and sensitive discrimination of amino acids with functionalized silver nanoparticles, ACS Comb. Sci. 17 (2015) 409–412.

[110] Y. He, B. Xu, W. Li, H. Yu, Silver nanoparticle-based chemiluminescent sensor array for pesticide discrimination, J. Agric. Food Chem. 63 (2015) 2930–2934.

[111] C. Li, P. Wu, X. Hou, Plasma-assisted quadruple-channel optosensing of proteins and cells with Mn-doped ZnS quantum dots, Nanoscale. 8 (2016) 4291–4298.

[112] W. Sun, Y. Lu, J. Mao, N. Chang, J. Yang, Y. Liu, Multidimensional sensor for pattern recognition of proteins based on DNA–gold nanoparticles conjugates, Anal. Chem. 87 (2015) 3354–3359.

[113] Y. Lu, H. Kong, F. Wen, S. Zhang, X. Zhang, Lab-on-graphene: Graphene oxide as a triple-channel sensing device for protein discrimination, Chem. Commun. 49 (2013) 81–83.

[114] Y. Gui, Y. Wang, C. He, Z. Tan, L. Gao, W. Li, et al., Multi-optical signal channel gold nanoclusters and their application in heavy metal ions sensing arrays, J. Mater. Chem. C 9 (2021) 2833–2839.

[115] J.-Y. Yang, X.-D. Jia, X.-Y. Wang, M.-X. Liu, M.-L. Chen, T. Yang, et al., Discrimination of antibiotic-resistant Gram-negative bacteria with a novel 3D nano sensing array, Chem. Commun. 56 (2020) 1717–1720.

[116] J.-Y. Yang, T. Yang, X.-Y. Wang, Y.-T. Wang, M.-X. Liu, M.-L. Chen, et al., A novel three-dimensional nanosensing array for the discrimination of sulfur-containing species and sulfur bacteria, Anal. Chem. 91 (2019) 6012–6018.

[117] L. Tan, Z. Chen, Y. Zhao, X. Wei, Y. Li, C. Zhang, et al., Dual channel sensor for detection and discrimination of heavy metal ions based on colorimetric and fluorescence response of the AuNPs-DNA conjugates, Biosens. Bioelectron. 85 (2016) 414–421.

CHAPTER 7

Synthesis, characterization, and applications of Ag and Au nanoparticles in obtaining electrochemical bio/sensors

Mônika Grazielle Heinemann, Caroline Pires Ruas, Daiane Dias

Laboratório de Eletroespectro Analítica (LEEA), Escola de Química e Alimentos, Universidade Federal do Rio Grande, Rio Grande, Brazil

7.1 Introduction

There is an increasing interest in nanoscience and nanotechnology in the recent years due to the impact of nanostructured materials in the improvement of life quality [1]. The term 'nano' was firstly mentioned in 1914 by Richard Adolf Zsigmondy. In 1959 during the American Physical Society's annual event, the physicist Richard Feynman introduced the concept of nanotechnology, being considered the "father" of modern nanotechnology [2,3]. In 1974, the Japanese engineer Norio Taniguchi was the first to use the term nanotechnology to classify this new technology that uses nanostructured materials [3]. Currently, the meaning is closer to what was described by Eric Drexler, which corresponds to the processing method involving atom-by-atom manipulation [4]. Therefore, nanotechnology is the study of the phenomenon and manipulation of physical systems that produce significant information with noticeable differences on a nanoscale (10^{-9} m) focusing on the design, characterization, production, and applications of systems and components at nanoscale [5]. At the end of the 20th century, the study of nanomaterials achieved significant advancement [6]. Thus, it is possible to define nanomaterials as materials that have at least one dimension in the nanometer range and different properties (size, distribution, phase, morphology, and crystallinity) when compared to the same material in the macroscopic scale (bulk), highlighting the fact that material properties are strongly dependent on particle size [6].

Among the various nanomaterials, the metallic nanoparticles (MNPs) was first synthesized by Michael Faraday in the 19th century [7]. Among MNPs, gold (Au) and silver (Ag) nanoparticles (NPs) present the most applications, as they have excellent optical, electronic, magnetic, chemical, physical, biological and electrical properties [8]. In addition, they showed good biocompatibility (when synthesized via green route), large surface area and easy surface functionalization to improve the electronic transfer processes [9].

Gold and Silver Nanoparticles.
DOI: https://doi.org/10.1016/B978-0-323-99454-5.00009-3

Nanoparticles are thermodynamically unstable and tend to aggregate. Therefore, the preparation of stable form of NPs is a great challenge [10]. Stability control is essential to maintain the important properties of the size-dependent nanoparticles for efficient applications in different areas [11]. Therefore, researchers are constantly looking for new synthesis or functionalization methods to increase the chemical stability of NPs, which will be discussed in this chapter. Several methods are employed for the synthesis of Au and Ag based NPs, which can be broadly divided into chemical (bottom-up), physical (top-down) and biogenic [12]. The top-down method consists of physical/mechanical processes (laser, ablation, microwave, lithography, etc.) that allows to manipulate materials from macro to nanometer scale [13]. The bottom-up method involves a chemical processes (chemical, reduction, solution evaporation, sol-gel process, etc.) in which the growth of nanoparticles can be controlled atom by atom [12]. The combination of both methods can favor the formation of stable NPs.

However, regardless of the methods used, the characterization techniques like Scanning Electron Microscopy (SEM), Field Emission Scanning Electron Microscopy (FE-SEM), Transmission Electron Microscopy (TEM), Fourier transform infrared spectroscopy (FTIR), X-ray diffraction (XRD), Energy Dispersive X-ray (EDX), ultraviolet visible (UV-Vis) spectroscopy, Dynamic light Scattering (DLS), Atomic force microscopy (AFM), X-ray photoelectron spectroscopy (XPS), etc. are used to characterize the morphology, composition and other important properties of the nanostructure formed.

Despite the many advantages of using NPs, negative impacts on the environment and human health can occur due to the toxic reagents normally used in chemical and physical methods (like sodium borohydride, amines, hydrazine hydrate, thiols etc.) besides the use of high temperatures, high energy consumption, radiation and pressure [14]. Thus, ecologically friendly and biocompatible reagents can reduce the toxicity of the NPs formed, causing a smaller impact [15]. In this way, green nanotechnology that can be related with the use of biogenic sources aims to reduce consequences linked to the production of NPs via chemical (using toxic solvents) and physical route (with the reduction of synthesis time and energy used) [16]. Among these biogenic sources plant and algae extracts, fungi and bacteria can be employed as reducing and stabilizing agents in the production of AuNPs and AgNPs [14].

Biogenic synthesis (or green synthesis) has several advantages when compared to traditional synthesis because of the use of natural compounds and/or extracts, and more economical procedures that enabling the formation of more stable NPs with a greater reproducibility [17]. The biogenic sources are composed by biomolecules such as proteins, polysaccharides, phenolic compounds, among others, that are able to act simultaneously in the reduction and stabilization of NPs (in only one step), without the need of other reagents [18]. Furthermore, the use of biogenic sources can minimize eventual toxic effects of NPs and also favor the interaction with biological systems such as cells, proteins or tissues [19].

Thanks to the excellent properties of AuNPs and AgNPs have present several applications in different areas such as antifungal [20], antimicrobial [21], antibacterial, insecticidal, antioxidant and membrane damage [22] and catalysis [23]. Additionally, they can be used in the removal of contaminants [24,25], as antioxidant function [26] and photocatalytic dye degradation activity [27]. In addition, these nanomaterials are used in the preparation of electrochemical sensors [28] or biosensors [29] due to the large surface area and their electro-active, improving the electrocatalytic effects, which enables the dimmish of the over potential reaction of many analytes and leads to the obtention of methods with optimal detection limits [28]. NPs also help to increase the sensitivity, selectivity, conductivity, signal/noise ratio and electron transfer in these devices [29].

Electrochemical techniques have great advantages over chromatographic and/or spectrometric techniques because of the low cost and sensitive, enabling the obtention of determination in a short time of analysis and with detection limits comparable to those before mentioned [30]. The modification of electrochemical bio/sensors from MNPs represents a valuable strategy for the development of selective devices, with high sensitivity enabling the detection or monitoring of drugs [28,31], cancer [29], toxins [30], nitro compounds [32], hydrogen peroxide [33], pesticides [34], thiourea [35], condensed tannins [36], dye degradation [37], SARS-CoV-2 spike protein [38,39], and others. Therefore, this chapter was narrated with the aim to address general aspects of the physical, chemical and biogenic synthesis of AuNPs and AgNPs, the main characterization methods applied and the applications of these NPs in the development of bio/sensors followed by the future perspectives about this important topic.

7.2 Synthetic approaches for nanoparticles

7.2.1 Bottom-up or chemical methods

The bottom-up method or chemical method used in general for the preparation of MNPs, including AuNPs and AgNPs. This method is able to produce nanomaterials with fewer defects, known chemical composition, size and shape [40]. Furthermore, in this type of reaction, it is possible to control experimental parameters, such as reaction time, temperature, concentration of the precursor and the solvent used by chemical reduction (sol-gel method, ultrasound, plasma process and others) that directly influence the morphology of MNPs [41]. These experimental parameters can be controlled in order to obtain MNPs with known morphology, presenting advantages during the application of the formed nanomaterials, as well as in their physical and chemical properties [42]. Table 7.1 depict some works described in the literature that used different methods by the bottom-up to prepare AuNPs and AgNPs.

In the chemical method, the interaction between metallic ion and reducing agent as well as the adsorption of the stabilizing agent with the metallic surface directly influences the chemical and physical properties, the morphology (shape and size) and the stability of MNPs [54]. Using chemical reduction of metallic precursors is possible to obtain AuNPs and AgNPs with high reproducibility, well-defined morphology and high surface area [55]. However, this process presents the disadvantage of impurities of the medium [54].

Table 7.1: Bottom-up synthesis of AuNPs and AgNPs.

Metallic precursor	Method of synthesis	Size (nm)/Shape	Characterization techniques	Refs.
$HAuCl_4$	Chemical reduction	20/spherical	TEM	[43]
$HAuCl_4$/		10 ± 3/spherical	UV-vis, XRD, TEM	[44]
$AgNO_3$		7 ± 3/spherical		
$HAuCl_4$		5 – 50/spherical	UV-vis, SEM	[45]
$HAuCl_4$		10 ± 1.4/spherical	UV-vis, TEM, FTIR	[46]
$HAuCl_4$		20 – 40/spherical	UV-vis, SEM, DLS, Zeta potential	[47]
$AgNO_3$		5/spherical	UV-vis, TEM	[48]
$HAuCl_4$	Sol gel process	15/spherical	UV-vis, TEM	[49]
$AgNO_3$		20 – 45 /[a]	AFM	[50]
$HAuCl_4$	Plasma process	20/anisotropic	TEM, SAED	[51]
$HAuCl_4$		4/spherical	UV-vis, EDX, TEM	[52]
$AgNO_3$	Ultrasonic irradiation	3/spherical	UV-vis, EDX, XRD, TEM	[53]

[a] not described.

The nucleation and growth of MNPs by chemical reduction can be performed by methods of Turkevich, Brust-Schiffrin and seed growth [56]. In the nucleation step, the reduction of the metallic precursor occurs, originating atoms with an oxidation state equal to zero [55]. MNPs are kinetically stable, but thermodynamically unstable tending to agglomerate without size control during the growth process, resulting in the formation of metallic bulks [57]. Therefore, during the nucleation step, the reaction rate must be slow to avoid the MNPs agglomeration [58].

In the literature, many works used the bottom-up method to form MNPs where experimental parameters are modified to obtain more stable materials. Metallic precursors of Au (such as $HAuCl_4$) and of Ag (such as $AgNO_3$) are used in most cases for the preparation of AuNPs and AgNPs, under agitation and reflux, in the presence of stabilizers and reducing agents. Dendrimers, ionic liquids, surfactants and/or polymers are reported as good stabilizers because of their ability to control size and prevent MNPs clumping through steric hindrance and/or electrostatic repulsion [59]. In the stabilization of MNPs by steric hindrance, the coordination of bulky molecules on the metallic surface occurs with the use of polymers (e.g., polyvinyl-pyrrolidone and polyvinyl alcohol). On the other hand, in the stabilization induced by electro-static repulsion, Coulomb repulsion occurs between the particles caused by the double layer of charges, formed by cations and anions present on the metallic surface [57,58,60,61].

The first synthesis by means of chemical reduction was described in 1857 by Michael Faraday, where he prepared AuNPs by the reduction of an aqueous solution of $HAuCl_4$ with phosphorus dissolved in carbon disulfide [62]. In the end of 19th and in the beginning of 20th century, researchers described AuNPs preparation methods using reducing agents (hydrogen peroxide, formaldehyde) where the solutions stability were linked to the charge of the colloidal particles [40]. In this period, mathematical calculus emerged stipulating the distance that the particles had to in order to avoid agglomeration.

A well-known synthesis is the Turkevich where the metallic precursor $HAuCl_4$ is reduced with citric acid in an aqueous medium, verifying the acid's performance as a reducing and stabilizing agent, obtaining spherical AuNPs with an average size of 1–2 nm [62]. Over the years, this method has undergone adaptations, making it possible to change the experimental parameters, nowadays known as the "Turkevich reduction method" [63].

As reducers, hydrazines, molecular hydrogen (H_2), sodium citrate ($Na_3C_6H_5O_7$), ascorbic acid ($C_6H_8O_6$) and sodium borohydride ($NaBH_4$) can be highlighted, because they act as electron donor substances during the reduction of metallic precursors [64]. The reduction can be observed by the change of solution color containing the metallic precursor, reducing and stabilizing agent, which is the first indication of MNPs formation. For AuNPs, the color change is quite characteristic. Usually, the solution changes from yellow to "reddish purple" indicating the formation of these particles. However, other colors can be observed, according to the variation in their shape and size [65]. Furthermore, due to the optical properties of AuNPs and AgNPs, they can be characterized by spectroscopy in the visible ultraviolet region, where is possible to observe the characteristic plasma bands [66].

A very common application of MNPs is as catalysts, mainly due to their high surface area, resulting from the surface-volume ratio. However, surface active atoms generally tend to clump, decreasing the activity and the selectivity during the catalytic process. Therefore, it is important to use a stabilizing agent that does not interact with the surface [67]. Additionally, AuNPs and AgNPs, for example, are applied in the manufacture of biosensors due to their excellent physical, chemical, mechanical and optical properties, in addition to the high stability in different environments [68,69].

The sol-gel method is also widely described in the literature for the nanomaterial's preparation [70]. The term "sol" is used to define a dispersion of colloidal particles with an average size between 1 and 100 nm, while the term "gel" is related to the rigid structure of the colloidal particles. Colloidal chemistry, as the method involving the sol-gel can be called, is a promising route for the preparation of nanomaterials which has many advantages, especially when combined with MNPs. This method involves polymerization reactions, through hydrolysis and condensation, catalyzed by an acid or a base [71]. The synthesis process of this method can be divided into a few steps. The first is the hydrolysis of metal oxide, which occurs in the presence of water and/or alcohol, to form the gel. The following step is condensation, where the formation of the polymeric network and the aging process occurs with changes in structure, properties and porosity [3]. After the aging step, drying step takes place (elimination of water and organic solvent) followed by calcination to assist the nanomaterials formation. An example of this method modification, was reported in 2011 by Yao and co-authors [72], where they prepare silica nanowires modified with AuNPs by the sol-gel method.

Sol-gel synthesis is a chemical reaction that involves the presence of several intermediaries, which can generate MNPs by gelation, precipitation and hydrothermal treatment, being a promising alternative for the nanomateriais preparation [73]. According to the literature, one of the main advantages of the sol-gel method is its versatility, as well as its reproducibility, allowing the incorporation of MNPs during the process to prepare nanomaterials with greater applicability [74].

Another reported and widely used method for the formation of MNPs is the ultrasound bath. The ultrasound process is based on acoustic cavitation and, due to the intensity of the cycles in the stages of bubble generation, waves are produced increasing the collisions between the particles [75]. In the work developed by Ganguly and co-authors [75], graphene sheets were prepared impregnated with AgNPs by the top-down method, with the sonicator. The process was completely "green", as it did not use organic solvent or stabilizers in the reaction medium.

However, it is noteworthy that the use of the ultrasound bath, as well as other methods described in the literature, allows the combination of the two methods (bottom-up and top-down) during the process of the MNPs formation. In this case, several factors such as power and ultrasonic frequency, temperature and solvent, directly influence the size, physical and chemical properties of NPs obtained [76].

MNPs obtained by plasma processes generally requires the use of few chemical additives compared to methods involving the bottom-up method [41]. According to Yonezawa and co-authors [41], this process produces NPs with a wide size distribution and the particles tend to agglomerate, which would not be an ideal condition. Therefore, the method chosen for the synthesis of MNPs is very important, as it will directly influence the desired application of the nanomaterial, mainly considering the stability and morphology of the MNPs.

7.2.2 Top-down or physical methods

The top-down method consists of physical/mechanical processes that allow the material to be manipulated from a macrometric scale to the desired scale. The preparation of MNPs is based on the reduction of the starting material by different chemical and physical processes [54]. Many techniques can be highlighted in the nanomaterials preparation, such as lithographic process, laser ablation, chemical vapor deposition and laser pyrolysis. Among the methods mentioned, laser ablation stands out for being a method that can be easily controllable in the MNPs or bimetallic preparation [77]. Table 7.2 depicts some works that used different methods by the top-down approaches to prepare AuNPs and AgNPs.

Table 7.2: Top-down synthesis of AuNPs and AgNPs.

Metallic Precursor	Method of synthesis	Size (nm)/shape	Characterization techniques	Refs.
Precursor metallic of Ag	Evaporation/ condensation	4 – 150/[a]	TEM, XRD	[78]
AgNO$_3$ and Ag(S$_2$O$_3$)$_2$ $^{3-}$	Sonochemical	2 – 8/[a]	UV-vis, SEM, AFM	[79]
HAuCl$_4$.3H$_2$O	Lithography	2 – 6/[a]	SEM, AFM	[80]
Ag(fod)(PEt$_3$)	CVD	4/[a]	SEM, XPS, AFM, XRD	[81]
[a]	Laser ablation (AuNPs)	1 – 2/spherical	UV-vis, HRTEM	[82]
[a]		2 – 12/spherical	UV-vis, TEM	[83]
[a]		40 – 60/spherical	SEM	[84]
[a]	Laser ablation (AgNPs)	1 – 9/spherical	UV-vis, HRTEM	[82]
[a]		10 – 20/spherical	UV-vis, TEM	[85]

[a] Not informed.

The top-down approach involves the preparation of MNPs in either a liquid, a gas or vapor phase. The method involved is able to prepare NPs based on specific conditions (such as temperature) in order to obtain greater control over size. In the liquid phase, specifically, the size of NPs depends on the control over the separation or precipitation of phases, through the change of parameters such as pH or temperature, during the reaction [40]. The top-down approach, when used, favors the control of the size and shape of the NPs, forming monodisperse materials with well-defined morphologies, however the cost of the equipment needed for the synthesis is high [86].

Fu and coauthors [40] present a comprehensive review of the application of MNPs prepared by the top-down method using the lithography process (photolithography, electron beam lithography, nanoimprint lithography) generating NPs with well-defined shapes and sizes. However, other methods have been reported by several researchers. In 1995, Esumi and coauthors [87] reported a physical method of reducing the metallic precursor (HAuCl$_4$) under an excitation of ultraviolet light forming anisotropic AuNPs.

Other important method applied to MNPs synthesis is laser ablation that involves the vaporization of the metallic precursor due to the high energy of laser irradiation resulting in the formation of the nanomaterial [88]. In this method, a metallic precursor is mixed in a solvent containing a stabilizing agent (usually a surfactant) and during the irradiation process, the metallic atoms are vaporized and solvated by surfactant molecules forming the MNPs [89]. However, it is also possible to perform the reaction without any chemical additives (stabilizers or organic solvents) [90]. Zhang and coauthors reported the synthesis of AuNPs by the laser ablation technique resulting in the formation of nanomaterials with well-defined morphology and high catalytic efficiency [91].

Chemical vapor deposition (CVD) is also widely reported in the literature as top-down method to prepare nanomaterials based on carbon (such as graphene and carbon nanotubes) with excellent quality, especially due to the high surface area [3]. It is important to highlight that in the CVD, the particle size distribution is controlled by mixing the cold and hot gas that carries the evaporated metal, resulting in particles with well-defined morphology [92]. This method is based on the deposition of solid films on the substrate surface through chemical reactions. During this process, the choice of a good catalyst is important, as it plays a significant role in the morphology of the formed nanomaterials [3]. To the best of our knowledge, this method has not been applied to synthetize AgNPs and AuNP. However, Bhaduri, and co-authors [93] reported the study about synthetic carbon nanofibers (CNFs) and CVD to obtain the Cu-CNFs composite by calcination, reduction and deposition by chemical vapor. Firstly, the $Cu(NO_3)_2$ salt was dissolved in deionized water and impregnated into graphite powder. During the calcination step, $Cu(NO_3)_2$ was adhered to the graphite surface and converted into copper oxides (Cu_2O and CuO) followed by the reduction process with H_2.

Another reported process is the synthesis by lasers pyrolysis which consists of the decomposition of metallic precursors in the presence of inert gases (argon or helium). The control of the MNPs morphology is made by the gas pressure, triggering the nucleation and growth, influencing the size and distribution of the nanostructures [89,94].

The methods described above are flexible and quite versatile techniques being able to be applied not only in the preparation of MNPs, but also of other nanomaterials (nanoceramics, for example) with the possibility of experimental adjustments before being applied in several areas [95]. There are many methods developed for the formation of MNPs using the top-down method, but this method has disadvantages that need to be taken into account. One of them is the impurities that may be present on the surface of the NPs and the other is the difficulty in controlling the shape and size which will influence the desired application of the nanomaterial [96].

Nonetheless, there are many examples also using the combination of the bottom-up and top-down methods to obtain materials with "clean" and greater surface area. The high surface area directly influences the catalytic activity, as it is associated with several material properties, such as strong surface reactivity [97]. Both methods are capable of producing MNPs with high surface area and well-defined chemical and physical properties. However, their combination can be considered for the production of nanomaterials with interesting properties [98].

7.2.3 Biogenic synthesis

Biogenic nanoparticles (BNPs) have more advantages when compared to those synthesized by chemical and physical route, since in the biogenic synthesis, chemical reagents are not used (only the metallic precursor), the BNPs present greater stability and biocompatibility,

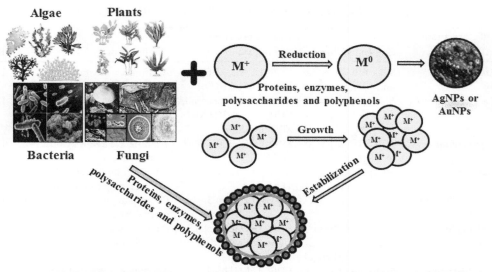

Fig. 7.1: Mechanism of formation of AuNPs and AgNPs synthesized by biogenic route.

reducing their toxicity and not occurring the formation of other substances [99]. In addition, in the biogenic synthesis, toxic and aggressive products are not used, high energy consumption is not demanded, the processes are considered low cost, water is used as a solvent, the reaction occurs in only one step, it can be used on a large scale and dispenses the use of stabilizer and coating chemicals [100].

In the formation of BNPs, microorganisms and/or biomolecules from organisms that act as reducing, capping and coating agents [101]. The use of bioactive compounds and organisms allows less use of chemical reagents and synthesis steps, which reduces NPs of low toxicity bringing new perspectives for their use in humans [25]. The biogenic synthesis of AuNPs and AgNPs using plants, algae, bacteria and fungi is widely discussed in the literature [102]. A schematic illustrating the mechanism of biogenic synthesis is shown in Fig. 7.1. As it can be observed, the biomolecules present in these biogenic sources are able to reduce metallic ions (Ag^+ to Ag^0 or Au^{3+} to Au^0) and controlled their growth by stabilizing the BNPs formed [22]. Next, the biosynthesis of AuNPs and AgNPs will be discussed in detail.

7.2.3.1 Plants

Biogenic synthesis from plant extracts or from the living organism itself (phytosynthesis) is the most used in the production of MNPs [17,18,103,104]. Plants contain metabolites such as flavonoids, terpenoids, soponins and polyphenols, as well as proteins for several aspects related to their physiology [17]. These biomolecules are responsible for the reduction and stabilization of MNPs, which making plants extracts as one of the main suppliers of molecules for biosynthesis [103]. Extracts from the root, leaf, bark, seeds, flowers and fruits are mixed

Table 7.3: Biogenic synthesis of AgNPs and AuNPs with plants.

Metallic precursor/ Plant	Size (nm)/shape	Characterization techniques	Applications	Refs.
Au/*Linum usitatissimum*	3.4 – 6.9/spherical and triangular	FTIR, UV-vis, XRD, XPS, EDX and TEM	Antitumor activity	[105]
Au/*Persicaria salicifolia*	5 – 23/spherical	UV-vis, Zeta potential, FTIR, EDX, HRTEM, XRD and XPS	Medical	[106]
Au/*Nigella sativa*	20 – 50/polygonal or elliptical	XRD, FTIR, EDX, TEM and DLS	Therapeutic agent against gastric cancer	[107]
Au/*Kaempferia parviflora*	44/spherical	UV-vis, FTIR, XRD and TEM	Antimicrobial, antioxidant and catalytic degradation agent	[108]
Au/*Camellia Sinensis* (green tea)	43/[a]	UV-vis, EDX, DLS	Photocatalytic degradation of methylene blue	[109]
Au and Ag/*Nypa fruticans*	10 – 15 (AgNPs) and 15 – 20 (AuNPs)/spherical	FTIR, XRD, EDX, DLS, UV-vis, SEM and TEM	Bacterial activity and catalytic degradation	[103]
Au/*Curcuma wenyujin*	100/spherical	FTIR, TEM, UV-vis, DLS and AFM	Anticancer potential against breast cancer cells	[110]
Au/*Citrus limonum*	4 – 90/spherical	EDS, FE-SEM, UV–vis and FTIR.	Detection of dichlorvos residue	[111]
Au/*Simarouba glauca*	10/prism and spherical	XRD, FTIR, UV-vis and HRTEM	Antimicrobial activity	[112]
Ag/*Malva parviflora*	50.6/spherical	UV-vis, FTIR, TEM, FE-SEM and Zeta potential	Antifungal activity	[18]
Ag/*Kalanchoe pinnata*	40.8/spherical	UV-vis, XRD, Zeta potential, FESEM, HRTEM	Antibacterial activity and photocatalytic activity	[19]
Ag/*Eucalyptus camaldulensis*	28/spherical	UV-vis, FTIR, XRD, SEM, EDX, DLS and Zeta potential	[a]	[17]
Ag/jasmine flower	22/fiber shape	XRD, FTIR, UV-vis, SEM and TEM	biological activities and photocatalytical degradation	[104]
Ag/*Moringa oleifera*	8/spherical	TEM, UV-vis, XRD, EDS and FTIR	Antimicrobial activities and copper detection	[113]
Ag/*Delonix regia*	72.77/nonuniform spherical	TEM, SEM, XRD, UV-vis and FTIR	Anticancer activity	[114]

[a] Not informed.

with a solution of the metallic precursor (usually $AgNO_3$ for AgNPs and $HAuCl_4$ for AuNPs) and kept in the conditions for the synthesis of NPs [102]. A large number of plant species are found in nature and many can be used in the biosynthesis of MNPs, as summarized in Table 7.3. Important information regarding size and shape of AgNPs and AuNPs, the characterization techniques applied and some applications are described.

In the work of Hosny and Fawzy [106], the use of aqueous extract of the leaf of *Persicaria salicifolia* plant was reported for the first time to obtain spherical AuNPs with a diameter around 5–23 nm and plasmon peak at 535 nm indicating the success in phytosynthesis in the formation of NPs. Several molecules such as quercetin glycosides, myricetin glycosides, polyphenols (e.g., tannin and anthraquinone), acylated flavonoids, apigenin, and luteolin 7-O-glucoside are found in the chemical composition of *Persicaria salicifolia* [115,116]. Rutin and quercetin are the flavonoids most common found in this species. Thus, these two compounds play a dual role as a reducing and stabilizing agent once they proved to be capable of donating electrons or hydrogen atoms to gold ions and reducing to Au^0 [117]. The FTIR analysis showed the characteristic bands corresponding to glycosides and flavonoids that are present in the extract of *Persicaria salicifolia* [106].

AuNPs were biosynthesized with *Kaempferia parviflora*, which is a medicinal plant commonly known as black ginger and has extensive pharmacological activity. Several biomolecules with volatile oil, phenolic glycosides and numerous flavonoids reduced and stabilized the AuNPs [118]. The reaction for the formation of AuNPs required 30 minutes at 30°C. The surface plasmon resonance (SPR) band at 530 nm confirmed the AuNPs formation and the TEM analysis supported that the spherical-shaped nanomaterials were well dispersed with high crystal in nature and size of 44±3 nm [108].

Doan and co-authors [103] used the *Nypa fruticans* fruit peel to biosynthesize NPs of Au and Ag. With the optimization of parameters such as concentration, temperature and time, the authors obtained spherical NPs with size ranging from 10–15 nm for the AgNPs and 15–20 nm for AuNPs. In addition, the biomolecules present in this plant, such as phenolic compounds and proteins play an important role in the reduction of the metal ions and in the stabilization of the NPs formed due to the fact that unstable phenoxyl radicals formed from the oxidation of phenolic groups reduced the metallic ions and the metallic nuclei produced from flocculation [119].

Thangamani and Bhuvaneshwari [112] evaluated different volumes of leaf extract from *Simarouba glauca* tree and different volumes of metallic precursor ($HAuCl_4$) to synthesize AuNPs with small diameters. Among the volumes used, 0.3 mL of extract and 2.7 mL of $HAuCl_4$ (0.4 mg mL^{-1}) were optimized to generate NPs with a smaller diameter of 10 nm. Additionally, it was also possible to observe by high-resolution transmission electron microscope (HR-TEM) images that the AuNPs showed different morphologies with a mixture of prism and sphere-like particles with a decrease of the size when there was an increase in the concentration of the leaf extract [112].

Delonix regia leaf extract was used for the biogenic synthesis of AgNPs and the UV-Vis spectrum showed a band at 455 nm indicating the formation of AgNPs [114]. The DLS analysis showed a hydrodynamic diameter of 72.77 nm to the NPs, the XRD analysis confirmed the crystalline nature of AgNPs and the FTIR showed peaks corresponding to the phenols and amines that are present in the *Delonix regia* extract, which probably acted in the reduction and stabilization of AgNPs.

7.2.3.2 Algae (phycosynthesis)

Algae are living organisms that have a large number of species, which make them very attractive due to the possibility of synthesize different types of NPs providing opportunities for new scientific discoveries [120]. They can be divided into microalgae and macroalgae. These two groups being phylogenetically different, but they share a mutual characteristic, the plastids, which are structures arising from an endosymbiotic event between a photosynthetic cyanobacterium and a eukaryotic cell [121].

Microalgae are unicellular and inhabit benthic zones and deep waters, while macroalgae are multicellular beings that can be found in coastal zones between seas and in the sub-tidal region [122]. They are classified according to the predominance of their pigments in brown algae (phylum *Phaeophyta*), red algae (phylum *Rhodophyta*) and green algae (phylum *Chlorophyta*) and have bioactive compounds with structures different from those found in other living beings and with functional activities selected [123]. Algae, in their chemical compositions have polysaccharides, proteins, polyunsaturated fatty acids, and pigments with bioactive compounds that are able to reduce metallic ions and stabilize the AuNPs and AgNPs formed [124]. Several factors such as easy availability, nontoxic nature, presence of excellent bioactive compounds make algae a great option for the biogenic synthesis of NPs [10], as it can be seen in the works cited in Table 7.4.

AgNPs were biosynthesized using *Spirogyra hyalina* (green algae) as reducing and covering agent. The synthesized AgNPs showed a characteristic SPR peak at 451 nm. The SEM analyzes supported the formation of spherical shape AgNPs with average size of 52.7 nm [22]. Furthermore, the results obtained by FTIR showed peaks corresponding to amines and alcohols, CH bond stretching of alkene and alkane, C=C bond stretching of alkene, NO bond stretching of nitro compound and OH bond stretching of carboxylic acid [22]. These different functional groups associated with AgNPs came from *Spirogyra hyalina* extract, which participated in the capping and reduction of Ag ions. In another study, four green algae, two chlorophytes (*Coelastrum astroideum* and *Desmodesmus armatu*) and two charophytes (*Cosmarium punctulatum* and *Klebsormidium flaccidum*) were used in the preparation of AgNPs. The TEM images supported the formation of spherical shape AgNPs with size ranging from 1.8 to 5.4 nm [128]. Additionally, the FTIR analysis demonstrated the presence of hydroxyl groups of peptidoglycan nature acting as stabilizing agents on the surface of AgNPs. With this study, the authors discovered that there was no pattern for AgNPs in terms of microorganism phyla, once the characteristics of AgNPs change depending on the structures of the cellular machinery of the microorganisms [128]. Considering the different size of biosynthesized AgNPs, they can be applied in catalytic and biological processes as their physical and chemical properties strongly depend on size and shape, which will directly influence their structure and the reactivity [27,128].

Table 7.4: Biogenic synthesis of AgNPs and AuNPs with algae.

Metallic precursor/Plant	Size (nm)/shape	Characterization techniques	Applications	Refs.
Au/*Sargassum plagiophyllum* (brown)	65.87/poly-dispersed	UV-vis, FTIR, HR-TEM, Zeta potential and XRD	Antimicrobial effect	[125]
Au/*Padina tetrastromatica* (brown)	10 – 70/spherical	UV-vis, XRD, SEM, TEM and FTIR	Anticancer activity	[126]
Au/*Gracilaria verrucosa* (red)	20 – 80/different shapes	UV-vis, TEM, XRD, FTIR and Zeta Potential	Biocompatibility against normal human embryonic kidney (HEK-293) cell lines.	[124]
Au/*Caulerpa racemosa* (green)	13.7 – 85.4/spherical-to-oval	UV-vis, XRD, FTIR, FE-SEM and HR-TEM,	Controlled the growth of human colon adenocarcinoma (HT-29) cells	[127]
Ag/*Spirogyra hyalina* (green)	52.7/spherical	UV-vis, FTIR, SEM, XRD and EDX	Antibacterial, antifungal, insecticidal and antioxidant	[22]
Ag/chlorophytes and charophytes	1.8 – 5.4/spherical	UV-vis, FTIR, TEM and EDS	[a]	[128]
Ag/*Fucus gardeneri* (brown)	19.39/spherical	UV-vis, HRTEM, XRD, FTIR and EDX	Reduction of nitrophenol	[129]
Ag/*Chlorella vulgaris* (green)	55.06 ± 9.67/spherical	FTIR, FESEM, UV-vis and XRD	Photocatalytic dye degradation	[27]
Ag/*Gelidium corneum* (red)	20 – 40/spherical or angular	UV-vis, TEM, EDS, XRD and FTIR	Antimicrobial and Antibiofilm Activity	[130]
Ag/*Spyridia filamentosa* (red)	20 – 30/spherical	UV-vis, TEM, FTIR and XRD	Antibacterial activity and anticancer activity	[131]
Ag/*Ulva flexuosa*	4.93 – 6.70/spherical	HR-TEM, UV-vis, FTIR, SEM and EDX	Water disinfection	[132]
Ag/*Enteromorpha compressa*	14/spherical	UV-vis, XRD, FTIR, HRTEM and EDX	Biomedical and Pharmaceutical	[133]
Ag/*Ecklonia cava*	43/spherical	UV-vis, TEM, DLS,	Pharmaceutical, nutraceutical and cosmeceutical	[134]
Ag/*Gracilaria birdiae* (red)	20.2- 94.9/spherical	UV-vis, FTIR, DLS and TEM	Antimicrobial (*Escherichia coli* - Gram-negative) and (*Staphylococcus aureus* - Gram-positive)	[135]

[a] Not informed.

AuNPs can also be biosynthesized with different types of algae, such as the green seaweed *Caulerpa racemosa*. In this synthesis, an aqueous solution of $HAuCl_4$ (1 mol L^{-1}) was used in a 1:1 v/v ratio with the algae extract and the reaction was kept under stirring for 24 hours [127]. An absorption band was observed at 528 nm (characteristic of AuNPs), the zeta potential analysis confirmed the stability of the NPs formed (–27.4 mV) and the HR-TEM measurement showed a diameter for NPs from 13.7 to 85.4 nm.

Chellapandian and co-authors [124] evaluated the extract of the red *alga Gracilaria verrucosa* in the green synthesis of AuNPs. In this synthesis, first the *alga* extract was heated at 60°C for 20 minutes and then the $HAuCl_4$ solution was added by dripping, until reach 0.0199 mol L^{-1}

and the mixture was remained under stirring vigorously for another 20 min. The TEM results revealed several morphologies such as spherical, oval, pentagonal, rhombohedral and triangular, with a diameter between 20 and 80 nm. The zeta potential analysis showed the anionic character of the formed nanoparticles with a value of –21.3 mV. The results of FTIR showed that the biomolecules present (proteins, phenolic and aromatic compounds) played an important role in the formation and stabilization of AuNPs [124].

7.2.3.3 Bacteria

Different mechanisms can act in the formation of AuNPs and AgNPs using bacteria as a biogenic source. These mechanisms can occur through the internalization of some ions, which undergo reduction in the cytoplasmic environment or it can be reduced outside the cell by the action of secreted metabolites or by the action of molecules existing on the outer surface of the bacteria [136]. The use of bacteria as a biogenic source has several advantages such as easy handling, manipulation of genetic material and adaptability to a different kind of environmental conditions [137]. However, the production of these microorganisms is more expensive than the production of plant and algae extracts [137]. Several bacteria can be used to biosynthesize Au and Ag NPs (Table 7.5) and some works will be discussed below.

Biogenic synthesis using bacteria can produce the tailored functional AuNP due to the coating of different biomolecules present in the bacteria [148]. However, many problems still exist in the production of biocompatible AuNPs using harmful bacteria [148]. In the study by Qiu and coauthors [138], gradient centrifugation was used to remove the toxic part of AuNPs biosynthesized with *Staphylococcus aureus*. The obtained NPs are spherical with sizes ranging from 5 to 20 nm and the face cubic crystal (FCC) structure carrying clear lattice fringes spacing of 0.14 nm in AuNPs, which represented gold (220) planes. The average zeta potential value was –42.50 mV, which means the conjugates of AuNPs are stable. In addition, the position of the bands obtained by FTIR indicates that the secondary structure of the enzyme is present on the surface of AuNPs.

Elbahnasawy and co-authors [140] used *rothia endophytica* as a new endophytic bacterial strain isolated from healthy corn roots for the biogenic synthesis of AgNPs, in which they obtained cubic NPs, well dispersed with size ranging between 47 and 72 nm. The analysis by XRD showed face-centered cubic nanocrystal structured AgNPs and the FTIR analysis demonstrated the existence of functional groups of strain biomolecules that could be responsible for the fabrication, coating, and stabilization of AgNPs

Pseudomonas stutzeri was used for the biogenic synthesis of AgNPs in the presence of a solution of 2 mmol L^{-1} $AgNO_3$. The mixture was incubated at pH 9 and 60°C for 24 hours [144]. The FTIR spectra indicated that biomolecules such as amino acids, proteins or enzymes from the bacteria were involved in the synthesis of AgNPs and the TEM image showed the formation of NPs with sizes ranging from 10 to 50 nm.

Table 7.5: Biogenic synthesis of AgNPs and AuNPs with bacteria.

Metallic precursor/ Bacteria	Size (nm)/shape	Characterization techniques	Applications	Refs.
Au/*Staphylococcus aureus*	5 – 20/spherical	TEM, DLS, zeta potential UV-vis, FTIR and EDX	Muscle tissue engineering	[138]
Au/*Vibrio alginolyticus*	100 – 150/irregular	SEM, TEM, UV-vis and FTIR	Anti-inflammatory and anticancer	[139]
Ag/*Rothia endophytica*	47 – 72/cubic	UV-vis, TEM, XRD and FTIR	Antifungal activity	[140]
Ag/*Nostoc muscorum*	30/polycrystalline	UV-vis, SEM, EDS, TEM, AFM, DLS and XRD.	Antibacterial activity	[141]
Ag/*Bacillus cereus*	21.5/spherical	UV-vis, FTIR, XRD and SEM	Antibacterial activity	[142]
Ag/*Klebsiella pneumoniae*	40 – 80/pseudo-spherical	AFM, UV-vis, TEM and SEM	a	[143]
Ag/Bacillus cereus	6 – 50/spherical	Uv-vis, FTIR, XRD, TEM, SEM, EDX, DLS and Zeta potential	Cytotoxic effect against cancer cell line and larvicidal activity	[136]
Ag/*Pseudomonas stutzeri*	10 – 50/a	UV-vis, FTIR, EDX, EDS, XRD and TEM	Detection of platinum and antibacterial activity	[144]
Ag/*Bacillus thuringiensis*	42/spherical	UV-vis, XRD, FE-SEM, TEM and EDX	Antibacterial and anti-biofilm activities	[145]
Ag/*Ureibacillus thermosphaericus*	10 – 100/spherical	UV-vis, DLS, TEM, XRD and FTIR	a	[146]
Ag/Escherichia coli, Bacillus megaterium, Acinetobacter sp. and Stenotrophomonas maltophilia	15 – 50/spherical	UV-vis, XRD, TEM, SEM and EDS	a	[147]

[a] Not informed.

7.2.3.4 Fungi

Fungi are micro or macroscopic beings that are capable of exerting different impacts on the environment and on human beings [149]. They have several biological components in their cellular structures, which are involved in their metabolic processes and, because of this, there may be potential for bioaccumulation and tolerance to metals [150]. Due to these factors, fungi are used in the production of MNPs, as they can be obtained on a large scale and provide economic viability in the production [150].

Biosynthesis from fungi (or mycosynthesis) is widely used and more viable, when compared to the use of bacteria, as they are easy to culture on a large scale, present high tolerance to metals and high wall binding capacity [151]. The process and manipulation of biomass can be simple and it can be extended to the synthesis of NPs of different chemical compositions,

Table 7.6: Biogenic synthesis of AgNPs and AuNPs with fungi.

Metallic precursor/Fungi	Size (nm)/shape	Characterization techniques	Applications	Refs.
Au/*Aspergillus flavus*	12/spherical	UV-Vis, TEM, FTIR, zeta potential, EDX and XRD	Cytotoxic and catalytic activities	[155]
Au/*Glomus aureum*	250/round but irregular in shape	FTIR, EDX, SEM and UV-Vis	Anti-bacterial potential and photocatalytic degradation	[156]
Au/*Trichoderma harzianum*	30/spherical	XRD, FTIR, SEM, EDS, TEM, DLS and zeta potential	Biosorption of metals	[157]
Au/*Flammulina velutipes*	74.32/triangular, spherical, and irregular	UV–vis, FESEM, AFM, DLS, XRD, Zeta Potential and FTIR	Catalysts in reducing the methylene blue dye	[158]
Ag/*Aspergillus oryzae*	109.6/spherical	TEM, DLS, FTIR and SEM	Antitumor activity and anti-microbial activity	[159]
Ag/*Aspergillus terreus*	13 – 49/spherical	UV-Vis, FTIR, SEM and EDX	Antitumor activity	[160]
Ag /*Trichoderma* strains	5 – 35/spherical	UV-Vis, XRD, TEM and FTIR	Antibacterial activity against *Enterococcus pernyi*	[161]
Ag/*Penicillium verrucosum*	10 – 12/irregular	UV-Vis, TEM, SEM, EDX and XRD	Antifungal activity (*Fusarium chlamydosporum* and *Aspergillus flavus*)	[162]
Ag/*Aspergillus niger*	10.31/spherical	UV-Vis, TEM and FTIR	Evaluate its effect against the newly identified *Allovahlkampfia spelaea*	[163]
Ag/*Cladosporium cladosporioides*	30 – 60/spherical	UV–vis, FTIR, AFM, FE-SEM, XRD and DLS	Antioxidant and antimicrobial activity	[164]
Ag/*Humicola sp*	5 – 25/spherical	UV-Vi-s, TEM, XRD, HR-TEM, EDX and SEM	Cytoxicity using normal and cancer cell lines	[165]

shapes and sizes having an economic feasibility and possibility of covering large surfaces by adequate mycelium growth [152].

In biogenic synthesis with fungi, NPs of Au and Ag can be obtained intracellularly and extracellularly. In the case of intracellular production, Ag^+ or Au^{3+} ions are reduced to their metallic form and they are accumulated in the microorganism's biomass [148]. The size of the NPs formed is controlled through parameters such as pH, temperature, substrate concentration, and substrate exposure time [148]. In the extracellular production of NPs, a quantity of fungus biomass is kept for a time in water and then filtered, and the reduction of Ag^+ or Au^{3+} ions is made in the filtrate without the fungus presence [153]. In this type of process, it is not necessary to separate the fungus mycelia from the NPs produced, unlike the intracellular process [154]. Several works using different types of fungi are shown in Table 7.6 and some will be discussed below.

AuNPs were synthesized using the extracellular cell free colorless filtrate of fungus *Glomus aureum*. When mixed with the $HAuCl_4$ solution showed burgundy red color indicating the formation of AuNPs, which was confirmed from the appearance of excitation of SPR [148,156]. This change in color is often indicative of the change in the oxidation state of metallic gold to Au^0, which was reduced by the bioreductors present in extracellular cell free filtrate of *Glomus aureum*, through interaction with lipid and membrane protein [156].

Some studies have shown that many biosorbents are capable of biosynthesizing metal ions into nanoparticles from reduction, nucleation and stabilization [166]. As observed in the work by Do Nascimento and co-authors [157], in which they evaluated Au biosorption using the biomass of the fungus *Trichoderma harzianum* and the formation of AuNPs simultaneously. The results showed that 0.4 g L^{-1} of fungal biomass was added to a $HAuCl_4$ solution (400 mg L^{-1}) to carry out the biosorption and biogenic synthesis of AuNPs within 30 minutes of reaction time under constant agitation to obtain spherical shape NPs of size below 30 nm.

AgNPs were biosynthesized from the fungus *Aspergillus oryzae* by optimizing parameters such as $AgNO_3$ concentration, cell free filtrate (CFF) ratio, pH and reaction time [159,167]. The best results were obtained with 3.75 mL of CFF, 2.25 mmol L^{-1} of $AgNO_3$, 108 minutes of reaction time and pH of 12.25. In these conditions, the AgNPs showed a spherical shape with 50% of the biosynthesized sample with a size less than 109.6 nm. The FTIR analysis delineated that the proteins present in the fungus were responsible for the reduction and stabilization of AgNPs [159]. Hulikere and coauthors [164] reported the synthesis of AgNPs using the marine fungus *Cladosporium cladosporioides*, which was isolated from the brown alga *Sargassum wightii*. The AgNPs suspension showed a UV-Vis absorption band at 440 nm indicating the presence of the NPs formed. The images obtained by FE-SEM showed that the material presented spherical morphology and particle size ranging from 30 to 60 nm. XRD analysis demonstrated the crystalline nature of AgNPs exhibiting peaks related to the FCC structure characteristic of metallic silver.

7.3 Characterization techniques

7.3.1 SEM-EDS

The SEM coupled with energy dispersive spectroscopy (EDS) allows to analyze the morphology and the surface composition of materials by electron beam with small diameter (generated by tungsten filament, lanthanum hexaboride, field electron emission, cathode or anode) [168,169]. Specially, the EDS is important technique for the qualitative analysis of chemical elements by surface scanning high energy electrons [168]. The signal obtained by SEM-EDS can be by (1) secondary electrons (electrons of low energy) that provide information about the topography of samples or (2) backscattered electrons (electrons of high energy) that provide information related to the composition through the difference in contrast generated by the atomic number of chemical elements present in the samples [170].

Through the X-rays emitted by the atoms present on the surface, it is possible to determine the composition of certain regions of the materials and to identify chemical elements of the samples by the comparison between the peaks and the binding energy of the elements [171,172]. It should be noted that for the MNPs characterization by SEM, the colloidal solutions must be converted into powder form to be placed on a support and coated with a conductive metal after being placed in the high-vacuum column of the microscope, where the electron beam will scan the surface [173].

Despite SEM provides important information about the surface morphology and size of NPs, it also presents resolution limitations to analyze smaller particles, as it presents limited information about the size distribution [174].

In the case of the works developed by Abdullah and coauthors [22], the SEM was used to characterize spherical AgNPs from *Spirogyra hyalina* algae extract with an average size around 52 nm. A special remark highlighted by the authors was that the size of the nanomaterial depended on the concentration of the metallic precursor and the pH during the reaction. The authors also used the EDS technique to characterize the NPs formed. In another work, Ardakani and coauthors [29], prepared AuNPs by biosynthesis approach by using yeast from *Saccharomyces cerevisiae*. The biomass was mixed with the metallic precursor $HAuCl_4$ for 48 hours at room temperature. AuNPs were used as biosensors capable of detecting the presence of mutant genes, as well as functioning as cancer biomarkers. To characterize this nanomaterial, the SEM technique was used to determine the spherical and monodisperse structure of the AuNPs with unspecified size.

7.3.2 TEM

The TEM allows to analysis the shape, morphology, size and dispersion of nanomaterials. In this technique, an electron beam passes through the sample, causing a strong interaction, resulting in the information about the nanomaterial at nanometer scale [175]. Like in the SEM measurements, the electron beam can be generated by thermoionic heating of a filament. However, due to the short wavelength of the beam, which is inversely proportional to the energy of the beam incident on the sample (e.g., 0.0025 nm at 200 kV and 0.0017 nm at 300 kV), structures with a higher resolution can be observed [168]. The electron beam that strikes the ultra-thin sample is transmitted, originating the scattered electrons (elastically and inelastically) that are focused by electromagnetic lenses and projected onto a screen, originating morphological information of the nanomaterial [176]. The knowledge on the NPs size and distribution is essential for some areas once the particle influences the release of drugs in the body [136]. Additionally, it is known that NPs with small size tend to agglomerate more, and they must be well stabilized by some agents [177].

Bindhu and coauthors [178], prepared anisotropic AgNPs from beetroot extract and they determined the size of 16 nm for the spherical NPs and 5–20 nm for the triangle NPs by

TEM. Bindhu and coworkers [113] also used TEM to characterize the AgNPs obtained from the extract of *Moringa oleifera* flowers. According to the authors, the presence and increase of flavonoid and phenolic compounds helped in the reduction of the metallic Ag precursor. The presence of monodisperse nanoparticles was confirmed through TEM micrographs. From *Tectona Grandis* seed extract, which contains carbohydrates, lipids, enzymes, flavonoids, terpenoids, and alkaloids, Rautela and coworkers [179] synthetized AgNPs and characterized the shape, dispersion, and size of NPs (10–30 nm) by TEM.

It is important to highlight that compared to SEM, the TEM present higher resolution, enabling a more detailed study of the morphology of NPs. However, analysis by TEM present disadvantages related to the need of high vacuum, thorough preparation of the sample, and the conditions for obtaining images with good quality [69].

7.3.3 XRD

The XRD technique allows the identification of the crystalline phases present in the nanomaterials and provide the position of the peaks of Bragg, intensity and shape of MNPs, which become larger and absent in clusters [180]. This technique is considered the most appropriate method for determining the crystalline phases, in which a qualitative and quantitative analysis is possible [69]. For that, Rietveld analysis can provide the crystallite size of MNPs by broadening the diffractogram peaks. The nanoparticle crystallite size exhibits a linear correlation between particle size and crystallite size [181].

When an X-ray beam is focused on a crystalline sample, there is an interaction with the atoms present, originating the phenomenon of diffraction. This phenomenon is only possible due to the ordering of atoms in the crystalline planes, separated from each other by distances in the same order of magnitude as the wavelengths of X-rays [182]. XRD can analyze structural characteristics of MNPs through their diffraction patterns [69]. The composition of the particles can be determined by comparing the peaks position and intensity with the crystallographic sheets, available as a reference in the Center International for Diffraction (ICDD). However, for amorphous materials and diffractograms of materials with a size smaller than 3 nm, the crystallographic sheets are not indicated and adequate [183].

Bykkam and coauthors [184] prepared AgNPs through the coprecipitation of $AgNO_3$ using *Ocimum* leaf extract, which acts as a reducing and stabilizing agent leading to the formation of AgNPs with well-defined morphology. The authors performed a study using the XRD to confirm the formation of NPs. The crystallinity index, the unit cell parameters, the specific surface area were studied. In addition to this information, the crystallite size was also calculated, as well as the FCC structure of AgNPs was confirmed.

Another information that can be taken from the peak's characteristic of MNPs is the size of the nanomaterials that can be obtained by mathematical equations such as Debye-Scherrer, which can be described as [184]:

D = Kλ

B.cos θ

Where:

D = crystallite size; K = Scherrer's constant (ranging from 0.9 to 1); λ = wavelength of the X-ray source (usually 1.54 nm); B = total width of the "largest" diffraction peak (in radians) and θ = Bragg diffraction angle (in radians). All this information can be taken from the diffractogram and the calculation of crystallite size is not exact, but it can provide an idea of particle size.

Balasubramanian and coworkers [185] investigated the crystal structure of AuNPs through the XRD technique prepared by biogenic approach by using *Jasminum auriculatu* extract. The main diffraction peaks obtained were at 38.3°, 44.2°, 64.28° and 77.32° and they can be indexed to the crystallographic planes (111), (200), (220) and (311), respectively. Also, according to the diffractogram peaks, it was found that the NPs had a spherical shape and, they calculated the particle size (8–41 nm) by applying the Scherrer equation. This information was corroborated by the TEM results.

The XRD analysis of AuNPs obtained by Pandey and co-workers [186] using aqueous solution of *Cyamopsis tetragonaloba* known as guar gum, showed that the nanomaterial presents a FCC structure of crystalline with a size of 8 nm. This AuNPs was successfully used in the development of sensor to detect ammonia. The same researcher group also carried out a similar study, but using the metallic precursor $AgNO_3$ mixed with a solution of guar gum [187]. The composite formed was applied as an optical sensor for the detection of ammonia in biological fluids (sweat, saliva, among others). The XRD study was used to assess the purity of the AgNPs formed. The diffractogram presented characteristics peaks at 38.06°, 44.22°, 64.48°, and 77.32° corresponding to the crystallographic planes (111), (200), (220), and (311), respectively. Thus, using XRD, it is possible to observe whether there was a reduction in the metallic precursor and, consequently, the formation of NPs. Additionally, this technique enables the identification of peaks corresponding to the crystallographic planes of the MNPs, as well as the identification of the crystal lattice [188].

7.3.4 UV-Vis spectroscopy

The UV-Vis spectroscopy technique was commonly used to characterize the nanomaterials made of Au and Ag. In this technique, the light absorbed and scattered by a sample is quantified, being placed between the light source and a photodetector, so that the beam intensity is measured and then deducted from the blank [189]. The UV-Vis technique becomes an important tool to identify and characterize AuNPs and AgNPs due to their unique optical properties [190]. This technique is widely used to confirm the formation of AuNPs and AgNPs through characteristic light absorption, as it is fast, simple and low [191]. Furthermore, this technique is important to study the MNPs stability, being able to monitor the decrease in the plasmonic band of colloidal solutions since its preparation [185].

The UV-Vis analysis is very simple and uses a small volume of MNPs (~2 mL) which is added to a quartz cuvette that will be subjected to a wavelength scan to determine the characteristic SPR band of each nanomaterial [22]. AuNPs and AgNPs have very interesting optical properties, which lead to several applications of these nanomaterials. The plasmonic resonance of AuNPs and AgNPs can be associated with the phenomenon of interaction of the electromagnetic wave with the collective oscillations of the electronic cloud on the surface of these nanomaterials. Because of this, colloidal solutions have different coloring and strong absorption in the visible region [192].

Observing the coloration of the MNPs solution, it is possible to know about the dispersion and formation of nanomaterials [193]. The color is associated with the type of ligand and solvent (among others) that the MNPs are inserted. When there is a change in the color of the colloidal solution, an indication of agglomeration can be considered. This phenomenon can be confirmed by TEM once the surface area tends to increase [192]. The solutions containing NPs of Au and Ag have very characteristic colors, which are directly associated with the morphology (size, shape, and dispersion) [190].

AgNPs showed an intense absorption band from 400 to 500 nm, whereas AuNPs from 525 to 555 nm due to the SPR phenomenon [194]. The position and shape of the SPR band are strongly dependent on the dielectric medium of the NPs and on the adsorbed species on the surface. Furthermore, for spherical NPs the appearance of only one plasmonic band is expected, while anisotropic particles can present two or more bands depending on the different forms presented [195].

7.4 Electrochemical bio/sensing applications

7.4.1 Sensors

The Au and Ag NPs are widely used in the development of electrochemical sensors and biosensors due to their conductive properties, high surface area, ability to accelerate the transfer of electrons between electroactive substances and the electrode surface, and ability to include NPs in various configurations during the development of the bio/sensor [31]. An electrochemical sensor is a device that transforms chemical information from a chemical reaction, a biochemical process, or a physical property of the investigated system into an electrical signal [196,197]. These sensors can be divided into two basic units (Fig. 7.2A), the receptor being responsible for transforming the chemical information into a form of measurable energy and the transducer transforming the energy coming from the receiver into an observable analytical signal [196,197].

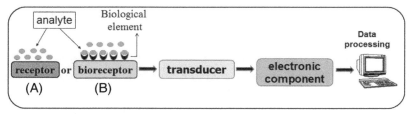

Fig. 7.2: Representative scheme of sensor (A) and biosensor (B).

In the electrochemical sensor, an electrochemical interaction between analyte and electrode happens whose signal can be electrically stimulated or occur spontaneously [198]. They have several advantages and, combined with NPs, they present faster analysis, higher detection and quantification limits, sensitivity and lower cost (depending on the synthesis used) [37]. Many works are reported in the literature using sensors modified with NPs of Au and Ag synthesized by biogenic route (Table 7.7) and chemical synthesis (Table 7.8).

CPE/AuNPs-PG - gold nanoparticles and pineapple gum modified glassy carbon paste electrode; AuNPs/GO/GCE - gold nanoparticles and graphite oxide film modified glassy carbon electrode; AuNPs-GCE: gold nanoparticles modified glassy carbon electrode; CPE/AuNPs-PFR - carbon paste electrodes modified with porphyran-capped gold nanoparticles; AgNPs-GCE: silver nanoparticles modified glassy carbon electrode; AgPdNPs-GCE: silver and palladium nanoparticles modified glassy carbon electrode; RGO-Ag-GCE: reduced graphene oxide-silver nanocomposite modified glassy carbon electrode; Pt-TCE-AgNPs: Pt-tipped carbon electrode - silver nanoparticles; AgNPs/CPE – silver nanoparticles modified carbon paste electrode; AgNS/GCE - Silver nanospheres modified glassy carbon electrode; AgNPs-xGnP/GCE – silver nanoparticles obtained and exfoliated graphite nanoplatelets applied to a glassy carbon electrode; CPE/AgNPs-PFR - carbon paste electrodes modified with porphyran-capped silver nanoparticles; DPV - differential pulse voltammetry; SWV - square wave voltammetry; CV – cyclic voltammetry.

Among the sensing studies using biogenic NPs, Lopes and coauthors [199] synthesized AuNPs for the first time with polysaccharides extracted from pineapple gum and used in the modification of the carbon paste electrode (CPE) to develop a sensitive electrochemical sensor for the determination of the antihistamine drug promethazine hydrochloride. The results obtained by the authors through CV and EIS, showed that AuNPs can increase the electron transfer properties of CPE due to its large surface area and high electrical conductivity. The results by SWV estimated the limit of detection (LD) (1.33 µmol L^{-1}) and quantification (LQ) (4.44 µmol L^{-1}), and applied successfully to quantify the analyte in different pharmaceutical products [199].

Taib and coauthors [200] used an aqueous extract of the leaves of the *Hibiscus sabdariffa* plant to biosynthesize AuNPs and applied it in the modification of a glassy carbon electrode (GCE) by dropping an aliquot of the AuNPs onto its surface and subsequently drying it in an

Table 7.7: Brief summary with information about electrochemical sensors developed with AgNPs and AuNPs synthesized with different biogenic sources.

Biogenic source	NPs size (nm) and shape	Sample	Analyte	Working electrode	Detection technique	LOD (μmol L^{-1})	Refs.
Pineapple gum (fruit)	10.3 ± 1.6/ spherical	Commercial pharmaceutical formulations	Antihistamine drug promethazine hydrochloride (PMZ)	CPE/AuNPs-PG	SWV	1.33	[199]
Hibiscus sabdariffa (plant)	7 ± 2/ spherical	Tap water and mineral water	Nitrite	Au-NPs/GCE	CV	110	[200]
Sueda fruciotosa (plant)	6 – 8/ spherical	Azo dyes	Methylene blue	AuNPs-GCE	CV	a	[201]
Bischofia javanica Blume (plant)	10 – 30/a	Milk, powdered milk, honey and eye drops	Chloramphenicol	AuNPs/GO/ GCE	Amperometry	0.25	[28]
Porphyra (red seaweed)	16.7 ± 2.9/ spherical	Pharmaceutical formulations	5-fluorouracil	CPE/AuNPs-PFR	DPV	0.66	[202]
Trachyspermum ammi (plant)	5 – 20/ cubic and rod	a	p-nitrophenol	Pt-TCE-AgNPs	CV	a	[203]
Camellia japônica (plant)	12 – 25/ spherical	a	Nitrobenzene and eosin-Y dye	AgNPs-GCE	CV and amperometry	0.012	[37]
Onion	6/spherical	Orange and Kiwi juices and apple	Ascorbic acid	AgNPs/CPE	SWV	0.1	[204]
Caruluma edulis (plant)	2 – 10/ spherical	a	Bromothymol blue	AgNPs-GCE	CV	a	[205]
Nilgiri (plant)	30/spherical	Tap water	Nitrite	AgNS/GCE	Amperometry	0.031	[206]
Lithodora hispidula (plant)	AgNPs (15.03/ spherical)	a	H$_2$O$_2$	AgPDNPs-GCE	CV and amperometry	0.52	[207]
Araucaria angustifólia (plant)	91.0 ± 0.5/ spherical	Pharmaceutical products	Paracetamol	AgNPs-xGnP/GCE	SWV, DPV and CV	0,00085	[208]
Plectranthus amboinicus (plant)	22.5/ spherical	a	H$_2$O$_2$	RGO-Ag-GCE	CV and amperometry	0,312	[209]
Porphyra (red seaweed)	24.4 ± 11.7/ spherical	Pharmaceutical formulation	5- fluorouracil	CPE/AgNPs-PFR	SWV	10.7	[210]

a not described.

Table 7.8: Brief summary with information about electrochemical bio/sensors developed with AgNPs and AuNPs synthesized by chemical route.

Precursor	Sample	Analyte	Working electrode	Detection technique	LOD	Refs.
Electrodeposition	Tap water, artificial saliva and artificial sweat	Cr (III) and Cr (VI)	Ag–Au-SPCE	DPV and SWV	0.1 µg L^{-1}	[213]
AgNPs synthesized with TACoPc and PANI	Tap water	HQ and CC	TACoPc/PANI/Ag-NPs/GCE	DPV	HQ (0,60 µmol L^{-1}) e CC (0,46 µmol L^{-1})	[214]
Electrodeposition	Syrup, plasma and urine	Guaifenesin	AgNPs/PL-cys/PGE	DPV	0.0061 µmol L^{-1}	[215]
Silicate sol–gel, hydrazine, ammonium chloride and nitric acid	a	H$_2$O$_2$ and nitrobenzene	TPDT-Ag-NPs/GCE	LSV, DPV and SWV	H$_2$O$_2$ (LSV = 0.5 and DPV 0.1 µmol L^{-1}) Nitrobenzene (1 µmol L^{-1})	[216]
Poly(acrylamide-co-diallyldimethylammoniumchloride) and N1-(3-trimethoxysilylpropyl) diethylenetriamine	a	H$_2$O$_2$	AgNPs/GCE	LSV and SWV	LSV (5 µmol L^{-1}) and SWV (0.2 µmol L^{-1})	[217]
Mercaptopropyl-trimethoxysilane, NaBH$_4$ and tetramethoxysilane	Industrial waste water	Cyanide	GE/sol–gel/AgNPs	CV	1.4×10^{-8} mol L^{-1}	[218]
Sodium citrate	Foodstuffs	Butylated hydroxyanisole	AuNPs/CS/SPCE	DPV	0.001 µg mL^{-1}	[219]
Na$_3$CA and NH$_2$OH·HCl	Natural water	Uranyl ions	AuNPs/GCE	DPASV	0.3 µg L^{-1}	[220]
Sodium citric,	Pharmaceutical formulations, synthetic urine and river water	Clindamycin	AuNPs-GO-CTS-ECH/GCE	SWV	0.29 µmol L^{-1}	[221]

^{a} not described.

oven for 6 hours. The developed sensor was used to study the nitrite electrooxidation process using CV, in which the nitrite ions oxidized into nitrate ions during this process, as it can be observed in the following reactions [211].

$$2NO_2 \leftrightarrow 2NO_2 + 2e^- \tag{7.1}$$

$$2NO_2 + H_2O \rightarrow NO_2^- + NO_3^- + 2H^+ \tag{7.2}$$

$$NO_2^- + H_2O \rightarrow NO_3^- + 2H^+ + 2e^- \tag{7.3}$$

Furthermore, AuNPs can participate in the electro-oxidation of nitrite through the formation of the Au^0-NO_2^- complex on the surface of the modified GCE in the presence of nitrite [212]. This complex is electrochemically oxidized to Au^+ and NO_2 and then converted to Au^0 and NO_3^-, as it shown in the following reactions:

$$Au^0 + NO_2^- \rightarrow Au^0 - NO_2^- \tag{7.4}$$

$$Au^0 - NO_2^- \rightarrow Au^+ - NO_2 + 2e^- \tag{7.5}$$

$$Au^+ - NO_2 + H_2O \rightarrow Au^0 + NO_3^- + 2H^+ \tag{7.6}$$

It should be noted that the GCE without modification and under the same conditions did not respond to the analyte [200].

In the study by Lima and coauthors [202], porphyran, which is a polysaccharide extracted from red seaweed (*Porphyra*) was used to reduce and stabilize the AuNPs. The porphyrin stabilized AuNPs was used to modify CPE and applied for the detection of anti-tumor agent 5-flurouracil (5-FU). Fromthe results obtained by CV and EIS, it was observed that the nanocomposite improved the electrochemical performance of the electrodes by increase in current and a decrease in the charge transfer resistance due to the high conductivity and large surface area of the AuNPs [28]. Using DPV, the LD and LQ value of 0.66 and 2.22 $\mu mol\ L^{-1}$ was obtained, respectively. In addition to AuNPs, AgNPs are also used in the development of sensors, as shown in the work by Zamarchi and coauthors [208]. They used pine nut extract (*Araucaria angustifolia*) as a reducing and stabilizing agent in the synthesis of AgNPs and applied it to a GCE with graphite nanoplates exfoliated for the determination of paracetamol. The results obtained with the sensor demonstrated good repeatability, reproducibility, stability with the estimated LD of 0.850 $nmol\ L^{-1}$. It was applied successfully for the quantification of paracetamol in pharmaceutical products.

In the study by Khalilzadeh and Borzoo [204], the AgNPs was synthesized by using the aqueous extract of onion. The CPE was developed by mixing AgNPs with a known mass of powdered graphite and paraffin (which acted as a binder). This sensor was used to determine ascorbic acid with greater sensitivity.

Bojko and coauthors [210] developed a work very similar to that described by Lima and coauthors [80]. The sulfated polysaccharide porphyran (PFR) obtained from the red alga *Porphyra* was used in the stabilization of AgNPs and D-glucose as a reducing agent. AgNPs were used as a modifier in the EPC, improving the electrochemical properties and the electroanalytical performance of the electrode, obtaining a LD of 10.7 $\mu mol\ L^{-1}$, confirming its potential as a promising new analytical detection tool for quality control purposes of fluorouracil [83].

It should be noted that, so far, no works have been found in the literature that use fungi and bacteria in the development of electrochemical sensors. NPs synthesized via chemical route are widely used in the development of electrochemical sensors, as it can be seen in Table 7.8, in which several applications for sensors modified with NPs are reported.

Ag–Au-SPCE – silver and gold screen-printed carbon electrode; TACoPc/PANI/AgNPs/ GCE - tetra amino cobalt phthalocyanine/polyaniline/silver nanoparticles modified glassy carbon electrode; AgNPs/PLcys/PGE –silver nanoparticles and poly(L-cysteine) modified pencil graphite electrode; TPDT-AgNPs/GCE - N-[3-(trimethoxysilyl)propyl]diethylenetriamine - silver nanoparticles modified glassy carbon electrode; AgNPs/GCE - silver nanoparticles modified glassy carbon electrode; GE/sol–gel/AgNPs – gold electrode and sol-gel and silver nanoparticles;

AuNPs/CS/SPCE – gold nanoparticles, chitosan and screen-printed carbon electrode; AuNPs/ GCE - gold nanoparticles modified glassy carbon electrode; AuNPs-GO-CTS-ECH/GCE - glassy carbon electrode modifiedwith graphene oxide and gold nanoparticles within a film of crosslinked chitosan with epichlorohydrin; HQ – hydroquinone; CC – catechol; DPV - differential pulse voltammetry; SWV - square wave voltammetry; CV – cyclic voltammetry; DPASV - differential pulse anodic stripping voltammetry.

Screen-printed carbon electrodes (SPCEs) modified with bimetallic Au and Ag NPs or Ag-Au bimetallic oxide NPs were applied to speciation of chromium (Cr(III) and Cr(VI)) in environmental samples [213]. The results obtained showed low LDs (0.1 μg L^{-1} for both species) and the sensors were successful applied to speciation in the samples.

In several works, Au is used in the composition of the working electrode for the electrochemical detection of Cr(VI) to obtain excellent results. However, for the quantification of Cr(III), its response is not sufficient [222–224]. Therefore, studies were done with Au oxides and it was found that they could electrocatalyze Cr(III) oxidation to produce a better electrochemical response. However, Au oxides are unstable and it can be reduced easily [213,225,226]. These studies indicated that the silver segregation can stabilize one-dimensional surface gold oxides and this strategy can be used for the development of sensors applied to Cr speciation.

The determination of hydroquinone (HQ) and catechol (CC) was performed by the production of an electrocatalyst from tetra amino cobalt phthalocyanine (TACoPc) in conjunction with the polyaniline conducting (PANI) and which was subsequently used to decorate AgNPs using Ag$^+$ solution on the surface of TACoPc/PANI composite to form a supramolecular hybrid system. This process was carried out in an aqueous medium at room temperature and confirmed by several spectroscopic techniques. The electrocatalyst was added to the surface of a GCE and successfully used for simultaneous detection of HQ and CC, exhibiting excellent electrocatalytic activity with good stability, selectivity, reproducibility and a low LD for HQ (0.60 μmol L^{-1}) and CC (0.46 μmol L^{-1}) [214].

Table 7.9: Brief summary with information about electrochemical biosensors developed with AgNPs and AuNPs synthesized with different biogenic sources and chemical route.

Precursor	Sample	Analyte	Working electrode	Detection technique	LOD	Refs.
Saccharomyces cerevisiae (fungi)	a	Acute lymphoblastic leukemia	AuNPs/CD/GCE	DPV, CV and chronoamperometry	1.5 pmol L^{-1} (target DNA) and 0.26 pg mL^{-1} (target antigen)	[29]
Bacillus stearo-thermophilus (bacteria)	Clostridium difficile	Toxin A (TOA)	Nafion/thionine/AuNPs/SPE	CV	1 nmol L^{-1}	[30]
Trisodium citrate	Serum	microRNA-21	AuNS/capture probe/TB/GCE	DPV	78 amol L^{-1}	[231]
Trisodium citrate	Purified water, milk, grapefruit juice and green tea	Lipopolysaccharide	GFLC/LPS/PBA/GE	DPV	0.51×10^{-10} µg mL^{-1}	[232]
Porous graphene oxide and ascorbic acid	Human serum	COVID-19 protein	Anti-COVID-19 antibody/AuNPs@rPGO/ITO	SWV	39.5 fmol L^{-1}	[39]
Single-layer graphene, dimethyl formamide, dioctyl sulfosuccinate sodium and ethylene glycol	Fluid from an oilfield	Dissimilatory sulfite reductase (DsrAB) gene from sulfate-reducing bacteria (SRB)	3D G AuNPs/GCE	DPV	9.41×10^{-15} mol L^{-1}	[229]
Electrodeposition	a	Bisphenol A	Tyr/AuNPs/T-NH$_2$/AuE	CV	0.133 µmol L^{-1}	[233]
Trisodium citrate, tetrathoxysilane, chitosan and ethanol	a	Monocrotophos, methyl parathion and carbaryl	AChE–AuNPs-SiSG/GCE	CV	a	[234]
Trisodium citrate and NaBH$_4$	Spiked human serum and cell lysates	Wild-type p53 protein	dsDNA/p53 protein/AgNPs/BDT/AuE	LSV	0.1 pmol L^{-1}	[235]
Trisodium citrate and NaBH$_4$	Serum	T4 polynucleotide kinase	ligase/AgNPs/AuE	LSV	0.01 U mL^{-1}	[236]

a not described.

7.4.2 Biosensors

Biosensors are devices capable of identifying specific substances qualitatively and quantitatively. They have several advantages such as low analysis time, possibility of miniaturization, application online and in the field of measurement, portability, low operating, project and final consumer costs [227]. The biosensor has the same composition as a sensor (Fig. 7.2B), however in the receptor (bioreceptor in this case) an element of biological recognition is added (enzymes, cells, animal or plant tissue, DNA, antibodies) [228]. Electrochemical biosensors are widely used and have several advantages such as high selectivity, sensitivity, low cost and have a variety of measurable reaction products in catalytic processes, including protons, electrons, light, and heat [227].

The number of applications of biosensors in conjunction with metals such as Au and Ag is increasing (Table 7.9) due to the improved performance of the fabricated devices because of the improvement in its electrical, optical and dielectric properties because these metals have excellent catalytic activity, electrical and thermal conductivity [227,229,230].

A very interesting application of the biosensor constructed by a nanocomposite of quantum dots of carbon and AuNPs (biosynthesized with the fungus *Saccharomyces cerevisiae*) was applied for the detection of acute lymphoblastic leukemia cancer by detection of mutant genes and cancer biomarkers [29]. AuNPs synthesized using trisodium citrate was used as a support material in the preparation of a microRNA biosensor. In this approach, the conducting polymer polypyrrole was used to form the support material AuNPs superlattice and the redox indicator toluidine blue (TB) for obtaining amplified signal. After that, the immobilization of the single-stranded RNA and hybridization with the microRNA sequence was confirmed by VC and DPV. From the oxidation peak current of TB, the microRNA can be detected down to 78 aM [231].

The biosensor showed high sensitivity, selectivity, reproducibility, showing an excellent response in the analysis of real serum samples, in addition to being useful in the early detection of breast cancer by direct detection of the serum microRNA-21 in real clinical samples [231]. Additionally, the AuNPs surface can be modified for the conjugation of biomolecules for the development of biosensors [229]. Biofunctional groups such as copolymers, nucleic acids and proteins can also be used to functionalize the biosensor surface to detect different enzymes and neurotransmitters [237]. The AuNPs was synthesized with sodium citrate and conjugated with gabapentin, which is a derivative of the inhibitory neurotransmitter GABA [238,239]. Through UV-Vis analysis, it was possible to observe the formation of the conjugate, which is formed due to the electrostatic attraction between them. Additionally, the results by fluorescence microscopy showed the interaction of neural cells with AuNPs/gabapentin, indicating that this compound can be used as a basis for neurotransmitter-based biosensor development [239].

AuNPs/CD/GCE – gold nanoparticles and carbon quantum dots modified glassy carbon electrode; Nafion/thionine/AuNPs/SPE – Nafion/thionine/gold nanoparticles/modified screen-printed electrode; AuNS/capture probe/TB/GCE – gold nanoparticles and toluidine blue modified glassy carbon electrode; GFLC/LPS/PBA/GE - Gold nanoparticles/ferrocene/liposome cluster/lipopolysaccharide/pyrene-1-boronic acid/graphite electrode; dsDNA/p53 protein/AgNPs/BDT/AuE – dsDNA/p53 protein/silver nanoparticles modified gold electrode;

Anti-COVID-19 antibody /AuNPs@rPGO/ ITO – Anti-COVID-19 antibody immobilized on gold nanoparticles/reduced porous graphene oxide modified indium tin oxide; Ligase/AgNPs/ AuE – ligase/ silver nanoparticles/gold electrode; 3D G/AuNPs/GCE - 3-dimensional graphene functionalized with gold nanoparticles modified glassy carbon electrode; Tyr/AuNPs/ T-NH$_2$/AuE – sensor based on the immobilization of tyrosinase onto gold nanoparticles and thioctic acid amide self-assembled monolayers modified gold electrode; AChE–AuNPs-SiSG/ GCE – acetylcholinesterase combined to silica sol–gel (SiSG) assembling AuNPs composite modified glass carbon electrode; DPV - differential pulse voltammetry; SWV - square wave voltammetry; CV – cyclic voltammetry; LSV - linear-sweep voltammetry.

7.5 Conclusions and future perspectives

Gold and silver are the most used metals for the preparation of NPs by different routes, as they produce NPs with excellent properties such as optical, electrical, mechanical, chemical, biological. In addition, NPs also present a large surface area, which enables their application in several fields. The chosen route and the control of experimental parameters will directly influence the MNPs morphology and consequently, their applications. However, despite the advantages of these syntheses, biogenic synthesis presents itself as an alternative method, as it does not require toxic reagents and the synthesis can be performed from natural sources such as plants, algae, bacteria and fungi, Additionally, it is a simple and low-cost production method, which allows its application in several sectors, such as pharmaceuticals, environmental, in the removal and degradation of contaminants, in the development of bio/sensors, among others. Regardless of the MNPs preparation method, a detailed study about its morphology and characterization is important. The knowledge of the morphology (size, shape, and dispersion) of nanomaterials allows a better result in the desired application.

However, it is important to emphasize that some challenges (such as greater control over their size, shape and monodispersity) still need to be overcome for an industrial applicability of NPs. Most of these studies are carried out on a laboratory scale, however for an industrial scale the costs become higher and the process tedious. The identification of specific molecules that are present in the biogenic sources used is another challenge demanded, once they are used as reducers and stabilizers in the formation of NPs. Thus, further researches should focus on understanding the exact mechanism of biogenic NPs formation.

References

[1] M.C. Daniel, D. Astruc, Gold nanoparticles: Assembly, supramolecularchemistry, quantum-size-related properties, and applications toward, Chem. Rev. 104 (2004) 293–346.

[2] R.P. Feynman, There's Plenty of Room at the Bottom, Eng. Sci. 23 (1960) 22–36, https://resolver.caltech. edu/CaltechES. Accessed date: September 7, 2022 23.5.1960Bottom.

[3] N. Baig, I. Kammakakam, W. Falath, I. Kammakakam, Nanomaterials: A review of synthesis methods, properties, recent progress, and challenges, Mater. Adv. 2 (2021) 1821–1871, doi:10.1039/d0ma00807a.

[4] K.E. Drexler, Molecular engineering: An approach to the development of general capabilities for molecular manipulation, Proc. Natl. Acad. Sci. U. S. A. 78 (1981) 5275–5278, doi:10.1073/pnas.78.9.5275.

[5] C. Buzea, I.I. Pacheco, K. Robbie, Nanomaterials and nanoparticles: Sources and toxicity, Biointerphases. 2 (2007) MR17–MR71. doi:10.1116/1.2815690.

[6] S. Hasan, A review on nanoparticles : Their synthesis and types, Res. J. Recent Sci. Res . J . Recent . Sci 4 (2014) 1–3 Uttar Pradesh (Lucknow Campus).

[7] X. M. Faraday, The Bakerian Lecture. —Experimental relations of gold (and other metals) to light, Philos. Trans. R. Soc. London. 147 (1857) 145–181, doi:10.1098/rstl.1857.0011.

[8] C. Shi, N. Zhu, Y. Cao, P. Wu, Biosynthesis of gold nanoparticles assisted by the intracellular protein extract of Pycnoporus sanguineus and its catalysis in degradation of 4-nitroaniline, Nanoscale Res. Lett. (2015) 10, doi:10.1186/s11671-015-0856-9.

[9] C.G.A. Das, V.G. Kumar, T.S. Dhas, V. Karthick, K. Govindaraju, J.M. Joselin, J. Baalamurugan, Antibacterial activity of silver nanoparticles (biosynthesis): A short review on recent advances, Biocatal. Agric. Biotechnol. 27 (2020) 101593, doi:10.1016/j.bcab.2020.101593.

[10] N. Kulkarni, U. Muddapur, Biosynthesis of metal nanoparticles: A review, J. Nanotechnol. 2014 (2014), doi:10.1155/2014/510246.

[11] L. Carlini, C. Fasolato, P. Postorino, I. Fratoddi, I. Venditti, G. Testa, C. Battocchio, Comparison between silver and gold nanoparticles stabilized with negatively charged hydrophilic thiols: SR-XPS and SERS as probes for structural differences and similarities, Colloids Surfaces A Physicochem. Eng. Asp. 532 (2017) 183–188, doi:10.1016/j.colsurfa.2017.05.045.

[12] N.T. Khan, M.J. Khan, Metallic nanoparticles fabrication methods– A brief overview, SunKrist Nanotechnol. Nanosci. J. 2 (2020) 1–6, doi:10.46940/snnj.02.1002.

[13] R.N. Wijesena, N. Tissera, Y.Y. Kannangara, Y. Lin, G.A.J. Amaratunga, K.M.N. De Silva, A method for top down preparation of chitosan nanoparticles and nanofibers, Carbohydr. Polym. 117 (2015) 731–738, doi:10.1016/j.carbpol.2014.10.055.

[14] S. Kumar, V. Lather, D. Pandita, Green synthesis of therapeutic nanoparticles: An expanding horizon, Nanomedicine 10 (2015) 2451–2471, doi:10.2217/nnm.15.112.

[15] V. Sanchez-Mendieta, A. Rafael, Green synthesis of noble metal (Au, Ag, Pt) nanoparticles, assisted by plant-extracts, Noble Met. (2012). doi:10.5772/34335.

[16] H. Duan, D. Wang, Y. Li, Green chemistry for nanoparticle synthesis, Chem. Soc. Rev. 44 (2015) 5778–5792, doi:10.1039/c4cs00363b.

[17] I. Alghoraibi, C. Soukkarieh, R. Zein, A. Alahmad, J.G. Walter, M. Daghestani, Aqueous extract of Eucalyptus camaldulensis leaves as reducing and capping agent in biosynthesis of silver nanoparticles, Inorg. Nano-Metal Chem. 50 (2020) 895–902, doi:10.1080/24701556.2020.1728315.

[18] F. Al-Otibi, K. Perveen, N.A. Al-Saif, R.I. Alharbi, N.A. Bokhari, G. Albasher, R.M. Al-Otaibi, M.A. Al-Mosa, Biosynthesis of silver nanoparticles using Malva parviflora and their antifungal activity, Saudi J. Biol. Sci. 28 (2021) 2229–2235, doi:10.1016/j.sjbs.2021.01.012.

[19] Ruby Aryan, M.S. Mehata, Green synthesis of silver nanoparticles using Kalanchoe pinnata leaves (life plant) and their antibacterial and photocatalytic activities, Chem. Phys. Lett. 778 (2021) 138760, doi:10.1016/j.cplett.2021.138760.

[20] T.M. Dawoud, M.A. Yassin, A.R.M. El-Samawaty, A.M. Elgorban, Silver nanoparticles synthesized by Nigrospora oryzae showed antifungal activity, Saudi J. Biol. Sci. 28 (2021) 1847–1852, doi:10.1016/j. sjbs.2020.12.036.

[21] A.A. Heflish, A.E. Hanfy, M.J. Ansari, E.S. Dessoky, A.O. Attia, M.M. Elshaer, M.K. Gaber, A. Kordy, A.S. Doma, A. Abdelkhalek, S.I. Behiry, Green biosynthesized silver nanoparticles using Acalypha wilkesiana extract control root-knot nematode, J. King Saud Univ. - Sci. 33 (2021) 101516, doi:10.1016/j.jksus.2021.101516.

[22] N.S.Al-Radadi Abdullah, T. Hussain, S. Faisal, S. Ali Raza Shah, Novel biosynthesis, characterization and bio-catalytic potential of green algae (Spirogyra hyalina) mediated silver nanomaterials, Saudi J. Biol. Sci. 29 (2022) 411–419, doi:10.1016/j.sjbs.2021.09.013.

[23] A.R.M. Abd El-Aziz, A. Gurusamy, M.R. Alothman, S.M. Shehata, S.M. Hisham, A.A. Alobathani, Silver nanoparticles biosynthesis using Saussurea costus root aqueous extract and catalytic degradation efficacy of safranin dye, Saudi J. Biol. Sci. 28 (2021) 1093–1099, doi:10.1016/j.sjbs.2020.11.036.

[24] M.G. Heinemann, C.H. Rosa, G.R. Rosa, D. Dias, Biogenic synthesis of gold and silver nanoparticles used in environmental applications: A review, Trends Environ. Anal. Chem. (2021) 30, doi:10.1016/j.teac.2021. e00129.

[25] A. Saravanan, P.S. Kumar, S. Karishma, D.V.N. Vo, S. Jeevanantham, P.R. Yaashikaa, C.S. George, A review on biosynthesis of metal nanoparticles and its environmental applications, Chemosphere 264 (2021) 128580, doi:10.1016/j.chemosphere.2020.128580.

[26] C. Li, D. Chen, H. Xiao, Green synthesis of silver nanoparticles using Pyrus betulifolia Bunge and their antibacterial and antioxidant activity, Mater. Today Commun. 26 (2021) 102108, doi:10.1016/j. mtcomm.2021.102108.

[27] R. Rajkumar, G. Ezhumalai, M. Gnanadesigan, A green approach for the synthesis of silver nanoparticles by Chlorella vulgaris and its application in photocatalytic dye degradation activity, Environ. Technol. Innov. 21 (2021) 101282, doi:10.1016/j.eti.2020.101282.

[28] R. Karthik, M. Govindasamy, S.M. Chen, V. Mani, B.S. Lou, R. Devasenathipathy, Y.S. Hou, A. Elangovan, Green synthesized gold nanoparticles decorated graphene oxide for sensitive determination of chloramphenicol in milk, powdered milk, honey and eye drops, J. Colloid Interface Sci. 475 (2016) 46–56, doi:10.1016/j.jcis.2016.04.044.

[29] M. Mazloum-Ardakani, B. Barazesh, A distinguished cancer-screening package containing a DNA sensor and an aptasensor for early and certain detection of acute lymphoblastic leukemia, Clin. Chim. Acta. 497 (2019) 41–47, doi:10.1016/j.cca.2019.07.009.

[30] P. Luo, Y. Liu, Y. Xia, H. Xu, G. Xie, Aptamer biosensor for sensitive detection of toxin A of Clostridium difficile using gold nanoparticles synthesized by Bacillus stearothermophilus, Biosens. Bioelectron. 54 (2014) 217–221, doi:10.1016/j.bios.2013.11.013.

[31] A. Wong, A.M. Santos, O. Fatibello-Filho, Simultaneous determination of paracetamol and levofloxacin using a glassy carbon electrode modified with carbon black, silver nanoparticles and PEDOT:PSS film, Sensors Actuators, B Chem 255 (2018) 2264–2273, doi:10.1016/j.snb.2017.09.020.

[32] C.A. De Lima, P.S. Da Silva, A. Spinelli, Chitosan-stabilized silver nanoparticles for voltammetric detection of nitrocompounds, Sensors Actuators, B Chem 196 (2014) 39–45, doi:10.1016/j.snb.2014.02.005.

[33] H.V. Tran, C.D. Huynh, H.V. Tran, B. Piro, Cyclic voltammetry, square wave voltammetry, electrochemical impedance spectroscopy and colorimetric method for hydrogen peroxide detection based on chitosan/silver nanocomposite, Arab. J. Chem. 11 (2018) 453–459, doi:10.1016/j.arabjc.2016.08.007.

[34] C.A. De Lima, E.R. Santana, J.V. Piovesan, A. Spinelli, Silver nanoparticle-modified electrode for the determination of nitro compound-containing pesticides, Anal. Bioanal. Chem. 408 (2016) 2595–2606, https://doi.org/10.1007/s00216-016-9367-5.

[35] M. Rohani Moghadam, S. Akbarzadeh, N. Nasirizadeh, Electrochemical sensor for the determination of thiourea using a glassy carbon electrode modified with a self-assembled monolayer of an oxadiazole derivative and with silver nanoparticles, Microchim. Acta. 183 (2016) 1069–1077, https://doi.org/10.1007/s00604-015-1723-1.

[36] J.V. Piovesan, C.A. de Lima, E.R. Santana, A. Spinelli, Voltammetric determination of condensed tannins with a glassy carbon electrode chemically modified with gold nanoparticles stabilized in carboxymethylcellulose, Sensors Actuators, B Chem 240 (2017) 838–847, doi:10.1016/j.snb.2016.09.057.

[37] R. Karthik, M. Govindasamy, S.M. Chen, Y.H. Cheng, P. Muthukrishnan, S. Padmavathy, A. Elangovan, Biosynthesis of silver nanoparticles by using Camellia japonica leaf extract for the electrocatalytic reduction of nitrobenzene and photocatalytic degradation of Eosin-Y, J. Photochem. Photobiol. B Biol. 170 (2017) 164–172, doi:10.1016/j.jphotobiol.2017.03.018.

[38] K.Y.P.S. Avelino, G.S. dos Santos, I.A.M. Frías, A.G. Silva-Junior, M.C. Pereira, M.G.R. Pitta, B.C. de Araújo, A. Errachid, M.D.L. Oliveira, C.A.S. Andrade, Nanostructured sensor platform based on organic polymer conjugated to metallic nanoparticle for the impedimetric detection of SARS-CoV-2 at various stages of viral infection, J. Pharm. Biomed. Anal. 206 (2021) 114392, doi:10.1016/j.jpba.2021.114392.

[39] W.A. El-Said, A.S. Al-Bogami, W. Alshitari, Synthesis of gold nanoparticles@reduced porous graphene-modified ITO electrode for spectroelectrochemical detection of SARS-CoV-2 spike protein, Spectrochim. Acta - Part A Mol. Biomol. Spectrosc. 264 (2022) 120237, doi:10.1016/j.saa.2021.120237.

[40] P. Slepička, N.S. Kasálková, J. Siegel, Z. Kolská, V. Švorčík, Methods of gold and silver nanoparticles preparation, Materials (Basel). 13 (2020) 1. https://doi.org/10.3390/ma13010001.

[41] T. Yonezawa, D. Čempel, M.T. Nguyen, Microwave-induced plasma-in-liquid process for nanoparticle production, Bull. Chem. Soc. Jpn. 91 (2018) 1781–1798, https://doi.org/10.1246/bcsj.20180285.

[42] B.R. Karimadom, H. Kornweitz, Mechanism of producing metallic nanoparticles, with an emphasis on silver and gold nanoparticles, using bottom-up methods, Molecules 26 (2021), https://doi.org/10.3390/molecules26102968.

[43] J. Turkevich, P. Stevenson, J. Hillier, A study of the nucleation and growth processes in the synthesis of colloidal gold, (1951).

[44] M.M. Kemp, A. Kumar, S. Mousa, T.J. Park, P. Ajayan, N. Kubotera, S.A. Mousa, R.J. Linhardt, Synthesis of gold and silver nanoparticles stabilized with glycosaminoglycans having distinctive biological activities, Biomacromolecules 10 (2009) 589–595, https://doi.org/10.1021/bm801266t.

[45] J. Wagner, J.M. Köhler, Continuous synthesis of gold nanoparticles in a microreactor, Nano Lett. 5 (2005) 685–691, https://doi.org/10.1021/nl050097t.

[46] J.J. Yuan, A. Schmid, S.P. Armes, A.L. Lewis, Facile synthesis of highly biocompatible poly(2-(methacryloyloxy)ethyl phosphorylcholine)-coated gold nanoparticles in aqueous solution, Langmuir 22 (2006) 11022–11027, https://doi.org/10.1021/la0616350.

[47] Z.N. Jameel, Synthesis of the gold nanoparticles with novel shape via chemical process and evaluating the structural, morphological and optical properties, Energy Procedia 119 (2017) 236–241, doi:10.1016/j.egypro.2017.07.075.

[48] R. Stiufiuc, C. Iacovita, C.M. Lucaciu, G. Stiufiuc, A.G. Dutu, C. Braescu, N. Leopold, SERS-active silver colloids prepared by reduction of silver nitrate with short-chain polyethylene glycol, Nanoscale Res. Lett. 8 (2013) 1–5, https://doi.org/10.1186/1556-276X-8-47.

[49] Y. Kobayashi, M.A. Duarte, L.M. Marzán, Sol - gel processing of silica-coated gold nanoparticles, Langmuir 17 (2001) 6375–6379.

[50] N. Lkhagvajav, I. Yaşa, E. Çelik, M. Koizhaiganova, O. Sari, Antimicrobial activity of colloidal silver nanoparticles prepared by sol-gel method, Dig. J. Nanomater. Biostructures. 6 (2011) 149–154.

[51] N. Saito, J. Hieda, O. Takai, Synthesis process of gold nanoparticles in solution plasma, Thin. Solid. Films 518 (2009) 912–917, doi:10.1016/j.tsf.2009.07.156.

[52] A. Izadi, R.J. Anthony, A plasma-based gas-phase method for synthesis of gold nanoparticles, Plasma Process. Polym. 16 (2019) 1–7, https://doi.org/10.1002/ppap.201800212.

[53] C. He, L. Liu, Z. Fang, J. Li, J. Guo, J. Wei, Formation and characterization of silver nanoparticles in aqueous solution via ultrasonic irradiation, Ultrason. Sonochem. 21 (2014) 542–548, doi:10.1016/j.ultsonch.2013.09.003.

[54] P.G. Jamkhande, N.W. Ghule, A.H. Bamer, M.G. Kalaskar, Metal nanoparticles synthesis: An overview on methods of preparation, advantages and disadvantages, and applications, J. Drug Deliv. Sci. Technol. 53 (2019) 101174, doi:10.1016/j.jddst.2019.101174.

[55] M.M. Seitkalieva, D.E. Samoylenko, K.A. Lotsman, K.S. Rodygin, V.P. Ananikov, Metal nanoparticles in ionic liquids: Synthesis and catalytic applications, Coord. Chem. Rev. 445 (2021) 213982. doi: 10.1016/j.ccr.2021.213982.

[56] C. Daruich De Souza, B. Ribeiro Nogueira, M.E.C.M. Rostelato, Review of the methodologies used in the synthesis gold nanoparticles by chemical reduction, J. Alloys Compd. 798 (2019) 714–740, doi:10.1016/j.jallcom.2019.05.153.

[57] J. Dupont, G.S. Fonseca, A.P. Umpierre, P.F.P. Fichtner, S.R. Teixeira, Transition-metal nanoparticles in imidazolium ionic liquids: Recyclable catalysts for biphasic hydrogenation reactions, J. Am. Chem. Soc. 124 (2002) 4228–4229, https://doi.org/10.1021/ja025818u.

[58] H. Bönnemann, R.M. Richards, Nanoscopic metal particles - Synthetic methods and potential applications, Eur. J. Inorg. Chem. (2001) 2455–2480, https://doi.org/10.1002/1099-0682(200109)2001:10<2455::aid-ejic2455>3.0.co;2-z.

[59] C.J. Jia, F. Schüth, Colloidal metal nanoparticles as a component of designed catalyst, Phys. Chem. Chem. Phys. 13 (2011) 2457–2487, https://doi.org/10.1039/c0cp02680h.

[60] J. Cookson, The preparation of palladium nanoparticles, Platin. Met. Rev. 56 (2012) 83–98, doi:10.1595/147106712×632415.

[61] M. Rakap, PVP-stabilized Ru-Rh nanoparticles as highly efficient catalysts for hydrogen generation from hydrolysis of ammonia borane, J. Alloys Compd. 649 (2015) 1025–1030, doi:10.1016/j.jallcom.2015.07.249.

[62] S. Li, Z. Dai, Y. Han, Synthesis and applications of gold nanoparticles, Spec. Petrochemicals. 36 (2019) 66–70.

[63] L.F. Gorup, E. Longo, E.R. Leite, E.R. Camargo, Moderating effect of ammonia on particle growth and stability of quasi-monodisperse silver nanoparticles synthesized by the Turkevich method, J. Colloid Interface Sci. 360 (2011) 355–358, doi:10.1016/j.jcis.2011.04.099.

[64] Z. Khan, S.A. Al-Thabaiti, A.Y. Obaid, A.O. Al-Youbi, Preparation and characterization of silver nanoparticles by chemical reduction method, Colloids Surfaces B Biointerfaces 82 (2011) 513–517, doi:10.1016/j.colsurfb.2010.10.008.

[65] M. Rao, W. Li, M. Das, Synthesis, characterization and applications of gold nanoparticles: An introduction on synthesis of gold nanoparticles using Actinobacteria, Int. J. Pharma Bio Sci. 10 (2016), https://doi.org/10.22376/ijpbs.2019.10.2.b8-14.

[66] V. Borse, A.N. Konwar, Synthesis and characterization of gold nanoparticles as a sensing tool for the lateral flow immunoassay development, Sensors Int 1 (2020) 100051, doi:10.1016/j.sintl.2020.100051.

[67] F.A. Harraz, S.E. El-Hout, H.M. Killa, I.A. Ibrahim, Palladium nanoparticles stabilized by polyethylene glycol: Efficient, recyclable catalyst for hydrogenation of styrene and nitrobenzene, J. Catal. 286 (2012) 184–192, doi:10.1016/j.jcat.2011.11.001.

[68] M. Zhang, S. Shao, H. Yue, X. Wang, W. Zhang, F. Chen, L. Zheng, J. Xing, Y. Qin, High stability au NPS: From design to application in nanomedicine, Int. J. Nanomed. 16 (2021) 6067–6094, https://doi.org/10.2147/IJN.S322900.

[69] X.F. Zhang, Z.G. Liu, W. Shen, S. Gurunathan, Silver nanoparticles: Synthesis, characterization, properties, applications, and therapeutic approaches, Int. J. Mol. Sci. (2016) 17, https://doi.org/10.3390/ijms17091534.

[70] M. Shahjahan, Synthesis and characterization of silver nanoparticles by sol-gel technique, Nanosci. Nanometrology. 3 (2017) 34, https://doi.org/10.11648/j.nsnm.20170301.16.

[71] S. Ramesh, Sol-gel synthesis and characterization of Ag3(2+x)AlxTi4-xO11+s nanoparticles, J. Nanosci. 2013 (2013) 9, https://www.researchgate.net/publication/258400614_Sol-Gel_Synthesis_and_Characterization_of_Nanoparticles.

[72] Z.F. Yao, K.L. Huang, S.Q. Liu, X.Z. Song, Y.H. Li, J. Guo, J.G. Yu, A novel method to prepare gold-nanoparticle-modified nanowires and their spectrum study, Chem. Eng. J. 166 (2011) 378–383, doi:10.1016/j.cej.2010.06.015.

[73] D.S. Ahlawat, R. Kumari, I.Yadav Rachna, Synthesis and characterization of sol-gel prepared silver nanoparticles, Int. J. Nanosci. 13 (2014) 1–8, doi:10.1142/S0219581×14500045.

[74] S. Ganachari, N. Banapurmath, B. Salimath, J. Yaradoddi, A. Shettar, A. Hunashyal, A. Venkataraman, P. Patil, H. Shoba, G. Hiremath, Handbook of ecomaterials synthesis techniques for preparation of nanomaterials, Handb. Ecomater. 1 (2019) 85–88, https://doi.org/10.1007/978-3-319-68255-6.

[75] S. Ganguly, P. Das, M. Bose, T.K. Das, S. Mondal, A.K. Das, N.C. Das, Sonochemical green reduction to prepare Ag nanoparticles decorated graphene sheets for catalytic performance and antibacterial application, Ultrason. Sonochem. 39 (2017) 577–588, doi:10.1016/j.ultsonch.2017.05.005.

[76] K. Okitsu, M. Ashokkumar, F. Grieser, Sonochemical synthesis of gold nanoparticles: Effects of ultrasound frequency, J. Phys. Chem. B. 109 (2005) 20673–20675, https://doi.org/10.1021/jp0549374.

[77] K. Loza, M. Heggen, M. Epple, Synthesis, structure, properties, and applications of bimetallic nanoparticles of noble metals, Adv. Funct. Mater. 30 (2020), https://doi.org/10.1002/adfm.201909260.

[78] J.H. Jung, H. Cheol Oh, H. Soo Noh, J.H. Ji, S. Soo Kim, Metal nanoparticle generation using a small ceramic heater with a local heating area, J. Aerosol Sci. 37 (2006) 1662–1670, doi:10.1016/j.jaerosci.2006.09.002.

[79] J.P. Zhang, P. Chen, C.H. Sun, X.J. Hu, Sonochemical synthesis of colloidal silver catalysts for reduction of complexing silver in DTR system, Appl. Catal. A Gen. 266 (2004) 49–54, doi:10.1016/j.apcata.2004.01.025.

[80] M.K. Corbierre, J. Beerens, R.B. Lennox, Gold nanoparticles generated by electron beam lithography of gold(I)-thiolate thin films, Chem. Mater. 17 (2005) 5774–5779, https://doi.org/10.1021/cm051085b.

[81] S. Wack, P. Lunca Popa, N. Adjeroud, J. Guillot, B.R. Pistillo, R. Leturcq, Large-scale deposition and growth mechanism of silver nanoparticles by plasma-enhanced atomic layer deposition, J. Phys. Chem. C. 123 (2019) 27196–27206, https://doi.org/10.1021/acs.jpcc.9b06473.

[82] V. Amendola, S. Polizzi, M. Meneghetti, Free silver nanoparticles synthesized by laser ablation in organic solvents and their easy functionalization, Langmuir 23 (2007) 6766–6770, https://doi.org/10.1021/la0637061.

[83] N.V. Tarasenko, A.V. Butsen, E.A. Nevar, N.A. Savastenko, Synthesis of nanosized particles during laser ablation of gold in water, Appl. Surf. Sci. 252 (2006) 4439–4444, doi:10.1016/j.apsusc.2005.07.150.

[84] R. Kadhim, M. Noori, A. Ali, Preparation of gold nanoparticles by pulsed laser ablation in NaOH solution, J. Babylon Univ. Pure Appl. Sci. 22 (2012) 547–551.

[85] T. Tsuji, D.H. Thang, Y. Okazaki, M. Nakanishi, Y. Tsuboi, M. Tsuji, Preparation of silver nanoparticles by laser ablation in polyvinylpyrrolidone solutions, Appl. Surf. Sci. 254 (2008) 5224–5230, doi:10.1016/j.apsusc.2008.02.048.

[86] X. Fu, J. Cai, X. Zhang, W. Di Li, H. Ge, Y. Hu, Top-down fabrication of shape-controlled, monodisperse nanoparticles for biomedical applications, Adv. Drug Deliv. Rev. 132 (2018) 169–187, doi:10.1016/j.addr.2018.07.006.

[87] K. Esumi, K. Matsuhisa, K. Torigoe, Preparation of rodlike gold particles by UV irradiation using cationic micelles as a template, Langmuir 11 (1995) 3285–3287, https://doi.org/10.1021/la00009a002.

[88] T.M. Rashid, U.M. Nayef, M.S. Jabir, F.A.H. Mutlak, Synthesis and characterization of Au:ZnO (core:shell) nanoparticles via laser ablation, Optik (Stuttg) 244 (2021) 167569, doi:10.1016/j.ijleo.2021.167569.

[89] D.K. Golhani, B.G. Krishna, A. Khare, S.A.H. Zaidi, Techniques for nanoparticle synthesis, Int. J. Adv. Res. Ideas Innov. Technol. (2018) 443–451.

[90] A. Pyatenko, K. Shimokawa, M. Yamaguchi, O. Nishimura, M. Suzuki, Synthesis of silver nanoparticles by laser ablation in pure water, Appl. Phys. A Mater. Sci. Process. 79 (2004) 803–806, https://doi.org/10.1007/s00339-004-2841-5.

[91] J. Zhang, M. Chaker, D. Ma, Pulsed laser ablation based synthesis of colloidal metal nanoparticles for catalytic applications, J. Colloid Interface Sci. 489 (2017) 138–149, doi:10.1016/j.jcis.2016.07.050.

[92] S. Majidi, F.Z. Sehrig, S.M. Farkhani, M.S. Goloujeh, A. Akbarzadeh, Current methods for synthesis of magnetic nanoparticles, Artif. Cells, Nanomed. Biotechnol 44 (2016) 722–734, https://doi.org/10.3109/21691401.2014.982802.

[93] J.K. George, N. Verma, B. Bhaduri, Hydrophilic graphitic carbon nitride-supported Cu-CNFs: An efficient adsorbent for aqueous cationic dye molecules, Mater. Lett. 294 (2021) 129762, doi:10.1016/j.matlet.2021.129762.

[94] S. Wang, L. Gao, Chapter 7 - Laser-Driven Nanomaterials and Laser-Enabled Nanofabrication for Industrial Applications, Editor(s): Sabu Thomas, Yves Grohens, Yasir Beeran Pottathara, In Micro and Nano Technologies, Industrial Applications of Nanomaterials, Elsevier, 2019, Pages 181-203, ISBN 9780128157497, https://doi.org/10.1016/B978-0-12-815749-7.00007-4.

[95] E. Borsella, R. D'Amato, G. Terranova, M. Falconieri, F. Fabbri, Synthesis of nanoparticles by laser pyrolysis: From research to applications, J. Anal. Appl. Pyrolysis. 104 (2013) 461–469, doi:10.1016/j.jaap.2013.05.026.

[96] E.M. Modan, A.G. Plăiaşu, Advantages and disadvantages of chemical methods in the elaboration of nanomaterials, Ann. "Dunarea Jos" Univ. Galati. Fascicle IX, Metall. Mater. Sci. 43 (2020) 53–60, https://doi.org/10.35219/mms.2020.1.08.

[97] M. Auffan, J. Rose, J.Y. Bottero, G.V. Lowry, J.P. Jolivet, M.R. Wiesner, Towards a definition of inorganic nanoparticles from an environmental, health and safety perspective, Nat. Nanotechnol. 4 (2009) 634–641, https://doi.org/10.1038/nnano.2009.242.

[98] A. Majdalawieh, M.C. Kanan, O. El-Kadri, S.M. Kanan, Recent advances in gold and silver nanoparticles: Synthesis and applications, J. Nanosci. Nanotechnol. 14 (2014) 4757–4780, https://doi.org/10.1166/jnn.2014.9526.

[99]   A. Schröfel, G. Kratošová, I. Šafařík, M. Šafaříková, I. Raška, L.M. Shor, Applications of biosynthesized metallic nanoparticles - A review, Acta Biomater. 10 (2014) 4023–4042, doi:10.1016/j.actbio.2014.05.022.

[100] R. Singh, U.U. Shedbalkar, S.A. Wadhwani, B.A. Chopade, Bacteriagenic silver nanoparticles: Synthesis, mechanism, and applications, Appl. Microbiol. Biotechnol. 99 (2015) 4579–4593, https://doi.org/10.1007/s00253-015-6622-1.

[101] Y. Park, A new paradigm shift for the green synthesis of antibacterial silver nanoparticles utilizing plant extracts, Toxicol. Res. 30 (2014) 169–178, https://doi.org/10.5487/TR.2014.30.3.169.

[102] P. Singh, Y.J. Kim, D. Zhang, D.C. Yang, Biological synthesis of nanoparticles from plants and microorganisms, Trends Biotechnol. 34 (2016) 588–599, doi:10.1016/j.tibtech.2016.02.006.

[103] V.D. Doan, M.T. Phung, T.L.H. Nguyen, T.C. Mai, T.D. Nguyen, Noble metallic nanoparticles from waste Nypa fruticans fruit husk: Biosynthesis, characterization, antibacterial activity and recyclable catalysis, Arab. J. Chem. 13 (2020) 7490–7503, doi:10.1016/j.arabjc.2020.08.024.

[104] M. Aravind, A. Ahmad, I. Ahmad, M. Amalanathan, K. Naseem, S.M.M. Mary, C. Parvathiraja, S. Hussain, T.S. Algarni, M. Pervaiz, M. Zuber, Critical green routing synthesis of silver NPs using jasmine flower extract for biological activities and photocatalytical degradation of methylene blue, J. Environ. Chem. Eng. 9 (2021) 104877, doi:10.1016/j.jece.2020.104877.

[105] N.S. Al-Radadi, Green biosynthesis of flaxseed gold nanoparticles (Au-NPs) as potent anti-cancer agent against breast cancer cells, J. Saudi Chem. Soc. 25 (2021) 101243. doi: 10.1016/j.jscs.2021.101243.

[106] M. Hosny, M. Fawzy, Instantaneous phytosynthesis of gold nanoparticles via Persicaria salicifolia leaf extract, and their medical applications, Adv. Powder Technol. 32 (2021) 2891–2904, doi:10.1016/j.apt.2021.06.004.

[107] S. Dhandapani, X. Xu, R. Wang, A.M. Puja, H. Kim, H. Perumalsamy, S.R. Balusamy, Y.J. Kim, Biosynthesis of gold nanoparticles using Nigella sativa and Curtobacterium proimmune K3 and evaluation of their anticancer activity, Mater. Sci. Eng. C. 127 (2021) 112214, doi:10.1016/j.msec.2021.112214.

[108] B.A. Varghese, R.V.R. Nair, S. Jude, K. Varma, A. Amalraj, S. Kuttappan, Green synthesis of gold nanoparticles using Kaempferia parviflora rhizome extract and their characterization and application as an antimicrobial, antioxidant and catalytic degradation agent, J. Taiwan Inst. Chem. Eng. 126 (2021) 166–172, doi:10.1016/j.jtice.2021.07.016.

[109] T. Singh, A. Jayaprakash, M. Alsuwaidi, A.A. Madhavan, Green synthesized gold nanoparticles with enhanced photocatalytic activity, Mater. Today Proc 42 (2020) 1166–1169, doi:10.1016/j.matpr.2020.12.531.

[110] N. Zhang, J. Yu, P. Liu, J. Chang, D. Ali, Gold nanoparticles synthesized from Curcuma wenyujin inhibits HER-2 /neu transcription in breast cancer cells (MDA-MB-231/HER2), Arab. J. Chem. 13 (2020) 7264–7273, doi:10.1016/j.arabjc.2020.08.007.

[111] D.S. Bhagat, W.B. Gurnule, S.G. Pande, M.M. Kolhapure, A.D. Belsare, Biosynthesis of gold nanoparticles for detection of dichlorvos residue from different samples, Mater, Today Proc 29 (2019) 763–767, doi:10.1016/j.matpr.2020.04.589.

[112] N. Thangamani, N. Bhuvaneshwari, Green synthesis of gold nanoparticles using Simarouba glauca leaf extract and their biological activity of micro-organism, Chem. Phys. Lett. 732 (2019) 136587, doi:10.1016/j.cplett.2019.07.015.

[113] M.R. Bindhu, M. Umadevi, G.A. Esmail, N.A. Al-Dhabi, M.V. Arasu, Green synthesis and characterization of silver nanoparticles from Moringa oleifera flower and assessment of antimicrobial and sensing properties, J. Photochem. Photobiol. B Biol. 205 (2020) 111836, doi:10.1016/j.jphotobiol.2020.111836.

[114] M.A. Siddiquee, M. ud din Parray, S.H. Mehdi, K.A. Alzahrani, A.A. Alshehri, M.A. Malik, R. Patel, Green synthesis of silver nanoparticles from Delonix regia leaf extracts: In-vitro cytotoxicity and interaction studies with bovine serum albumin, Mater. Chem. Phys. 242 (2020) 122493, doi:10.1016/j.matchemphys.2019.122493.

[115] S. Hussein, U. EL-Magly, M. Tantawy, S. Kawashty, N. Saleh, Phenolics of selected species of Persicaria and Polygonum (Polygonaceae) in Egypt, Arab, J. Chem. 10 (2017) 76–81, doi:10.1016/j.arabjc.2012.06.002.

[116] A.M.M. Youssef, Z.A.S. El-Swaify, Anti-tumour effect of two persicaria species seeds on colon and prostate cancers, Biomed. Pharmacol. J. 11 (2018) 635–644, https://doi.org/10.13005/bpj/1416.

[117] M.S. Brewer, Natural antioxidants: Sources, compounds, mechanisms of action, and potential applications, Compr. Rev. Food Sci. Food Saf. 10 (2011) 221–247. https://doi.org/10.1111/j.1541-4337.2011.00156.x.

[118] M.A. Awad, N.E. Eisa, P. Virk, A.A. Hendi, K.M.O.O. Ortashi, A.A.S.A. Mahgoub, M.A. Elobeid, F.Z. Eissa, Green synthesis of gold nanoparticles: Preparation, characterization, cytotoxicity, and anti-bacterial activities, Mater. Lett. 256 (2019) 126608, doi:10.1016/j.matlet.2019.126608.

[119] M. Mosaviniya, T. Kikhavani, M. Tanzifi, M. Tavakkoli Yaraki, P. Tajbakhsh, A. Lajevardi, Facile green synthesis of silver nanoparticles using Crocus Haussknechtii Bois bulb extract: Catalytic activity and antibacterial properties, Colloids Interface Sci. Commun. 33 (2019) 100211, doi:10.1016/j.colcom.2019.100211.

[120] E. Ibañez, A. Cifuentes, Benefits of using algae as natural sources of functional ingredients, J. Sci. Food Agric. 93 (2013) 703–709, https://doi.org/10.1002/jsfa.6023.

[121] M.B. Cooper, A.G. Smith, Exploring mutualistic interactions between microalgae and bacteria in the omics age, Curr. Opin. Plant Biol. 26 (2015) 147–153, doi:10.1016/j.pbi.2015.07.003.

[122] L. O'Sullivan, B. Murphy, P. McLoughlin, P. Duggan, P.G. Lawlor, H. Hughes, G.E. Gardiner, Prebiotics from marine macroalgae for human and animal health applications, Mar. Drugs. 8 (2010) 2038–2064, https://doi.org/10.3390/md8072038.

[123] S. Yende, U. Harle, B. Chaugule, Therapeutic potential and health benefits of Sargassumspecies, Pharmacogn. Rev. 8 (2014) 1–7, https://doi.org/10.4103/0973-7847.125514.

[124] C. Chellapandian, B. Ramkumar, P. Puja, R. Shanmuganathan, A. Pugazhendhi, P. Kumar, Gold nanoparticles using red seaweed Gracilaria verrucosa: Green synthesis, characterization and biocompatibility studies, Process Biochem. 80 (2019) 58–63, doi:10.1016/j.procbio.2019.02.009.

[125] T.S. Dhas, P. Sowmiya, V.G. Kumar, M. Ravi, K. Suthindhiran, J.F. Borgio, G. Narendrakumar, V.R. Kumar, V. Karthick, C.M.V. Kumar, Antimicrobial effect of Sargassum plagiophyllum mediated gold nanoparticles on Escherichia coli and Salmonella typhi, Biocatal. Agric. Biotechnol. 26 (2020) 101627, doi:10.1016/j.bcab.2020.101627.

[126] S. Rajeshkumar, M.H. Sherif, C. Malarkodi, M. Ponnanikajamideen, M.V. Arasu, N.A. Al-Dhabi, S.M. Roopan, Cytotoxicity behaviour of response surface model optimized gold nanoparticles by utilizing fucoidan extracted from padina tetrastromatica, J. Mol. Struct. 1228 (2021) 129440, doi:10.1016/j.molstruc.2020.129440.

[127] M. Manikandakrishnan, S. Palanisamy, M. Vinosha, B. Kalanjiaraja, S. Mohandoss, R. Manikandan, M. Tabarsa, S.G. You, N.M. Prabhu, Facile green route synthesis of gold nanoparticles using Caulerpa racemosa for biomedical applications, J. Drug Deliv. Sci. Technol. 54 (2019) 101345, doi:10.1016/j.jddst.2019.101345.

[128] L.C. Moraes, R.C. Figueiredo, R. Ribeiro-Andrade, A.V. Pontes-Silva, M.L. Arantes, A. Giani, C.C. Figueredo, High diversity of microalgae as a tool for the synthesis of different silver nanoparticles: A species-specific green synthesis, Colloids Interface Sci. Commun. 42 (2021), doi:10.1016/j.colcom.2021.100420.

[129] K.F. Princy, A. Gopinath, Green synthesis of silver nanoparticles using polar seaweed Fucus gardeneri and its catalytic efficacy in the reduction of nitrophenol, Polar Sci. 30 (2021) 100692, doi:10.1016/j.polar.2021.100692.

[130] B. Yılmaz Öztürk, B. Yenice Gürsu, İ. Dağ, Antibiofilm and antimicrobial activities of green synthesized silver nanoparticles using marine red algae Gelidium corneum, Process Biochem. 89 (2020) 208–219, doi:10.1016/j.procbio.2019.10.027.

[131] N. Valarmathi, F. Ameen, A. Almansob, P. Kumar, S. Arunprakash, M. Govarthanan, Utilization of marine seaweed Spyridia filamentosa for silver nanoparticles synthesis and its clinical applications, Mater. Lett. 263 (2020) 127244, doi:10.1016/j.matlet.2019.127244.

[132] D. Dixit, D. Gangadharan, K.M. Popat, C.R.K. Reddy, M. Trivedi, D.K. Gadhavi, Synthesis, characterization and application of green seaweed mediated silver nanoparticles (AgNPs) as antibacterial agents for water disinfection, Water Sci. Technol. 78 (2018) 235–246, https://doi.org/10.2166/wst.2018.292.

[133] V.S. Ramkumar, A. Pugazhendhi, K. Gopalakrishnan, P. Sivagurunathan, G.D. Saratale, T.N.B. Dung, E. Kannapiran, Biofabrication and characterization of silver nanoparticles using aqueous extract of seaweed

Enteromorpha compressa and its biomedical properties, Biotechnol. Reports. 14 (2017) 1–7, doi:10.1016/j. btre.2017.02.001.

[134] J. Venkatesan, S.K. Kim, M.S. Shim, Antimicrobial, antioxidant, and anticancer activities of biosynthesized silver nanoparticles using marine algae ecklonia cava, Nanomaterials 6 (2016), https://doi.org/10.3390/ nano6120235.

[135] A.P. de Aragão, T.M. de Oliveira, P.V. Quelemes, M.L.G. Perfeito, M.C. Araújo, J. de A.S. Santiago, V.S. Cardoso, P. Quaresma, J.R. de Souza de Almeida Leite, D.A. da Silva, Green synthesis of silver nanoparticles using the seaweed Gracilaria birdiae and their antibacterial activity, Arab. J. Chem. 12 (2019) 4182–4188, doi:10.1016/j.arabjc.2016.04.014.

[136] S.M. Alsharif, S.S. Salem, M.A. Abdel-Rahman, A. Fouda, A.M. Eid, S. El-Din Hassan, M.A. Awad, A.A. Mohamed, Multifunctional properties of spherical silver nanoparticles fabricated by different microbial taxa, Heliyon 6 (2020) e03943, doi:10.1016/j.heliyon.2020.e03943.

[137] A.S. Dakhil, Biosynthesis of silver nanoparticle (AgNPs) using Lactobacillus and their effects on oxidative stress biomarkers in rats, J. King Saud Univ. - Sci. 29 (2017) 462–467, doi:10.1016/j.jksus.2017.05.013.

[138] R. Qiu, W. Xiong, W. Hua, Y. He, X. Sun, M. Xing, L. Wang, A biosynthesized gold nanoparticle from Staphylococcus aureus – as a functional factor in muscle tissue engineering, Appl. Mater. Today. 22 (2021) 100905, doi:10.1016/j.apmt.2020.100905.

[139] R. Shunmugam, S. Renukadevi Balusamy, V. Kumar, S. Menon, T. Lakshmi, H. Perumalsamy, Biosynthesis of gold nanoparticles using marine microbe (Vibrio alginolyticus) and its anticancer and antioxidant analysis, J. King Saud Univ. - Sci. 33 (2021) 101260, doi:10.1016/j.jksus.2020.101260.

[140] M.A. Elbahnasawy, A.M. Shehabeldine, A.M. Khattab, B.H. Amin, A.H. Hashem, Green biosynthesis of silver nanoparticles using novel endophytic Rothia endophytica: Characterization and anticandidal activity, J. Drug Deliv. Sci. Technol. 62 (2021) 102401, doi:10.1016/j.jddst.2021.102401.

[141] S. Husain, S.K. Verma, M.Azam Hemlata, M. Sardar, Q.M.R. Haq, T. Fatma, Antibacterial efficacy of facile cyanobacterial silver nanoparticles inferred by antioxidant mechanism, Mater. Sci. Eng. C. 122 (2021) 111888, doi:10.1016/j.msec.2021.111888.

[142] N. Mujaddidi, S. Nisa, S. Al Ayoubi, Y. Bibi, S. Khan, M. Sabir, M. Zia, S. Ahmad, A. Qayyum, Pharmacological properties of biogenically synthesized silver nanoparticles using endophyte Bacillus cereus extract of Berberis lyceum against oxidative stress and pathogenic multidrug-resistant bacteria, Saudi J. Biol. Sci. 28 (2021) 6432–6440, doi:10.1016/j.sjbs.2021.07.009.

[143] N.H. Sayyid, Z.R. Zghair, Biosynthesis of silver nanoparticles produced by Klebsiella pneumoniae, Mater. Today Proc. 42 (2021) 2045–2049, doi:10.1016/j.matpr.2020.12.257.

[144] M.P. Desai, R.V. Patil, K.D. Pawar, Selective and sensitive colorimetric detection of platinum using Pseudomonas stutzeri mediated optimally synthesized antibacterial silver nanoparticles, Biotechnol. Reports. 25 (2020) e00404, doi:10.1016/j.btre.2019.e00404.

[145] M. Khaleghi, S. Khorrami, H. Ravan, Identification of Bacillus thuringiensis bacterial strain isolated from the mine soil as a robust agent in the biosynthesis of silver nanoparticles with strong antibacterial and anti-biofilm activities, Biocatal. Agric. Biotechnol. 18 (2019) 101047, doi:10.1016/j.bcab.2019.101047.

[146] M.M. Juibari, S. Abbasalizadeh, G.S. Jouzani, M. Noruzi, Intensified biosynthesis of silver nanoparticles using a native extremophilic Ureibacillus thermosphaericus strain, Mater. Lett. 65 (2011) 1014–1017, doi:10.1016/j.matlet.2010.12.056.

[147] S. Zaki, M.F. El Kady, D. Abd-El-Haleem, Biosynthesis and structural characterization of silver nanoparticles from bacterial isolates, Mater. Res. Bull. 46 (2011) 1571–1576, doi:10.1016/j. materresbull.2011.06.025.

[148] A. Ahmad, P. Mukherjee, S. Senapati, D. Mandal, M.I. Khan, R. Kumar, M. Sastry, Extracellular biosynthesis of silver nanoparticles using the fungus Fusarium oxysporum, Colloids Surfaces B Biointerfaces 28 (2003) 313–318, doi:10.1016/S0927-7765(02)00174-1.

[149] S. Kaminskyj, K. Jilkine, A. Szeghalmi, K. Gough, High spatial resolution analysis of fungal cell biochemistry - Bridging the analytical gap using synchrotron FTIR spectromicroscopy, FEMS Microbiol. Lett. 284 (2008) 1–8, https://doi.org/10.1111/j.1574-6968.2008.01162.x.

[150] K.N. Thakkar, S.S. Mhatre, R.Y. Parikh, Biological synthesis of metallic nanoparticles, Nanomed. Nanotechnol., Biol. Med. 6 (2010) 257–262, doi:10.1016/j.nano.2009.07.002.

[151] D. Mandal, M.E. Bolander, D. Mukhopadhyay, G. Sarkar, P. Mukherjee, The use of microorganisms for the formation of metal nanoparticles and their application, Appl. Microbiol. Biotechnol. 69 (2006) 485–492, https://doi.org/10.1007/s00253-005-0179-3.

[152] M. Gericke, A. Pinches, Biological synthesis of metal nanoparticles, Hydrometallurgy 83 (2006) 132–140, doi:10.1016/j.hydromet.2006.03.019.

[153] N. Durán, P.D. Marcato, O.L. Alves, G.I.H. De Souza, E. Esposito, Mechanistic aspects of biosynthesis of silver nanoparticles by several Fusarium oxysporum strains, J. Nanobiotechnology. 3 (2005) 1–7, https://doi.org/10.1186/1477-3155-3-8.

[154] P. Mohanpuria, N.K. Rana, S.K. Yadav, Biosynthesis of nanoparticles: Technological concepts and future applications, J. Nanoparticle Res. 10 (2008) 507–517, https://doi.org/10.1007/s11051-007-9275-x.

[155] M.A. Abu-Tahon, M. Ghareib, W.E. Abdallah, Environmentally benign rapid biosynthesis of extracellular gold nanoparticles using Aspergillus flavus and their cytotoxic and catalytic activities, Process Biochem. 95 (2020) 1–11, doi:10.1016/j.procbio.2020.04.015.

[156] S. Ali, H. Ali, M. Siddique, H. Gulab, M.A. Haleem, J. Ali, Exploring the biosynthesized gold nanoparticles for their antibacterial potential and photocatalytic degradation of the toxic water wastes under solar light illumination, J. Mol. Struct. 1215 (2020) 128259, doi:10.1016/j.molstruc.2020.128259.

[157] J.M. do Nascimento, N.D. Cruz, G.R. de Oliveira, W.S. Sá, J.D. de Oliveira, P.R.S. Ribeiro, S.G.F. Leite, Evaluation of the kinetics of gold biosorption processes and consequent biogenic synthesis of AuNPs mediated by the fungus Trichoderma harzianum, Environ. Technol. Innov. 21 (2021) 101238, doi:10.1016/j.eti.2020.101238.

[158] M.A. Rabeea, M.N. Owaid, A.A. Aziz, M.S. Jameel, M.A. Dheyab, Mycosynthesis of gold nanoparticles using the extract of Flammulina velutipes, Physalacriaceae, and their efficacy for decolorization of methylene blue, J. Environ. Chem. Eng. 8 (2020) 103841, doi:10.1016/j.jece.2020.103841.

[159] A.M. Elshafei, A.M. Othman, M.A. Elsayed, N.G. Al-Balakocy, M.M. Hassan, Green synthesis of silver nanoparticles using Aspergillus oryzae NRRL447 exogenous proteins: Optimization via central composite design, characterization and biological applications, Environ. Nanotechnology, Monit. Manag. 16 (2021) 100553, doi:10.1016/j.enmm.2021.100553.

[160] A.M. Othman, M.A. Elsayed, A.M. Elshafei, M.M. Hassan, Application of central composite design as a strategy to maximize the productivity of Agaricus bisporus CU13 laccase and its application in dye decolorization, Biocatal. Agric. Biotechnol. 14 (2018) 72–79, doi:10.1016/j.bcab.2018.02.008.

[161] M. Qu, W. Yao, X. Cui, R. Xia, L. Qin, X. Liu, Biosynthesis of silver nanoparticles (AgNPs) employing Trichoderma strains to control empty-gut disease of oak silkworm (Antheraea pernyi), Mater. Today Commun. 28 (2021) 102619, doi:10.1016/j.mtcomm.2021.102619.

[162] M.A. Yassin, A.M. Elgorban, A.E.R.M.A. El-Samawaty, B.M.A. Almunqedhi, Biosynthesis of silver nanoparticles using Penicillium verrucosum and analysis of their antifungal activity, Saudi J. Biol. Sci. 28 (2021) 2123–2127, doi:10.1016/j.sjbs.2021.01.063.

[163] H.M.M. Farrag, F.A.A.M. Mostafa, M.E. Mohamed, E.A.M. Huseein, Green biosynthesis of silver nanoparticles by Aspergillus niger and its antiamoebic effect against Allovahlkampfia spelaea trophozoite and cyst, Exp. Parasitol. 219 (2020) 108031, doi:10.1016/j.exppara.2020.108031.

[164] M. Manjunath Hulikere, C.G. Joshi, Characterization, antioxidant and antimicrobial activity of silver nanoparticles synthesized using marine endophytic fungus- Cladosporium cladosporioides, Process Biochem. 82 (2019) 199–204, doi:10.1016/j.procbio.2019.04.011.

[165] A. Syed, S. Saraswati, G.C. Kundu, A. Ahmad, Biological synthesis of silver nanoparticles using the fungus Humicola sp. and evaluation of their cytoxicity using normal and cancer cell lines, Spectrochim. Acta - Part A Mol. Biomol. Spectrosc. 114 (2013) 144–147, doi:10.1016/j.saa.2013.05.030.

[166] C.F. Carolin, P.S. Kumar, A. Saravanan, G.J. Joshiba, M. Naushad, Efficient techniques for the removal of toxic heavy metals from aquatic environment: A review, J. Environ. Chem. Eng. 5 (2017) 2782–2799, doi:10.1016/j.jece.2017.05.029.

[167] A.M. Othman, M.A. Elsayed, N.G. Al-Balakocy, M.M. Hassan, A.M. Elshafei, Biosynthesized silver nanoparticles by Aspergillus terreus NRRL265 for imparting durable antimicrobial finishing to polyester cotton blended fabrics: Statistical optimization, characterization, and antitumor activity evaluation, Biocatal. Agric. Biotechnol. 31 (2021) 101908, doi:10.1016/j.bcab.2021.101908.

[168]   W.A. Mannheimer, Microscopia dos materiais: Uma introdução, Microsc. Eletrônica Transm. (2002) 1–213.

[169]   B.A. Dedavid, C.I. Gomes, G. Machado, Microscopia Eletrônica de Varredura -Aplicações e Preparação de Amostras, Dados Int. Cat. Na Publicação. (2007) 60.

[170]   A.E. Vladár, V.D. Hodoroaba, Characterization of Nanoparticles by Scanning Electron Microscopy, Elsevier Inc., Amsterdam, 2019, doi:10.1016/B978-0-12-814182-3.00002-X.

[171]   A. Von Bohlen, M. Brücher, B. Holland, R. Wagner, R. Hergenröder, X-ray standing waves and scanning electron microscopy - Energy dispersive X-ray emission spectroscopy study of gold nanoparticles, Spectrochim. Acta - Part B At. Spectrosc. 65 (2010) 409–414, doi:10.1016/j.sab.2010.04.017.

[172]   M. Puchalski, P. Dąbrowski, W. Olejniczak, P. Krukowski, P. Kowalczyk, K. Polański, The study of silver nanoparticles by scanning electron microscopy, energy dispersive X-ray analysis and scanning tunnelling microscopy, Mater. Sci. Pol. 25 (2007) 473–478.

[173]   K. Jores, W. Mehnert, M. Drechsler, H. Bunjes, C. Johann, K. Mäder, Investigations on the structure of solid lipid nanoparticles (SLN) and oil-loaded solid lipid nanoparticles by photon correlation spectroscopy, field-flow fractionation and transmission electron microscopy, J. Control. Release. 95 (2004) 217–227, doi:10.1016/j.jconrel.2003.11.012.

[174]   S.L. Pal, U. Jana, P.K. Manna, G.P. Mohanta, R. Manavalan, Nanoparticle: An overview of preparation and characterization, J. Appl. Pharm. Sci. 01 (2011) 228–234.

[175]   E. Buhr, N. Senftleben, T. Klein, D. Bergmann, D. Gnieser, C.G. Frase, H. Bosse, Characterization of nanoparticles by scanning electron microscopy in transmission mode, Meas. Sci. Technol. (2009) 20, https://doi.org/10.1088/0957-0233/20/8/084025.

[176]   P. Heera, S. Shanmugam, Nanoparticles characterization and applications: An overview, Int. J. Curr. Microbiol. App. Sci. 08 (2015) 379–386, https://doi.org/10.35652/igjps.2019.92s44.

[177]   L. Betancor, H.R. Luckarift, Bioinspired enzyme encapsulation for biocatalysis, Trends Biotechnol. 26 (2008) 566–572, doi:10.1016/j.tibtech.2008.06.009.

[178]   M.R. Bindhu, M. Umadevi, Antibacterial activities of green synthesized gold nanoparticles, Mater. Lett. 120 (2014) 122–125, doi:10.1016/j.matlet.2014.01.108.

[179]   A. Rautela, J. Rani, M. Debnath (Das), Green synthesis of silver nanoparticles from Tectona grandis seeds extract: characterization and mechanism of antimicrobial action on different microorganisms, J. Anal. Sci. Technol. 10 (2019), https://doi.org/10.1186/s40543-018-0163-z.

[180]   A.P.F. Albers, F.G. Melchiades, R. Machado, J.B. Baldo, A.O. Boschi, Um método simples de caracterização de argilominerais por difração de raios X, Cerâmica 48 (2002) 34–37, https://doi.org/10.1590/s0366-69132002000100008.

[181]   S. Banerjee, K. Loza, W. Meyer-Zaika, O. Prymak, M. Epple, Structural evolution of silver nanoparticles during wet-chemical synthesis, Chem. Mater. 26 (2014) 951–957, https://doi.org/10.1021/cm4025342.

[182]   D. Zanchet, M.B.D. Hall, D. Ugarte, Structure population in thioi-passivated gold nanoparticles, J. Phys. Chem. B. 104 (2000) 11013–11018, https://doi.org/10.1021/jp0017644.

[183]   S. Mourdikoudis, R.M. Pallares, N.T.K. Thanh, Characterization techniques for nanoparticles: Comparison and complementarity upon studying nanoparticle properties, Nanoscale. 10 (2018) 12871–12934. https://doi.org/10.1039/c8nr02278j.

[184]   S. Bykkam, M. Ahmadipour, S. Narisngam, V.R. Kalagadda, S.C. Chidurala, RETRACTED: Extensive studies on X-Ray diffraction of green synthesized silver nanoparticles, Adv. Nanoparticles. 04 (2015) 1–10, https://doi.org/10.4236/anp.2015.41001.

[185]   S. Balasubramanian, S.M.J. Kala, T.L. Pushparaj, Biogenic synthesis of gold nanoparticles using Jasminum auriculatum leaf extract and their catalytic, antimicrobial and anticancer activities, J. Drug Deliv. Sci. Technol. 57 (2020) 101620, doi:10.1016/j.jddst.2020.101620.

[186]   S. Pandey, G.K. Goswami, K.K. Nanda, Green synthesis of polysaccharide/gold nanoparticle nanocomposite: An efficient ammonia sensor, Carbohydr. Polym. 94 (2013) 229–234, doi:10.1016/j.carbpol.2013.01.009.

[187]   S. Pandey, G.K. Goswami, K.K. Nanda, Green synthesis of biopolymer-silver nanoparticle nanocomposite: An optical sensor for ammonia detection, Int. J. Biol. Macromol. 51 (2012) 583–589, doi:10.1016/j.ijbiomac.2012.06.033.

[188] M. Mahdieh, A. Zolanvari, A.S. Azimee, M. Mahdieh, Green biosynthesis of silver nanoparticles by Spirulina platensis, Sci. Iran. 19 (2012) 926–929, doi:10.1016/j.scient.2012.01.010.

[189] F. Schmid, Biological Macromolecules: UV-visible Spectrophotometry, ELS. (2001) 1–4. https://doi.org/10.1038/npg.els.0003142.

[190] D. Grasseschi, D. dos Santos, Nanomateriais plasmônicos: parte i. fundamentos da espectroscopia de nanopartículas e sua relação com o efeito SERS, Quim. Nova. 43 (2020) 1463–1481.

[191] M. Nasrollahzadeh, S. Mahmoudi-Gom Yek, N. Motahharifar, M.Ghafori Gorab, Recent developments in the plant-mediated green synthesis of Ag-based nanoparticles for environmental and catalytic applications, Chem. Rec. 19 (2019) 2436–2479, https://doi.org/10.1002/tcr.201800202.

[192] S. Eustis, H.Y. Hsu, M.A. El-Sayed, Gold nanoparticle formation from photochemical reduction of Au 3+ by continuous excitation in colloidal solutions. A proposed molecular mechanism, J. Phys. Chem. B. 109 (2005) 4811–4815, https://doi.org/10.1021/jp0441588.

[193] Mohamed Anwar K. Abdelhalim, The bioaccumulation and toxicity induced by gold nanoparticles in rats in vivo can be detected by ultraviolet-visible (UV-visible) spectroscopy, African J. Biotechnol. 11 (2012) 9399–9406, https://doi.org/10.5897/ajb11.3102.

[194] A.G. Ingale, A.N. Chaudhari, Biogenic synthesis of nanoparticles and potential applications: An eco-friendly approach, J. Nanomed. Nanotechnol. 4 (2013) 7, https://doi.org/10.4172/2157-7439.1000165.

[195] L. Breggin, R. Falkner, N. Jaspers, J. Pendergrass, R. Porter, C. House, Securing the Promise of Nanotechnologies: Towards Transatlantic Regulatory Cooperation, (2009) 122. http://www.chathamhouse.org/sites/default/files/public/Research/Energy, Environment and Development/r0909_nanotechnologies.pdf.

[196] A. Hulanicki, S. Glab, F. Ingman, Chemical sensors definitions and classification, Pure Appl. Chern. 63 (1991) 1247–1250.

[197] D. Baratella, E. Bonaiuto, M. Magro, J. de Almeida Roger, Y. Kanamori, G.P.P. Lima, E. Agostinelli, F. Vianello, Endogenous and food-derived polyamines: Determination by electrochemical sensing, Amino Acids 50 (2018) 1187–1203, https://doi.org/10.1007/s00726-018-2617-4.

[198] M. Akhoundian, A. Rüter, S. Shinde, Ultratrace detection of histamine using a molecularly-imprinted polymer-based voltammetric sensor, Sensors (Switzerland) 17 (2017), https://doi.org/10.3390/s17030645.

[199] L.C. Lopes, D. Lima, A.C. Mendes Hacke, B.S. Schveigert, G.N. Calaça, F.F. Simas, R.P. Pereira, M. Iacomini, A.G. Viana, C.A. Pessôa, Gold nanoparticles capped with polysaccharides extracted from pineapple gum: Evaluation of their hemocompatibility and electrochemical sensing properties, Talanta 223 (2021), doi:10.1016/j.talanta.2020.121634.

[200] S.H. Mohd Taib, K. Shameli, P. Moozarm Nia, M. Etesami, M. Miyake, R. Rasit Ali, E. Abouzari-Lotf, Z. Izadiyan, Electrooxidation of nitrite based on green synthesis of gold nanoparticles using Hibiscus sabdariffa leaves, J. Taiwan Inst. Chem. Eng. 95 (2019) 616–626, doi:10.1016/j.jtice.2018.09.021.

[201] Z.U.H. Khan, A. Khan, Y. Chen, A. ullah Khan, N.S. Shah, N. Muhammad, B. Murtaza, K. Tahir, F.U. Khan, P. Wan, Photo catalytic applications of gold nanoparticles synthesized by green route and electrochemical degradation of phenolic Azo dyes using AuNPs/GC as modified paste electrode, J. Alloys Compd. 725 (2017) 869–876, doi:10.1016/j.jallcom.2017.07.222.

[202] D. Lima, G.N. Calaça, A.G. Viana, C.A. Pessôa, Porphyran-capped gold nanoparticles modified carbon paste electrode: A simple and efficient electrochemical sensor for the sensitive determination of 5-fluorouracil, Appl. Surf. Sci. 427 (2018) 742–753, doi:10.1016/j.apsusc.2017.08.228.

[203] N. Chouhan, R. Ameta, R.K. Meena, Biogenic silver nanoparticles from Trachyspermum ammi (Ajwain) seeds extract for catalytic reduction of p-nitrophenol to p-aminophenol in excess of NaBH4, J. Mol. Liq. 230 (2017) 74–84, doi:10.1016/j.molliq.2017.01.003.

[204] M.A. Khalilzadeh, M. Borzoo, Green synthesis of silver nanoparticles using onion extract and their application for the preparation of a modified electrode for determination of ascorbic acid, J. Food Drug Anal. 24 (2016) 796–803, doi:10.1016/j.jfda.2016.05.004.

[205] Z.U.H. Khan, A. Khan, A. Shah, Y. Chen, P. Wan, A.U. Khan, K. Tahir, N. Muhamma, F.U. Khan, H.U. Shah, Photocatalytic, antimicrobial activities of biogenic silver nanoparticles and electrochemical degradation of water soluble dyes at glassy carbon/silver modified past electrode using buffer solution, J. Photochem. Photobiol. B Biol. 156 (2016) 100–107, doi:10.1016/j.jphotobiol.2016.01.016.

[206] M. Shivakumar, K.L. Nagashree, S. Manjappa, M.S. Dharmaprakash, Electrochemical detection of nitrite using glassy carbon electrode modified with silver nanospheres (AgNS) obtained by green synthesis using pre-hydrolysed liquor, Electroanalysis 29 (2017) 1434–1442, https://doi.org/10.1002/elan.201600775.

[207] E. Turunc, R. Binzet, I. Gumus, G. Binzet, H. Arslan, Green synthesis of silver and palladium nanoparticles using Lithodora hispidula (Sm.) Griseb. (Boraginaceae) and application to the electrocatalytic reduction of hydrogen peroxide, Mater. Chem. Phys. 202 (2017) 310–319, doi:10.1016/j.matchemphys.2017.09.032.

[208] F. Zamarchi, I.C. Vieira, Determination of paracetamol using a sensor based on green synthesis of silver nanoparticles in plant extract, J. Pharm. Biomed. Anal. 196 (2021) 113912, doi:10.1016/j.jpba.2021.113912.

[209] Y. Zheng, A. Wang, W. Cai, Z. Wang, F. Peng, Z. Liu, L. Fu, Hydrothermal preparation of reduced graphene oxide–silver nanocomposite using Plectranthus amboinicus leaf extract and its electrochemical performance, Enzyme Microb. Technol. 95 (2016) 112–117, doi:10.1016/j.enzmictec.2016.05.010.

[210] L. Bojko, G. de Jonge, D. Lima, L.C. Lopes, A.G. Viana, J.R. Garcia, C.A. Pessôa, K. Wohnrath, J. Inaba, Porphyran-capped silver nanoparticles as a promising antibacterial agent and electrode modifier for 5-fluorouracil electroanalysis, Carbohydr. Res. 498 (2020), doi:10.1016/j.carres.2020.108193.

[211] M. Etesami, N. Mohamed, Preparation of Pt/MWCNTs catalyst by taguchi method for electrooxidation of nitrite, J. Anal. Chem. 71 (2016) 185–194, https://doi.org/10.1134/S1061934816020040.

[212] J. Losada, M.P. García Armada, E. García, C.M. Casado, B. Alonso, Electrochemical preparation of gold nanoparticles on ferrocenyl-dendrimer film modified electrodes and their application for the electrocatalytic oxidation and amperometric detection of nitrite, J. Electroanal. Chem. 788 (2017) 14–22, doi:10.1016/j.jelechem.2017.01.066.

[213] K. Zhao, L. Ge, T.I. Wong, X. Zhou, G. Lisak, Gold-silver nanoparticles modified electrochemical sensor array for simultaneous determination of chromium(III) and chromium(VI) in wastewater samples, Chemosphere 281 (2021) 130880, doi:10.1016/j.chemosphere.2021.130880.

[214] S. Malali Sudhakara, M. Chattanahalli Devendrachari, H. Makri Nimbegondi Kotresh, F. Khan, Silver nanoparticles decorated phthalocyanine doped polyaniline for the simultaneous electrochemical detection of hydroquinone and catechol, J. Electroanal. Chem. 884 (2021) 115071, doi:10.1016/j.jelechem.2021.115071.

[215] A. Dehnavi, A. Soleymanpour, Silver nanoparticles/poly(L-cysteine) nanocomposite modified pencil graphite for selective electrochemical measurement of guaifenesin in real samples, Meas. J. Int. Meas. Confed. 175 (2021) 109103, doi:10.1016/j.measurement.2021.109103.

[216] P. Rameshkumar, P. Viswanathan, R. Ramaraj, Silicate sol-gel stabilized silver nanoparticles for sensor applications toward mercuric ions, hydrogen peroxide and nitrobenzene, Sensors Actuators, B Chem 202 (2014) 1070–1077, doi:10.1016/j.snb.2014.06.069.

[217] P. Viswanathan, R. Ramaraj, Polyelectrolyte assisted synthesis and enhanced catalysis of silver nanoparticles: Electrocatalytic reduction of hydrogen peroxide and catalytic reduction of 4-nitroaniline, J. Mol. Catal. A Chem. 424 (2016) 128–134, doi:10.1016/j.molcata.2016.08.001.

[218] A. Taheri, M. Noroozifar, M. Khorasani-Motlagh, Investigation of a new electrochemical cyanide sensor based on Ag nanoparticles embedded in a three-dimensional sol-gel, J. Electroanal. Chem. 628 (2009) 48–54, doi:10.1016/j.jelechem.2009.01.003.

[219] S. Motia, B. Bouchikhi, N. El Bari, An electrochemical molecularly imprinted sensor based on chitosan capped with gold nanoparticles and its application for highly sensitive butylated hydroxyanisole analysis in foodstuff products, Talanta 223 (2021) 121689, doi:10.1016/j.talanta.2020.121689.

[220] S. Shi, H. Wu, L. Zhang, S. Wang, P. Xiong, Z. Qin, M. Chu, J. Liao, Gold nanoparticles based electrochemical sensor for sensitive detection of uranyl in natural water, J. Electroanal. Chem. 880 (2021) 114884, doi:10.1016/j.jelechem.2020.114884.

[221] A. Wong, C.A. Razzino, T.A. Silva, O. Fatibello-Filho, Square-wave voltammetric determination of clindamycin using a glassy carbon electrode modified with graphene oxide and gold nanoparticles within a crosslinked chitosan film, Sensors Actuators, B Chem 231 (2016) 183–193, doi:10.1016/j.snb.2016.03.014.

[222] A.K. Das, C.R. Raj, Shape-controlled growth of surface-confined Au nanostructures for electroanalytical applications, J. Electroanal. Chem. 717–718 (2014) 140–146, doi:10.1016/j.jelechem.2014.01.019.

[223] W. Jin, G. Wu, A. Chen, Sensitive and selective electrochemical detection of chromium(VI) based on gold nanoparticle-decorated titania nanotube arrays, Analyst 139 (2014) 235–241, https://doi.org/10.1039/c3an01614e.

[224] J. Wei, Z. Guo, X. Chen, D.D. Han, X.K. Wang, X.J. Huang, Ultrasensitive and ultraselective impedimetric detection of cr(vi) using crown ethers as high-affinity targeting receptors, Anal. Chem. 87 (2015) 1991–1998, https://doi.org/10.1021/ac504449v.

[225] C.M. Welch, M.E. Hyde, O. Nekrassova, R.G. Compton, The oxidation of trivalent chromium at polycrystalline gold electrodes, Phys. Chem. Chem. Phys. 6 (2004) 3153–3159, https://doi.org/10.1039/b404243c.

[226] S. Hoppe, Y. Li, L.V. Moskaleva, S. Müller, How silver segregation stabilizes 1D surface gold oxide: A cluster expansion study combined with: Ab initio MD simulations, Phys. Chem. Chem. Phys. 19 (2017) 14845–14853, https://doi.org/10.1039/c7cp02221b.

[227] J. Kirsch, C. Siltanen, Q. Zhou, A. Revzin, A. Simonian, Biosensor technology: Recent advances in threat agent detection and medicine, Chem. Soc. Rev. 42 (2013) 8733–8768, https://doi.org/10.1039/c3cs60141b.

[228] M. Campàs, R. Carpentier, R. Rouillon, Plant tissue-and photosynthesis-based biosensors, Biotechnol. Adv. 26 (2008) 370–378, doi:10.1016/j.biotechadv.2008.04.001.

[229] S. Chen, Y.Frank Cheng, G. Voordouw, Three-dimensional graphene nanosheet doped with gold nanoparticles as electrochemical DNA biosensor for bacterial detection, Sensors Actuators, B Chem 262 (2018) 860–868, doi:10.1016/j.snb.2018.02.093.

[230] J. Wang, From DNA biosensors to gene chips, Nucleic. Acids. Res. 28 (2000) 3011–3016, https://doi.org/10.1093/nar/28.16.3011.

[231] L. Tian, K. Qian, J. Qi, Q. Liu, C. Yao, W. Song, Y. Wang, Gold nanoparticles superlattices assembly for electrochemical biosensor detection of microRNA-21, Biosens. Bioelectron. 99 (2018) 564–570, doi:10.1016/j.bios.2017.08.035.

[232] T. Chen, A. Sheng, Y. Hu, D. Mao, L. Ning, J. Zhang, Modularization of three-dimensional gold nanoparticles/ferrocene/liposome cluster for electrochemical biosensor, Biosens. Bioelectron. 124–125 (2019) 115–121, doi:10.1016/j.bios.2018.09.101.

[233] N. Wang, H.Y. Zhao, X.P. Ji, X.R. Li, B.B. Wang, Gold nanoparticles-enhanced bisphenol A electrochemical biosensor based on tyrosinase immobilized onto self-assembled monolayers-modified gold electrode, Chinese Chem. Lett. 25 (2014) 720–722, doi:10.1016/j.cclet.2014.01.008.

[234] D. Du, S. Chen, J. Cai, A. Zhang, Electrochemical pesticide sensitivity test using acetylcholinesterase biosensor based on colloidal gold nanoparticle modified sol-gel interface, Talanta 74 (2008) 766–772, doi:10.1016/j.talanta.2007.07.014.

[235] L. Hou, Y. Huang, W. Hou, Y. Yan, J. Liu, N. Xia, Modification-free amperometric biosensor for the detection of wild-type p53 protein based on the in situ formation of silver nanoparticle networks for signal amplification, Int. J. Biol. Macromol. 158 (2020) 580–586, doi:10.1016/j.ijbiomac.2020.04.271.

[236] Y. Jiang, J. Cui, T. Zhang, M. Wang, G. Zhu, P. Miao, Electrochemical detection of T4 polynucleotide kinase based on target-assisted ligation reaction coupled with silver nanoparticles, Anal. Chim. Acta. 1085 (2019) 85–90, doi:10.1016/j.aca.2019.07.072.

[237] I. Pantic, J. Paunovic, I. Dimitrijevic, A. Blachnio, A. Przepiorka, S. Pantic, Novel strategies for neurotransmitter detection and measurement using advanced materials, Rev. Adv. Mater. Sci. 42 (2015) 1–5.

[238] K. Cai, R.P.R. Nanga, L. Lamprou, C. Schinstine, M. Elliott, H. Hariharan, R. Reddy, C.N. Epperson, The impact of gabapentin administration on brain gaba and glutamate concentrations: A 7T 1H-MRS study, Neuropsychopharmacology. 37 (2012) 2764–2771. https://doi.org/10.1038/npp.2012.142.

[239] J.S. Boruah, P. Kalita, D. Chowdhury, M. Barthakur, Conjugation of citrate capped gold nanoparticles with gabapentin to use as biosensor, Mater. Today Proc. 46 (2019) 6404–6408, doi:10.1016/j.matpr.2020.06.422.

CHAPTER 8

Silver and gold nanoparticles: Potential cancer theranostic applications, recent development, challenges, and future perspectives

Swapnali Londhe[a,b], **Shagufta Haque**[a,b], **Chitta Ranjan Patra**[a,b]

[a]*Department of Applied Biology, CSIR-Indian Institute of Chemical Technology, Hyderabad, Telangana, India* [b]*Academy of Scientific and Innovative Research (AcSIR), Ghaziabad, Uttar Pradesh, India*

8.1 Introduction

Nanotechnology involving noble metal nanoparticles especially silver (Ag) and gold (Au) has emerged as one of the most indispensable tools for various biomedical applications worldwide [1–6]. Apart from biomedical applications, in the sector of science and technology, noble metal nanoforms such as gold and silver successfully imparted solutions to several challenges in industrial fields (catalysis, environmental, and agriculture, etc) [7–11]. Owing to their unique fundamental properties, it influences excellent chemical reactivity, catalytic behavior as well as biological activity as compared to bulk materials of the same composition. As a therapeutic agent, noble metals (Au & Ag) and their derivatives have a history in biomedical research [1–6,12–14]. Gold and silver-based nanoparticles act as a novel drug delivery system because of special characteristics such as large surface area, less toxicity, and enhanced stability within the living system [15–17]. Currently, gold and silver nanomaterials are considered as an alternative approach for cancer therapeutics and diagnostics [18–21].

As per reports, silver is a noble metal with a soft and lustrous texture having high electrical as well as thermal conductivity [22,23]. The silver in the nanoscale range has a greater surface area that makes it a suitable candidate for absorption and catalysis [24,25]. Recently, silver nanoparticles (AgNPs) have attracted great attention because of their remarkable conductivity as well as chemical stability in several medicinal science fields, including antimicrobial and wound healing. When AgNPs are placed in the dielectric environment, it permits surface plasmon resonance leading to higher optical absorption. It is extensively used as an antibacterial agent in food industries [24,26,27]. Apart from this, AgNPs has potent anticancer, antiinflammatory, antiangiogenic, and antioxidant properties [28–30].

Gold and Silver Nanoparticles.
DOI: https://doi.org/10.1016/B978-0-323-99454-5.00006-8

Similarly, gold is a versatile and multifunctional agent used in different biomedical applications over centuries due to its anticancer, antibacterial, anticorrosive, and antioxidant properties. The modern science is making it more advantageous by manufacturing gold at nanoscale range. The gold nanoparticles (AuNPs) are functionalized with different thiol and amine groups for desired activities that could be precisely used in drug delivery [31,32]. The smaller size of the AuNPs increases its interaction for binding with biomolecules located on either side of the cell cytoplasm [33]. Several reports demonstrated the versatility and multifunctionality of AuNPs as a potent drug delivery agent as well as significant candidate in transferring drugs without any issue [15,16,34]. Moreover, these properties can be altered for imaging, photodynamic, as well as photothermal therapy [21,33]. The upcoming section encompasses potential applications of silver and Au nanoparticles in several fields with more emphasis on cancer therapeutics and diagnostics.

8.1.1 Historical and medicinal background of silver and gold compounds

Silver and AuNPs are developed for several applications in the sector of pharmaceuticals, biomedical field such as for diagnostic tool, antibiofilm, antibacterial, drug delivery systems, cosmetics, personal care products, etc. [1,3,12]. For thousands of years, silver and its derivatives have been utilized for antibacterial and therapeutic applications. The use of silver has been first recorded in 750 AD for manufacturing of liquid container, vessels for water, and food storage to prevent spoilage. It has been documented that different civilization of primordial Romans, Greeks, and Egyptians used silver for food preservation [13,35]. In 1850 BC, silver was used to treat wounds; hence the use of silver in wound healing can be dated back for several centuries. Moreover, in the present days due to the wide antibacterial spectrum, silver-containing ointments, creams, biomedical products like wound dressings are readily used [24].

Similarly, the use of gold colloids has been fascinating over the last several centuries before the advancement of nanomedicine [36]. In ancient times, the usage of gold by Romans has a long history for decorative purposes. Over 150 years ago, the AuNPs synthesis era begin with the work of Michael Faraday in 1857, who first observed the red color of gold colloidal solution and its different properties than bulk gold. After 50 years, Maxwell's electromagnetic equations explained AuNPs visible absorption phenomenon [36–38]. Since, ancient time, gold has been enormously used in several fields of sciences (chemistry, biology, medicine) and engineering with modified chemical functional groups [37]. Hence, the gold and silver are readily used as nanomaterials in the current scenario for various biomedical applications.

8.2 Nanoparticles and its various synthetic approaches of silver and gold nanoparticles (brief overview)

The route of synthesis of gold and AgNPs is categorized into three types namely:

- Physical method [39–47]
- Chemical method [43,46, 48–54]
- Biological method [29,46,55–59]

Specifically, the particle size, shape, and monodispersity are essential parameters for nanoparticle synthesis [43,52]. The upcoming section briefly highlights the different synthesis procedures for the synthesis of gold and AgNPs.

8.2.1 Physical method

Physical methods are free from use of any chemical or biological agents to synthesize gold and AgNPs other than the precursors of silver and gold salts. The physical method includes laser ablation [39], microwave [40], ball milling [41], thermal evaporation [42], vapor phase [43] and evaporation-condensation [44] approaches for synthesis of AgNPs. While, the AuNPs are synthesized by gamma irradiation [43], pyrolysis, electrochemical method, and laser ablation from corresponding source [45].

8.2.2 Chemical method

Chemical method includes synthesis of metallic nanoparticles by reduction of their respective salts using chemical agents and stabilizing agent that prevents the aggregation of nanoparticles [43]. Chemical synthesis includes thermal reduction [48], electrochemical synthetic method, microemulsion techniques [49], photoreduction method [50], and microwave assisted synthesis [51]. The most common method for synthesis of AgNPs is the chemical reduction of silver salts using inorganic or organic reducing agent (sodium citrate, sodium borohydride, ascorbate, polyethylene glycol) [43,46,52,53]. Similarly, preparation of AuNPs using chemical reduction includes two steps such as, (1) reduction by agents like hydrazine, formaldehyde, hydroxylamine, polyols, citric, carbon monoxide, and oxalic acid, (2) stabilization of nanoparticles by sulfur ligands, phosphorus ligands, dendrimers, surfactants, and polymers [54]. Chemical method is advantageous as it requires less time for synthesis of nanoparticles with high amount of yield.

8.2.3 Biological method

The conventional nanoparticle synthesis methods are costly, hazardous, and nonecofriendly. Hence, researchers are attracted towards eco-friendly biological precursors for nanoparticle synthesis [29]. There are various reports of gold and AgNPs synthesized through biological routes using different plants including their parts as well as microorganism [55,56].The synthesis of nanoparticles by biological method includes bottom-up approach carried out with the help of these biological sources that acts as a capping, stabilizing, and reducing agents [29].

Usually, plant extract contains several compounds (polyphenols, proteins, carbohydrates) that acts as reducing, capping, and stabilizing agent during the synthesis of metal nanoparticles. [29,46,57,58]. For example, during the synthesis of silver and gold nanoparticles, the first step involves collection of plant material and dissolving it in appropriate solvent. Secondly, gold/

silver salts of different concentration are mixed with plant extracts at various temperatures and pH [29,43].

8.3 Background of cancers

Cancer is one of the deadly diseases of 21st century which mainly affects the social, economic, and ethnic characteristics of the population. The several cancer-causing factors include harmful radiations, exposure to pollutants, environmental factors, lack of physical activity, smoking, and stress. Overall, 5–10% cases of cancer are associated with heredity [38,60]. For proliferation and survival, cancer cells gain differentiated capabilities as compared to normal cells [61]. Cancer is characterized by random cell division, unrestricted growth of cells, and invasiveness. The cancer development is mostly due to alternation and mutation of proto-oncogenes as well as tumor suppressor genes. Cancer cells exhibit various biologic hallmarks resulting from changes in genetic and epigenetic level. These changes convert normal cells into cancerous, having the capability to uncontrollable growth and movement [38,60,62,63]. The metabolic characteristics and pathways of cancer cells are different from normal cells in order to fulfill their needs. Current research mainly focuses on anticancer drug targeting that are nontoxic to the normal cells and specifically target proliferating cancer cells. Thus, the recent treatment strategies available commercially are focusing more on the metabolism of cancer cells for designing and development of anticancer drugs as well as inhibitors of the metabolic pathways [63,64].

8.3.1 Global statistics and market for cancer

The cancer is majorly responsible for the mortality and morbidity worldwide. As per GLOBO-CAN 2008 estimates, about 7.6 million cancer deaths and 12.7 million cancer cases was predicted in 2008 [65]. As reported by GLOBOCAN, 8.2 million deaths of cancer patients and 14.1 million new cases were found in 2012 [66]. If one goes by statistical analysis given by WHO, about 1.15 million new patients were estimated in 2018 and it predicted to almost two-fold by 2040 [67]. It is becoming prime reason for the premature deaths of young and middle-aged people. The specific types of cancers like lung, colorectal, and female breast cancer are responsible for one-third of total cancer incidence as well as global mortality burden [68]. According to WHO, cancer causes more death than cardiovascular diseases [69]. Global demography predicted that, by 2025, 420 million new cancer cases will increase annually. In India, the cancer accounts for 9% of all deaths and leading among the noncommunicable diseases [70].

8.3.2 Various treatment strategies for cancers and their limitations

The traditional treatment of cancer includes local and systemic approaches. The local treatment contains oncological surgery and radiation therapy. Systemic treatment includes hormone therapy, biological therapy, and chemotherapy [71–73]. Surgery plays a key role in the treatment of cancer as it gives complete clearance, lowers tumor mass that enhances systemic treatment effects [71,73]. Whereas, radiation therapy to the tumor leads to impaired metabolic function and cell division. In systemic treatment, chemotherapy is one of the most popular treatments of cancer. It uses drugs that blocks cell division, inhibit signaling pathways. Hormone therapy includes targeting the expression of the receptor of a specific hormone related to the regression of tumor [61,73]. But the currently available chemotherapeutic therapies are associated with acute as well as delayed toxic effects, as it cytotoxic to both cancer and normal cells. These treatments are insignificant as it lacks specificity, affects both normal and cancer cells, exhibit noncompliance, inconvenience to patient as well as side effects to the healthy cells [74]. But novel designed therapies contain greater therapeutic index along with reduction of side effect on normal cells. The main objective of the designed treatments is to induce cancer cell cytotoxicity by the combination of targeted metabolic inhibitors along with chemotherapeutic agents [63,74]. Among the recent designed technologies, nanotechnology is readily explored for cancer therapeutics. Therefore, scientists are focusing on the development of nanotechnology as an alternative approach to overcome the drawbacks of the conventional medicine for cancer therapeutics.

8.4 Role of metal nanoparticles in cancer therapy

Several anticancer drugs are not able to achieve their target with sufficient concentration. It does not provide pharmacological effect efficiently as it injures healthy cells and tissues. Nanomaterials provide plenty of tools for the treatment of cancer by crossing biological barriers and delivering therapeutic targets directly [75]. The unique properties of metal nanoparticles include easy synthesis, surface functionalization, wide optical properties, and greater surface area to volume ratio that provides novel opportunities for therapeutics of cancer [76,77].The metal nanoparticles such as gold, iron, zinc, titanium, and silver have significant potential as an anticancer agent based on their surface modification [78].They are one of the significant candidates to reduce side effects related to the conventional treatments of cancer. They play key role in cancer therapy by gene silencing, drug delivery, and better targeting. Metal nanoparticle functionalization along with targeting ligand provides significant energy deposition in cancer cells. Along with therapeutic advantages, metal nanoparticles are significant diagnostic tool for cancer imaging [79]. Due to their smaller size and shape, they can simply interact with biological molecules both inside as well as at the surface of cells providing target specificity along with better signaling. The novel physical properties (fluorescence enhancement, surface plasmonic resonance) and chemical properties (catalytic activity) make them a significant candidate for imaging along with therapeutic [77,80]. The upcoming section highlights the different biomedical applications of gold and silver-based nanoparticles depending upon their conjugation, targeting agents as well as mode of action.

8.5 Therapeutic applications of silver and gold nanoparticles for cancer theranostics (detailed discussion)

8.5.1 Silver nanoparticles (AgNPs)

Based on the distinctive physical properties of AgNPs like size, shape, and optical properties (light scattering) they can be explored as an in vitro and in vivo therapeutic and diagnostic tool for cancer [81,82].

8.5.1.1 Biosynthesized AgNPs

The biological synthesis of metal nanoparticles is an emerging branch of nanotechnology for several biomedical applications. Generally, plants contain various phytochemicals which acts as stabilizing and reducing agents for synthesis of nanoparticle making it a safe and cheap option [83]. In this context, Haque et al. synthesized silver nanoparticles (AgZE) by *Zinnia elegans* (ZE) ethanolic leaf extract that was characterized using several physicochemical techniques. The cell viability was checked by various (HEK-293T, CHO, H9c2, and EA. hy926) normal cell lines which supported the biocompatible nature of AgZE. Further, the anticancer activity of AgZE was exhibited in U-87 (glioblastoma cell line) by ROS generation and through apoptosis. Moreover, the in vivo study demonstrates the noninvasive in vivo bioimaging activity of AgZE at the NIR region [28].The Fig. 8.1: Panel I column III and Fig. 8.1: Panel II, column III shows fluorescence images of C57BL6/J mice organs at for 4 and 24 hours, respectively. The data exhibited accumulation of AgZE in different organs (kidney, spleen, liver, colon, and lung) due to their ability to penetrate deep in the tissues. The untreated animals did not exhibit fluorescence except autofluorescence in some organs. Further, all the treated and untreated mice was sacrificed and organs were taken out for ex vivo imaging in order to observe fluorescence and biodistribution of AgZE at 4hr (Fig. 8.1: Panel I, row III, column III) and 24 hours (Fig. 8.1: Panel II, row III, column III). There was preferential accumulation of AgZE NPs in brain, kidney, liver, colon, spleen, and lung which were further confirmed by ICPOES analysis. The above finding confirmed that the synthesized AgZE can act as significant therapeutic and NIR bioimaging agent for cancer treatment [28].

In another work, Venugopal et al. synthesized AgNPs using *Beta vulgaris* extract as a reducing agent. The biosynthesized AgNPs exhibited better cytotoxicity against A549, Hep2, and MCF-7 cells as compared to normal cells. The AO/EtBr staining results indicated the apoptotic nature of AgNPs for inducing cancer cell death [84].

Jadhav et al. investigated the anticancer activity of AgNPs synthesized using an aqueous extract of *Salacia chinensis* bark. The absence of cytotoxicity against blood erythrocytes and human fibroblasts confirms their biocompatible nature. The in vitro assay demonstrated anticancer activity against lungs, liver, pancreas, prostate, breast, and cervical cancer cells [85]. Kummara et al. manufactured AgNPs through biological method using *Azadirachta indica* as well as by chemical method. Further, the toxicity of biologically and chemically synthesized AgNPs was

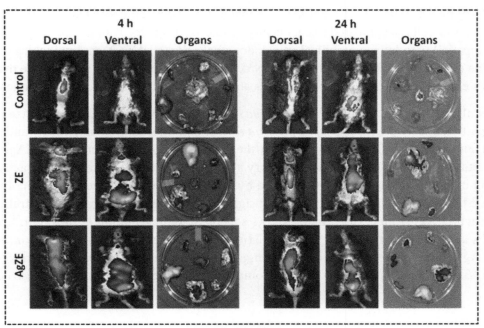

Fig. 8.1: Representative images of in vivo and ex vivo biodistribution of ZE and AgZE in C57BL6/J female mice using a noninvasive in vivo imager at 4–24-hour time points. Panel I: 4 hours, Panel II: 24 hours; Column I: dorsal side, Column II: ventral side, and Column III: organs of the C57BL6/J mice. Row I: Control mice; Row II: i.p. injected ZE mice; and Row III: i.p. injected AgZE mice. The images of the mice are taken at 710 nm excitation and 820 nm emission. The brain, heart, lung, liver, colon, kidney, and spleen show fluorescence upon intraperitoneal injection with ZE and AgZE, indicating their maximum distribution into those respective sites. These experiments are performed thrice. *Reprinted with permission from Haque et al. [Ref. 28] Biosynthesized Silver Nanoparticles for Cancer Therapy and In Vivo Bioimaging. Cancers. 2021; 13(23): 6114.Copyright © 2021 MDPI.*

compared against (normal cell) human skin dermal fibroblast along with NCI-H460 (lung cancer cells). The green synthesized AgNPs showed a significant increase in ROS (Reactive oxygen species) generation, decrease in cell viability, and cellular apoptosis induction in NCI-H460 cells. Moreover, the biosynthesized AgNPs has not shown any cytotoxic effect on human skin dermal fibroblast cells while chemical synthesized AgNPs showed significant toxicity against human skin dermal fibroblast cells. Furthermore, green synthesized AgNPs exhibits less toxicity towards human red blood cells than chemically synthesized nanoparticles. Thus, green synthesized nanoparticles impart selective toxicity to the cancer cells [86].

Mukherjee et al. designed and developed an efficient green approach for the synthesis of AgNPs using silver nitrate and leaf extract of *Olax scandens* as a reducing agent. The biological synthesized AgNPs showed enhanced antibacterial and anticancer activity against

(B16F10) mouse melanoma cells, (A549) human lung cancer cell lines, and (MCF7) human breast cancer cells as compared to chemically synthesized AgNPs. Along with this, it showed biocompatibility to human umbilical vein endothelial cells (HUVEC), cardiomyoblast normal cell line (H9C2), and Chinese hamster ovary cells (CHO). These results demonstrate that AgNPs can acts as a drug delivery as well as an anticancer agent for cancer therapeutics [87].

Gul et al. synthesized AgNPs by (*Poa annua*) annual meadow grass. The synthesized nanoparticles loaded with (EDL) *Euphorbia dracunculoides Lam.* plant extract (anticancer drug), and starch was used for coating. Further, the synthesized nanocomposite (AgNPs-EDL-Starch) was explored for drug delivery application. The in vitro anticancer activity and biocompatibility was demonstrated by MTT (by SCC7 cell lines) and hemolytic assay. Moreover, AgNPs did not show any significant changes after oral administration in Sprague Dawley rats. The Fig. 8.2 shows schematic representation of biosynthesized AgNPs and their in vitro and in vivo evaluation for anticancer therapy. The biosynthesized AgNPs are loaded with starch (coating and stabilizing agent) as well as anticancer agent *(E. dracunculiodes Lam.)*. The biocompatibility and cytotoxicity of synthesized

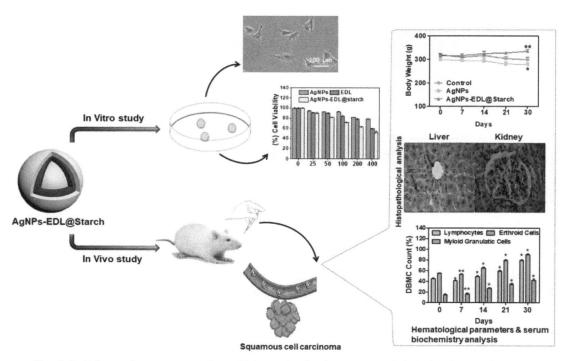

Fig. 8.2: Schematic representation of biosynthesized AgNPs and their in vitro and in vivo evaluation for anticancer therapy. *Reprinted with permission from Gul et al. [Ref. 88] (2021). Grass-mediated biogenic synthesis of silver nanoparticles and their drug delivery evaluation: a biocompatible anti-cancer therapy. Chem. Eng. J. 2021; 407: 127202. Copyright © 2021 Elsevier.*

nanoparticles was carried out by several in vitro (MTT) and in vivo (hematological parameter and biochemistry analysis) experiments as discussed above. Hence, AgNPs plays promising role in cancer therapeutics [88].

8.5.1.2 Polymeric AgNPs

Polymeric nanocompounds are bioengineered macromolecules used in several areas of biomedicine, where the field of cancer theranostics have attracted immense attention [89]. In this study, Boca et al. synthesized silver nanotriangles (AgNTs) coated with chitosan for high resonance in the NIR (near-infrared) region in order to act as a photothermal agent against (human lung cancer cell line) NCI-H460. The excitation was conducted from Ti:Sapphire laser at 800 nm wavelength. Moreover, it was found that the cell mortality rate is higher in presence of chitosan coated AgNTs as compared to thiolated polyethylene glycol coated (AuNRs) gold nanorods (commonly used hyperthermic agent). Fig. 8.3 shows the viability of cells using Hoechst stain (used to label cells nuclei) where it was performed at increasing concentration of two polymer coated nanoparticle (AuNRs and AgNTs) at 24 hours incubation in both normal (HEK) and cancer cells (NCI-H460). Figs. 8.3A and 8.3B demonstrated cell viability plots with nanoparticle concentration. The plot shows biocompatibility and 95% viability of chitosan- AgNTs towards HEK cells within 24 hours. In contrast, chitosan-AgNTs are cytotoxic to NCI-H460 cancer cells in same condition (Fig. 8.3A). The Fig. 8.3B shows that the PEG-AuNRs are less toxic for NCI-H46 cancer cells and slightly more towards HEK cells. The Fig. 8.3C and 8.3D shows the fluorescent images of viable and nonviable NCI-H460 cells. The red coloration and fragmented nuclei support late apoptosis of NCI-H460 cancer cells. Therefore, the above results shows, AgNTs is significant phototherapeutic agent against cancer [90].

Sanpui et al. reported delivery system for AgNPs using chitosan nanocarrier for apoptosis induction. The cytotoxic efficacy of AgNPs was analyzed by biochemical and morphological analysis in human colon cancer cell lines (HT29). The silver-chitosan nanocarrier induced apoptosis through mitochondrial pathway by depolarization of mitochondria membrane potential. The intracellular ROS generation was induced with silver chitosan nanocarrier treatment. Hence the AgNPs could be a potential candidate for cancer therapeutic approach [91].

In another work, Capanema et al. designed the hybrid hydrogel containing AgNPs enclosed in CMC (carboxymethylcellulose) polymer and conjugated it with doxorubicin drug. This system was developed using in situ where Ag^+ is reduced by CMC polymer which functions as a capping ligand. The electrostatic conjugation of CMC polymer with doxorubicin forms aqueous colloidal nanocomplexes. Further, citric acid is crosslinked with nanoconjugate by covalent bonds under mild temperature and pH conditions. This synergistic effect of AgNPs exhibits melanoma cancer cell killing. The hybrid nanostructure is proven to be effective against Gram negative and Gram positive bacteria. Hence, hybrid silver nanoconjugate was

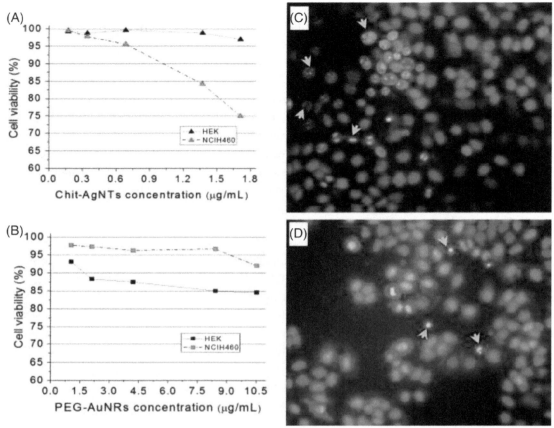

Fig. 8.3: Cytotoxic profiles of Chit-AgNTs (A) and PEG-AuNRs (B) towards HEK (black symbols) and NCI-H460 (red symbols) cells. The cells viability is expressed as function of nanoparticles concentrations. NCI-H460 cells double-stained with Hoechst-viability and Propidium Iodide-mortality indicators in the presence of Chit-AgNTs (C) and PEG-AuNRs (D). The arrows indicate condensed and fragmented nuclei typical of apoptotic cells. (For interpretation of the references to color in this figure legend, the reader is referred to the web version of this article.) *Reprinted with permission from Boca et al. [Ref. 90] Chitosan-coated triangular silver nanoparticles as a novel class of biocompatible, highly effective photothermal transducers for in vitro cancer cell therapy. Cancer Lett. 2011; 311(2): 131-140. Copyright © 2011 Elsevier.*

used against melanoma skin cancer, which demonstrates the significant release of doxorubicin against in vitro melanoma cells [89].

8.5.1.3 Silver nanocomplexes

The silver nanocomplexes attained sustainable attraction of researchers due to their ability to avoid Ag^+ leaching and nonspecific accumulation in other organs [92,93]. In this context, Mukherjee et al. synthesized silver Prussian blue analogue nanoparticles (SPBNPs) by incorporation of Prussian blue and silver nitrate for anticancer activity [92]. The designed SPBNPs showed significant in vitro inhibition of cancer cells (MCF7, B16F10, SKOV3, and A549). Further, in vivo result demonstrated that, synthesized AgNPs shows in vivo bio-compatibility in C57BL6/J mice and melanoma tumor inhibition in mouse model. Fig. 8.4 shows the histopathological changes within the melanoma tumor tissue after treatment with SPBNPs. The SPBNPs containing mice exhibited larger diameter and aggregating nests as compared to untreated mice and only AgNPs (Fig. 8.4A; Row 1). The black arrow indicated tumor cell lysis area and yellow arrow shows tumor necrotic cells in SPBNPs treated mice, which demonstrates significant tumor regression of SPBNPs. The immunohistochemistry analysis proven that the tumor of SPBNPs treated mice suppresses SOX2 and Ki-67 expression (decreased red fluorescence) as compared to AgNPs-treated and control untreated mice. It supports significant anticancer activity (Fig. 8.4A; Row 2 and-3). Further, the upregulation of tumor suppressor protein p53 in SPBNPs treated mice was evidenced by

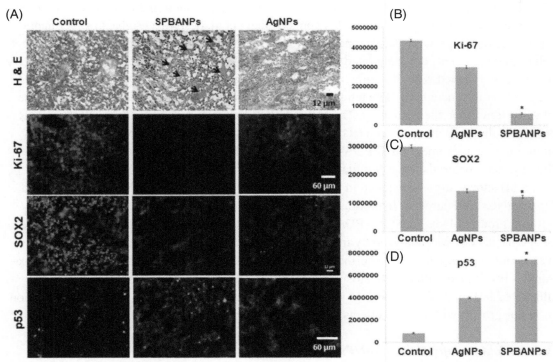

Fig. 8.4: (A) Histopathology and immunohistochemistry (Ki-67, SOX2, and p53) data of tumor tissues of mice treated with SPBANPs or AgNPs (black arrow indicates the area of tumor cell lysis and yellow arrow indicates tumor necrotic cells); (B–D) quantification of immunofluorescence data (B: Ki-67; C: SOX2 & D: p53) using Image J software [* represents $P < .05$]. *Reprinted with permission from Mukherjee et al. [Ref. 92]. Silver Prussian blue analogue nanoparticles: rationally designed advanced nanomedicine for multifunctional biomedical applications. ACS Biomater. Sci. Eng. 2019; 6(1): 690-704. Copyright © 2019 American Chemical Society.*

immunohistochemistry analysis. The above results support that, SPBNPs follows p53 mediated pathway for suppression of tumor as in Fig. 8.4A: Row 4. The (Fig. 8.4B–D) shows quantification of (b) Ki67 (c) SOX-2 (d) p53[92]. In another work, Kavinkumar et al. synthesized composites containing AgNPs, (GO) graphene oxide and (rGO) reduced graphene oxide using $AgNO_3$ and graphite where vitamin C is used as a reducing agent. The various GO and AgNPs composites (rGO, GO, AgNPs, rGO-AgNPs, and GO-AgNPs) were evaluated for cytotoxicity against human lung cancer cell line (A549). The result demonstrated that rGO-AgNPs exhibited better cytotoxicity against A549 cells as compared to rGO, GO-AgNP, and GO. Along with this, rGO-AgNPs exhibited AgNPs partial agglomeration on rGO sheet which lowers interaction between A549 cells and rGO-AgNPs. It lowers anticancer activity of rGO-AgNPs as compared to AgNPs. Hence, the interaction between AgNPs and rGO leads to better biocompatibility and less toxicity [94].

8.5.1.4 Chemically synthesized AgNPs

The chemically synthesized AgNPs provides novel and multifunctional nanoplatform for cancer therapeutics and diagnostics. In this context, Jiang et al. designed (AgNCs) silver nanocluster for tumor growth supresion. Firstly, folic acid is modified on (PDA) polydopamine (photothermal agent) that illustrated NIR absorbance. Further, AgNCs are developed on DNA template and loaded with doxorubicin (anticancer drug). Upon excitation by NIR light, PDA creates heat in cancer cells. The temperature increment breaks AgNCs leading to the release of doxorubicin for chemotherapy. The Fig. 8.5 shows infrared images of mice. Without NIR radiation, the temperature of mice is about 35°C. The temperature is increased significantly in tumor area after injection of PDA-folic acid. It confirms that, in targeted photothermal therapy, nanomaterials are accumulated close to tumor tissues for enhanced efficacy. The Fig. 8.5B shows tumor size taken in 15 days. The size of tumor did not decline for NIR and Dox-AgNCs -NIR groups. In contrast, the tumor growth is totally suppressed in the combinatorial group of Dox-AgNCs + PDA-folic acid + NIR as in Figs. 8.5C and 5D. The number of apoptotic cells was analyzed by immunohistochemical staining by cleaved caspase-3. It was found that, targeted photothermal therapy by PDA-folic acid and chemotherapy by Dox-AgNCs result in significant apoptosis in tumor cells as compared to control group which is illustrated in Fig. 8.5E. The both in vitro and in vivo result demonstrates significant reduction in tumor growth with no side effects [95].

8.5.2 Gold nanoparticles (AuNPs)

AuNPs are specified by better biocompatibility, smallest size, high atomic number, and remarkable contrasting agent [96]. Since, tumor cells are targeted by active and passive targeting. Therefore, active targeting involves AuNPs coupled with targeting agents which are specific to the tumor while passive targeting involves the osmotic tension effect to connect the tumor tissue for enhanced imaging [96]. AuNPs are valuable in addressing obstacles in the treatment of cancer due to their characteristics such as surface Plasmon resonance, enhanced

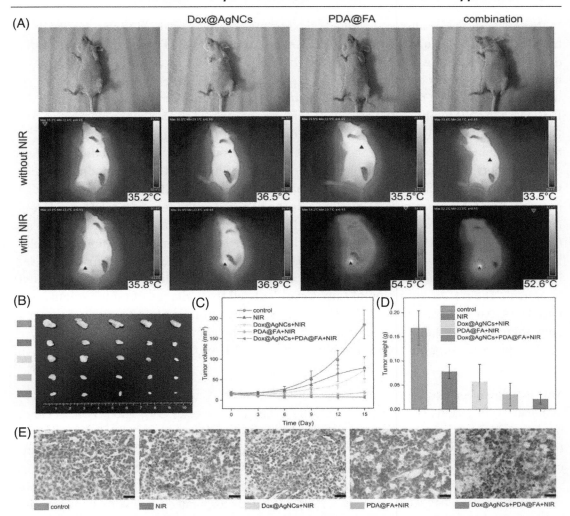

Fig. 8.5: Tumor growth of tumor-bearing mice after different treatments with and without NIR irradiation: Dox@AgNCs, PDA@FA, and both Dox@AgNCs and PDA@FA. (A) IR thermal images of tumor-bearing mice before and after 808 nm laser irradiation. (B) Photographs of harvested subcutaneous tumors (15 days). (C) Time-dependent tumor volume growth curves. (D) The weights of tumors harvested from mice after treatments (15 days). (E) Immunohistochemical staining of the apoptotic marker cleaved caspase-3, from different groups. Scale bars are 20 μm. *Reprinted with permission from Jiang et al. [Ref. 95] Synergistic Chemo-thermal Therapy of Cancer by DNA-Templated Silver Nanoclusters and Polydopamine Nanoparticles. ACS Appl. Mater. Interfaces. 2021; 13(18): 21653-21660. Copyright © 2021 American Chemical Society.*

permeation as well as retention in cancer cells, absorbance in NIR light and ability to conjugate with other drugs [96, 97]. AuNPs have been used to enhance localized hyperthermia for cancer radiotherapy, in a computed tomography imaging, photodynamic therapy and as a drug carrier [33]. The upcoming section highlights the different applications of AuNPs for cancer therapeutics.

In this context, Lee et al. demonstrated the anticancer activity of doxorubicin loaded (nanocarrier) DNA-AuNPs for treatment of ovarian cancer [98]. The anticancer effect of designed nanocarrier was observed in vitro by the EZ-Cytox cell viability assay using human ovarian cancer cell lines (SK-OV-3, A2780, and A2780). Further, in vivo study was evaluated in SK-OV-3 xenograft mice model. The superior anticancer effects of Dox-DNA-AuNP inhibited the growth of tumor as compared to Dox alone. Moreover, Dox-DNA-AuNP showed 2.5 times greater tumor growth inhibition rate than Dox. **Fig. 8.6(A–C)** demonstrates fluorescence images of tumor containing mice treated with different treatments. PBS buffer (control), free Dox, and Dox-DNA-AuNPs were administered into mice containing SK-OV-3 xenograft tumors which carries (GFP) green fluorescent proteins. The GFP represents green fluorescence in xenografted tumor mice. Dox-DNA-AuNP demonstrated more reduction in size of tumor compared to free Dox (Fig. 8.6B) and PBS buffer (Fig. 8.6A). The Fig. 8.6D clearly exhibited significant suppression effect of Dox-DNA-AuNP on growth of tumor than PBS and free Dox. The size of tumor increased in free Dox-treated and PBS-treated groups while tumor was declined in Dox-DNA-AuNP treatment. The rate of inhibition of tumor was depicted from change of tumor volume in Dox-DNA-AuNP treatment and free Dox. Fig. 8.6E demonstrated body weight of tumor containing mice with all the treatments, which shows no significant changes in all days. Hence, the result supports tumor inhibition efficacy of Dox-DNA-AuNP in mice model (in vivo) [98].

Valente et al. evaluated cellular uptake and diffusion of charged AuNPs (positive, negative, and neutral) in various breast cancer model by hydrogel composite [99]. The 3D breast cancer model was loaded in hydrogel composites to demonstrate cellular uptake of AuNPs quantitatively and qualitatively. The result demonstrated that positive AuNPs exhibited more cellular uptake and slow transport by matrix of hydrogel. On the other hand, neutral charged particles showed higher diffusion but slow cellular uptake. The Fig. 8.7 shows 3D model of two different breast cancer cells (MCF-7 and MDAMB-231) are incorporated into hydrogel matrix. The live or dead cell assay (PrestoBlue assay) confirms viability of encapsulated cells in hydrogel matrix. Fig. 8.7(A) for MCF-7 and Fig. 8.7(B) for MDA-MB-231 exhibited cluster formation. The MDA-MB-231 displays high concentration of gold which might be related to distribution of cells inside the hydrogel matrix. The Fig. 8.7A shows MCF-7 clusters inside the hydrogel matrix. Fig. 8.7B shows greater distribution of MDA-MB-231 because of its stellate shape. Hence, due to greater distribution of MDAMB-231cells, it facilitates more penetration of AuNPs [99].

8.5.2.1 Biosynthesized AuNPs

The biosynthesized AuNPs are cheaper, nontoxic, and eco-friendly in nature and their applications in cancer therapeutics are readily explored. In this context, Rajeshkumar et al. synthesized AuNPs using bacteria *Enterococus sp.* and exhibited anticancer activity against A549 (lung cancer cell lines) and HepG2 (liver cancer cell lines) [100]. In another study, Mukherjee

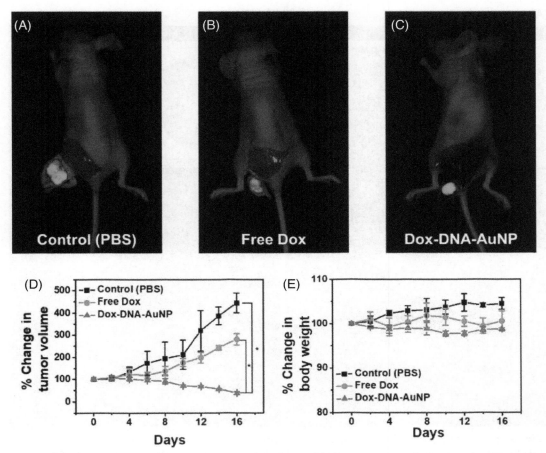

Fig. 8.6: The merged images of optical and fluorescent images obtained from the xenografted SK-OV-3-tumor bearing mice model treated with (A) PBS buffer, (B) free Dox, and (C) Dox-DNA-AuNP at the end of the test. SK-OV-3 tumor contains green fluorescence proteins (GFP). The time dependent changes in the (D) tumor volumes and (E) body weights in each drug treated tumor bearing mice model. Statistical comparisons were made using *t*-test; * *P* < .001. *Reprinted with permission from Lee C et al.[Ref. 98]. In Vivo and In vitro anticancer activity of doxorubicin-loaded DNA-AuNP nanocarrier for the ovarian cancer treatment. Cancers. 2020; 12(3):634. Copyright © 2020 MDPI.*

et al. reported in vitro and in vivo delivery of (anticancer drug) doxorubicin using biologically synthesized AuNPs by leaf extract of *Peltophorum pterocarpum*. It showed biocompatibility in both in vitro as well as in vivo systems and stability in various physiological buffers. Further, drug delivery system designed using AuNPs loaded doxorubicin resulted in a remarkable inhibition of cancer cells A549 and B16F10 in vitro. In the in vivo study, it did not exhibit any toxicity in C57BL6/J mice after peritoneal injection. Hence it can be a cost-effective strategy for cancer therapeutics [16]. Qian et al. synthesized AuNPs using an aqueous extract

(A)

(B)

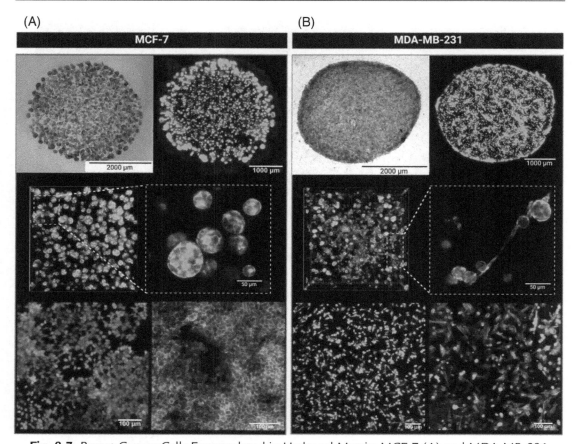

Fig. 8.7: Breast Cancer Cells Encapsulated in Hydrogel Matrix: MCF-7 (A) and MDA-MB-231 (B) breast cancer cells were encapsulated in a hydrogel composite matrix composed of 1 mg/mL Collagen type I and 11.25% (w/v) Gelatin Methacryloyl. Pellets of hydrogel loaded with cells were formed by dispensing 10 μL of hydrogel/cell mixture and exposing it to 365 nm UV light for 4 minutes. Cells were allowed to grow in the hydrogel for 1 week, forming clusters within matrix Bright field images [top left, (A) and (B)] illustrate the growth and cluster formation of cells, while fluorescent images [top right, (A) and (B)] display live cells in green and dead cells in red, after staining with Calcein AM and Ethidium Homodimer1. Fluorescent labeling of F-actin and nuclei [middle (A) and (B) images] with Alexa Fluor 488 and DAPI, respectively, illustrates the morphology and cluster formation of the breast cancer cells. Cell monolayers images [bottom (A) and (B) images] illustrate the morphology of breast cancer cell lines in 2D models. *Reprinted with permission from Valente et al. [Ref. 99]. Exploring Diffusion and Cellular Uptake: Charged Gold Nanoparticles in an in Vitro Breast Cancer Model. ACS Appl. Bio Mater. 2020; 3(10): 6992-7002. Copyright © 2020 American Chemical Society.*

of *Alternanthera sessilis*. The synthesized AuNPs induced a significant cytotoxic effect and modulated intrinsic apoptosis mechanism to enhance apoptosis in cervical cancer cells [101]. Mukherjee et al. synthesized AuNPs by *Lantanamontevidensis* aqueous leaf extract. The biosynthesized AuNPs showed biocompatibility in both in vitro and in vivo. Moreover, it demonstrated a significant anticancer activity by the generation of oxidative stress and ROS that induces caspase-3 upregulation. Further, the cell cycle analysis exhibited cell cycle arrest which resulted in apoptosis [102]. Khandanlou et al. synthesized AuNPs using leaf extract of *Backhousia citriodora*. It exhibited in vitro anticancer activity against (liver cancer cell line) HepG2 and (breast cancer cell line) MCF-7 in a dose-dependent manner [103]. Hence, biosynthesized AgNPs act as a promising approach in cancer therapeutics.

8.6 Other applications of gold and silver nanoparticles
8.6.1 Biomedical applications

The emerging advancement in nanotechnology alters the way of diagnosis, treatment, and prevention of various diseases in all aspects of life [1,104]. By considering significant interaction of molecules within outer or inner surface, the metal-based nanoparticles (AgNPs and AuNPs) have been developed to remediate disease effectively. These are considered as potential therapeutic agent for biomedical application due to their significant properties such as specificity to target cells, low toxicity, and biocompatibility [105]. Recently for disease theranostics, AgNPs, and AuNPs are the most versatile nanoparticles [21,22].They have depicted promising results in biomedical applications including cancer theranostics. Gold and AgNPs are conjugated with other metal, polymers, or biological molecule to enhance specificity against different diseases [106].

8.6.1.1 Antimicrobial (antibacterial, antifungal, antiviral)

Microbial infection is one of the prime reasons of accelerating mortality and morbidity, which weakens the immune system. The available conventional therapies have certain limitation like production cost, antibiotic resistance etc. Therefore, scientists have been exploiting different technologies including nanotechnology for microbial treatment [13,107]. Recently, silver and AuNPs are widely used for antimicrobial therapies.

8.6.1.1.1 Silver nanoparticles (AgNPs)

Silver has received remarkable attention in antimicrobial activity due to the emergence of antibiotic-resistant strains and its less capacity to permit resistance. Because of intrinsic therapeutic properties, AgNPs manifests a wide antibacterial capability for micro-organisms [13, 24]. AgNPs and its derivatives have been currently recognized in the biomedical research for its antibacterial, antioxidant, and antifungal properties [13].

8.6.1.1.1.1 Biosynthesized AgNPs

The antimicrobial actions of AgNPs includes microbial surface membrane adhesion, penetration into cells for disruption of biomolecules, intracellular damage, and destabilization, induction of cellular cytotoxicity by ROS generation, inhibiting cell signal transduction pathway leading to death of organism [13,107]. For example, Rafique et al. synthesized AgNPs by using *Albiziaprocera* leaf extract which showed remarkable antibacterial activity against both Gram-positive (*Staphylococcus aureus*) and Gram-negative bacteria (*Escherichia coli)* [108]. In another work, Ramesh et al. synthesized AgNPs from aqueous leaf extract of *Ficushispida Linn. f.* It demonstrated significant antibacterial activity by zone of inhibition against *E.coli* and *B. subtilis* [109].

8.6.1.1.1.2 Polymeric AgNPs

The AgNPs are coated with polymeric material in order to attain stability and controlled release. The polymer coated AgNPs have gained much focus in antibacterial activity due to more surface binding potential and low toxicity [110]. In this context, Xia et al. demonstrated the development of AgNPs stabilized by gallic acid and coated onto the surface of leather as well as immobilized it on skin collagen by chromium crosslinking [111]. The chemical attachment of AgNPs on leather surface enhanced the hydrophobicity and altered surface potential of leather surface from positive to negative charge. It demonstrated significant antiadhesive ability of microbes. Furthermore, it exhibited wide antibacterial activity against *S. aureus, C. albicans, E. coli,* and *methicillin-resistant S. aureus.* Fig. 8.8A shows the antiadhesive ability of gallic acid modified AgNPs, which was investigated by taking pristine leather as a control. As compared to pristine leather, almost no microbe was attached to the gallic acid modified AgNPs coated leather even at high microbial concentration. The antiadhesion rate of microbes for gallic acid modified AgNPs was checked against *E. coli, MRSA, S. aureus, and C. albicans* which exhibits its significant antiadhesive activity (Fig. 8.8B). Further, to understand antiadhesive mechanism of microbes, the antiadhesive model was described in Fig. 8.8C. For pristine leather surface, water is prime nutrient source for growth of microbes, due to easy absorption on surface. Therefore, the microbes are readily attached with hydrophilic pristine leather in atmosphere aerosols with humidity. The surface zeta potential of pristine leather carries positive charge under aqueous environment due to zwitterions crosslinkings, results in easy microbial adhesion on leather surface. On other hand, the water repellency of gallic acid-AgNPs coated leather avoids absorption of water due to its hydrophobic nature. Along with this, gallic acid-AgNPs coating changes surface of leather from positive to negative charge, which repels microbe adhesion due to electrostatic repulsive force [111]. Tarani et al. reported the biodegradable green synthesis of combined chitosan AgNPs using *Lactobacillusreuteri.* The biosynthesized AgNPs have significant antibacterial activity against *B. subtilis* and *E. coli* exhibited through disc diffusion method. Moreover, the hybrid AgNPs showed zone of inhibition against *B. subtilis.* Hence, results demonstrated that the biosynthesized nanoparticles are effective against bacterial pathogens [112].

8.6.1.1.2 Gold nanoparticles (AuNPs)

AuNPs are widely used in several fields such as drug delivery, photodynamic therapy as well as antibacterial activity [33,113]. Currently, the AuNPs is becoming a significant antibacterial agent for complementing antibiotics in order to overcome antimicrobial resistance [114].

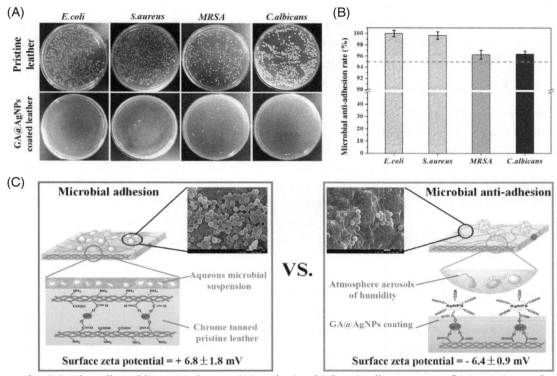

Fig. 8.8: The adhered live microbes on (A) and microbial anti-adhesion rate of (B) pristine and GA@AgNPs coated leather; Comparison of the microbially adhesive and antiadhesive action model of the pristine and GA@AgNPs coated leather (C). *Reprinted with permission from Xia et al. [Ref. 111]. Chromium cross-linking based immobilization of silver nanoparticle coating on leather surface with broad-spectrum antimicrobial activity and durability. ACS Appl. Mater. Interfaces. 2018; 11(2): 2352-2363. Copyright © 2018 American Chemical Society.*

8.6.1.1.2.1 *Biosynthesized AuNPs*

The biosynthesized AuNPs exhibits several biomedical applications depending on type of plant extract used for the synthesis [29]. In this context, Bindhu et al. synthesized AuNPs by using fruit extract of *A. comosus* which acts as a reducing agent. The synthesized AuNPs exhibited better antibacterial activity for both Gram negative and positive bacteria [115]. In another work, Abdel-Raouf et al. developed AuNPs with the ethanolic extract of *Galaxaura elongate* which illustrated antibacterial activity by zone of inhibition assay against *Klebsiella pneumonia*, MRSA (Methicillin-resistant *S. aureus*), *Pseudomonas aeruginosa, E. coli,* and *S. aureus*. Moreover, *Galaxaura elongate* extracted nanoparticles showed high effectiveness against *K. pneumonia* and *E. coli* [116]. In another work, Velmurugan et al. exhibited AuNPs synthesized in-situ on the silk, leather, and cotton fabrics by three different methods which are green synthesis, chemical synthesis, as well as combination of green and chemical synthesis. The synthesized AuNPs were tested against *Brevibacterium linens* (skin disease causing pathogen) by using live/dead BacLight bacterial viability test [117]. Hence, biosynthesized AuNPs could be alternative approach for antimicrobial activity.

8.6.1.1.2.2 Chemically synthesized AuNPs

Nanoparticles synthesis by chemical method have attracted much attention because of its significant features (high reaction rate, energy saving, low reaction time). In this context, Arshi et al. developed AuNPs by microwave irradiation method using citric acid as a reducing agent and CTAB (cetyltrimethyl ammonium bromide) as a binding agent. The antibacterial activity was demonstrated using AuNPs against the Gram negative bacteria *E. coli* by use of standard agar well diffusion method. The result demonstrates that the obtained nanoparticles exhibit high antibacterial activity with a zone of inhibition [118]. Therefore, silver and AuNPs are effective for antimicrobial activity.

8.6.1.2 Drug and gene delivery

Drug delivery includes the release of biologically active agents at the specific target location. Nanomedicine improves drug delivery of active pharmaceutical ingredient with the help of engineered nanomaterials [119,120]. Nanomaterials such as silver and AuNPs allow multiple delivery modes for clinical approach, due to their ability to penetrate across small capillaries .

8.6.1.2.1 Silver nanoparticles (AgNPs)

The AgNPs conjugation with biomolecules has significant applications in the development as well as design of biocompatible pharmaceuticals, as potent drug delivery agent, biosensors and for theranostics. The AgNPs are widely explored due to their plasmonic, optoelectronic, and biological characteristics for future pharmaceutical perspectives [6,121,122]. In this context, Yadollahi et al. illustrated the development of AgNPs with physical crosslinking of chitosan hydrogel beads (where sodium tripolyphosphate is used as crosslinker) for drug delivery application. The resulting formulation demonstrated prolonged and more controlled release of AgNPs containing chitosan beads [123]. In another example, Benyettou et al. designed AgNPs-based drug delivery system for intracellular delivery of (Dox) doxorubicin and (Ald) alendronate (where AgNPs were coated with bisphosphonate Ald) to improve anticancer therapeutics of both drugs [15]. Mukherjee et al. exhibited formation of AgNPs using *Olax scandens* leaf extract. The biosynthesized AgNPs acted as an efficient antibacterial, anticancer, drug delivery vehicle as well as imaging agent [87]. In another work, Patra et al. demonstrated anticancer drug doxorubicin delivery through biosynthesized gold and AgNPs produced from *Butea monosperma* leaf extract. The **Fig. 8.9** shows the schematic representation of biosynthesized AgNPs and AuNPs as well as their drug delivery application. The silver and AuNPs are synthesized from $AgNO_3$ and $HAuCl_4$ respectively, using *Butea monosperma* extract. Further, the anticancer drug (Doxorubicin) is loaded on the nanoparticles for targeted drug delivery system. The gold and silver biosynthesized nanoparticles conjugated with doxorubicin drug exhibited higher anticancer activity than free drug [124]. Sarkar et al. engineered AgNPs modified with (PEG) polyethylene glycol and chitosan-*g*-polyacrylamide. Further, gene transfection efficiency was enhanced by immobilization of Arg–Gly–Asp–Ser (RGDS) peptide. This study demonstrated that the engineered AgNPs can be used as significant nonviral carrier for delivery of gene [125]. Overall, AgNPs can act as an efficient drug deliver agent.

8.6.1.2.2 Gold nanoparticles (AuNPs)

Gene delivery is one of the newly emerging approaches for acquired and inherited disease treatment which occurs due to incorrect gene expression. It involves exogenous genetic material delivery to the targeted cells using specified vectors [126]. Among broad range of carrier systems, AuNPs are considered as a major metal nanoparticle for gene as well as drug

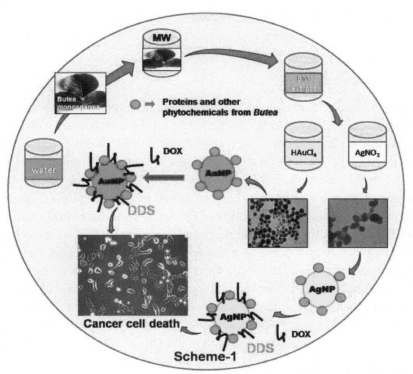

Fig. 8.9: Over all schematic representation for the green synthesis, characterization of biosynthesized b-AuNP and b-AgNP and their probable applications in drug delivery. *Reprinted with permission from Patra et al. [Ref. 124]. Green synthesis, characterization of gold and silver nanoparticles and their potential application for cancer therapeutics. Mater. Sci. Eng., C. 2015; 53: 298-309. Copyright © 2015 Elsevier.*

delivery [16,126]. AuNPs have been explored in these applications for their enormous electronic structure, highly significant magnetic, electronic, and optical properties [126]. Owing to novel physiochemical properties, it is mostly involved in targeted drug delivery to decrease side effects of healthy cells [127]. In this study, Lajunen et al. developed AuNPs encapsulated into liposomes (drug carrier) for drug release. The liposome was developed with AuNPs using pH and heat-sensitive composition of lipid. The pH-sensitive design demonstrates more release of drug in an acidic situation as compared to neutral. It did not exhibit any cellular toxicity to (HUVECs) human umbilical vein endothelial cells and human retinal pigment epithelial cells. Thus, it is a promising approach for the site and time-specific delivery of drugs towards the targeted site [128].

Thambiraj et al. synthesized AuNPs by chemical reduction method where the anticancer drug docetaxel was noncovalently conjugated while folic acid was conjugated covalently. Furthermore, the cytotoxicity of nanoparticles was demonstrated using the lung cancer cell line, H520 [129]. In another example, Shan et al. developed nonviral gene delivery vectors based on entrapment of dendrimer with AuNPs which show highly efficient gene transfection ability. The AuNPs entrapped dendrimers have the ability for the efficient compaction of pDNA. It demonstrated enhanced significant gene transfection in selected cell lines as shown by fluorescence microscopic imaging and Luc assay [130]. Therefore, silver and AuNPs can emerge as a potential candidate in drug and gene delivery.

8.6.1.2.2.1 Biosynthesized AuNPs

AuNPs exhibited a high capacity for delivery of several drug molecules, vaccines, proteins up to their target site and also control drug release [131]. Biosynthesized AuNPs are nontoxic, biocompatible drug vehicle for carrying anticancer drug. In this context, Mukherjee et al. reported in vitro and in vivo delivery of anticancer drug (doxorubicin) by using green synthesized AuNPs. The administration of AuNPs-doxorubicin demonstrated inhibition of different cancer cells such as A549 (lung cancer) and B16F10 (melanoma). The in vivo toxicity study demonstrated that, after IP injection of the AuNPs conjugated doxorubicin, there were no changes in serum clinical biochemistry and hematology. These results demonstrated that gold-based delivery system could be a cost-effective approach in near future [16]. Pooja et al. synthesized AuNPs using *gum karaya* for the delivery of anticancer drug (Gemcitabine hydrochloride). It exhibited high biocompatibility as well as cell survival in normal cell line (CHO). The anticancer drug gemcitabine was loaded onto the nanoparticle surface to improve its loading and targeting efficiency. The result demonstrated that gemcitabine loaded nanoparticles exhibited greater inhibition of tumor growth than alone gemcitabine [34].

8.6.1.2.2.2 AuNPs with polymer

In another study, Elbialy et al. synthesized magnetic AuNPs functionalized with thiol terminated PEG (polyethylene glycol) which were entrapped with anticancer drug doxorubicin.

Further, in vivo studies revealed that this targeted magnetic drug delivery system exhibited a greater drug accumulation all over the tumor as compared to passive targeting drug delivery. To check the in vivo toxic effect, various biochemical parameters were measured. This drug delivery vehicle reduces the nonspecific distribution of chemotherapeutic drugs as well as lowers side effects to healthy tissues [132]. Hence it shows significant therapeutic anticancer activity in presence of magnetic field.

8.6.1.3 Wound healing

Wound healing is one of the complicated biological processes which involve stepwise molecular and cellular interactions to direct repair and restoration of damaged tissue [133,134]. Currently, silver and AuNPs imparted significant attention in wound healing products.

8.6.1.3.1 Silver nanoparticles (AgNPs)

AgNPs are exploited as one of the most significant nanostructures known to be a promising candidate in various clinical applications. From long decades ago, silver compounds have been widely used for both hygienic and healing purposes, because of their high bactericidal as well as antimicrobial activity. Hence, different silver-containing compounds are used for wound healing [135]. Due to such significant biological properties, it could be explored for dressing of wounds and ulcers [134].

8.6.1.3.1.1 Polymeric AgNPs

Several biopolymers (gelatin, sodium alginate) are promising materials due to their biocompatible and biodegradable nature. Various reports have been established for synthesis of AgNPs. But polymer containing AgNPs have attracted great attention, because of its ability to stabilize aggregation of nanoparticles [136]. In this context, Liu et al. investigated the cell mediated response, dermal contraction event, and epidermal re-epithelization at wound healing site under influence of AgNPs. The proliferation event was performed in presence of AgNPs and silver sulfadiazine (control). The result demonstrated that AgNPs could accelerate wound closure rate by inducing keratinocytes migration [137]. In a different type of work, Tang et al. prepared skin adhesive using methacrylated hyaluronan–polyacrylamide (MHA–PAAm) hydrogel combined with AgNPs which were further attached to gelatin. It demonstrated hemostatic and antibacterial activities for wound healing. The in vivo results demonstrated that antibacterial and tissue adhesive properties of silver-based compounds were able to recover rat tail amputation and liver injury [138].

8.6.1.3.1.2 Silver nanocomplexes

In another study, Rao et al. designed silver nitroprusside complex (AgNNPs) by sodium nitroprusside and silver nitrate solutions for efficient antibacterial as well as in vivo wound healing activity. The designed nanoparticles showed remarkable antibacterial activity against both Gram-positive and Gram-negative bacteria through damaging the cell membrane which leads to cell death. Moreover, the AgNNPs exhibited efficient wound healing activity through enhanced antimicrobial activity as well as macrophage activation [93]. The wound was made in C57BL6 mice on their dorsal surface and treated with AgNNPs (1% and 0.1%) which was prepared in Vaseline and applied topically for alternative days (day 1–7). Images were taken by camera. The result showed that wound closure was enhanced in AgNNPs (0.1%) treated mice as compared to control (Fig. 8.10A). The wound closure is measured in mm scale which is represented in histogram that supports same results (Fig. 8.10B). The depicted results support that AgNNPs as significant wound healing properties [93].

8.6.1.3.2 Gold nanoparticles (AuNPs)

It was recently reported that AuNPs are significant for the cure of cutaneous wound and exert positive influence on wound healing [133,139]. In this context, Wang et al. designed a novel gene delivery system containing LL37 antimicrobial peptide for the treatment of diabetic wounds using AuNPs. AuNPs-LL37 was further conjugated with pro-angiogenic VEGF plasmids, which significantly improved the efficiency of gene transfection as compared to pristine AuNP or pDNAs. Furthermore, the result demonstrated that the designed joint system is responsible for acceleration of angiogenesis, reduction of bacterial infection, rapid re-epithelization, increasing expression of VEGF, and enhancing granulation of tissue formation [140].

8.6.1.3.2.1 Polymeric AuNPs

Various nanotechnology-based approaches were developed to induce wound healing. They can be used to deliver antiinflammatory, antibacterial agents to the wounded sites. For example, Leu et al. reported the wound healing effect of AuNPs along with EGCG (epigallocatechingallate), ALA (α-lipoic acid) for proliferation of Hs68, HaCaT cells. The AuNPs, EGCG, and ALA significantly increased Hs68 and HaCaT cells migration, proliferation, and accelerated wound healing on the skin of a mouse. Immunoblotting studies demonstrated that wound tissue exhibits increased expression of (Vascular Endothelial Growth Factor) VEGF and angiopoietin-1. This study supports the acceleration of mouse wound healing at the cutaneous site by anti-oxidation and anti-inflammatory effect [139]. In another study, Mahmoud et al. developed rod and spherical shaped AuNPs whose surface was modified with a cationic, anionic as well as neutrally charged polymer by loading it on the thermosensitive hydrogel poly ethylene glycol (PEG). In vitro study demonstrated that hydrogel has remarkable colloidal stability and extended-release, over the period of 48 hours. Hydrogels of PEG-gold nanorods-cationic poly allyl amine hydrochloride exhibited significant wound healing behavior upon tropical treatment in an in vivo model. Cationic charged and PEGylated gold nanorods hydrogel have accelerated deposition of collagen and re-epithelization of skin after 14 days of treatment as compared to control [141].

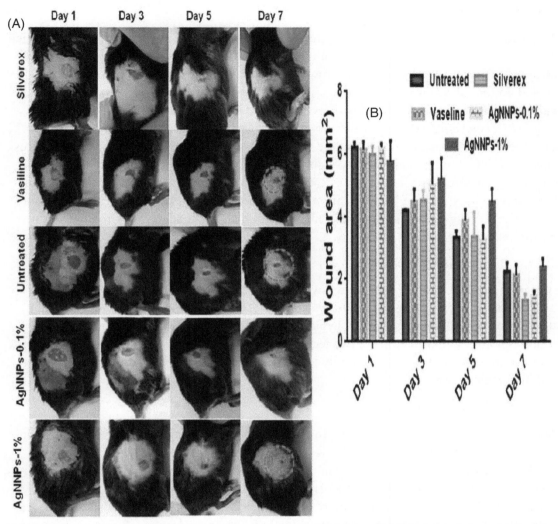

Fig. 8.10: Wound healing study in C57BL6 mice after treatment with AgNNPs. (A) Representative images of wound healing in mice. First row: mouse treated with SilverX; positive control experiment, Second row: vehicle control: Vaseline, Third row: untreated control group, fourth, and fifth row: Mouse treated with AgNNPs at different concentrations (0.1 and 1%). (B) Measurement of wound closure in mice after treatment with positive control (silverex), vehicle control, untreated and AgNNPs at various time points (Day 1, 3, 5, 7). *Reprinted with permission from Rao et al. [Ref. 93]. Ag2 [Fe (CN) 5NO] nanoparticles exhibit antibacterial activity and wound healing properties. ACS Biomater. Sci. Eng. 2018; 4(9): 3434-3449. Copyright © 2018 American Chemical Society.*

8.6.1.3.2.2 Biosynthesized AuNPs

The plant extract contains various phytochemicals for biogenic synthesis of AuNPs, which act as stabilizing and capping agent. In this context, Boomi et al. synthesized AuNPs by aqueous extract of *Acalyphaindica* for antibacterial and antioxidant properties. The AuNPs coated cotton fabrics were checked against bacterial strains *E. coli* and *Staphylococcus epidermidis* for antibacterial activity. It showed a significant inhibition zone against *Staphylococcus epidermidis*. Furthermore, wound healing activity was demonstrated in diabetic BALB/c mice model where the wound was entirely re-epithelized due to the presence of AuNPs [142]. In another work, Naraginti et al. reported the effect of silver and AuNPs synthesized by root extract of *Coleus forskohlii*. The assessment of in vivo wound healing activity was done by application of nanoparticles on dorsal side in albino Wistar rats illustrated significant reduction in inflammation and enhancing re-epithelialization during the healing process [143].

8.6.1.4 Biosensor and bioimaging

Biosensing and bioimaging are progressively advancing in the diagnosis of disease due to their nondestruction nature, reliability, and accuracy. Optical biosensing provides constant supervision of health metabolites by hygienic and wearable devices. The development of nanotechnology empowered these two techniques which remarkably enhance their specificity, contrast, multiplexibility, and sensitivity [5]. Silver and AuNPs are also used for biosensing and bioimaging purposes.

8.6.1.4.1 Silver nanoparticles (AgNPs)

AgNPs have become an efficient agent for biosensing and bioimaging activities due to their effective plasmonic characteristics [5,144]. Silver carries greater electrical and thermal conductivities as well as it is effective in the transfer of electrons along with sharp extinction bands. Moreover, modified AgNPs is much stable in air and water. Considering these features, AgNPs are an enormous tool in several fields such as diagnosis, environmental, and drug delivery along with biosensor and imaging [4,9,15]. In this context, Hai et al. designed multifunctional nanocomplex by AuNPs, tannic acid, and (GQD) graphene quantum dots for cancer therapeutics and biosensing. This nanocomplex demonstrates H_2O_2 responsive degradation as well as release of GQD and tannic acid. Thus, nanocomplex increases H_2O_2 in cancer cells, which exhibits better performance for recognition of cancer cell. The synergistic effect of tannic acid and Ag^+ provides death of cancer cell while GQD examine therapeutic efficacy by imaging [145].

8.6.1.4.2 Gold nanoparticles (AuNPs)

Among the long list of nanomaterials, AuNPs are mostly investigated for optical imaging and biosensing because of their chemical stability, biocompatibility, versatile properties, and

Table 8.1: Silver nanoparticles.

Sr no.	Application	Conjugation and synthesis	Function	Refs.
1.	Anticancer activity	AgNPs-GO and AgNPs-rGO	Cytotoxicity against human lung cancer cell line (A549)	[94]
		Beta vulgaris extract	Cytotoxicity against A549, MCF-7, and Hep2 cells	[84]
		Zinnia elegans ethanolic leaf extract	Anticancer activity against U-87 (glioblastoma cell line) by ROS generation	[28]
		Azadirachtaindica extract	Induction of cellular apoptosis in NCI-H460 cells	[86]
		AgNPs-chitosan	Photothermal agent against NCI-H460	[90]
		AgNPs-Chitosan nanocarrier	Cytotoxic efficacy of AgNPs against human colon cancer cell lines (HT29)	[91]
		Leaf extract of *Olaxscandens*	Anticancer activity against B16F10, A549, and MCF7	[87]
		AgNPs-CMC (carboxymethylcellulose) polymer	Against melanoma skin cancer	[89]
		AgNPs-folic acid-PDA	NIR breaks AgNC leading to the release of doxorubicin for chemotherapy	[95]
2.	Antimicrobial activity	Aqueous leaf extract of *Ficushispida Linn. f.*	Zone of inhibition of *E. coli* and *B. subtilis*	[109]
		AgNPs - gallic acid	Antibacterial activity against *Staphylococcus aureus, E. coli, Candida albicans,* and *methicillin-resistant S. aureus*	[111]
3.	Drug and gene delivery	AgNPs-chitosan hydrogel beads	More controlled release of AgNPs containing chitosan beads	[123]
4.	Wound healing	AgNPs-(MHA–PAAm) hydrogel	Recover rat tail amputation and liver injury	[138]
		Silver nitroprusside complex	Against both Gram-positive and Gram-negative bacteria through damaging the cell membrane	[93]

easy fabrication [146]. Andreescu et al. engineered glucose receptor on AuNPs to develop an efficient electrochemical glucose biosensor for detection of glucose [147].

Table 8.1 and Table 8.2 illustrate the different biomedical applications of gold and AgNPs as discussed above. [16,28,84,86,87,89,90,91,93,94,95,98,101,102,109,111,116–118,123,130,128,132,138,139–141,143]

8.6.2 Other industrial applications

Recently gold and silver-based nanomaterials are readily explored for different other industrial applications like catalysis, environmental purposes, agricultural, etc. due to their low toxicity, biocompatibility, and photo-optical distinctiveness [7–9,11,148–150]. The following sections highlight the several applications of the gold and AgNPs in the several field. [148].

Table 8.2: Gold nanoparticles.

Sr no.	Application	Conjugation and synthesis	Function	Refs.
1.	Anticancer activity	Leaf extract of *Peltophorumpterocarpum*AuNPs	Inhibition of cancer cells A549 and B16F10	[16]
		Aqueous extract of *Alternanthera sessilis*	Significant cytotoxic effect and modulate intrinsic apoptosis mechanism	[101]
		Lantanamontevidensis aqueous leaf extract	Generation of oxidative stress and ROS production	[102]
		Doxorubicin-loaded DNA-AuNPs	Delivery of doxorubicin at tumor site	[98]
2.	Antimicrobial activity	Ethanolic extract of *Galaxaura elongate*	Zone of inhibition against *Klebsiella pneumonia*, MRSA, *E. coli*, *P. aeruginosa* and *S. aureus*	[116]
		In-situ synthesis on the silk, leather and cotton fabrics	Against *Brevibacterium linens* (skin disease causing pathogen)	[117]
		AuNPs-citric acid-CTAB	Against the Gram-negative bacteria *E. coli* by use of standard agar well diffusion method	[118]
3.	Drug and gene delivery	AuNPs-dendrimer	Efficient compaction of pDNA and gene transfection	[130]
		AuNPs-doxorubicin	Inhibition of cancer cells A549 and B16F10	[16]
		AuNPs-doxorubicin-PEG	Greater drug accumulation all over the tumor	[132]
		AuNPs -liposomes	More release of drug in an acidic situation	[128]
4.	Wound healing	AuNPs-EGCG -ALA	Increased expression of VEGF and angiopoietin-1.	[139]
		AuNPs -polymer-PEG	Accelerated deposition of collagen and re-epithelization of skin	[141]
		AuNPs-LL37	Improved the efficiency of gene transfection	[140]
		Root extract of *Coleus forskohlii*.	Reducing inflammation and enhancing re-epithelialization	[143]

8.6.2.1 Electronics

Nanotechnology provides ultrafine particle size, structure, and crystallinity to the electronic devices which gives better performance over existing devices [150].

8.6.2.1.1 Silver nanoparticles (AgNPs)

AgNPs have efficient electronic, optical, and chemical properties utilized for microelectronics, catalytic, sensor materials, etc [151,152]. In this context, Jiang et al. designed AgNPs and silver ion composite ink for conductive printing on the flexible substrate [153]. The combination of AgNPs and silver ion carried composite ink act as conductive fillers. Along with this, a tape test was performed to observe the adherence of printed pattern while mechanical stability was investigated by cyclic bending test. The AgNPs and silver ions present within the composite ink mutually reduced curing temperature [153]. In another work, Tavakoli et al. used AgNPs coating of inkjet-printed traces along with gallium indium for increasing electrical conductivity, magnitude as well as tensile strain [154]. In another study, Chen et al. reported the use of AgNPs to apply on the electrode as a silver paste for higher conductivity and conductive adhesives [151]. Lu et al. developed an inexpensive process for a conductive silver nano network made up of silver nanowires along with AgNPs. It was observed that efficient conductivity was observed in this nano network which exhibits superior electrical stability under bias operation than thermally annealing silver nanowires [155]. Sangaonkar et al. synthesized kokum fruits mediated AgNPs for a more sensitive, colorimetric, and memristor-based assay to detect Hg^{2+} in water samples [156]. Matsukatova et al. developed Poly (p-xylylene) based memristor embedded with AgNPs to induce the characteristics of memristors [157]. Hence, AgNPs could be promising in development of electrical devices by decreasing their size.

8.6.2.1.2 Gold nanoparticles (AuNPs)

AuNPs have gathered admirable attention in several applications such as biosensors, photonics, electronics, etc.[158]. Minari et al. developed π junction AuNPs as material of electrode for room temperature deposition on the conductive metal layer [159]. These methods of patterning are designed for gold ink electrode and organic semiconductor layer for the creation of organic thin-film transistors by room temperature printing [159]. Zhou et al. manufactured air-stable small voltage flexible nonvolatile memory transistor by immersing AuNPs in poly (methyl methacrylate) (PMMA) as an element of charge storage for manufacturing low voltage and economically flexible electronics [160].

Overall, gold and AgNPs pose as a potential candidate in the world of electrical applications.

8.6.2.2 Catalysis

Over the last few years, selective synthesis of nanoparticles concerning shape, size, and their catalytic applications are gaining remarkable commercial interest [161].

8.6.2.2.1 Silver nanoparticles (AgNPs)

The AgNPs use in catalysis is beneficial as it prevents the use of ligands. The high surface area of AgNPs results in better catalytic activity [161]. Referring to this, Saha et al. explored the AgNPs for catalytic degradation and reduction of methylene blue dye [7]. In another work, Vidhu et al. emphasized the effect of AgNPs on the degradation of harmful dyes such as methylene blue and methyl orange by $NaBH_4$. This demonstrated capacity of AgNPs as a promising candidate for catalysis [162].

8.6.2.2.2 Gold nanoparticles (AuNPs)

The nanoparticulates such as Au catalysts are vital under mild conditions. They are novel candidate for lowering cost of chemical plants and increases the selectivity of the reaction

[163]. For example, Zinchenko et al. used DNA cross linked hydrogel as a matrix for AuNPs synthesis. DNA has the strongest affinity towards transition metals like gold. The result demonstrates that AgNPs containing DNA hydrogel showed greater catalytic activity in the nitrophenol to aminophenol hydrogenation reaction. Thus, the gold-based nanomaterial acts as a potential agent for catalytic application [164].

8.6.2.3 Environmental (wastewater treatment/water disinfection etc.)

Recently, the nanomaterials exploration is increased in wastewater treatment. For water purification, silver ions are used in hospitals, community water systems as it replaces chlorine ions to minimize adverse effects to the health. Some research demonstrated that the catalytic action of silver together with oxygen provides sanitizer effect that removes corrosive chlorine use [165].

8.6.2.3.1 Silver nanoparticles (AgNPs)

The AgNPs have significant antibacterial as well as efficient adsorption capacities which make them a precious alternative strategy to remove wastewater contamination [166]. Because of high stability, low cost, and controlled release rate of silver ions from nanoforms leads to the removal of inorganic anions, pollutants, bacteria, heavy metals etc. Moreover, the higher concentration of micro-organisms leads to increased use of disinfectant but, direct use of AgNPs instead of disinfectant exhibits better filtration [165]. For the antibacterial activity of AgNPs, Rus et al. developed filters impregnated with silver solutions for the treatment of drinking water. The retention of silver in the filters resulted in removal of *E. coli* bacteria, enhanced water quality as well as water flow rate [167]. To minimize biofouling microfiltration, Ferreira et al. developed AgNPs by chemical reduction with further integration into polyethersulfone (PES) microfiltration membranes (synthesized by phase inversion method) to form PES-AgNPs. The silver-based membrane suppressed the action of microorganism *Pseudomonas fluorescens*. These studies evaluated the bactericide properties of PES-AgNPs for wastewater treatment [168]. Hence, AgNPs are effective approach for wastewater management.

8.6.2.3.2 Gold nanoparticles (AuNPs)

AuNPs based sensor plays an enormous role in environmental monitoring by detecting heavy metals, toxins, organic, and inorganic pollutants of water with large sensitivity [169]. For example, Anwar et al. synthesized biogenic AuNPs on cotton cloths. The hydroxyl group of cellulose macromolecules is found abundantly on cotton which reduces Au ions into its nanoparticles. The AuNPs-cotton cloth acts as catalysts to degrade pollutants with a high reaction rate [170]. In another work, Das et al. synthesized nanogold bioconjugate by green chemical method for purification of contaminated water. The AuNPs were produced on the surface of a fungal strain, *Rhizopus oryzae*. This nano-bioconjugate exhibits strong adsorption capacity to several organo-phosphorous pesticides and could be utilized to achieve pesticides as well as pathogen-free water [171].

8.6.2.4 Textile industry

The nanotechnology utilization in textile is rapidly accelerating due to valuable and novel properties of nanoparticles. The nanomaterials provide multifunctional fabrics with several functions (easy cleaning, UV protection, hydrophobicity, antibacterial) [172].

8.6.2.4.1 Silver nanoparticles (AgNPs)

Currently, silver-based nanomaterials are explored in the textile sector due to their antimicrobial property. The nanomoieties can play a significant role to functionalize fabrics. The AgNPs are relatively preferred more over traditional antimicrobial agents due to bacterial resistance, stability, and ecofriendly nature [173]. In this context, Zhang et al. demonstrated that nanosilver based solution treated cotton fabric that exhibits significant antibacterial properties. The silver-treated fabric resulted in bacterial population reduction of *S. aureus* and *E. coli* [174]. In another study, Butola et al. used green chemistry approach for the synthesis of chitosan based AgNPs on the surface of linen fabric. This fabrication was done in the presence of pineapple crown extract containing biomolecules such as fructose, glucose, and sucrose. The chitosan polysaccharide was responsible to transfer antibacterial and antioxidant activity of AgNPs on the surface of linen fabric. The chitosan along with AgNPs imparted more antioxidant and antibacterial activity (against *S. aureus* and *E. coli*) along with stabilization of silver ions. These results support chitosan based AgNPs could be preferred as an alternative functional chemical agent in the textile industry [175].

8.6.2.4.2 Gold nanoparticles (AuNPs)

The metallic nanoparticles have high UV light stability hence; they act as an alternative agent for coloring of textiles. In this context, Tang et al. synthesized AuNPs with ligands (malate, tartarate, and citrate) using photochemistry and heating, where ligands act as both stabilizing as well as reducing agents. The synthesized AuNPs with ligands exhibited efficient coloring of silk, nylon fabrics, and wool fabrics [176]. In another study, Bindhu et al. synthesized β-cyclodextrin functionalized AuNPs for photocatalytic degradation, detection of heavy metals, and inhibition of bacterial growth. The β-cyclodextrin demonstrates better antibacterial activity by inhibiting bacterial growth that illustrates efficient sensing activity for toxic metals present in water. The photocatalytic degradation of textile dyeing waste-water using β-cyclodextrin functionalized AuNPs was more as compared to only AuNPs [177].

Further, Tang et al. prepared AuNPs using white silk fabric samples in the presence of heated solution where the color of the silk fabrics changed to brown and red due to the localized surface plasmon resonance (LSPR) feature of gold. The silk fabrics impregnated with AuNPs exhibited efficient antibacterial activity, thermal conductivity, and UV protection ability

[178]. Pisitsak et al. synthesized AuNPs using bark of *Xylocarpus granatum* which is rich in tannin and acts as both capping as well as reducing agent. The AuNPs demonstrated catalytic reduction of congo red dye (also known as carcinogenic azo dye) in the aqueous solution of sodium borohydride where the degradation efficiency was linked to the AuNPs loading. The AuNPs exhibited significant reduction in tearing strength [179].

8.6.2.5 Agriculture and food

Agriculture is considered as a backbone of economy globally. But certain microbes develop resistance against antibiotics that affects crops [180]. The disease existence in cultivars has significant losses in food production. The application of nanobiotechnology is useful to obtain new products to control pathogens in the cultivar [181].

8.6.2.5.1 Silver nanoparticles (AgNPs)

In the field of agriculture, nanotechnology focuses on target farming that involves the use of nanoparticles such as AgNPs which are found to be effective against fungal and bacterial infections [180]. The AgNPs due to their efficient antimicrobial properties are used in several agricultural activities. In this context, Terra et al. reported the microalgae based biosynthesized AgNPs use in agriculture to control pathogens [8].

The package obtained with the addition of silver nanocompounds with antibacterial properties provides efficient features. AgNPs might be incorporated into biodegradable polymers such as cellulose, starch, agarose, and nonbiodegradable polymers like polyethylene, vinyl alcohol for packaging of food [27].

8.6.2.5.2 Gold nanoparticles (AuNPs)

The AuNPs are being explored in the industries of nanopackaging due to their antibacterial, nontoxic, and inert nature [182].

8.7 Toxicological issues of silver and gold nanoparticles

The risk exerted by nanoparticles is one of the challenging tasks for industrial, academics, and regulatory communities. Scientific evidence focuses on the toxicity of nanoparticles in both in vitro and in vivo conditions [183]. The production of nanoparticles results in greater exposure to the environment and human life, leading to rising issues of toxicity. Nanoparticles are accumulated in various organs like the liver, spleen, kidney, brain, and heart after injection or skin contact. It led to damage of mitochondria, DNA mutation, and cell death [184]. The lack of reliability regarding the safety of nanoparticles is limiting their application in the biomedical area, particularly in diagnosis and cancer treatment [185]. Hence, efforts are aimed to understand the underlying this mechanism of toxicity [183]. Among several nanoparticles, silver, and AuNPs exhibited wide applications in various sector. The larger utilization of these nanoparticles also enhances risk of exposure. In this context, Shrivastava et al. reported toxic effect of silver and AuNPs on tissue and erythrocyte of mouse after exposure for 14 days. The result demonstrated that, the ROS generation was enhanced and antioxidant enzyme was lowered upon nanoparticle application in both tissue and erythrocytes. Along with that, interleukin-6, nitric oxide synthase, and inflammatory markers were enhanced after nanoparticle exposure. This suggested that oxidative stress induced by silver and AuNPs is the major mechanism of toxicity enhancement [186].

Bar-Ilan et al. demonstrated toxicity assessment of silver and AuNPs in zebrafish model. It was found that, at 120 hours postfertilization, the AgNPs exhibited 100% mortality while AuNPs produced less than 3% mortality at same condition. Moreover, AuNPs produced minimum toxic effects whereas AgNPs exhibited toxicity by Ag^+ formation [187]. In another work, Martirosyan et al. investigated acute toxicity of AgNPs followed by administration with phenolic compounds. It demonstrated that AgNPs toxicity is mainly due to release of Ag^+ ions and generation of oxidative stress [188].

Yang et al. demonstrated toxicological examination for biodistributed silver and AuNPs in mice. The result suggested that the silver and AuNPs were mainly deposited in mononuclear phagocytes (spleen and liver). Specifically, the AuNPs are deposited on liver while the high presence of AgNPs is found in organs like lung, heart, and kidney. qRTPCR study exhibited that AgNPs is responsible for changes in gene expression. This study evaluated that the chemical composition of nanoparticles is the primary factor for biodistribution and toxicity [189].

Mannerstrom et al. reported the cytotoxicity of silver, AuNPs and mesoporous silica nanoparticles (MSNP) in the NR8383 macrophages, BALB/c 3T3 fibroblasts, and U937 monocytes. The result exhibited that the AuNPs and MSNP are least toxic to all cells while AgNPs shows higher toxicity to all cell types [190]. By considering the toxicity exerted by the silver and gold nanoparticles, they must be properly treated before being applied into the nature. But there are several reports are established which exhibited biocompatible nature of silver and gold nanoparticles. For examples, Jadhav et al. formulated AgNPs gel against bacterial infection and found to be biocompatible and effective against almost all microorganisms. Along with this, many silver and gold nanoparticle formulations are FDA approved (e.gSilverX) [191].

8.8 Challenges and future perspectives

From the last decades, nanotechnology is utilized in several biomedical areas due to their advantages like specificity, controlled release of drugs, improvement in drug bioavailability, etc. [192]. Among metallic nanomaterials, silver, and AuNPs are mostly explored due to their wide use in various sectors. The advancement of silver and AuNPs in cancer therapeutics as well as diagnostic have been accelerating in the last few years [193]. It shows novel opportunities for diagnosis and early detection of cancer [194].

Despite the advancement described within this chapter, still some challenges prevail which are needed to be addressed in this field. Although silver and gold nano-carriers have significant potential as a cytotoxic and drug delivery agent for cancer treatment, but for enhancement of efficacy certain approaches are to be considered. In order to analyze these problems, primarily there should be well planned and optimized synthesis procedure which can enhance clinical benefits, yield along with stability of silver and AuNPs [29,194]. Also, synthesis and characterization need to be performed carefully for drug delivery to prevent unwanted toxicity to the normal cells [195]. Along with that, more in vitro and in vivo studies should be accomplished. Moreover, appropriate surface functionalization is required in order to attach specific complex ligands, targeting agents, antibodies, and small molecules to achieve more efficacy [194]. Another key challenge is the FDA approval for nanomaterials. Thus, the design and development approaches of nanomaterial-based therapeutic agents are needed to be employed before they are used in the medicinal field [195].

The proper evaluation of nanomaterials to withstand long-term stability, biosafety, significant efficacy, and briefly in vivo pharmacokinetics study is essential before going to clinical trials. The key challenges in the clinical phase are biocompatibility, biodegradability, amount of dose, route of administration, uptake, and clearance [196]. After multiple ongoing and approved clinical trials, vigorous efforts are still required to deliver nanoparticles within physiological barriers to prevent the growth of secondary malignancy. To facilitate cancer nanomedicine development, the regulatory agencies should improve the nanotheranostic platforms regulatory guidelines to facilitate safety and efficacy [197]. The nanotechnology-based strategy will be assured for future success and development of personalized medicine [193]. If these challenges and future perspective are met then it would be an excellent treatment strategy for biomedical applications and cancer theranostics.

8.9 Conclusions

The emergence of nanoparticles in biomedical field has positive impact on human health. Metal nanoparticles have emerged as one of the indispensable tools in the biomedical field. Owing to their inherent properties, noble metal nanoparticles have important role in therapeutics and diagnostics of several diseases especially for cancer treatment. Gold and AgNPs have revolutionized cancer therapeutics through different mechanism of action along with targeting agents. Also, noble metal nanoparticles, silver and gold are readily used in drug delivery, gene silencing, wound healing as well as antimicrobial activity. Apart from these therapeutic benefits, gold and AgNPs are extensively used for several industrial applications. Considering all these properties and applications of gold and AgNPs, they can emerge as an efficient therapeutic as well as diagnostic agent for cancer and other diseases in the near future.

Acknowledgment

CRP is thankful to CSIR, New Delhi, for funding support from PAN CSIR CANCER RESEARCH PROGRAM (HCP0040). We thank the Director of the CSIR - Indian Institute of Chemical Technology (Ms. No. IICT/Pubs./2022/091dated March 21, 2022) for providing all the required facilities to carry out the work. SL and SH are thankful to the CSIR, New Delhi for supporting their Research Fellowship.

Abbreviations

AgNPs	silver nanoparticles
ALA	α-lipoic acid
AuNPs	gold nanoparticles
CTAB	cetyltrimethyl ammonium bromide
EGCG	Epigallocatechingallate
HUVEC	human umbilical vein endothelial cells
IP	intraperitoneal
LSPR	localized surface plasmon resonance
MHA–PAAm	methacrylatedhyaluronan–polyacrylamide
MRSA	methicillin-resistant *S. aureus*
NIR	near-Infrared
PCL	polycaprolactone
PEG	polyethylene glycol
PES	polyethersulfone
PMMA	poly(methyl methacrylate)
ROS	reactive oxygen species
VEGF	vascular endothelial growth factor
WHO	World Health Organization

References

[1] R.A. Sperling, P.R. Gil, F. Zhang, M. Zanella, WJ. Parak, Biological applications of gold nanoparticles, Chem. Soc. Rev. 37 (9) (2008) 1896–1908.

[2] E.C. Dreaden, A.M. Alkilany, X. Huang, C.J. Murphy, MA. El-Sayed, The golden age: Gold nanoparticles for biomedicine, Chem. Soc. Rev. 41 (7) (2012) 2740–2779.

[3] K.K. Wong, L. Xuelai, Silver nanoparticles—The real "silver bullet" in clinical medicine? Med. Chem. Comm 1 (2) (2010) 125–131.

[4] V.K. Chaturvedi, A. Singh, V.K. Singh, MP. Singh, Cancer nanotechnology: A new revolution for cancer diagnosis and therapy, Curr. Drug Metab. 20 (6) (2019) 416–429.

[5] G.A. Sotiriou, SE. Pratsinis, Engineering nanosilver as an antibacterial, biosensor and bioimaging material, Curr. Opin. Chem. Eng. 1 (1) (2011) 3–10.

[6] A.C. Burduşel, O. Gherasim, A.M. Grumezescu, L. Mogoantă, A. Ficai, E. Andronescu, Biomedical applications of silver nanoparticles: An up-to-date overview, Nanomaterials 8 (9) (2018) 681.

[7] J. Saha, A. Begum, A. Mukherjee, S. Kumar, A novel green synthesis of silver nanoparticles and their catalytic action in reduction of Methylene Blue dye, Sustain. Environ. Res. 27 (5) (2017) 245–250.

[8]   A.L.M. Terra, R.D.C. Kosinski, J.B. Moreira, J.A.V. Costa, MGD. Morais, Microalgae biosynthesis of silver nanoparticles for application in the control of agricultural pathogens, J. Environ. Sci. Health. Part B 54 (8) (2019) 709–716.

[9]   M.G. Heinemann, C.H. Rosa, G.R. Rosa, D. Dias, Biogenic synthesis of gold and silver nanoparticles used in environmental applications: A review. Trends Environ, Anal. Chem. 30 (2021) e00129.

[10] L. Castillo-Henríquez, K. Alfaro-Aguilar, J. Ugalde-Álvarez, L. Vega-Fernández, G. Montes de Oca-Vásquez, J.R. Vega-Baudrit, Green synthesis of gold and silver nanoparticles from plant extracts and their possible applications as antimicrobial agents in the agricultural area, Nanomaterials 10 (9) (2020) 1763.

[11] M.C. Daniel, D. Astruc, Gold nanoparticles: assembly, supramolecular chemistry, quantum-size-related properties, and applications toward biology, catalysis, and nanotechnology, Chem. Rev. 104 (1) (2004) 293–346.

[12] S.S. Abdalla, H. Katas, F. Azmi, MFM. Busra, Antibacterial and anti-biofilm biosynthesised silver and gold nanoparticles for medical applications: Mechanism of action, toxicity and current status, Curr. Drug Deliv. 17 (2) (2020) 88–100.

[13] S. Haque, Potential application of silver nanocomposites for antimicrobial activity, Biomedical Composites: Perspectives and Applications (2021) 93.

[14] M. Azharuddin, G.H. Zhu, D. Das, E. Ozgur, L. Uzun, A.P. Turner, HK. Patra, A repertoire of biomedical applications of noble metal nanoparticles, Chem. Commun. 55 (49) (2019) 6964–6996.

[15] F. Benyettou, R. Rezgui, F. Ravaux, T. Jaber, K. Blumer, M. Jouiad, et al., Synthesis of silver nanoparticles for the dual delivery of doxorubicin and alendronate to cancer cells, J. Mater. Chem. B. 3 (36) (2015) 7237–7245.

[16] S. Mukherjee, S. Sau, D. Madhuri, V.S. Bollu, K. Madhusudana, B. Sreedhar, et al., Green synthesis and characterization of monodispersed gold nanoparticles: Toxicity study, delivery of doxorubicin and its bio-distribution in mouse model, J. Biomed. Nanotech. 12 (1) (2016) 165–181.

[17] D.C. Lekha, R. Shanmugam, K. Madhuri, L.P. Dwarampudi, M. Bhaskaran, D. Kongara, et al., Review on silver nanoparticle synthesis method, antibacterial activity, drug delivery vehicles, and toxicity pathways: Recent advances and future aspects, J. Nanomater. 2021 (2021) 4401829.

[18] S. Mukherjee, CR. Patra, Biologically synthesized metal nanoparticles: Recent advancement and future perspectives in cancer theranostics, Future Sci 3(3) (2017) FSO203.

[19] T.Q. Huy, P. Huyen, A.T. Le, M. Tonezzer, Recent advances of silver nanoparticles in cancer diagnosis and treatment, Anti-Cancer Agents Med. Chem. 20 (11) (2020) 1276–1287.

[20] M. Fan, Y. Han, S. Gao, H. Yan, L. Cao, Z. Li, et al., Ultrasmall gold nanoparticles in cancer diagnosis and therapy, Theranostics 10 (11) (2020) 4944.

[21] N. Elahi, M. Kamali, MH. Baghersad, Recent biomedical applications of gold nanoparticles: A review, Talanta 184 (2018) 537–556.

[22] X.F. Zhang, Z.G. Liu, W. Shen, S. Gurunathan, Silver nanoparticles: Synthesis, characterization, properties, applications, and therapeutic approaches, Int. J. Mol. Sci. 17 (9) (2016) 1534.

[23] M. Jouni, A. Boudenne, G. Boiteux, V. Massardier, B. Garnier, A. Serghei, Electrical and thermal properties of polyethylene/silver nanoparticle composites, Polym. Compos. 34 (5) (2013) 778–786.

[24] F. Paladini, M. Pollini, Antimicrobial silver nanoparticles for wound healing application: Progress and future trends, Materials 12 (16) (2019) 2540.

[25] M.A. Zakaria, A.A. Menazea, A.M. Mostafa, Al-A EA, Ultra-thin silver nanoparticles film prepared via pulsed laser deposition: Synthesis, characterization, and its catalytic activity on reduction of 4-nitrophenol, Surf. Interfaces. 19 (2020) 100438.

[26] A.A. Yaqoob, K. Umar, MNM. Ibrahim, Silver nanoparticles: Various methods of synthesis, size affecting factors and their potential applications–A review, Appl. Nanosci. 10 (5) (2020) 1369–1378.

[27] E.O. Simbine, L.D.C. Rodrigues, J. Lapa-Guimaraes, E.S. Kamimura, C.H. Corassin, et al., Application of silver nanoparticles in food packages: A review, Food Science and Technology 39 (4) (2019) 793–802.

[28] S. Haque, C.C. Norbert, R. Acharyya, S. Mukherjee, M. Kathirvel, CR. Patra, Biosynthesized silver nanoparticles for cancer therapy and in vivo bioimaging, Cancers 13 (23) (2021) 6114.

[29] S. Haque, CR. Patra, Biosynthesized nanoparticles (gold, silver and platinum): Therapeutic role in angiogenesis, Compr. Anal. Chem 94 (2021) 471–505.

[30] V. Kumar, S. Singh, B. Srivastava, R. Bhadouria, R. Singh, Green synthesis of silver nanoparticles using leaf extract of Holoptelea integrifolia and preliminary investigation of its antioxidant, anti-inflammatory, antidiabetic and antibacterial activities, J. Environ. Chem. Eng. 7 (3) (2019) 103094.

[31] J.B. Vines, J.H. Yoon, N.E. Ryu, D.J. Lim, H. Park, Gold nanoparticles for photothermal cancer therapy, Front. Chem 7 (2019) 167.

[32] SA. Bansal, V. Kumar, J. Karimi, A.P. Singh, S. Kumar, Role of gold nanoparticles in advanced biomedical applications, Nanoscale Adv 2 (9) (2020) 3764–3787.

[33] X. Zhang, Gold nanoparticles: Recent advances in the biomedical applications, Cell Biochem. Biophys. 72 (3) (2015) 771–775.

[34] D. Pooja, S. Panyaram, H. Kulhari, B. Reddy, S.S. Rachamalla, R. Sistla, Natural polysaccharide functionalized gold nanoparticles as biocompatible drug delivery carrier, Int. J. Biol. Macromol. 80 (2015) 48–56.

[35] J.Y. Maillard, P. Hartemann, Silver as an antimicrobial: Facts and gaps in knowledge, Crit. Rev. Microbiol. 39 (4) (2013) 373–383.

[36] A. Balfourier, J. Kolosnjaj-Tabi, N. Luciani, F. Carn, F. Gazeau, Gold-based therapy: From past to present, Proc. Natl. Acad. Sci 117 (37) (2020) 22639–22648.

[37] D.A. Giljohann, D.S. Seferos, W.L. Daniel, M.D. Massich, P.C. Patel, CA. Mirkin, Gold nanoparticles for biology and medicine, Angew. Chem. Int. Ed. 49 (19) (2010) 3280–3294.

[38] K. Sztandera, M. Gorzkiewicz, Klajnert-Maculewicz B. Gold nanoparticles in cancer treatment, Mol. Pharm. 16 (1) (2018) 1–23.

[39] V. Amendola, S. Polizzi, M. Meneghetti, Laser ablation synthesis of silver nanoparticles embedded in graphitic carbon matrix, Sci. Adv. Mater. 4 (3-4) (2012) 497–500.

[40] X. Zhao, Y. Xia, Q. Li, X. Ma, F. Quan, C. Geng, Z. Han, Microwave-assisted synthesis of silver nanoparticles using sodium alginate and their antibacterial activity, Colloids Surf. A: Physicochem. Eng. Asp. 444 (2014) 180–188.

[41] G. Khayati, K. Janghorban, The nanostructure evolution of Ag powder synthesized by high energy ball milling, Adv. Powder Technol. 23 (3) (2012) 393–397.

[42] L.S. Kibis, A.I. Stadnichenko, E.M. Pajetnov, S.V. Koscheev, V.I. Zaykovskii, AI. Boronin, The investigation of oxidized silver nanoparticles prepared by thermal evaporation and radio-frequency sputtering of metallic silver under oxygen, Appl. Surf. Sci. 257 (2) (2010) 404–413.

[43] K. Alaqad, TA. Saleh, Gold and silver nanoparticles: synthesis methods, characterization routes and applications towards drugs, J. Environ. Anal. Toxicol. 6 (4) (2016) 525–2161.

[44] M. Yusuf, Silver nanoparticles: Synthesis and applications, Handbook of Ecomaterials (2020) 2343–2356.

[45] I.A. Mohammed, Al-G FJ, Gold nanoparticle: Synthesis, functionalization, enhancement, drug delivery and therapy: A review, Sys Rev Pharm 11 (6) (2020) 888–910.

[46] V. Pareek, A. Bhargava, R. Gupta, N. Jain, J. Panwar, Synthesis and applications of noble metal nanoparticles: A review, Adv. Sci. Eng. Med. 9 (7) (2017) 527–544.

[47] J.P. Abid, A.W. Wark, P.F. Brevet, HH. Girault, Preparation of silver nanoparticles in solution from a silver salt by laser irradiation, Chem. Commun (7) (2002) 792–793.

[48] D. Parida, P. Simonetti, R. Frison, E. Bülbül, S. Altenried, Y. Arroyo, et al., Polymer-assisted in-situ thermal reduction of silver precursors: A solventless route for silver nanoparticles-polymer composites, Chem. Eng. Sci. 389 (2020) 123983.

[49] Y. Han, J. Jiang, S.S. Lee, Ying JY Reverse microemulsion-mediated synthesis of silica-coated gold and silver nanoparticles, Langmuir 24 (11) (2008) 5842–5848.

[50] G.N. Xu, X.L. Qiao, X.L. Qiu, JG. Chen, Preparation and characterization of stable monodisperse silver nanoparticles via photoreduction, Colloids Surf. A: Physicochem. Eng. Asp. 320 (1-3) (2008) 222–226.

[51] A.V. Rane, K. Kanny, V.K. Abitha, S. Thomas, Methods for synthesis of nanoparticles and fabrication of nanocomposites. In: Synthesis of Inorganic Nanomaterials, Elsevier, 2018, pp. 121–139.

[52] S. Iravani, H. Korbekandi, S.V. Mirmohammadi, B. Zolfaghari, Synthesis of silver nanoparticles: Chemical, physical and biological methods, Res Pharm Sci 9 (6) (2014) 385.

[53] J. Natsuki, T. Natsuki, Y. Hashimoto, A review of silver nanoparticles: Synthesis methods, properties and applications, Int. J. Mater. Sci. Appl. 4 (5) (2015) 325–332.

[54] R. Herizchi, E. Abbasi, M. Milani, A. Akbarzadeh, Current methods for synthesis of gold nanoparticles, Artif Cells Nanomed Biotechnol 44 (2) (2016) 596–602.

[55] S. Pirtarighat, M. Ghannadnia, S. Baghshahi, Green synthesis of silver nanoparticles using the plant extract of Salvia spinosa grown in vitro and their antibacterial activity assessment, J. nanostructure chem. 9 (1) (2019) 1–9.

[56] K. Luo, S. Jung, K.H. Park, YR. Kim, Microbial biosynthesis of silver nanoparticles in different culture media, J. Agric. Food Chem. 66 (4) (2018) 957–962.

[57] M. Rafique, I. Sadaf, M.S. Rafique, MB. Tahir, A review on green synthesis of silver nanoparticles and their applications, Artif Cells Nanomed Biotechnol 45 (7) (2017) 1272–1291.

[58] N. Pantidos, LE. Horsfall, Biological synthesis of metallic nanoparticles by bacteria, fungi and plants, J. Nanomed. Nanotechnol. 5 (5) (2014) 1.

[59] A. Roy, O. Bulut, S. Some, A.K. Mandal, MD. Yilmaz, Green synthesis of silver nanoparticles: Biomolecule-nanoparticle organizations targeting antimicrobial activity, RSC Adv. 9 (5) (2019) 2673–2702.

[60] W. Kaplan, Background paper 6.5 cancer and cancer therapeutics. In: Priority medicines for Europe and the world: update, World Health Organization, 2013, pp. 5–6.

[61] R.G. de Oliveira Júnior, A.F.C. Adrielly, J.R.G. da Silva Almeida, R. Grougnet, V. Thiery, L. Picot, Sensitization of tumor cells to chemotherapy by natural products: A systematic review of preclinical data and molecular mechanisms, Fitoterapia 129 (2018) 383–400.

[62] B. Kalyanaraman, Teaching the basics of cancer metabolism: Developing antitumor strategies by exploiting the differences between normal and cancer cell metabolism, Redox. Biol. 12 (2017) 833–842.

[63] A.M. Lewandowska, M. Rudzki, S. Rudzki, T. Lewandowski, B. Laskowska, Environmental risk factors for cancer-review paper, Ann. Agric. Environ. Med. 26 (1) (2019) 1–7.

[64] A. Luengo, D.Y. Gui, Vander Heiden MG. Targeting metabolism for cancer therapy, Cell Chem. Biol. 24 (9) (2017) 1161–1180.

[65] A. Jemal, F. Bray, M.M. Center, J. Ferlay, E. Ward, D. Forman, Global cancer statistics. Ca-Cancer, J. Clin. 61 (2) (2011) 69–90.

[66] J. Zugazagoitia, C. Guedes, S. Ponce, I. Ferrer, S. Molina-Pinelo, L. Paz-Ares, Current challenges in cancer treatment, Clin. Ther. 38 (7) (2016) 1551–1566.

[67] R.D. Smith, MK. Mallath, History of the growing burden of cancer in India: From antiquity to the 21st century, J. glob. oncol. 5 (2019) 1–15.

[68] F. Bray, J. Ferlay, I. Soerjomataram, R.L. Siegel, L.A. Torre, A. Jemal, Global cancer statistics 2018: GLOBOCAN estimates of incidence and mortality worldwide for 36 cancers in 185 countries, Ca-Cancer J. Clin. 68 (6) (2018) 394–424.

[69] Y. Mao, et al., Epidemiology of lung cancer, Surg. Oncol. Clin 25 (3) (2016) 439–445.

[70] P. Mathur, K. Sathishkumar, M. Chaturvedi, P. Das, K.L. Sudarshan, S. Santhappan, et al., Cancer statistics, 2020: Report from national cancer registry programme, India. JCO Glob. Oncol. 6 (2020) 1063–1075.

[71] K. Łukasiewicz, M. Fol, Microorganisms in the treatment of cancer: Advantages and limitations, J. Immunol. Res. (2018).

[72] V. Dilalla, G. Chaput, T. Williams, K. Sultanem, Radiotherapy side effects: Integrating a survivorship clinical lens to better serve patients, Curr Oncol 27 (2) (2020) 107–112.

[73] V. Schirrmacher, From chemotherapy to biological therapy: A review of novel concepts to reduce the side effects of systemic cancer treatment, Int. J. Oncol. 54 (2) (2019) 407–419.

[74] S. Senapati, A.K. Mahanta, S. Kumar, P. Maiti, Controlled drug delivery vehicles for cancer treatment and their performance, Signal Transduct Target Ther 3 (1) (2018) 1–19.

[75] M. Chidambaram, R. Manavalan, K. Kathiresan, Nanotherapeutics to overcome conventional cancer chemotherapy limitations, J. Pharm. Pharm. Sci. 14 (1) (2011) 67–77.

[76] M. Buttacavoli, N.N. Albanese, G. Di Cara, R. Alduina, C. Faleri, M. Gallo, et al., Anticancer activity of biogenerated silver nanoparticles: An integrated proteomic investigation, Oncotarget 9 (11) (2018) 9685.

[77] R.R. Arvizo, S. Bhattacharyya, R.A. Kudgus, K. Giri, R. Bhattacharya, P. Mukherjee, Intrinsic therapeutic applications of noble metal nanoparticles: Past, present and future, Chem. Soc. Rev 41 (7) (2012) 2943–2970.

[78] S. Jain, N. Saxena, M.K. Sharma, S. Chatterjee, Metal nanoparticles and medicinal plants: Present status and future prospects in cancer therapy, Mater. Today: Proc. 31 (2020) 662–673.

[79] A. Sharma, A.K. Goyal, G. Rath, Recent advances in metal nanoparticles in cancer therapy, J. Drug Target. 26 (8) (2018) 617–632.

[80] H. Sharma, P.K. Mishra, S. Talegaonkar, B. Vaidya, Metal nanoparticles: A theranostic nanotool against cancer, Drug Discov. Today. 20 (9) (2015) 1143–1151.

[81] AM. Grumezescu, Design of Nanostructures for Theranostics Applications, William Andrew, Elsevier, Amsterdam, Netherlands, 2017.

[82] V. Mirabello, D.G. Calatayud, R.L. Arrowsmith, H. Ge, SI. Pascu, Metallic nanoparticles as synthetic building blocks for cancer diagnostics: From materials design to molecular imaging applications, J. Mater. Chem. B. 3 (28) (2015) 5657–5672.

[83] C.R. Patra, S. Mukherjee, R. Kotcherlakota, Biosynthesized silver nanoparticles: A step forward for cancer theranostics? Nanomedicine 9 (10) (2014) 1445–1448.

[84] K. Venugopal, H. Ahmad, E. Manikandan, K.T. Arul, K. Kavitha, M.K. Moodley, et al., The impact of anticancer activity upon Beta vulgaris extract mediated biosynthesized silver nanoparticles (ag-NPs) against human breast (MCF-7), lung (A549) and pharynx (Hep-2) cancer cell lines, J. Photochem. Photobiol. B: Biol. 173 (2017) 99–107.

[85] K. Jadhav, S. Deore, D. Dhamecha, R. Hr, S. Jagwani, S. Jalalpure, R. Bohara, Phytosynthesis of silver nanoparticles: Characterization, biocompatibility studies, and anticancer activity, ACS Biomater. Sci. Eng. 4 (3) (2018) 892–899.

[86] S. Kummara, M.B. Patil, T. Uriah, Synthesis, characterization, biocompatible and anticancer activity of green and chemically synthesized silver nanoparticles–A comparative study, Biomed. Pharmacother. 84 (2016) 10–21.

[87] S. Mukherjee, D. Chowdhury, R. Kotcherlakota, S. Patra, Potential theranostics application of bio-synthesized silver nanoparticles (4-in-1 system), Theranostics 4 (3) (2014) 316.

[88] A.R. Gul, F. Shaheen, R. Rafique, J. Bal, S. Waseem, TJ. Park, Grass-mediated biogenic synthesis of silver nanoparticles and their drug delivery evaluation: A biocompatible anti-cancer therapy, Chem. Eng. J. 407 (2021) 127202.

[89] N.S. Capanema, I.C. Carvalho, A.A. Mansur, S.M. Carvalho, A.P. Lage, HS. Mansur, Hybrid hydrogel composed of carboxymethylcellulose–silver nanoparticles–doxorubicin for anticancer and antibacterial therapies against melanoma skin cancer cells, ACS Appl. Nano Mater. 2 (11) (2019) 7393–7408.

[90] S.C. Boca, M. Potara, A.M. Gabudean, A. Juhem, P.L. Baldeck, S. Astilean, Chitosan-coated triangular silver nanoparticles as a novel class of biocompatible, highly effective photothermal transducers for in vitro cancer cell therapy, Cancer Lett. 311 (2) (2011) 131–140.

[91] P. Sanpui, A. Chattopadhyay, SS. Ghosh, Induction of apoptosis in cancer cells at low silver nanoparticle concentrations using chitosan nanocarrier, ACS Appl. Mater. Interfaces. 3 (2) (2011) 218–228.

[92] S. Mukherjee, R. Kotcherlakota, S. Haque, S. Das, S. Nuthi, D. Bhattacharya, et al., Silver Prussian blue analogue nanoparticles: rationally designed advanced nanomedicine for multifunctional biomedical applications, ACS Biomater. Sci. Eng. 6 (1) (2019) 690–704.

[93] B.R. Rao, R. Kotcherlakota, S.K. Nethi, N. Puvvada, B. Sreedhar, A. Chaudhuri, CR. Patra, Ag2 [Fe (CN) 5NO] nanoparticles exhibit antibacterial activity and wound healing properties, ACS Biomater. Sci. Eng. 4 (9) (2018) 3434–3449.

[94] T. Kavinkumar, K. Varunkumar, V. Ravikumar, S. Manivannan, Anticancer activity of graphene oxide-reduced graphene oxide-silver nanoparticle composites, J. Colloid Interface Sci. 505 (2017) 1125–1133.

[95] Y. Jiang, M. Sun, N. Ouyang, Y. Tang, P. Miao, Synergistic chemo-thermal therapy of cancer by DNA-templated silver nanoclusters and polydopamine nanoparticles, ACS Appl. Mater. Interfaces. 13 (18) (2021) 21653–21660.

[96] J. Peng, X. Liang, Progress in research on gold nanoparticles in cancer management, Medicine (Baltimore). 98 (18) (2019).

[97] J. Lee, D.K. Chatterjee, M.H. Lee, S. Krishnan, Gold nanoparticles in breast cancer treatment: promise and potential pitfalls, Cancer Lett. 347 (1) (2014) 46–53.

[98] C.S. Lee, T.W. Kim, D.E. Oh, S.O. Bae, J. Ryu, H. Kong, TH. Kim, *In Vivo* and *In Vitro* anticancer activity of doxorubicin-loaded DNA-AuNP nanocarrier for the ovarian cancer treatment, Cancers 12 (3) (2020) 634.

[99] K.P. Valente, A. Suleman, AG. Brolo, Exploring diffusion and cellular uptake: Charged gold nanoparticles in an in vitro breast cancer model, ACS Appl. Bio Mater. 3 (10) (2020) 6992–7002.

[100] S. Rajeshkumar, Anticancer activity of eco-friendly gold nanoparticles against lung and liver cancer cells, J. Genet. Eng. Biotechnol. 14 (1) (2016) 195–202.

[101] L. Qian, W. Su, Y. Wang, M. Dang, W. Zhang, C. Wang, Synthesis and characterization of gold nanoparticles from aqueous leaf extract of Alternanthera sessilis and its anticancer activity on cervical cancer cells (HeLa), Artif Cells Nanomed Biotechnol 47 (1) (2019) 1173–1180.

[102] S. Mukherjee, M. Dasari, S. Priyamvada, R. Kotcherlakota, V.S. Bollu, CR. Patra, A green chemistry approach for the synthesis of gold nanoconjugates that induce the inhibition of cancer cell proliferation through induction of oxidative stress and their in vivo toxicity study, J. Mater. Chem. B. 3 (18) (2015) 3820–3830.

[103] R. Khandanlou, V. Murthy, D. Saranath, H. Damani, Synthesis and characterization of gold-conjugated Backhousia citriodora nanoparticles and their anticancer activity against MCF-7 breast and HepG2 liver cancer cell lines, J. Mater. Sci. 53 (5) (2018) 3106–3118.

[104] S. Ahmed, M. Ahmad, B.L. Swami, S. Ikram, A review on plants extract mediated synthesis of silver nanoparticles for antimicrobial applications: A green expertise, J. Adv. Res. 7 (1) (2016) 17–28.

[105] S.B. Yaqoob, R. Adnan, R.M. Rameez Khan, M. Rashid, Gold, silver, and palladium nanoparticles: A chemical tool for biomedical applications, Front. Chem. 8 (2020) 376.

[106] H. Chugh, D. Sood, I. Chandra, V. Tomar, G. Dhawan, R. Chandra, Role of gold and silver nanoparticles in cancer nano-medicine, Artif Cells Nanomed. Biotechnol. 46 (2018) 1210–1220.

[107] A. Salleh, R. Naomi, N.D. Utami, A.W. Mohammad, E. Mahmoudi, N. Mustafa, MB. Fauzi, The potential of silver nanoparticles for antiviral and antibacterial applications: a mechanism of action, Nanomaterials 10 (8) (2020) 1566.

[108] M. Rafique, I. Sadaf, M.B. Tahir, M.S. Rafique, G. Nabi, T. Iqbal, K. Sughra, Novel and facile synthesis of silver nanoparticles using Albizia procera leaf extract for dye degradation and antibacterial applications, Mater. Sci. Eng., C. 99 (2019) 1313–1324.

[109] A.V. Ramesh, D.R. Devi, G. Battu, KA. Basavaiah, Facile plant mediated synthesis of silver nanoparticles using an aqueous leaf extract of Ficus hispida Linn. f. for catalytic, antioxidant and antibacterial applications, S. Afr. J. Chem. Eng. 26 (2018) 25–34.

[110] F. Masood, Polymeric nanoparticles for targeted drug delivery system for cancer therapy, Mater. Sci. Eng., C. 60 (2016) 569–578.

[111] Q. Xia, L. Yang, K. Hu, K. Li, J. Xiang, G. Liu, Y. Wang, Chromium cross-linking based immobilization of silver nanoparticle coating on leather surface with broad-spectrum antimicrobial activity and durability, ACS Appl. Mater. Interfaces. 11 (2) (2018) 2352–2363.

[112] S. Tharani, D. Bharathi, R. Ranjithkumar, Extracellular green synthesis of chitosan-silver nanoparticles using Lactobacillus reuteri for antibacterial applications, Biocatal. Agric. Biotechnol. 30 (2020) 101838.

[113] B. Lee, DG. Lee, Synergistic antibacterial activity of gold nanoparticles caused by apoptosis-like death, J. Appl. Microbiol. 127 (3) (2019) 701–712.

[114] S. Sathiyaraj, G. Suriyakala, A.D. Gandhi, R. Babujanarthanam, K.S. Almaary, T.W. Chen, K. Kaviyarasu, Biosynthesis, characterization, and antibacterial activity of gold nanoparticles, J. Infect. Public Health. 14 (12) (2021) 1842–1847.

[115] M. Bindhu, M. Umadevi, Antibacterial activities of green synthesized gold nanoparticles, Mater. Lett. 120 (2014) 122–125.

[116] N. Abdel-Raouf, N.M. Al-Enazi, IB. Ibraheem, Green biosynthesis of gold nanoparticles using Galaxaura elongata and characterization of their antibacterial activity, Arab. J. Chem. 10 (2017) S3029–S3039.

[117] P. Velmurugan, J. Shim, K.S. Bang, BT. Oh, Gold nanoparticles mediated coloring of fabrics and leather for antibacterial activity, J. Photochem. Photobiol. B: Biol. 160 (2016) 102–109.

[118] N. Arshi, F. Ahmed, S. Kumar, M.S. Anwar, J. Lu, B.H. Koo, CG. Lee, Microwave assisted synthesis of gold nanoparticles and their antibacterial activity against Escherichia coli (E. coli), Curr. Appl. Phys. 11 (1) (2011) S360–S363.

[119] J.K. Patra, G. Das, L.F. Fraceto, E.V.R. Campos, M.D.P. Rodriguez-Torres, L.S. Acosta-Torres, et al., Nano based drug delivery systems: Recent developments and future prospects, J. Nanobiotechnology. 16 (1) (2018) 1–33.

[120] L. Wang, M. Zheng, Z. Xie, Nanoscale metal–organic frameworks for drug delivery: A conventional platform with new promise, J. Mater. Chem. B. 6 (5) (2018) 707–717.

[121] A.G. Nene, M. Galluzzi, H. Luo, P. Somani, S. Ramakrishna, XF. Yu, Synthetic preparations and atomic scale engineering of silver nanoparticles for biomedical applications, Nanoscale 13 (33) (2021) 13923–13942.

[122] P. Prasher, M. Sharma, H. Mudila, G. Gupta, A.K. Sharma, D. Kumar, K. Dua, Emerging trends in clinical implications of bio-conjugated silver nanoparticles in drug delivery, Colloids Interface Sci. Commun. 35 (2020) 100244.

[123] M. Yadollahi, S. Farhoudian, H. Namazi, One-pot synthesis of antibacterial chitosan/silver bio-nanocomposite hydrogel beads as drug delivery systems, Int. J. Biol. Macromol. 79 (2015) 37–43.

[124] S. Patra, S. Mukherjee, A.K. Barui, A. Ganguly, B. Sreedhar, CR. Patra, Green synthesis, characterization of gold and silver nanoparticles and their potential application for cancer therapeutics, Mater. Sci. Eng., C. 53 (2015) 298–309.

[125] K. Sarkar, S.L. Banerjee, P.P. Kundu, G. Madras, K. Chatterjee, Biofunctionalized surface-modified silver nanoparticles for gene delivery, J. Mater. Chem. B. 3 (26) (2015) 5266–5276.

[126] R.B. KC, B. Thapa, N. Bhattarai, Gold nanoparticle-based gene delivery: Promises and challenges, Nanotechnol. Rev. 3 (3) (2014) 269–280.

[127] M. Das, K.H. Shim, S.S.A. An, D.K. Yi, Review on gold nanoparticles and their applications, Toxicol. Environ. Health Sci. 3 (4) (2011) 193–205.

[128] T. Lajunen, L. Viitala, L.S. Kontturi, T. Laaksonen, H. Liang, E. Vuorimaa-Laukkanen, A. Urtti, Light induced cytosolic drug delivery from liposomes with gold nanoparticles, J Control Release 203 (2015) 85–98.

[129] S. Thambiraj, S. Shruthi, R. Vijayalakshmi, DR. Shankaran, Evaluation of cytotoxic activity of docetaxel loaded gold nanoparticles for lung cancer drug delivery, Cancer Treat. Res. Commun. 21 (2019) 100157.

[130] Y. Shan, T. Luo, C. Peng, R. Sheng, A. Cao, X. Cao, X. Shi, Gene delivery using dendrimer-entrapped gold nanoparticles as nonviral vectors, Biomaterials 33 (10) (2012) 3025–3035.

[131] F.Y. Kong, J.W. Zhang, R.F. Li, Z.X. Wang, W.J. Wang, W. Wang, Unique roles of gold nanoparticles in drug delivery, targeting and imaging applications, Molecules 22 (9) (2017) 1445.

[132] N.S. Elbialy, M.M. Fathy, WM. Khalil, Doxorubicin loaded magnetic gold nanoparticles for in vivo targeted drug delivery, Int. J. Pharm 490 (1-2) (2015) 190–199.

[133] M. Ovais, I. Ahmad, A.T. Khalil, S. Mukherjee, R. Javed, M. Ayaz, ZK. Shinwari, Wound healing applications of biogenic colloidal silver and gold nanoparticles: recent trends and future prospects, Appl. Microbiol. Biotechnol. 102 (10) (2018) 4305–4318.

[134] T. Gunasekaran, T. Nigusse, MD. Dhanaraju, Silver nanoparticles as real topical bullets for wound healing, J Am Coll Clin Wound Spec: Vols 3 (4) (2011) 82–96.

[135] C. Rigo, L. Ferroni, I. Tocco, M. Roman, I. Munivrana, C. Gardin, B. Zavan, Active silver nanoparticles for wound healing, Int. J. Mol. Sci. 14 (3) (2013) 4817–4840.

[136] F.R. Diniz, R.C.A. Maia, L. Rannier Andrade, L.N. Andrade, M. Vinicius Chaud, C.F. da Silva, P. Severino, Silver nanoparticles-composing alginate/gelatine hydrogel improves wound healing in vivo, Nanomaterials 10 (2) (2020) 390.

[137] X. Liu, P.Y. Lee, C.M. Ho, V.C. Lui, Y. Chen, C.M. Che, KK. Wong, Silver nanoparticles mediate differential responses in keratinocytes and fibroblasts during skin wound healing, ChemMedChem 5 (3) (2010) 468–475.

[138] Q. Tang, C. Chen, Y. Jiang, J. Huang, Y. Liu, P.M. Nthumba, J. Ren, Engineering an adhesive based on photosensitive polymer hydrogels and silver nanoparticles for wound healing, J. Mater. Chem. B. 8 (26) (2020) 5756–5764.

[139] J.G. Leu, S.A. Chen, H.M. Chen, W.M. Wu, C.F. Hung, Y.D. Yao, YJ. Liang, The effects of gold nanoparticles in wound healing with antioxidant epigallocatechin gallate and α-lipoic acid, Nanomed.: Nanotechnol. Biol. Med. 8 (5) (2012) 767–775.

[140] S. Wang, C. Yan, X. Zhang, D. Shi, L. Chi, G. Luo, J. Deng, Antimicrobial peptide modification enhances the gene delivery and bactericidal efficiency of gold nanoparticles for accelerating diabetic wound healing, Biomater. Sci. 6 (10) (2018) 2757–2772.

[141] N.N. Mahmoud, S. Hikmat, D.A. Ghith, M. Hajeer, L. Hamadneh, D. Qattan, EA. Khalil, Gold nanoparticles loaded into polymeric hydrogel for wound healing in rats: Effect of nanoparticles' shape and surface modification, Int. J. Pharm. 565 (2019) 174–186.

[142] P. Boomi, R. Ganesan, G.P. Poorani, S. Jegatheeswaran, C. Balakumar, H.G. Prabu, M. Saravanan, Phyto-engineered gold nanoparticles (AuNPs) with potential antibacterial, antioxidant, and wound healing activities under in vitro and in vivo conditions, Int. J. Nanomedicine. 15 (2020) 7553.

[143] S. Naraginti, P.L. Kumari, R.K. Das, A. Sivakumar, S.H. Patil, VV. Andhalkar, Amelioration of excision wounds by topical application of green synthesized, formulated silver and gold nanoparticles in albino Wistar rats, Mater. Sci. Eng., C. 62 (2016) 293–300.

[144] P. Tan, H. Li, J. Wang, SC. Gopinath, Silver nanoparticle in biosensor and bioimaging: Clinical perspectives, Biotechnol. Appl. Biochem. 68 (6) (2021) 1236–1242.

[145] X. Hai, Y. Li, K. Yu, S. Yue, Y. Li, W. Song, X. Zhang, Synergistic in-situ growth of silver nanoparticles with nanozyme activity for dual-mode biosensing and cancer theranostics, Chin. Chem. Lett. 32 (3) (2021) 1215–1219.

[146] P. Si, N. Razmi, O. Nur, S. Solanki, C.M. Pandey, et al., Gold nanomaterials for optical biosensing and bioimaging, Nanoscale Adv 3 (10) (2021) 2679–2698.

[147] S. Andreescu, LA. Luck, Studies of the binding and signaling of surface-immobilized periplasmic glucose receptors on gold nanoparticles: A glucose biosensor application, Anal. Biochem. 375 (2) (2008) 282–290.

[148] A. Majdalawieh, M.C. Kanan, O. El-Kadri, SM. Kanan, Recent advances in gold and silver nanoparticles: Synthesis and applications, J. Nanosci. Nanotechnol. 14 (7) (2014) 4757–4780.

[149] G.V. Lowry, A. Avellan, LM. Gilbertson, Opportunities and challenges for nanotechnology in the agri-tech revolution, Nat. Nanotechnol. 14 (6) (2019) 517–522.

[150] A. Rae, Real life applications of nanotechnology in electronics, OnBoard Technol (2006) 28.

[151] D. Chen, X. Qiao, X. Qiu, J. Chen, Synthesis and electrical properties of uniform silver nanoparticles for electronic applications, J. Mater. Sci. 44 (4) (2009) 1076–1081.

[152] I. Saini, MY. Himanshi Optoelectronic and sensing applications of plasmonic silver nanoparticles (2020).

[153] H. Jiang, C. Tang, Y. Wang, L. Mao, Q. Sun, L. Zhang, C. Zuo, Low content and low-temperature cured silver nanoparticles/silver ion composite ink for flexible electronic applications with robust mechanical performance, Appl. Surf. Sci. 564 (2021) 150447.

[154] M. Tavakoli, M.H. Malakooti, H. Paisana, Y. Ohm, D. Green Marques, P. Alhais Lopes, C. Majidi, EGaIn-assisted room-temperature sintering of silver nanoparticles for stretchable, inkjet-printed, thin-film electronics, Adv. Mater. 30 (29) (2018) 1801852.

[155] H. Lu, D. Zhang, X. Ren, J. Liu, WC. Choy, Selective growth and integration of silver nanoparticles on silver nanowires at room conditions for transparent nano-network electrode, Acs Nano 8 (10) (2014) 10980–10987.

[156] G.M. Sangaonkar, M.P. Desai, T.D. Dongale, KD. Pawar, Selective interaction between phytomediated anionic silver nanoparticles and mercury leading to amalgam formation enables highly sensitive, colorimetric and memristor-based detection of mercury, Sci. Rep. 10 (1) (2020) 1–12.

[157] A.N. Matsukatova, A.V. Emelyanov, A.A. Minnekhanov, D.A. Sakharutov, A.Y. Vdovichenko, R.A. Kamyshinskii, PK. Kashkarov, Memristors based on poly (p-xylylene) with embedded silver nanoparticles, Tech. Phys. Lett. 46 (1) (2020) 73–76.

[158] W. Zhao, M.A. Brook, Y. Li, Design of gold nanoparticle-based colorimetric biosensing assays, ChemBioChem 9 (15) (2008) 2363–2371.

[159] T. Minari, Y. Kanehara, C. Liu, K. Sakamoto, T. Yasuda, A. Yaguchi, M. Kanehara, Room-temperature printing of organic thin-film transistors with π-junction gold nanoparticles, Adv. Funct. Mater 24 (31) (2014) 4886–4892.

[160] Y. Zhou, S.T. Han, Z.X. Xu, VAL. Roy, Low voltage flexible nonvolatile memory with gold nanoparticles embedded in poly (methyl methacrylate), Nanotech 23 (34) (2012) 344014.

[161] M. A Bhosale, B.M. Bhanage, Silver nanoparticles: Synthesis, characterization and their application as a sustainable catalyst for organic transformations, Curr. Org. Chem. 19 (8) (2015) 708–727.

[162] V. Vidhu, D. Philip, Catalytic degradation of organic dyes using biosynthesized silver nanoparticles, Micron 56 (2014) 54–62.

[163] DT. Thompson, Using gold nanoparticles for catalysis, Nano Today 2 (4) (2007) 40–43.

[164] A. Zinchenko, Y. Miwa, L.I. Lopatina, V.G. Sergeyev, S. Murata, DNA hydrogel as a template for synthesis of ultrasmall gold nanoparticles for catalytic applications, ACS Appl. Mater. Interfaces. 6 (5) (2014) 3226–3232.

[165] T. Esakkimuthu, D. Sivakumar, S. Akila, Application of nanoparticles in wastewater treatment, Pollut. Res 33 (03) (2014) 567–571.

[166] M. Zahoor, N. Nazir, M. Iftikhar, S. Naz, I. Zekker, J. Burlakovs, F. Ali Khan, A review on silver nanoparticles: Classification, various methods of synthesis, and their potential roles in biomedical applications and water treatment, Water. 13 (16) (2021) 2216.

[167] A. Rus, V.D. Leordean, P. Berce, Silver nanoparticles (AgNP) impregnated filters in drinking water disinfection. In: MATEC Web Conf, EDP Sciences, Modern Technologies in Manufacturing, France, 2017, 07007.

[168] A.M. Ferreira, É.B. Roque, F.V.D. Fonseca, CP. Borges, High flux microfiltration membranes with silver nanoparticles for water disinfection, Desalination Water Treat 56 (13) (2015) 3590–3598.

[169] C. Wang, C. Yu, Detection of chemical pollutants in water using gold nanoparticles as sensors: A review, Rev. Anal. Chem. 32 (1) (2013) 1–14.

[170] Y. Anwar, I. Ullah, M. Ul-Islam, K.M. Alghamdi, A. Khalil, T. Kamal, Adopting a green method for the synthesis of gold nanoparticles on cotton cloth for antimicrobial and environmental applications, Arab. J. Chem. 14 (9) (2021) 103327.

[171] S.K. Das, A.R. Das, AK. Guha, Gold nanoparticles: microbial synthesis and application in water hygiene management, Langmuir 25 (14) (2009) 8192–8199.

[172] J.K. Patra, S. Gouda, Application of nanotechnology in textile engineering: An overview, Int. J. Eng. Res. 5 (5) (2013) 104–111.

[173] S.P. Deshmukh, S.M. Patil, S.B. Mullan, SD. Delekar, Silver nanoparticles as an effective disinfectant: A review, Mater. Sci. Eng., C. 97 (2019) 954–965.

[174] F. Zhang, X. Wu, Y. Chen, H. Lin, Application of silver nanoparticles to cotton fabric as an antibacterial textile finish, Fibers Polym. 10 (4) (2009) 496–501.

[175] S. Salam, D. Verma, B. Butola, Facile synthesis of chitosan-silver nanoparticles onto linen for antibacterial activity and free-radical scavenging textiles, Int. J. Biol. Macromol. 133 (2019) 1134–1141.

[176] B. Tang, J. Tao, S. Xu, J. Wang, C. Hurren, W. Xu, X. Wang, Using hydroxy carboxylate to synthesize gold nanoparticles in heating and photochemical reactions and their application in textile colouration, Chem. Eng. J. 172 (1) (2011) 601–607.

[177] M.R. Bindhu, P. Saranya, M. Sheeba, C. Vijilvani, T.S. Rejiniemon, A.M. Al-Mohaimeed, MS. Elshikh, Functionalization of gold nanoparticles by β-cyclodextrin as a probe for the detection of heavy metals in water and photocatalytic degradation of textile dye, Environ. Res 201 (2021) 111628.

[178] B. Tang, L. Sun, J. Kaur, Y. Yu, X. Wang, In-situ synthesis of gold nanoparticles for multifunctionalization of silk fabrics, Dyes Pigm. 103 (2014) 183–190.

[179] P. Pisitsak, K. Chamchoy, V. Chinprateep, W. Khobthong, P. Chitichotpanya, S. Ummartyotin, Synthesis of gold nanoparticles using tannin-rich extract and coating onto cotton textiles for catalytic degradation of Congo red, J. Nanotech. (2021) 2021.

[180] A.M. Partila, Bioproduction of silver nanoparticles and its potential applications in agriculture. In: Nanotechnology for Agriculture, Springer, Singapore, 2019, pp. 19–36.

[181] T. Singh, S. Shukla, P. Kumar, V. Wahla, V.K. Bajpai, IA. Rather, Application of nanotechnology in food science: Perception and overview, Front. microbiol. 8 (2017) 1501.

[182] S. Paidari, S.A. Ibrahim, Potential application of gold nanoparticles in food packaging: A mini review, Gold Bull. (2021) 1–6.

[183] S. Sabella, R.P. Carney, V. Brunetti, M.A. Malvindi, N. Al-Juffali, G. Vecchio, P.P. Pompa, A general mechanism for intracellular toxicity of metal-containing nanoparticles, Nanoscale 6 (12) (2014) 7052–7061.

[184] H.A. Jeng, J. Swanson, Toxicity of metal oxide nanoparticles in mammalian cells, J. Environ. Sci. Health 41 (12) (2006) 2699–2711.

[185] S. Medici, M. Peana, A. Pelucelli, MA. Zoroddu, An updated overview on metal nanoparticles toxicity. In: Semin. Cancer Biol., Elsevier, Amsterdam, Netherlands, 2021.

[186] R. Shrivastava, P. Kushwaha, Y.C. Bhutia, SJS. Flora, Oxidative stress following exposure to silver and gold nanoparticles in mice, Toxicol. Ind. Health. 32 (8) (2016) 1391–1404.

[187] O. Bar-Ilan, R.M. Albrecht, V.E. Fako, Furgeson DY. Toxicity assessments of multisized gold and silver nanoparticles in zebrafish embryos, Small 5 (16) (2009) 1897–1910.

[188] A. Martirosyan, A. Bazes, YJ. Schneider, In vitro toxicity assessment of silver nanoparticles in the presence of phenolic compounds–Preventive agents against the harmful effect? Nanotoxicology 8 (5) (2014) 573–582.

[189] L. Yang, H. Kuang, W. Zhang, Z.P. Aguilar, H. Wei, H. Xu, Comparisons of the biodistribution and toxicological examinations after repeated intravenous administration of silver and gold nanoparticles in mice, Sci. Rep. 7 (1) (2017) 1–12.

[190] M. Mannerström, J. Zou, T. Toimela, I. Pyykkö, T. Heinonen, The applicability of conventional cytotoxicity assays to predict safety/toxicity of mesoporous silica nanoparticles, silver and gold nanoparticles and multi-walled carbon nanotubes, Toxicol. in Vitro 37 (2016) 113–120.

[191] K. Jadhav, D. Dhamecha, D. Bhattacharya, M. Patil, Green and ecofriendly synthesis of silver nanoparticles: Characterization, biocompatibility studies and gel formulation for treatment of infections in burns, J. Photochem. Photobiol. 155 (2016) 109–115.

[192] H.I. Gomes, C.S. Martins, JA. Prior, Silver nanoparticles as carriers of anticancer drugs for efficient target treatment of cancer cells, Nanomaterials 11 (4) (2021) 964.

[193] V.S. Madamsetty, A. Mukherjee, S. Mukherjee, Recent trends of the bio-inspired nanoparticles in cancer theranostics, Front. pharmacol. 10 (2019) 1264.

[194] G. Bor, I.D. Mat Azmi, A. Yaghmur, Nanomedicines for cancer therapy: Current status, challenges and future prospects, Ther. Deliv. 10 (2) (2019) 113–132.

[195] G. Sanità, B. Carrese, A. Lamberti, Nanoparticle surface functionalization: How to improve biocompatibility and cellular internalization, Front. Mol. Biosci. 7 (2020) 381.

[196] R. Arvizo, R. Bhattacharya, P. Mukherjee, Gold nanoparticles: Opportunities and challenges in nanomedicine, Expert Opin. Drug Deliv. 7 (6) (2010) 753–763.

[197] Z. Mirza, S. Karim, Nanoparticles-based drug delivery and gene therapy for breast cancer: Recent advancements and future challenges. In: Semin. Cancer Biol. 69, Elsevier, Amsterdam, Netherlands, 2021, pp. 226–237.

CHAPTER 9

Recent progress in gold and silver nanoparticle mediated drug delivery to breast cancers

Parth Malik[a], **Gajendra Kumar Inwati**[b], **Rachna Gupta**[c], **Tapan Kumar Mukherjee**[d]

[a]*School of Chemical Sciences, Central University of Gujarat, Gandhinagar, Gujarat, India* [b]*Department of Physics, University of the Free State, Bloemfontein, ZA, South Africa* [c]*Department of Biotechnology, Vishwa-Bharati Shantiniketan, Bolpur, West Bengal, India* [d]*Institute of Biotechnology, Amity University, Noida, Uttar Pradesh, India*

9.1 Introduction

Arguably the most frequently prevailing cancer in women, breast cancer (BC) is a major health issue, with 2012 GLOBOCAN estimates listing 1.7 million female casualties, increasing by a staggering 18% from those of 2008 cases. The listing by American Cancer Society predicts almost one in every eight United States Women to develop BC in her lifetime. More seriously, the records present a gloomy future with global BC cases expected to add 3.2 million new cases on yearly basis till 2050 [1]. Such precarious and threatening figures are the evident reasons for increasing global scientific attention to screen effective preventive and treatment routes. On a molecular scale, BC is significantly heterogeneous in terms of its pathological distinctions, with some cases exhibiting slow growth and remarkable prognosis while others being significantly more aggressive. Though advances in the imaging techniques and multiple detection routes have aided in timely detection of this critical disease but unfolding queries pertaining to an accurate mechanistic understanding still presents manifold unending puzzles. Alarming rise to global cancer related deaths has been a big reason for increasing funding and investments in BC research, with the 2008 global contribution of 11% reaching to the 12% extent in 2012. On a clinical platform, the BC classification is made via morphological features, comprising infiltrating ductal carcinoma having no specific identification along with multiple special types, such as infiltrating lobular carcinoma, tubular, mucinous, medullary, and adenoid cystic carcinoma. Further subclassification is simplified on the basis of histological grading, including relative extent of cellular differentiation, nuclear pleomorphism, and the mitotic estimation. In general, the smaller tumors are typically recognized with an earlier stage at the presentation extent contrary to those of infiltrating ductal carcinomas. It is also relatively common that breast tumors with a high histological grading

Gold and Silver Nanoparticles.
DOI: https://doi.org/10.1016/B978-0-323-99454-5.00012-3

are often large at the presentation, exhibiting local, or distant metastasis. On the basis of mutually distinct estrogen receptors (ER), progesterone receptors (PRs), and Her2 oncogene expressions, breast tumors are classified into five distinct subtypes. The ER positive tumors are relatively more common with a smaller size, lower grading, and in being lymph node negative contrary to those of ER negative tumors. In nut shell, Luminal A and Luminal B are the two ER/PR positive subgroups while the ER-negative tumors are distinguished into three subtypes. Fig. 9.1 depicts the ER and PR expression dependent BC classification, wherein the ER negative tumors are characterized by an elevated HER2 expression along with the genes associated with myoepithelial cells and a varied gene expression exhibiting normal subtype. Quite intriguingly, the HER2 and basal like subtypes of ER negative BC pathologically develops a poorer outcome compared to those of luminal and normal-like groups [2]. These two subtypes commonly express a worse prognosis irrespective of their developmental stages, with a common trait being the advanced stage development at the presentation stage [3–4].

The HER2 and basal like BC typically possess a greater extent of stem cells, which could limit the treatment recourse by contributing to replenished tumor population and the gradual multidrug resistant responses. Of note, the HER2 oncogene is associated with epidermal growth factor receptor family and is overexpressed in nearly 20% of breast tumors [5]. Perhaps, PR negativity amongst all ER positive tumors is a standalone predictor of HER2 positivity [6]. Breast tumors which do not express ER, PR, or HER2; are called as triple negative breast cancers (TNBCs), contributing to nearly 15% of all BCs. Usually, the TNBCs phenotype is quite aggressive, with a grade III staging and a higher metastatic sensitivity. It occurs with a higher frequency in younger women, African-American women, and in the individuals screened with mutated BRCA1 [7–9]. The severity of TNBCs is presented more evidently by the fact that despite a 12% lower risk of developing ER positive tumors in post-menopausal women who mothered more than two children, the risk of developing TNBC is a whopping 46% greater [10]. A relatively less known BC subtype is the inflammatory breast cancer (IBC) even though it is clinically and biologically distinct. Distinct aspects of IBC

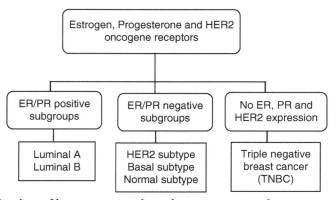

Fig. 9.1: Classification of breast cancers based on estrogen and progesterone expressions.
Most fatal are the ER and PR null regime, the triple negative breast cancer.

include the implicit prevalence of tumor emboli in the dermal lymphatic channels with presumably unusual breast tenderness, together presenting an advanced stage. On being detected, the IBC sufferers have axillary lymph node involvement in the diagnosis wherein up to 35% cases exhibit distinct metastases. Studies estimate as much as 50% of IBC with elevated Her2 levels compared to 20% in the non-IBC, BC which collectively contributes to a high extent of metastases at the presentation [11]. The faster and greater extent of metastases in IBC is due to uncharacteristically high angiogenic and angioinvasive nature of these tumors, arising from the elevated expression of proangiogenic factors. The enhanced expression of genes contributing to basal like phenotype, vascular associated genes, and immunologically related genes are the prominent traits of IBC. Apart from these, the genes controlling cell migration and invasion such as integrin β4, VASP, and ARNT, encoding the β-subunit of hypoxia inducible factor 1 (HIF-1) which together regulate the angiogenesis, are also overexpressed in IBC.

Inevitably though, the forefront challenge is to arrest the rising BC mortality, which is possible only by the enhanced reliability of timely diagnosis and safer (not patient sensitizing) drug delivery approaches, ensuing minimal side-effects. Delivering drugs to tumors is engrossed with multiple oppositions, the foremost being the difficulty in reaching the exact site of need. Physiological barriers engage the delivered drugs in nonspecific interactions owing to which the net amount of drug reaching the tumor site in conventional drug delivery is precariously inadequate. As a result, the sufferer is forced to consume greater amount of drug which in turn, has a debilitating effect on the overall coordination besides causing multiple other chronic side-effects. It is therefore, the rightful need of the hour to switch to safer and moderate drug binding assuring delivery vehicles having their exterior physical periphery being made of biologically native materials. The net aim is to enhance the bioavailability of rather more hydrophobic antitumor drugs, either alone or via combinatorial delivery with manifold natural and pronounced immunological modulators. In this context, nanoparticles (NPs) of insert metals have emerged as true assets whereby targeting to reach the needed tumor interior locations could be effectively accomplished. Additional benefit of a NPs mediated drug delivery is the moderated drug dosage which is rendered effective by assuring their stealth trafficking so that these are not rapidly degraded or excreted from within the physiological boundaries. Incentives of using inert metal constituted NPs as drug delivery vehicles stems from their robust, easy to reproduce and with immense room to self-optimize preparation methods. These carriers bind the drug via noncovalent interactions so that there is no structural deterioration and concurrent functional loss of native drug molecule. With such insights, the present chapter focuses on the emerging potential of Au and Ag NPs as drug carriers to deliver drugs and contain the BC mortality.

9.2 Pathological whereabouts of breast cancer

Development of BC is associated with a progression of a series of intermediated events, commencing from ductal hyperproliferation which is succeeded with an in situ evolution of carcinoma, invasive carcinoma and finally into a metastatic manifestation [12]. Considering the diversity in clinical progression of the disease, identifying reliable predictive tumor markers is the inevitable need of hour. This also argues for a significant ease in the clinical management of sufferers, assisting the diagnostic procedures, staging, therapeutic response evaluation, detection of recurrence, distant metastasis, and prognosis. Together, such steps become the platform for novel and accurate treatment modalities. Identifying the stage specific quantitative expression of implicit markers could be therefore a decisive route towards arresting the BC mortality besides reducing the length of treatment duration. As per the latest norms of US National Institutes of Health (NIH) Working Group and Biomarkers Consortium, a molecular marker is an indicator of pathogenic or normal biological events or a specified pharmacological response of therapeutic recourse [13]. Usually, most of these markers are proteins although in recent times, altered gene and DNA expression patterns have also been prominently included.

The heterogeneity and complexity of BC with implicit pathologies, histological specifications, and clinical outcomes has been well-recognized. Apart from this, the fact that the neoplasia is characterized by well-defined molecular subgroups on the basis of gene expression profiling having a close relation with the behavior of molecular subtypes has also been an awakening factor. In a noted contribution, Sotiriou and Pusztai mentioned that observations from gene expression profiling studies significantly changed the considerable view of BC, providing a novel basis for molecular diagnosis [14]. In fact, the expression intensity of estrogen receptor (ER), progesterone receptor (PR), and human epidermal growth factor receptor 2 (HER2) has been used as predictive marker towards identifying predictive markers for a high-risk phenotype and choosing the most suitable therapeutic option [15]. The association of BC heterogeneity with the corresponding subtypes is evidently outlined in the comparative genomic hybridization data of multiple reports [16]. The time postsequencing of human genome and the proportionate progress in protein identification has resulted in the establishment of an integrated genomics and proteomics platform towards a better understanding of BC molecular characteristics adjacent to the emerging therapeutic developments. It is therefore quite increasingly significant to know and recognize the well-established molecular markers of therapeutic rationale which comprise a familiar aspect of BC diagnosis. The following paragraphs therefore rightly discuss the prominent aspects of BC prognostic and therapeutic markers.

(1) Hormone Receptors (HR): Of the one million BC sufferers annually, ~70% are diagnosed with positive HR status [17]. These proteins are expressed both in epithelium and breast stroma and bind to circulating hormones to regulate their cell specific effects [18–19]. Of the most well-studied HR in BC, ER, and PR. The BCs classified via positive immunohistochemistry (IHC) through a regulated ER and PR expression exhibit distinct clinical, pathological, and molecular traits [20]. The risk factors associated with ER+ and PR+ BC involve mechanisms with reference to estrogen and progesterone exposure in contrast to the mechanisms associated with ER and PR negative BC (remaining independent towards the hormone exposure). Inevitably, the ER and PR both are highly associated with patient age at the time of diagnosis, increasing steadily with age.

In the 1980s, Tamoxifen, a selective estrogen receptor modulator (SERM), was reported and documented as first antiestrogenic therapy via adjuvant mode [21]. Structurally rich with delocalized pi-conjugation, this drug works via regulating estrogen signaling by interacting with ERs, progressive studies of its administration revealed significant prevention of rat mammary carcinogenesis and the clinical inhibition of contralateral BC (Fig. 9.2) [22–23].

Investigations have significantly established the benefit of tamoxifen to ER-positive BC sufferers, conveying a greater suitability of long-term adjuvant therapy over the short-term regime along with increasing occurrence of endometrial cancer. Nevertheless, at present more than four lakh females globally, are alive due to long-term adjuvant Tamoxifen treatment [24]. The declining BC mortalities in United States and United Kingdom over the past decade are majorly attributed to such traits of Tamoxifen [25]. Antagonist actions of Tamoxifen are the outcomes of its ability to bind the ligand binding ER domain, effectively preventing the possibility of estrogen stimulation. Such binding subsequently prevents the critical ER conformational changes which are needed for an association of coactivators [26]. This mode of therapy exhibited clinical remission in ER positive BC patients, unlike the tumors exhibiting low or undetectable levels of these receptors [27]. Besides, patients expressing hormone receptors on tumor cells exhibited a better response to hormone therapy, inferred via higher disease free and overall survival [28–29]. Despite improved outcomes of hormone therapy in BC patients, optimal management of chronological BC manifesting procedures remains a crucial challenge.

(2) Human Epidermal Growth Factor Receptor 2 (HER-2): The association of HER-2 with BC is in practice ever since 1987 when Slamon and colleagues reported an enhanced survival driven poor prognosis with amplified HER-2 expression. This is a trans membrane tyrosine kinase receptor belonging to the family of EGFR receptors, encoded by ERBB2/HER2 oncogene prevailing on chromosome 17q21 [30]. This oncogene is observed in amplified form typically in (20%–30%) of BC and is widely regarded as a prognosis weakening marker.

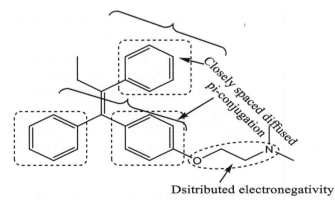

Fig. 9.2: Chemical structure of selective estrogen receptor modulator (SERM), Tamoxifen.
The drug serves as major curative source of breast cancer metastasis.

The major distinctions for such attributes of HER2 include its overexpression associated with a typically aggressive phenotype of tumor cells resulting in resistance to antihormonal, cytotoxic therapies with a concomitant low overall survival. Many other names are also frequently recognized for this gene, such as c-erb-2, cerbB-2, C-erbB-2, HER-2/neu, ERBB2, erbB2, erbB-2, neu/c-erbB-2/oncogene neu, neu protein, and neu [31].

In the cell signaling, the homo or hetero dimerization of HER family of receptors activates intracellular tyrosine kinase domain, promoting the autophosphorylation of tyrosine residues of cytoplasmic tail and consequently activates the cell survival and proliferation pathways [32]. Nevertheless, as per the crystallographic profiling, HER-2 exhibits a robust binding conformation even in the ligand's absence. Present use of HER-2 for BC treatment pertains to the administration of humanized monoclonal antibody (Mab), Trastuzumab directed against the HER-2 extracellular domains for the treatment of HER-2 positive BC cases. Antitumor potential of this Mab is reported in multiple clinical studies, with inhibitory effect on tumor growth and chemotherapy sensitizer [33]. Though the mechanisms through which Trastuzumab inhibits the HER-2 regulated cell signaling are not yet completely understood, its corresponding antitumor activities are believed to be controlled via inhibiting the receptor-receptor interaction, receptor action moderation via endocytosis, inactivation of extracellular domain cleavage of receptor and activation of antibody dependent cellular cytotoxicity [34–35]. Inadequacy of Trastuzumab therapy was resolved via development of more potent tyrosine kinase inhibitor, Lapatinib towards HER-2 protein. In general, the HER-2 status of a typical BC is estimated either via IHC analysis of HER-2 protein or fluorescent in situ hybridization (FISH) screening of gene copy number in primary tumor tissues. Studies on knowing the HER-2 expression inferred that its extracellular domain can be shed into circulation through proteolytic cleavage from the full-length HER-2 receptor. It is subsequently, detected in the serum of women with benign breast disease, primary and metastatic BC [36]. The soluble receptor can be quantified using ELISA, as also reported by Tan and group using an inexpensive and valid method, Dot blot method to detect serum HER-2 extents *vis-à-vis* BC progression [37].

Interestingly, the more threatening TNBC subtype of BC is characterized by missing ER, PR, and HER-2 expression and is associated with younger age at diagnosis [38]. Pulsating research for identifying the exclusive BC subtype continues due to concurrent inefficacy of antiendocrine and anti-HER2 targeted therapies along with the patient sensitizing and elaborate traditional version of cytotoxic chemotherapy. The main distinctions responsible for increasing interest in this BC subtype include aggressive clinical course, poor prognosis, and diminishing therapeutic options [39]. Clinical response of TNBC is typically more aggressive than luminal A and B molecular subtypes, often considered suitable for best and intermediate prognosis [40].

(3) Ki 67 Antigen: First described in 1983, the Ki-67 antigen is a labile, nonhistone nuclear protein that plays multiple prominent roles in cell cycle regulation and is expressed in mid-G1, S, G2, and M phases. The expression remains null in quiescent or resting cells of the

Go and early G1 phases. In general, the Ki-67 score is measured on histological sections by IHC methodology and is defined as percentage of stained invasive carcinoma cells. In a significant effort, Vielh and colleagues demonstrated a strong correlation between phase S and Ki-67, verifying that a quantitative evaluation of Ki-67 can offer a precise estimation of tumor proliferation index [41]. As per the 2011 St. Gallen Consensus, a 14% or lesser stained nuclei extent argues for a low or negative proliferation index while a greater than 14% stained extent corroborates for a positive or high proliferation index. Biological markers capable of predicting a pathological response to a primary systemic therapy of early form (a part of chemotherapy cycle) harbors significant clinical importance. Some noteworthy attempts are pertinent to mention here, regarding the Ki-67 index assessment and tumor progression. For instance, Patil and accomplices in 2011 evaluated the Ki-67 index and apoptotic index (AI) before, during, and after neoadjuvant chemotherapy using anthracycline in indigenous women suffering from BC but observed no significant distinctions [42]. Likewise, Tawfik and associates in 2013, demonstrated for the first time that a high Ki-67 expression in axillary lymph nodes is considerably correlated with a shorter patient survival [43]. It was hence concluded that the sufferers having a higher proliferative activity in lymph nodes indeed need an aggressive therapy alongside a closer clinical monitoring. In another worthwhile effort, Luprosi and associates discussed prognostic and therapeutic significance of Ki-67 biomarker and noted it as biomarker as a critical prognostic factor [44]. Still, routine assessment of this biomarker necessitates a standardization of assay techniques and scoring methods for its integration to erstwhile diagnostic assays.

(4) Tumor Protein p53 (TP53): This protein is involved in multiple critical pathways, including that of cell cycle arrest, apoptotic activation, DNA repair, and cellular senescence. All of which are essential for normal cellular homeostasis and genome integrity maintenance. Alteration of TP53 gene or a post-translational modification of p53 protein is often attempted to modulate its response to cellular stress. The molecular archaeology of TP53 mutation affects the etiology and molecular pathogenesis of human cancer. In BC, typically 30% patients exhibit TP53 gene mutation but the frequency varies from more than 80% in basal subtypes to below 15% in luminal A counterparts [2]. In the noted study by Allred and colleagues, expression of mutant p53 protein correlated with high tumor proliferation rate, early recurrence and death in node negative BC. With reference to BC, Dumay and associates screened the TP53 mutations in breast tumors from luminal, basal, and molecular apocrine subgroups and noticed considerable differences not only in TP53 mutation frequency but also in mutation type and their respective consequences. The investigators observed a high prevalence of mis-sensed mutations in luminal tumors and truncating mutations in basal tumors. In apocrine molecular tumors, even though insertions and deletions prevailed highly, p53 truncation showed no enhancement. These observations indicated differential mutational mechanisms, functional consequences, and selective pressures in the different BC subtypes. Mutations in TP53 gene result in altered molecular conformation and a prolonged protein half-life leading to nuclear accumulation of altered p53 protein. The IHC method screens this abnormal

accumulation, serving as indirect indicator of mutated TP53 gene [45]. The accumulation of this kind signifies a poor clinical outcome of BC sufferers. Unfortunately, in spite of its high prognostic value, there is no standardized treatment recourse that takes into account the expression status of this marker.

(5) Carbohydrate 15-3 and Carcinoembryonic Antigens (CA 15-3 and CEA): CA 15-3 peptides are shed or soluble forms of MUC-1, which prevails as a transmembrane protein constituted of two subunits together forming a stable dimer. The release is known to be regulated via two proteases, ADAM17 and MT-MMP1 [46]. This maker protein is heterogeneously expressed on the apical surface of normal epithelial cells but in BC, the expression is unexpectedly high in 90% cases [47]. The biomarker CEA is a glycoprotein known for its frequent expression in vast majority of human colorectal, gastric, and pancreatic cancers apart from non-small cell lung and breast carcinomas [48]. Screening the CEA expression in BC is a predictor of tumor size and nodal involvement. Typically, the CEA extents more than 7.5 $\mu g \cdot mL^{-1}$ are correlated with a high probability of subclinical metastases [49]. The prognosis of BC patients having CEA expression within the normal range at the time of diagnosis is comparatively better than those having higher CEA expressions [50].

The CA 15-3 in combination with CEA is a potential tumor marker in BC but the prognostic usefulness of these two marker proteins often meets a conflicting scenario. In the view of Geraghty and colleagues, the CA 15-3 has a superior prognostic significance over the CEA but the studies by Ebeling and accomplices suggest a higher CEA prognostic usefulness [51–52]. Two major studies examined the role of CA 15-3 for BC prognosis. In the first investigation, Sandri and colleagues observed the baseline CA 15-3 expression within subgroups of patients exhibiting luminal B and HER-2 positivity. The corresponding expression of CA 15-3 was noted as significant towards identifying a higher relapse risk, thereby inferring a necessity of adjuvant chemotherapy administration [53]. Putting simply, an abnormal CA 15-3 presurgical value is an indicative of increased risk of recurrence and death. Although this reflects an implicit positivity yet further studies via prospective trials and database cross-examination are needed to quantify a prognostic significance of presurgery CA 15-3 expression. In the event of conformity, elevated CA-15-3 expression could be included in the list of considering features before choosing the right course of treatment. The study by Mendes and colleagues demonstrates CA 15-3 as more efficient marker protein than CEA for detecting distant metastases. However, the investigators concluded that mere use of CA 15-3 (alone) postsurgical treatment is not sufficient, wherein CA 15-3 and CEA jointly ascertain an early metastasis in (60%–80%) BC patients [54–55].

(6) Breast Cancer Susceptibility Genes (BRCA1 and BRCA2): As many as 80% of familial BC cases are associated with one gene of hereditary susceptibility for breast and ovarian cancer, BRCA1 and BRCA2. These genes have been classified as tumor-suppressor genes due to a loss of wild type allele in the tumors of heterozygous carriers. BRCA proteins perform multiple important functions in diverse cellular processes, including the activation and

transcriptional regulation, repair of DNA damage, cellular proliferation, and differentiation [56]. The frequency and spectrum of mutations within BRCA1 or BRCA2 genes exhibit a significant variation between the ethnic groups and geographic regions, presumably due to interactions between different lifestyle and genetic characteristics. Studies have discussed the role of maternal or paternal inheritance of BRCA mutation affecting the BC risk. In this reference, Shapira and colleagues demonstrated a higher lifetime risk in paternal BRCA mutation. Contrary to this, in the studies of Senst and associates, the BC risk is apparently higher in woman with paternal BRCA1 mutation, the results of study were not significant [57]. Interestingly, the parental mutation also did not affect the risk in women having BRCA2 mutation. Thereby, the available figures are not adequate for justifying the various screening recommendations for the two subgroups. Prediction of BRCA1 or BRCA2 mutation can be made via family history profiles characterized by first degree relatives with ovarian or BC along with young age at the time of diagnosis, bilateral occurrence, and increased number of affected relatives. These predictors could be of significance in genetic counselling and decision making for a genetic test but these could be of limited usefulness as BC families exhibit a large number of BRCA1 or BRCA2 mutations in the absence of such risk factors. Over the past few years, genetic counselling and testing has been widely optimized to screen the BRCA1 and BRCA2 gene mutations in high-risk patients [58]. Individuals having undergone genetic testing via DNA sequencing for specific gene regions and become aware of carrying a BRCA mutation can get the BC diagnosis anticipated and thereby prevented on some occasions. In case of a high-risk status identification in these women, they do have an implicit option to undergo genetic counselling, testing more effective cancer surveillance, and related preventive options, such as prophylactic surgery and chemoprevention [59]. Amongst such options, bilateral mastectomy (despite being invasive), reduces nearly 90% BC risk in women with BRCA1/2 mutations [60].

The studies by Apostolou and Fostira elucidate BRCA1 and BRCA2 predispostion as a part of more susceptible genes in causing tumor. These findings include rare germline mutations in other highly penetrant genes, most important of which are TP53 mutations in Li-Fraumeni syndrome, serine/threonine kinase 11 mutations in Peutz-Jeghers syndrome, and PTEN (phosphate and tensin homolog on chromosome 10) mutations in Cowden syndrome [61]. The mutations that are rather more frequent but less penetrant have been identified in families having a history of BC clustering, in moderate or lowly penetrant genes such as CHEK2 (checkpoint kinase 2), ATM (ataxia telangiesctasis mutated), PALB2 (partner and localizer of BRCA2), and BRIPI (BRCA1-interacting protein C-terminal helicase 1).

Having known about the different biomarkers screened for accurate and efficient BC diagnosis, it becomes imperative to ascertain their expression on tumor cell surface. A quick glance at the overexpressed receptor proteins in the various BC configurations can be had in Table 9.1, whereby targeting of drugs using active mode is optimized [62]. This is because the conventional treatment of tumors is badly repetitive with limiting impacts of concurrent

Table 9.1: Major cell-surface receptor proteins overexpressed in breast cancer.

Sr. no.	Receptor protein	Associated tumor aggravating event
1.	Estrogen and progesterone	Age dependent complications, exclusively varying as per the menopausal and pregnancy conditions
2.	Prostaglandin E-receptor (EP-4)	Enhanced tumor migration, invasion, angiogenesis, and lymphangiogenesis, blocking killer functions of NK and T cells, immunological suppressing activity on macrophages
3.	Toll-like receptors (TLRs)	Affects adhesion and invasiveness via pro-modulating inflammatory and cytokine gene αvβ3 integrin expression, impaired TLR-4 activity reduces IL-6 and IL-8 functioning, TLR-3 initiates a tumor suppressing action via inhibiting cell proliferation and survival, TLR-9 induces invasion via sex steroid hormones and estrogen-α activity
4.	Human Epidermal Growth factor Receptor 2 (HER-2)	Overexpressed in nearly 20% noninflammatory and 50% inflammatory tumors, performs a vital role in angiogenesis, lymphangiogenesis, aromatase upregulation, and antiapoptotic action, contributes in frequent discordance in Trastuzumab treatment

therapies. Quite often, there has been a concern of delivering drugs at the needed locations effectively so that the undesirable toxicity and side-effects are minimized. This is how the nanoparticles (NPs) of Au and Ag make the tumor treatment more effective and thorough, that too by carrying minute drug quantities. It is pertinent to make clear here that Au and Ag NPs are not the only nanomaterials for delivering drugs to BC cells and a number of other options such as mesoporous and core-shell configured silica NPs, functionalized graphene assemblies and others, are in wide use for delivering drugs at the needed intra-tumor locations.

9.3 Fundamental aspects of gold and silver nanoparticles: Overview of formation methods

Recurrent resistant patterns and chronic patient sensitization has compelled a need to search for modified or alternate drug delivery vehicles, amongst which NPs of Au and Ag are the mainstays. Belonging to the 11th group of periodic tables and with an atomic number of 79, Au is positioned below Ag and has a larger size, Au has its outermost electron lying in the 6s-subshell contrary to that of 5s for Ag. This distinction confers a higher chemical stability to Au NPs, wherein removal of electron from outermost orbit is much easier than Ag. Physicochemically, the formation of NPs implies the attainment of a size within (1–100) nm corresponding to a manifestation of quantum confinement in one, two, or all three dimensions, depending on the characteristic size attained. Typical attainment of nanoscale dimensions is indicated by a change in optical (a different color than the bulk state), electronic, and chemical properties, as it involves the acquiring of zerovalent ionic state from that of parent salt precursor (can vary as per the suitability of chemical reduction to be attained). Easier removal of outermost electron from the native electronic arrangement of Au imparts a higher stability to the nanoscale Au owing to which it is viciously used in functionalization and biosensing applications. Contrary to this, Ag requires a higher energy to prevail in the monovalent (Ag^+) state, due to which the more toxic Ag NPs perform better towards the antimicrobial and antibacterial applications. Multiple studies elucidate the interaction mechanism of Ag NPs with the microbes via electrostatic interaction with the negatively charged plasma membrane phospholipids [63–64]. The major advantages of Au and Ag NPs as anticancer therapeutic agents include (1) their comparatively low chemical reactivity which allows them to stay monodispersed for a longer duration, (2) easier and robust preparation methods, (3) ability to display synergistic activities enabling additive chemo and photo thermal responses, (4) easier availability of different shapes, in line with the reports claiming better drug delivery with non-spherical NPs, (5) well-understood surface plasmon resonance artefacts, enabling accurate resonance attainment to monitor even slightest changes, and (6) knack for multiple functional activities, such as simultaneous diagnosis and treatment activities.

A number of methods are available for preparing Au and Ag NPs on a laboratory scale, the fundamental mechanism of which is unanimously aimed at the reduction of di or trivalent state of metal in the salt state to the zerovalent. Literature classifies the preparation methods broadly into physical, chemical, and biological regimes, typically distinguished in terms of the unique reducing agent(s) involved. The methods conceptually deal either with a reduction of bulky state to the nanoscale dimensions (the Top-down approach) or stopping the combination of atomic scale identities gradually to the nanoscale (the Bottom-up approach). The "Top-Down" principle is usually the physicists' choice wherein a lot of waste material is generated along with a requirement of steady external energy source. Contrary to this, the "Bottom-Up" principle is a green mechanism with much lower dependence on external energy and nearly negligible waste generation. Our own multiple contributions on this topic can be referred for an idea about the different preparation methods of Au and Ag NPs [62,65]. The availability of robust and efficient chemical reducing agents makes the chemical method of NPs preparation as the robust, reproducible, and easy to implement with comparatively less requirements of the infrastructure. Although biological methods (using plant components, such as leaf extract, stem extract etc. and those of microbial enzymes) are also significantly optimized for a pH dependent extracellular and intracellular formation of metal NPs but this method suffers in terms of long-term stability. Thereby, chemical method of NPs preparation is the standard choice for modern day biomedicine experts, wherein the aggregation stability is maintained till long durations and the capping agents exercise their operational control more strongly.

Mostly preferred method for making Au NPs on the laboratory scale is the one demonstrated by Turkevich and associates [66]. The method uses trisodium citrate as the reducing agent to form the monovalent Au from the precursor auric chloride, $AuCl_4$. A number of modifications have been optimized via varying the interacting stoichiometries and durations of the interactions between the metal salt precursor and reducing agent. It is pertinent to mention here that the reducing and stabilizing agents are the two fundamental requirements of all synthesis methods but it may be possible that a particular reducing agent acts equally well in terms of guarding against aggregation. This is accomplished through the specific interaction mechanism of reducing agent with that of metal salt precursor. Fig. 9.3 depicts the mechanism of Turkevich method for Au NPs preparation which makes use of trisodium citrate as reducing as well as stabilizing agent. The method is capable of forming (1–50) nm sized Au NPs, developing arrange of shapes such as rods, stars, ellipsoidal, and spherical, via varying the metal salt precursor and reducing agent mutual stoichiometries. A smaller size attributes to a higher quantum confinement and a more vigorous variation in physical and chemical properties. Several variations of this process are well-documented in multiple literature sources, the most prominent of which is Burst-Schiffrin thermal reduction [67]. In this method, the Au precursor salt, $HAuCl_4$ is transferred from an aqueous to an organic phase like toluene using tetraoctylammonium bromide (TOAB) as a phase-transfer agent. Thereafter, $NaBH_4$ is added as reducing agent in presence of dodecanethiol which develops deep brown appearance from the yester orange, confirming the Au-NPs formation. Several other green mechanisms are in active use for Au NPs formation, such as using ionic liquids, sonochemical method, microwave irradiation, reverse micellar driven reduction, and using biomolecules such as polysaccharides, nucleic acids, and amino acids as reducing agents for the salt precursor(s).

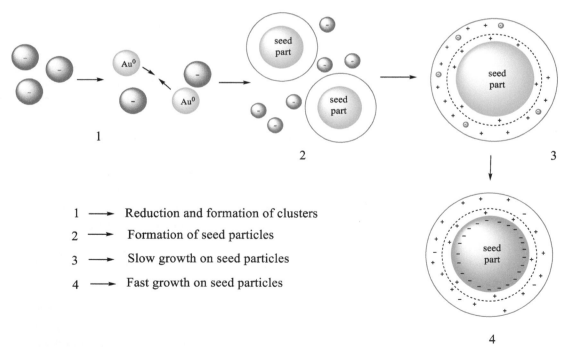

1   ⟶   Reduction and formation of clusters

2   ⟶   Formation of seed particles

3   ⟶   Slow growth on seed particles

4   ⟶   Fast growth on seed particles

Fig. 9.3: Mechanism of Turkevich method driven gold nanoparticle formation.
The method is a Bottom-Up approach and is regulated through a seeding patterned nanoparticle formation.

Table 9.2 lists the major features of these methods, which are all 'Bottom-Up" approaches and are the hands-on experiences of chemists [68]. Readers must note that besides these methods, a number of physical methods are also practiced for Au NPs formation, such as sputtering approaches, electrochemical reduction, and UV-irradiation accomplished metal salt precursor fragmentation. These methods are "Top-Down" approaches and inevitably needs a high maintenance infrastructure and that's why the use of these methods has been rather low on the laboratory scale. Similarly, a range of plant extracts and microbial species have also been used as green agents for reduction of precursor salts. These methods are indeed stead-fast and readily feasible but viciously suffer from maintaining long-term stability of the NPs, due to which these require additional inclusion of stabilizing agents. A number of literature sources could be referred for knowing details of these methods which are equipped for extra-cellular and intracellular feasibility, minimizing the environmental damage likely to be caused by heavy metal and recalcitrant rich sewage and industrial effluents [69–70].

Similar to Au, a number of formation methods are in practice for Ag NPs synthesis, such as evaporation-condensation and laser beam irradiation (physical), using reducing and capping agents like ascorbic acid, sodium citrate, $NaBH_4$, Tollens reagent, polyol process, biocompatible polymers such as polyvinyl alcohol (PVA), polyvinylpyrrolidone (PVP), and others (chemical). It should be noted that analogous to $HAuCl_4$ for Au NPs, $AgNO_3$ is the most used precursor

Table 9.2: Working principles of modified chemical methods for preparing gold nanoparticles on a laboratory scale. The methods are distinguished via varying reducing agents to reduce the corresponding metal salt precursors.

Sr. no.	Name of method	Working mechanism/distinct aspects/potential concerns
1.	Ionic liquids driven reduction of salt precursor(s)	Thermal stability, nonvolatility, and miscibility with cosolvents are the benefits, simultaneous activities of reducing and capping activities, capable of serving as templates, functionalization could form one-phased configured nanoparticles
2.	Sonochemical approach	Works through cavitation driven sono-chemolysis, involves high energy stimulus, robust, and less time consuming but may be vulnerable in long lasting stability
3.	Microwave irradiation	Green approach using the energy of microwaves, requires less time and minimal dependence on infrastructure, needs caution to guard from microwave interaction
4.	Reverse Micelle driven chemical reduction	Uses thermodynamically stable emulsions to form gold nanoparticles with controlled shape, diameter and spontaneous stability that minimizes random aggregation
5.	Biomolecule driven salt precursor reduction	Uses amino acids, lipids, carbohydrates, and proteins as reducing agents, works through van der Waal forces, London Dispersive Forces and larger surface area distributed hydrophilic-hydrophobic force gradients

for the synthesis of Ag NPs. Most of the synthesis mechanisms of Ag NPs use distinct reducing and stabilizing agents, whereby a range of particle sizes could be obtained by varying their mutual interaction stoichiometries. Table 9.3 lists the reducing and stabilizing agents along with the precursors and the optimized particle sizes of different methods used for preparing Ag NPs. Although wire and spherical morphologies are known for Ag NPs but the functional diversity of varying shapes is much more diverse in Au NPs with wires, stars, rod, sphere, and prism morphologies. The physicochemical stability of each configuration is highly vital for drug loading and studies have validated a higher drug loading and delivery suited potential of nonspherical morphologies. Due to a higher toxicity of Ag NPs, some studies have accomplished a combined use of Ag and Au NPs, using the latter as capping agent for an aggregation countering stability. It is highly befitting to know and understand the drug delivery mechanisms of Au and Ag NPs, whereby active and passive targeting of drug loaded NPs could be accomplished.

Table 9.3: A description of different synthesis methods of silver nanoparticles. The methods are distinguished *via* physical, chemical and photochemical means of accomplishing metal salt precursor reduction [68].

Source*	Method	Reducing agent	Stabilizing agent	Size (nm)
$AgNO_3$	CM	DMF		<25
$AgNO_3$	CM	$NaBH_4$	Surfactine	3–28
$AgNO_3$	CM	Tri sodium citrate	Trisodium citrate	<50
$AgNO_3$	CM	Tri sodium citrate	Trisodium citrate	30–60
$AgNO_3$	CM	Ascorbic acid		200–650
$AgNO_3$	CM	$NaBH_4$	DDA	7
$AgNO_3$	CM	Parafin	Oleylamine	10–14
$AgNO_3$	CM (Thermal)	Dextrose	PVP	22–25
$AgNO_3$	CM (Thermal)	Hydrazine		2–10
$AgNO_3$	CM (Oxidation of glucose)	Glucose	Cgluconic acid	40–80
$AgNO_3$	CM (polyol process)	Ethylene glycol	PVP	5–25
$AgNO_3$	CM (polyol process)	Ethylene glycol	PVP	50–115
$AgNO_3$	Electrochemical (polyol process)	Electrolysis: Titanium (cathode) and Platinum (anode)	PVP	11
$AgNO_3$	CM (Tollen)	m-Hydroxy benzaldehyde	SDS	15–260
Ag wires	PM	Electrical arc discharge, Water		10
$AgNO_3$	PM	Electric arc discharge	Sodium citrate	14–27
$AgNO_3$	CM (microemulsion)	Hydrazine hydrate	AOT	<2
$AgNO_3$	PR (microwave radiation)	Ethylene glycol		<2
$AgClO_4$	PR (pulse radiolysis)	Ethylene glycol	PVP	5–10
$AgNO_3$	PR (Photo reduction)	UV light		4–10
Ag_2SO_4	PR (X-ray radiation)	X-ray		2–8

* CM = chemical method, PM = physical method, PR = photochemical reduction.

9.4 Drug delivery mechanisms of nanoparticles

Delivering drugs through NPs to cancer cells is facilitated via active and passive targeting. The challenge in the overall process pertains to discrete identification of tumor cells which is further compounded by the concomitant risk of arresting the metastasis. The intent of drug consumption pertains to induce a characteristic change in the biological response subject to its known chemical structure. Typical hallmarks encountered by a trafficked drug in its passage to tumor cells comprise the troubled extracellular environment, the overexpressed cell surface receptor proteins, unusually faster growing pace and loss of cell-cell communication. Apart from this, the ability of forming new blood vessels for circulation of nutrients, named as angiogenesis is the significant feature of tumor cells for their distinction from normal cells. Since the challenge lies in adequate identification of tumor cells, the mechanisms are therefore recognized via site specific activities, termed as active and passive targeting. Another aspect worth noting is that there is an exclusive mention of these mechanisms with respect to tumor cells, primarily because interventions like NPs as drug delivery carriers are screened only in extreme cases or in terminal situations. Readers are suggested to refer 2014 contributions of Bertrand and accomplices (from Advanced Drug Delivery Reviews) and 2018 of ours (on Au and Ag NPs treated lung and breast cancers) for all aspects in the discussion on active and passive drug delivery [62,71].

9.4.1 Active targeting

Active targeting is attributed to identification of a nanoparticle surface ligand by the tumor cell specific over-expressed surface receptor(s). This approach involves considerable risk as any random misfired delivery is likely to harm the normal cells and the damage would be further manifolded due to functional cell-cell signaling in normal cells. Enhanced complexity of this delivery mechanism necessitates that the ligand specific antigen (or receptor) is not merely present on the tumor cells but also remains accessible to the administered NPs (Fig. 9.4). Table 9.4 provides some fundamental information about the various ligands used for targeted drug delivery using NPs along with their specific limitations, unique aspects, and current progress of the work [72–75]. Another essential factor regulating the success of active targeting is the chance inadequate targeted receptor localization and expression which could weaken the ligand binding. Frequently used ligands in active targeting of NPs include antibodies, peptides, nucleic acids, sugars, and a number of erstwhile small molecules such as vitamins and others. Likewise, the target moieties can be proteins, sugars, or lipids prevailing on the tumor cell surface.

Obtaining success in the active targeting is a multitude of multiple constraints aiming for multiple ligands conjugated NP moieties to surface receptors. Thereby, the target specificity and delivering efficacy emerge as decisive factors regulating the drug delivery efficacy. A major insolence of this mode pertains to a necessity of delivered NPs prevalence in the vicinity of

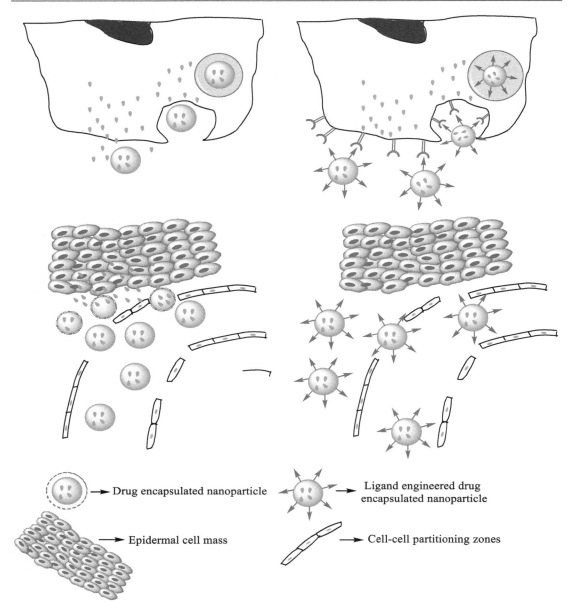

Fig. 9.4: Active and passive drug targeting mechanisms of nanoparticles.
The strategies elucidate a direct interaction with tumor (or injured) cell surface receptor (in active mode) and the intracellular entry via physicochemical drifting in passive mode.

targeted cell surface antigen for adequate recognition and interaction. As a result, the administered NPs could persist longer in the bloodstream and enable a homogenous biodistribution. This aspect is acutely affected by the restricted blood supply to the tumor cells because of which the NP affinity towards the targeted antigen cannot supersede the natural clearance

Table 9.4: The various ligands used for drug delivery via active targeting of drugs [72–75].

Name of ligand	Distinctive attributes/few examples	Performance affecting features
Monoclonal antibodies (Mabs)	Neutralizing molecules generated against specific antigen, reduce the damage to normal cells by their specificity of action, Trastuzumab, Pertuzumab and Bevacizumab are the main Mabs used for BC treatment, Trastuzumab has been more preferred for combinatorial use	Location of Mab recognition site on cell surface, blood stability of drug-Ab conjugate, facilitate drug release on interaction with cell, localization extent of Ab at the target site, host-toxicity of drug-Mab conjugate, biodistribution of drug-Mab conjugate relative to parent Ab
Aptamers	Less than 15 kDa single stranded DNA/RNA oligoucleotide sequences, moderate immunogenicity, high stability with improved biodistribution, can be produced at a low cost, easy to modify. S6 (for SK-BR-3 cells), XLX-1-A (for MDA-MB-231 cells), HBS (for HER2), SYL3c (for EpCAM)	Adequate modulation of immunogenicity, risk of unexpected toxicities due to capability of interaction with varied shapes of nanomaterials
Proteins and Peptide based targeting molecules	Nonantigenic, structurally simple, easy to engineer, numerous possibilities of application. Doxorubicin bound nonapeptide CD-CRGDCFC improved survival in mice having human MDA-MB-435 breast carcinomas, accumulation of cyclic RGD analog conjugated to fluorescent bacteriochlorophyll in tumor necrotic domain in MDA-MB-231-RFP bearing mice	Important to optimize immunogenicity by controlling interacting stoichiometry with antibodies, requires extreme caution in scaling up manufacture, high concentration of protein mandatory to confer a nonimmunogenic behavior
Small molecules	Low molecular weight organic compound capable of regulating a biological process, size in the order of 1 nm. Exhibit multiple biological functions: cell signaling, drugs etc. Can be natural (secondary metabolites) or artificial (antiviral drugs). Several combinations of polysaccharide and polymeric nanomaterials (gelatin, starch, chitosan, collagen) can be used in loaded configurations with chemotherapeutic drugs.	Therapeutic utility stems from the ability to inhibit a specific function of a protein or protein-protein interactions. So, it is important to regulate the action for abrupt responses and not to normal ones. Important to regulate the binding kinetics and ensure noncovalent binding with the drugs, so that no structural impairment prevails

mechanisms. The above constraints necessitate configuring NP architecture via opsonization to escape immune system detection with a long-lasting blood circulation time. These constraints culminate in the ultimate dependence on the facilitated transport (through Enhanced Permeation and Retention, EPR) and the relative inability to attenuate the NPs biodistribution. Several attempts have reported enhanced NP tumor internalization through which the corresponding therapeutic payload is manifolded. It is pertinent to mention here that EPR involves the physicochemically active intracellular drifting of drug loaded cargos and holds a greater relevance than passive targeting. Permeation implies entry within the cells while retention refers to the residence of a drug inside the tumor cell.

Modulating therapeutic efficacy of actively targeted drug loaded NPs is aimed to mediate an adequate endosomal escape or via encapsulating drugs which are impervious to the endosomal/lysosomal environment. For instance, anti-HER2 targeting moieties on liposomal surface are reported for increased nanoparticle uptake in HER2-expressing BC cells. Conjugation of ligand on NP surface significantly alters its native properties owing to which nanoparticle size (PS), geometry, surface charge, hydrophobicity, and composition are the key factors affecting the stability. Adequate binding forces (BF) with implicit redox environment and localized pH are the crucial factors affecting the in vivo performance of an active targeted delivery configuration of NPs. The net aim is to minimize the structural attenuation of delivered drugs along with its homogenous distribution over varied time durations. For instance, the significant opposition in the nuclease degradation for nucleic acid strands immobilized on NPs, necessitates a modulation of redox balance in the physiological environment of drug loaded NPs. A proper ligand selection in this regard with a counteraction of such opposition and null undue impact on loaded drugs is imperative and is the need of hour. Similarly, a too high surface hydrophobicity on the ligand surface with an excessive accumulation of one kind of charges is a limiting step as it could hamper the drug release when needed. So, the configuration of drug loaded NPs should be self-responsive to the localized intracellular conditions.

9.4.2 Passive targeting

Contrary to active mode, the passive drug targeting using NPs relies on the concentration gradient driven tumor cell internalization of certain macromolecules, first witnessed nearly 40 years ago. This was optimized through a therapeutic macromolecule accumulation inside a tumor cell in course of poly(Styrene-co-Maleic Acid)-Neo Carzino Statin (SMANCS), a 16 kDa polymer conjugate noncovalently supported binding to albumin, resulting in ~80 kDa molecular weight. Concurrent studies on the SMANCS preferential accumulation, deciphered its association with imperfect tumor blood vessels, and poor lymphatic tissue drainage. This scenario resulted in the coining of enhanced permeation and retention (EPR) terminology. The effect serves as a decisive link in the passive drug delivery, making it inevitable to understand the EPR dynamics and its consequences *vis-à-vis* tumor microenvironment.

The rationale for EPR is mediated via growth of a solid tumor to a threshold size after which no oxygen is furnished to it for further proliferation. As a result, the cells gradually perish and spontaneously secrete growth factors, forming new blood vessels from surrounding capillaries. This fresh blood vessel formation is termed as angiogenesis, characterized by a discontinuous epithelium devoid of basal membrane unlike the normal vascular structures. As a consequence, serious changes in normal capillary physiology leading to an attainment of >2000 nm sizes are observed, depending on tumor type, site, and its microenvironment. Such a typical and sporadic vascular organization cannot resist the transport of blood components to the tumor interstitium. This process comprises enhanced permeation domain of EPR effect. Unlike normal tissues, tumors are characterized by defective lymphatic functions, thereby forbidding the periodic draining and renewal of interstitial liquids, and the recovery of extravasated solutes and colloids with respect to circulation. While smaller molecules (<10 nm) can be reabsorbed through back diffusion to blood, the larger sized macromolecular cannot be trafficked. Ultimately, the macromolecules remain uncleared and get accumulated within the tumor interstitium. This comprises the retention part of EPR effect. On a developmental note, it is widely recognized that lymphatic functioning is nonuniform throughout the cancerous cells and the blood vessels in bulk are under higher mechanical stress. As a result, functional loss within the intratumoral regions is significantly higher than marginal cells. Thus, the dissimilar lymphatic function inside the tumor is a key factor for intracellular NP accrual. Despite a sound mechanistic understanding conferred by EPR effect, the restriction of its trials and investigations on humans has hindered its risk-free execution. Major factor contributing to this is the dissimilarity of optimal parameters needed for maximum distribution and retention in humans and animals. Apart from this, testing NPs administration in the animals at a preclinical stage is devoid of much predictive significance in humans chiefly due to a vulnerable PCPs in vivo (humans) unlike frequently studied mice models. To integrate the limitations, the localized conditions make it challenging to compare the humans and animal clinical data, such as that of PEGylated liposomal EPR, noted rather consistently in animals having high tumor to blood stoichiometry. Analogous generalizations on humans are not feasible, who have sarcomas as only cancers with a favored accumulation within liposomes. Similar to active mode, the various in vivo factors affecting the success of passive drug delivery are listed in Table 9.5, of which shape, surface charge, and hydrophobicity and the physicochemical environment surrounding tumors are the major concerns.

Table 9.5: Summary of various in vivo factors regulating the success in passive drug delivery approach.

Name of factor	Regulatory mechanism
Physicochemical properties	Noncovalent drug-carrier interactions, null use of in vivo energy, spontaneous release of drugs, optimize through in vitro analysis confirming adequate distribution.
Sizes/dimensions of drug loaded NPs cargos	Must be low than the diameter of blood vessels so that needed transport and trafficking happens smoothly. A larger size is more likely to result in aggregation and toxic responses.
Surface charge of drug loaded NPs complexes	Preferable to have a cationic sensitivity which facilitates an intracellular entry but a too high positive charge can immobilize the drug-carrier complex to cell periphery. Similarly, a negative surface polarity may result in random, nonspecific toxic response
Shape of the NPs used as drug carrier	In general, nonspherical morphologies are better suited for an adequate threshold of delivered drug content, affects the intracellular uptake by the targeted cells like those of nanorods
Tumor microenvironment	Redox status and aggravated signaling of cell-cell cross pathways (in tumors) may resist the passage of drugs into the tumor interior, indirectly enhances the toxicity and risk to normal and healthy cells

9.5 Recent attempts of gold and silver nanoparticles treated breast cancers

This section discusses select studies of recent past which used Au and Ag NPs for BC treatment. The major performance determining criteria in these attempts are characteristic shape and size of used NPs, nature of their capping agent, method of functionalization, and the stage of the tumor being treated. The NPs capped by chemical reducing agents retain their nanoscale dimensions for longer periods and are hence more stable towards a random aggregation. Similarly, functionalization using graphene and its derivatives is witnessing significant interest in recent times due to their selectivity of targeting tumor cells. A number of biocompatible polymers such as polyvinyl acetate (PVA), glutaraldehyde, polyethylene glycol, polycapralactone, chitosan is being used as stabilizers due to their green essence. The salient observations of past 3-year studies are discussed in subsequent paragraphs.

9.5.1 Gold nanoparticles for breast cancer treatment

The year 2018 witnessed several splendid attempts towards exploring the Au NPs BC elimination, with the very first study using photon and thermal sensitive Au NPs. The study reported by Calavia and associates comprises used 4 nm Au NPs (prepared using sodium borohydride as reducing agent), followed by PEG functionalization having varying carbon chain (CC) length connecting PEG to the Au comprising core. The chain comprised of either three (C3Pc) or 11 (C11Pc) carbons with the latter exhibiting a greater fluorescence emission while C3Pc did so in its bound state to the Au NPs surface despite the expected C3Pc fluorescence quenching due to a shorter CC. Apart from this, the C3Pc NP conjugate caused significant singlet oxygen generation which enabled an extraordinary photodynamic efficacy for human BC treatment [76].

Subsequent 2018 attempt demonstrated doxorubicin delivery via encapsulation on solid and hollow Au NPs adsorbed over the thermosensitive liposomes (TLs). The investigators observed that solid and hollow morphologies of NPs served as *nanoscale switches* to trigger the doxorubicin release within the BC cells. The NPs were screened for their photo-chemo-therapeutic efficacy, with hollow NPs being comparatively better for the doxorubicin uptake as well as lower systemic toxicity than solid ones. The doxorubicin loaded hollow and spherical NPs were 190 nm in diameter and had −29 mV ζ-potential. The hollow NPs revealed a nearly eight fold high anticancer efficacy compare to solid NPs, *through greater hypothermic capability* which transformed the incident NIR light into heat at the targeted site and induced the doxorubicin release [77].

Yet another noteworthy 2018 attempt used Au and Ag NPs prepared from *Phoenix dactylifera* pollen extract, followed by structural characterization and stability quantification of the corresponding phyto-constituents [78]. The as prepared NPs were subsequently screened for their cytotoxicity on MCF-7 BC cell line, where fluorescent microscopy along with altered p53 (pro-apoptotic) and Bcl_2 (anti-apoptotic) proteins, predicted dosage dependent cell-death. The 2.57 folds enhanced p53 and 1.64 folds reduced Bcl-2 protein expressions were witnessed for Au NPs whereas Ag NPs were less effective with 1.40 folds enhanced p53 and 1.08 folds reduced Bcl-2 expressions. Of note, p53 is pro-apoptotic protein whose transcriptional activation is associated with an apoptotic induction. Furthermore, this protein also modulates the expression of other apoptosis associated proteins such as Fas and Bax. Similarly, increased Bcl-2 functioning reduced the mitochondrial membrane permeability to release pro-apoptotic molecules and restrain the intrinsic mitochondria regulated cell death. So, both Au and Ag NPs caused cancer cell death via modulating the apoptosis associated p53 and Bcl-2 genes.

Another significant 2018 attempt used tumor cell derived extracellular vesicles (TDEV) (encouraged by their natural formation and cargo delivery abilities) constituted nanoplatform for multimodal miRNA delivery, photothermal therapy and magnetic resonance imaging (MRI) of 4T1 BC cells. The study reported from Stanford University featured in ACS Nano, loaded anti-miR-21 to the carrier vehicle which blocked the function of overexpressed endogenous oncogenic miR-21 in 4T1 cells. The investigators designed Cy5-anti-miR-21-loaded TDEVs using (liver cells) Hepg2 and SKB3 (for BC) cell lines, whereby significant homologous and heterologous transfection along with intracellular Cy5-anti-miR delivery. Monitoring the drug transport activity of TDEV loaded carriers in BC cells, the investigators noticed that a TDEV driven anti-miR-21 delivery decreased the doxorubicin resistance in BC cells, alongside triplicated cell death efficacy compared to an unaided doxorubicin delivery. Encouraged by these attributes of TDEVs, the investigators tried to use them as functionalization agents for the Au coated iron oxide NPs (Au coated FeO NPs). The TDEV functionalized Au coated FeO NPs exhibited splendid T2 contrast during the in vitro conducted MRI along with significant photothermal effect in 4T1 cells. It was noticed that TDEVs functioned well in combinations with Au coated FeO NPs as also demonstrated by tumor targeting and intracellular localization ability of TDEV-GIONs using MRI. So, this study established a compatibility of TDEV for enhanced tumor cell localization activity in combination with Au coated FeO NPs. A pictorial description of this study is depicted in Fig. 9.5, projecting a suitability of fabricating TDEVs using other organs and NPs combinations for a minimized intracellular resistance [79].

Moving to 2019, the first significant study used a nanoconjugate (NCg) (consisting of curcumin loaded Au-PVP (polyvinylpyrrolidone) NPs) conjugated with folic acid. Following preparation, the NCg was characterized by UV-Vis, FT-IR spectroscopy, X-ray powder diffraction, and thermogravimetic analysis. The in vitro anticancer and antimigratory attributes of NCg were studied using MTT and wound migration assays while the in vivo anticancer ability was studied in a preclinical BC orthotopic mouse model. The NCgs size and charge played a key role in curcumin release, analyzed using layer by layer assembly. Nearly 80% curcumin was released in an acidic pH and the as prepared NCg did not exhibit any aggregation on being incubated with human serum besides being *mimicked with intrinsic peroxidase*

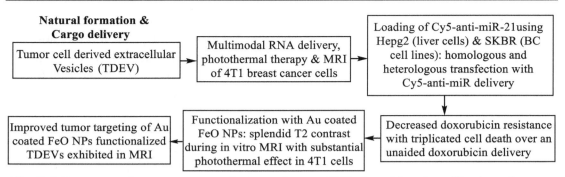

Fig. 9.5: Summary of steps corresponding to reference [79], describing the utilization of tumor cell derived extracellular vesicles for site specific doxorubicin delivery, in vitro MRI contrast agents and photothermal effect in 4T1 breast cancer cells.

behaviour in the presence of 3,3',5,5'-tetramethylbenzidine substrate. Results of anticancer studies of these NCgs using the MTT assay revealed an enhanced activity corresponding to lower doses in estrogen/progesterone receptor negative cells compared to ER/PR positive cells. Alongwith this, in vivo inspection of prepared NCgs inhibited the cell migration with high antitumor ability. The efficacy was reported specifically on tumor cells and no cytotoxicity was observed corresponding to the tested concentrations in normal human breast epithelial and mouse fibroblast cells. Thereby, PVP and folic acid aided in the tumor cell specific activity of curcumin which itself is well-known for its selective actions towards tumor cells. the observations create an interest towards using PVP as a support matrix for the delivery of many other polyphenols like curcumin [80].

The subsequent 2019 attempt screened the in vitro chemotherapeutic efficacy of ZnO NPs delivered via mesoporous silica nanolayer (MSN) to the drug sensitive (MCF-7, ER positive, CAL51, triple negative), and drug resistant (MCF-7TX, CALDOX) BC cells [81]. The ZnO NPs conjugated MSNs were coated on Au nanostars (shape diversity of Au NPs) for testing the NIR-II ranged imaging capabilities. Inspection using electron and confocal microscopy inferred an accumulation of MSN-ZnO-Au nanostars close to the plasma membrane and a parallel internalization by the BC cells. High resolution microscopy analysis of the MSN loaded ZnO NPs inferred a degradation of MSN coating outside the cells, releasing the ZnO NPs for interaction with plasma membranes. Together, the ZnO NPs loaded over the MSN reduced the viability of all tumor cell lines, of which the CAL51/CALDOX cells were more susceptible. Monitoring the targeted activity, the as prepared MSN loaded ZnO NPs were conjugated with Ab to Frizzled-7 (overexpressed receptor in several BC cells). This conjugation was made using disulphide linkage (-S-S-) which is easily cleaved by a high glutathione concentration inside the tumor cells. As a result of the cleavage, the Zn^{+2} were released into the cytoplasm. It was noticed that the Frizzled-7 targeting enhanced the MSN-ZnO-AuNs toxicity by more than three times, towards the drug resistant MCF-7 cell line (which exhibited a highest Frizzled-7 expression). Overall, the study illustrated MSN loaded ZnO NPs as promising anticancer agents to treat triple negative and drug resistant BCs besides highlighting a target specific ability of Frizzled-7. Fig. 9.6 represents a pictorial view of the steps of this study for an easier understanding.

The year 2020 continued the interest in Au-NPs aided drug delivery to BC cells, with the first attempt involving the use of Turkevich method prepared Au-NPs for a combinatorial delivery of doxorubicin and polo-like kinase (PLK1, a siRNA) to the SKBR3 cell line. The Au-NPs were coated with polyethyleneimine (PEI) which aided in PLK1 surface assembly. Doxorubicin was loaded on NPs via a pH sensitive linker, having a thiol group at one terminal, which aided in sustained drug release. Therapeutic activity of doxorubicin-PLK1 combine was assessed in 2D and 3D cultured systems, where reduced IC_{50} extent (0.06 ± 0.002) showed a synergistic response of doxorubicin and PLK1 (siRNA) gene delivery. The response was compared with unaided doxorubicin (0.805 ± 0.29), individual doxorubicin and PLOK1 delivery with Au NPs (0.122 ± 0.02 and 0.512 ± 0.3, respectively). The PEI stabilized Au NPs exhibited a core-shell morphology with *Au-NPs* forming the core and PEI conjugated to surface through a PEG linker (anticipated as biocompatible and non-toxic). Thiol bonds served as stability augmenting factor in the combined delivery which conjugated the doxorubicin simultaneously with Au-NPs. The optimized vehicle configuration also offered a provision to conjugate other drugs also to the Au-NPs surface for developing a multi-drug delivery response. The inclusion of Au-NPs as core in the vehicle always provides an option to incorporate photothermal therapy besides the imaging component for a higher localized toxicity and improved diagnosis of the tumor progress [82].

Subsequent 2020 comprises a slightly different than conventional attempt, using green nanotechnology as an initiative, whereby *mango peal constituent phytochemicals* were used as reducing agents to form Au-NPs. Briefly, 30 mg of dry mango peel powder was added to 6 mL of double distilled water in a 20 mL vial, followed by 10 minutes stirring at RT till a homogenous suspension is obtained. Thereafter, 100 µL, 0.1 M $NaAuCl_4$ solution was added following which its color changed to ruby-red within 2 minutes. Although polyphenols are frequently recommended and used as reducing agents for NPs capping activities but the mango peal phytochemicals (major ingredient is Mangiferin) create an interest in natural essence of the study. Retrieved Au-NPs were termed as Nano Swarna Bhasma (NSB drug), which was tested for (1) anticancer attributes in breast tumor cells, (2) preclinical therapeutic efficacy studies in BC bearing severe combined immunodeficiency syndrome (SCID) infected mice via oral delivery, and (3) first ever analysis of clinical translation from mice to human BC patients (using pilot human clinical trials. The study comprised a part of Ayurveda, Yoga, and Naturopathy, Unani, Sidha, and Homeopathy (AYUSH) initiative by the Government of India. Preclinical in vitro (MDA-MB-231 BC cell line) and in vivo investigations in breast tumor bearing mice unanimously established the NSB (the Ayurvedically fabricated Au-NPs) as being highly effective in controlling the BC growth, exhibiting a dose-dependent response.

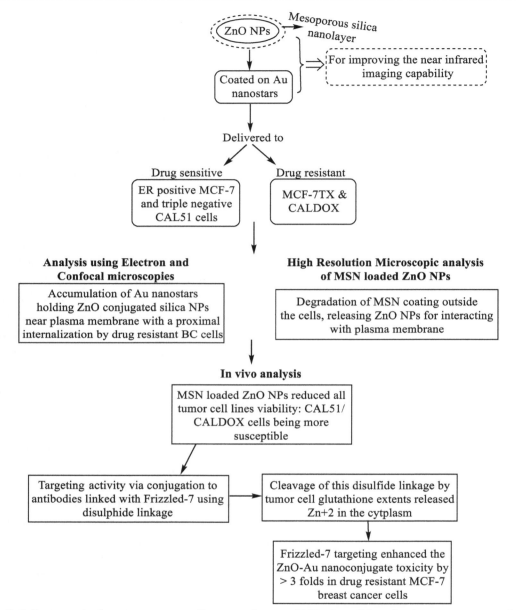

Fig. 9.6: Summary of steps corresponding to reference [81], describing the Au nanostar coating driven site-specific actions of silica coated ZnO NPs.

The engineered NPs were efficiently internalized in drug sensitive (MCF-7 cells) and drug resistant (MCF-7 TX and CALDOX) cells. Degradation of silica coating outside the cells facilitated the interaction of ZnO NPs with plasma membrane lipids.

This was the reason to pursue clinical trials in humans, wherein, patients treated with NSB drug capsules along the standard of care treatment (Arm B) expressed 100% clinical benefits compared to patients administered treatment with Arm A. This indicated significant clinical benefits of NSB drug in adjuvant therapy which could improve the therapeutic response of multiple chemotherapeutic drugs [83].

Another 2020 milestone attempt using Au-NPs as drug delivery carrier for BC treatment used Au-NPs to facilitate a significant uptake of hesperidin, a flavonoid glycoside being listed for its manifold therapeutic activities. The major hurdles in the action of this flavonoid include its below par aq solubility, which inhibits its trafficking to the desired tumor site. The study from Iraq and Saudi Arabia joint efforts, involved a chemical synthesis driven hesperidin (a flavanone glycoside prevailing in citrus fruits) loading on the Au-NPs. Manifold characterizations, including UV-V is spectroscopy, FT-IR, XRD, FESEM, TEM, EDX, and ζ-potential measurements, confirmed the chemical synthesis of hesperidin conjugated Au NPs (Hsp-Au NPs). The cytotoxic effect of Hsp-Au NPs on human BC cell line (MDA-MB-231) was ascertained using MTT and crystal violet assays. Analysis revealed a considerable decrease in the cell population along with an inhibition in the growth of treated cells when compared to normal human breast epithelial cell line (HBL-100).

Determination of apoptosis was made using Fluorescence microscope, using acridine orange propidium iodide dual staining assay. The in vivo study was conducted in mice to ascertain the Hsp-Au NPs toxicity, to ascertain which, the expressions of hepatic and renal functionality markers were assessed. No significant distinctions could be found for the tested indicators, complemented by null apparent damages shown in the histological images of liver, spleen, lung, and kidney, subsequent to Hsp-Au NPs treatment. The Hsp loaded Au NPs were found to ameliorate the functional activity of macrophages against Ehrlich ascites tumor cells bearing mice, complemented by the inhibited IL-1β, IL-6, and TNF secretions post Hsp-Au NPs treatment [84]. For an easier understanding, Fig. 9.7 summarizes the various aspects of this study. So, this study comprised another attempt focusing Au NPs as efficient carriers of natural polyphenols, which improved their structural expression along with an enhanced specificity.

Towards the end of 2020, another study similar to the previous attempt, reported improved anticancer potential of cisplatin (CDDP), through its Au-NPs mediated delivery to human breast cancer (SKBR3) and normal (MCF-10A) cells. The attempt used L-lysine as linker to conjugate cisplatin with Au NPs. The as developed nanodrug (Au-NPs-Lys-CDDP) was rigorously characterized using UV-V is spectroscopy, electron force microscopy, particle size, and ζ-potential measurements. Apoptosis and morphology variations were screened using Flow Cytometry and acridine orange/ethidium bromide (AO/EdBr) staining respectively. The CDDP loaded Au-NPs through lysine revealed 85 nm as particle size alongside a −25 mV

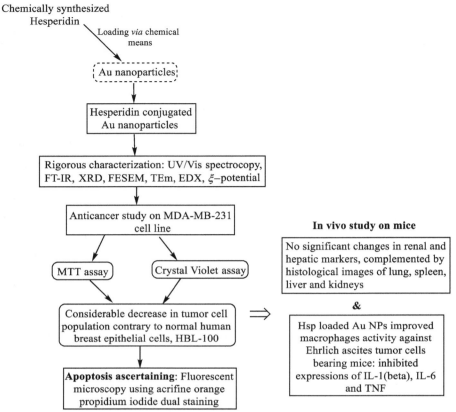

Fig. 9.7: Pictorial description of reference [84], wherein Au nanoparticles conjugated with hesperidin engineered the drug trafficking in MDA-MB-231 TNBC cell line.
The flavanone loaded with Au nanoparticles enhanced the apoptosis of tumor cells with successful replication in mice models.

ζ-potential, inferring an efficient uptake by tumor cells. The CDDP loaded Au-NPs were distinguished from Au-NPs by a strong absorption shift in the 525 nm region, besides eightfold lower IC_{50}, against the SKBR3 cell lines. The CDDP delivery on conjugation with Au-NPs caused a substantial enhancement in the SKBR3 apoptosis, with minimal cytotoxic effects on normal cells. So, this study is yet another evidence to prove that Au-NPs could efficiently function as sustained carriers of chemotherapeutic drugs, with reduced dosage, a sustained release profile, and a localized toxicity expression [85].

9.5.2 Silver nanoparticles for breast cancer treatment

Unlike Au, Ag NPs use BC treatment more focused towards utilization of a higher Ag^+ intracellular toxicity. It is because of this reason that contrary to the most Au-NPs studies using Au NPs as drug carriers, those with Ag NPs either use them in the native form as tumor treatment agents or just to enhance the cytotoxicity of a moderately toxic drug such as several polyphenols. Therefore, the exact intent of Au and Ag NPs treated BC is distinguished via carrier abilities of the former and the drug like functioning of Ag NPs. A careful look at the pathophysiology of BC cells reveals significant changes in sugar expression patterns on cell surfaces which play decisive roles in tumor progression and metastases. In general, two most common molecules used to target anticancer drugs at the designated locations in BC cell are, (1) albumin and (2) trastuzumab. While albumin is a protein recognized for its caveolae facilitated endocytosis, trastuzumab is a *humanized monoclonal antibody* binding specifically to HER2 receptor. Apart from these, the substantial drug targeting molecules in BC treatment include agglutinins, glycoproteins which bind specifically to sugars. Therefore, carbohydrates having an abundant expression on cell surfaces are often the targets of anticancer drugs due to their implicit controls on cell-cell and cell-extracellular matrix interactions.

Beginning from select 2019 studies, the first from United States was inspired by the ability of Ag NPs to treat aggressive cancers, based on which the investigators attempted to correlate the physicochemical attributes of Ag NPs with their biological activities. As a result, the study involved a thorough structure-function analysis, mechanistic, safety, and efficacy evaluation *vis-à-vis* Ag NPs ability to treat TNBCs. Of note, TNBCs are the most dreaded tumors in terms of their mortality contribution amongst the BC deaths. The findings revealed a cytotoxicity of Ag NPs to TNBC cells which was mediated independent of their sizes, shapes and used stabilizing agents. However, this toxicity is selective for tumor cells and towards benign breast epithelial cells, no significant toxicity was noticed. Analysis of cytotoxicity expressions in TNBCs and nonmalignant breast epithelial cells revealed their similar sensitivity, inferring a necessity for nanoparticle formulation for a TNBC specific toxicity induction. It was observed that despite being internalized by TNBC and nonmalignant BC cells, the Ag NPs undergo swift degradation only within TNBC cells. The distinct response of Ag NPs within the TNBC cells is due to the depleted intracellular antioxidants (such as glutathione) that causes endoplasmic reticulum (ER) stress. Such degradation is however not observed in nonmalignant breast epithelial cells. Separate studies revealed extensive DNA damage in 3D TNBC tumor nodules (in an in vitro setting) on being exposed to Ag NPs but no alterations were observed in the normal architecture of breast acini in 3D cell culture, nor any sort of DNA damage or apoptotic induction. The study also revealed an effectiveness of systematically administered Ag NPs at nontoxic dosages towards a reduced growth of TNBC tumor xenografts in mice. Thereby, on the whole, this study elucidates the TNBC treatment efficacy of Ag NPs as safe and specific moieties [86].

Another significant 2019 attempt matched the yester attempt in terms of ER stress interception of Ag NPs within the BC cells. Citing the repeated development of multidrug resistance (MDR) in BC cells, the study from Hungary observed size dependent cellular activities of Ag NPs towards modulated P-glycoprotein interception and the counteracting of MDR in the BC cells. Evaluating the antitumor efficacy of 5 and 75 nm Ag NPs in the TNBC cells, the investigators noted that 75 nm Ag NPs were more pronounced in decreasing the P-glycoprotein (Pgp) efflux activity which reduced the resistant response towards doxorubicin. Further, it was observed that these activities of 75 nm Ag NPs were due to depleted ER calcium levels which resulted in ER stress and decreased a plasma membrane Pgp positioning. Thereby, this study provided a mechanism to reduce the MDR through enhanced ER stress, which can be extended to other chemotherapeutic drugs too [87].

The year 2020 further witnessed intensified attempts using Ag NPs for BC treatment, wherein Ag NPs individually and with other chemotherapeutic agents, have induced sustained toxicity to moderate or eliminate tumorigenic activities. The first of these attempts comprised a collaborative attempt from Saudi Arabia and Egypt, which examined the effect of silver citrate NPs (AgNPs-CIT) on NF-κB activation alongside tumor necrosis factor (TNFα) mRNA/protein expressions in the phorbol myristate acetate (PMA) stimulated MCF-7 human BC cell line. It is pertinent to mention here that NF-κB is the prominent transcription factor responsible for the activation of several inflammatory mediators (including TNFα), having a lethal association with the tumor onset. The Ag NPs used herein were prepared using citrate reduction method and characterized for spherical morphology having (19.2 ± 0.1 to 277 ± 0.12) nm particle sizes and (−9.99 ± 0.8 to −34.63 ± 0.1) mV surface charges. The prepared Ag NPs exhibited a UV absorption wavelength within (381–452) nm range, produced >40% DPPH radical scavenging, were nontoxic to MCF-7 BC cells besides inhibiting PMA-induced NF-κB, p65 activation along with TNFα mRNA/protein expressions. Analysis revealed significantly inhibited TNFα expression upon being administered with AgNPs-CIT through suppressed NF-κB signaling in simulated BC cells. The study therefore explained the anticancer and anti-inflammatory attributes of silver citrate NPs. The results could be extended to other NPs through modulated TNFα activities and it would be a paradigm shift if Ag NPs prepared using plants, micro-organisms, and physical methods (such as electrodeposition, laser ablation) are compared for their TNFα driven anticancer response for tumors other than those of breast [88].

The second rigorous attempt is a study from China wherein nonspecific and below par tumor sensitization ability of Au NPs (due to tumor radio-resistance and nontarget tissue irradiation) was improved using hyaluronic acid modified Au-Ag alloy NPs (Au-Ag@HA NPs). The Au-Ag@HA NPs were prepared through $AgNO_3$ and $HAuCl_4$ co-reduction (using trisodium citrate as reducing agent) followed by HA surface modification. The intent of conferring HA modification to Ag-Au alloy NPs was to improve their preferential targeting to 4T1 BC cells that over-express the CD44 receptor. The introduction of Ag in the Au NPs imparted superior multienzyme like activities for efficient tumor catalytic therapy. Monitoring the antitumor activity of Au-Ag@HA NPs, investigators noted the ionizing radiation and peroxidase like activity which together boosted the generation of ˙OH radicals along with toxic Ag^+ release at the tumor sites. Overall, the results summarized a mechanism for improved nanozyme and

Ag^+ combined therapy against cancer and could advance the development of multifunctional nanoplatforms for synergistic antitumor activities. A role of stoichiometry optimization is important which keeps the monodispersed state of NPs intact [89].

Yet another significant 2020 attempt is a study from the collaborative attempts of Egypt and Saudi Arabia, wherein cefotaxime (a drug) conjugated Ag NPs (Cf-AgNPs) were synthesized to counter the MDR bacterial infections. The study analyzed the cell death status of Cf-AgNPs via apoptotic induction in human RPE-1 normal and MCF-7 BC cells. The exact purpose was to screen the antimicrobial efficacy of cefotaxime conjugated Ag NPs against the cefotaxime resistant *E. coli* and MRSA strains. Characterization of as prepared Ag NPs in neat state and cefotaxime conjugated states revealed a spherical morphology with (7.42–18.3) and (8.48–25.3) nm as particle sizes.

Compared to neat Ag NPs and pure antibiotic, the Cf-AgNPs were strongly antimicrobial, with an MIC ranging between (3 and 8) $\mu g \cdot mL^{-1}$ against *E. coli* and *MRSA* bacterial stains. Apart from this, the Cf-AgNPs exhibited no cytotoxic effect on normal cells even at 12 $\mu g \cdot mL^{-1}$ till 24 hours. The IC_{50} for the Ag NPs in neat state and of Cf-AgNPs was computed as 12 $\mu g \cdot mL^{-1}$ towards human RPE-1 normal and human MCF-7 BC cell lines. To counter the inferences of apoptosis mediated cell death, a monitoring of pro-apoptotic genes (p53, p21, and Bax) and anti-apoptotic genes (Bcl-2) was done, which revealed upregulated and suppressed activities (Fig. 9.8). These distinctions were noticed on 48 hours treatment of NPs at 24 $\mu g \cdot mL^{-1}$ dosage, which was subsequently chosen as most effective concentration. Thus, this study reported a mechanism using Ag NPs, to revive the anticancer activity of old and unresponsive cefotaxime, which were subsequently expressed as cefotaxime-CS-Ag NPs [90]. Similar trials should be extended for other metallic NPs and also for their possible combinations with individually non-suited chemotherapeutic drugs.

Another milestone attempt is a study from Turkey, wherein anticancer activities of *Cuminum cyminum L.* (Cumin) seed extract, chemically Cumin seeds) and biologically (using synthesized Ag NPs were evaluated on human breast adenocarcinoma cell line (MCF-7) and human breast adenocarcinoma metastatic cell line (AU565). The as synthesized NPs were thoroughly characterized for their nanoscale attributes using familiar spectroscopic techniques, dispersion profile assessment (using DLS) and morphology assessment using SEM. Cytotoxic and anticancer activities of chemically and biologically prepared Ag NPs were screened using MTT assay. It was noticed that biologically prepared Ag NPs (IC_{50} = 1.25 $\mu g \cdot mL^{-1}$) were less toxic against J774 macrophage cells compared to chemically prepared NPs (IC_{50} = 0.75 $\mu g \cdot mL^{-1}$). The biologically prepared Ag NPs also caused significant inhibition on human BC cells at non-toxic concentrations of 0.25 and 0.5 $\mu g \cdot mL^{-1}$. However, at elevated concentrations, the lethal effects of chemically prepared Ag NPs on BC cells were significantly higher than biologically prepared Ag NPs. Monitoring the antitumor activities of cumin extract, it was found that it killed half of the MCF-7 cells but did not show any inhibitory effect on AU565 cells. Contrary to this, chemically prepared Ag NPs caused 95% and 97% inhibitions in the MCF-7

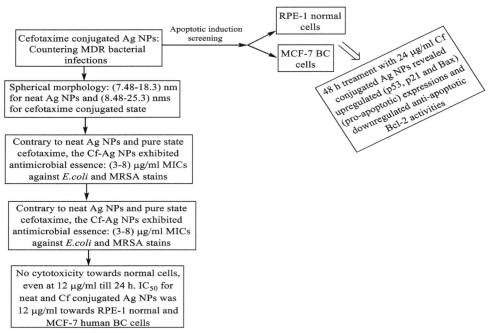

Fig. 9.8: Stepwise explanation of reference [90], wherein cefotaxime conjugated Ag nanoparticles exhibited enhanced antimicrobial activity against *E. coli* and MRSA stains.
The functional nanoparticles displayed spherical morphology with <20 nm particle sizes.

and AU565 cell lines corresponding to highest dosage, 12.5 μg·mL^{-1}. The similar extents for biologically prepared Ag NPs were estimated as 87.5% and 96%. Since the biologically prepared Ag NPs did not exhibit any toxicity in their native state, these could be pioneer agents as anticancer drugs for BC treatment. Similar prospects could be extended to related plant products, having composition similar to their respective extracts [91].

In a subsequent elegant attempt, Gonclaves and colleagues from China studied the toxicity and cell internalization of individual (Ag and Au), bimetallic (Ag-Fe, Au-Fe, and Ag-Au) aminolevulinic acid (ALA) NPs. Of note, ALA can regulate the generation of photoporphyrin IX, an intracellular photosensitizer commonly used in photodynamic therapy. The NPs were prepared using photoreduction method and characterized thoroughly for spectroscopic, dispersion, and morphology attributes. The amount of singlet oxygen generated by a yellow LED and ultrasound was studied for Au, Au-Fe, and Ag-Au NPs. To examine the characteristic impact on MCF-7 cells, the cytotoxicity assays of MCF-7 were performed in presence of NPs besides a quantification of PpIX fluorescence using high content screening (HCS). Inspection revealed a development of red fluorescence subsequent to 24 hour of NPs incubation with MCF-7 cells, suggested an efficient conversion of ALA on the NPs surface to PpIX. Best results for singlet oxygen generation with LED or ultrasound generation were retrieved for ALA:Ag-Au NPs. So, this study illustrated a significance of efficient ALA delivery to BC cells besides generation of singlet oxygen and ALA to PpIx conversion inside the cells, which paved way for photodynamic and sonodynamic therapies. The results could be amicably extended to the emerging studies on simultaneous NPs and photoradiation administration to cancer cells, whereby improved and sustained toxicities could be accomplished [92].

Another elegant effort of 2021 by Ulagesan and colleagues studied the cytotoxicity and anticancer activities of Ag NPs, prepared using aqueous extract of marine alga *Capsosiphon fulvescens* against the MCF-7 BC cells. The formed CfAgNPs were thoroughly characterized for their nanoscale attributes, revealing a (20–22) nm particle size, spherical morphology, and null aggregation. Analysis for cytotoxicity revealed an IC_{50} of 50 $\mu g \cdot mL^{-1}$, whereby the viability of treated MCF-7 cells decreased on increasing CfAgNPs concentrations. Morphology evaluation of the MCF-7 cells before and after treatment with CfAgNPs revealed irregular confluent aggregates with round polygonal cells on being treated with aqueous algal extract (devoid of CfAgNPs). Contrary to this, the Tamoxifen and CfAgNPs treatment resulted in significant morphological alterations. Analysis of apoptosis using 4',6-diamidino-2-phenylindole (DAPI) staining revealed CfAgNPs treated MCF-7 cells to develop a bright blue fluorescence with an evident, shortened, and disjointed chromatin. The control cells, on the contrary developed less bright fluorescence. Analysis using Flow Cytometry revealed a high percentage of cells in the late apoptosis, on being treated with Tamoxifen (77.2%) and CfAgNPs (74.6%). So, this study established another worthwhile analysis method to use CfAgNPs as BC treating agents with a sustained cytotoxicity profile (Fig. 9.9). An element of concern, however, is the in vitro conduct of the study which provides a caution to validate the results in the in vivo conditions using the same concentrations of CfAgNPs. It is well known that nearly 90% of in vitro results are not exactly replicated when examined in the in vivo conditions [93].

The above discussion amply illustrates the therapeutic significance of Au and Ag NPs *vis-à-vis* drug delivery to respective tumor sites in the lungs and breast. The success of NPs use as drug carriers is highly dependent on their targeted engineering for which one needs to have a thorough knowledge about the stage of the cancer cells being treated. This is because the intensity of cell surface receptors is not the same during different tumor stages and a distinct expression of these may result in totally different drug internalization extents within the tumor cells. Even though drug delivery using NPs is in practice for more than two decades (although its commercialization gained momentum in the last decade) and there is plentiful literature explaining the delivery modes along with corresponding pros and cons, still there are multiple myths which hinder a clear understanding of this subject. As also highlighted in some of the discussed studies, the efficacy of NPs mediated drug delivery does not depend on the implicit particle sizes and it is quite possible that larger sized particles are internalized more efficiently within the tumor cells. There are many reports of nanorods being more effective in localized toxicity induction within the tumor cells. So, the concern does not focus solely on particle size and in fact, it is the binding of drug with the NPs which is an important

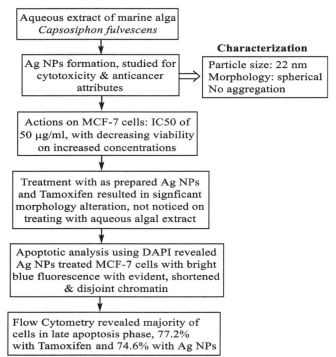

Fig. 9.9: Chronological steps of reference [93], with Ag nanoparticles prepared using *Capsosiphon fulvescens* (marine algae) were efficiently internalized in MCF-7 cells, resulting >70% population in late apoptotic phases, with Tamoxifen as well as in native configurations.
The particles exhibited a size of 22 nm with spherical morphology.

regulating factor. The delivery process mandates that once the drug loaded NPs are within the tumor cells, they should be capable of releasing the bound drug spontaneously due to the physicochemical dynamics of tumor cell activities. Alternatively, it necessitates that the body's own energy must not be expended for releasing the drug at the needed site. Secondly, the drug loaded carrier must be capable of requisite extent of opsonization, so that the detection of drug as a foreign entity does not metabolize it out before it reaches its inside the tumor cell. Therefore, surface engineering with respect to stealth delivery must be properly understood so that maximum amount of drug reaches within the tumor cell. Third and most important factor is the dosage moderation extent which implies that once we are using engineered NPs as delivery vehicles, a reduced amount of drug loading should be attempted contrary to the unaided drug delivery. Thereby, the acronym, ADME (i.e., Adsorption, Degradation, Metabolism, and Excretion) must be followed with highest caution in order to achieve success in the NPs aided drug delivery. The net conclusion is that binding of drug with the carrier must not be through drastic forces which could damage the native drug chemical structure and the task of prolonging intracellular stay of a drug must be accomplished. The latter part implies a prevention of prompt chemical degradation from the physiological environment which is prevented by the stealth delivery using NPs.

9.6 Conclusions and future prospects

Despite much progress in clinical and research phase, the NPs of Au and Ag have significantly reduced the BC mortality by improving the efficacy of drug delivery and reducing the toxic side effects. Availability of manifold reducing agents makes the chemical method of NPs preparation as the favored route with greater stability control compared to the physical and biological methods. Before NPs as drug delivery vehicles, the investigators must thoroughly ensure that the efficacy of drug delivery could be enhanced only through moderate interactions of drug-carrier binding. Thereby, the noncovalent interactions rightly emerge as the operational interfaces for a sustainable, ensuring minimal structural attenuation of native drug structure. In this context, the drug loaded vehicle must be capable of protecting the loaded drug from physiological detection and enzymatic degradation. Though time consuming but this criterion serves as a must to agree condition so that while releasing drug from bound carrier, no energy from body is spent and rather the release profile must be physiochemically induced or controlled. The binding interactions could be inspected on an in vitro scale using computational techniques and software. Reproducibility of this analysis can help an investigator control the drug performance via optimizing the carrier-drug binding stoichiometries. Besides, it must be ensured that the carrier does not chemically interact with the drug and merely function either as transporting medium or it would be an incentive to have a carrier which additively supports the drug activity. For instance, delivering a low concentration of chemotherapeutic drug with curcumin argues well for the site-specific activity. This is possible due to non-toxic and pleiotropic nature of curcumin, which guards it well against the possible side effects. Other important aspects which control the performance of NPs as drug carrier are their shape modulated activities and functionalization abilities. Through functionalization, the properties of a delivery vehicle could be varied as per the specific chemical structure of drug to be delivered. Specifically, Au NPs have been preferred either as capping agents or sensing the tumor regime, owing to their relatively high inertness and lower toxicity. A compelling issue limiting the trials is the ethical requirement of obtaining reproducible responses in different working environments. Secondly, there must be a rigorous summing up and inspection of reported trials (consulting the database) so that there is a solid hypothesis behind an intended choice of drug delivery vehicle including the carrier-drug binding activities. Future of Au and Ag NPs as BC drug delivery is indeed very bright but subject to wise selection of shape, size and functionalizing moiety with respect to chemical structure of native drug.

References

[1] Z. Tao, A. Shi, C. Lu, T. Song, Z. Zhang, J. Zhao, Breast cancer: Epidemiology and etiology, Cell Biochem. Biophys. 72 (2015) 333–338.

[2] T. Sorlie, C.M. Perou, R. Tibshirani, T. Aas, S. Geisler, H. Johnsen, et al., Gene expression patterns of breast carcinomas distinguish tumor subclasses with clinical implications, PNAS, United States of America 98 (2001) 10869–10874.

[3] K.R. Bauer, M. Brown, R.D. Cress, C.A. Parise, V. Caggiano, Descriptive analysis of estrogen receptor (er)-negative, progesterone receptor (pr)-negative, and her2-negative invasive breast cancer, the so-called triple-negative phenotype: A population-based study from the California cancer registry, Cancer 109 (2007) 1721–1728.

[4] H. Iwase, J. Kurebayashi, H. Tsuda, T. Ohta, M. Kurosumi, K. Miyamoto, et al., Clinicopathological analyses of triple negative breast cancer using surveillance data from the registration committee of the Japanese breast cancer society, Breast Cancer 17 (2010) 118–124.

[5] C.A. Hudis, Trastuzumab-mechanism of action and use in clinical practice, New Engl. J. Med. 357 (2007) 39–51.

[6] H.J. Huang, P. Neven, M. Drijkoningen, R. Paridaens, H. Wildiers, E. van Limbergen, et al., Association between tumor characteristics and HER-2/neu by immunohistochemistry in 1362 women with primary operable breast cancer, J. Clin. Pathol. 58 (2005) 611–616.

[7] K.A. Cadoo, M.N. Fornier, P.G. Morris, Biological subtypes of breast cancer: Current concepts and implications for recurrence patterns. Quart, J. Nucl. Med. & Mol. Imaging, 57 (2013) 312–321.

[8] G.J. Morris, S. Naidu, A.K. Topham, F. Guiles, Y. Xu, P. McCue, et al., Differences in breast carcinoma characteristics in newly diagnosed African-American and Caucasian patients: A single-institution compilation compared with the national cancer institute's surveillance, epidemiology, and end results database, Cancer 110 (2007) 876–884.

[9] R. Dent, M. Trudeau, K.I. Pritchard, W.M. Hanna, H.K. Kahn, C.A. Sawka, et al., Triple-negative breast cancer: Clinical features and patterns of recurrence, Clin. Cancer Res. 13 (2007) 4429–4434.

[10] A.I. Phipps, R.T. Chlebowski, R. Prentice, A. McTiernan, J. Wactawski-Wende, L.H. Kuller, et al., Reproductive history and oral contraceptive use in relation to risk of triple negative breast cancer, J. Natl. Cancer Inst. 103 (2011) 470–477.

[11] E. Turpin, I. Bieche, P. Bertheau, L.F. Plassa, F. Lerebours, A.de Roquancourt, et al., Increased incidence of erbb2 overexpression and TP53 mutation in inflammatory breast cancer, Oncogene 21 (2002) 7593–7597.

[12] K. Polyak, Breast cancer: Origins and evolution, J. Clin. Invest. 117 (2007) 3155–3163.

[13] A. Mishra, M. Verma, Cancer biomarkers: Are we ready for the prime time? Cancers 2 (2010) 190–208.

[14] L C.Sotiriou, Pusztai Gene-expression signatures in breast cancer, The New Engl. J. Med. 360 (2009) 752–800.

[15] B. Weigelt, J.S. Reis-Filho, Molecular profiling currently offers no more than tumor morphology and basic immunohistochemistry, Breast Cancer Res. 12 (2010) S5.

[16] E.H. van Beers, P.M. Nederlof, Array-CGH and breast cancer, Breast Cancer Res. 8 (2006) 210 Article.

[17] M.J. Piccart-Gebhart, New developments in hormone receptor-positive disease, Oncologist 16 (2011) 40–50.

[18] S.Z. Haslam, The ontogeny of mouse mammary gland responsiveness to ovarian steroid hormones, Endocrinol 125 (1989) 2766–2772.

[19] P.P. Rosen, Adenomyoepithelioma of the breast, Human Pathol. 18 (1987) 1232–1237.

[20] M.D. Althuis, J.H. Fergenbaum, M. Garcia-Closas, L.A. Brinton, M.P. Madigan, M.E. Sherman, Etiology of hormone receptor-defined breast cancer: A systematic review of the literature, Cancer Epidemiol. Biomarkers & Prevention 13 (2004) 1558–1561.

[21] E.V. Jensen, V.C. Jordan, The estrogen receptor: A model for molecular medicine, Clin. Cancer Res. 9 (2003) 1980–1989.

[22] V.C. Jordan, Effect of tamoxifen (ICI 46,474) on initiation and growth of DMBA induced rat mammary carcinomata, Eur. J. Cancer 12 (1976) 419–424.

[23] B. Fisher, J. Costantino, C. Redmond, R. Poisson, D. Bowman, J. Couture, et al., A randomized clinical trial evaluating tamoxifen in the treatment of patients with node-negative breast cancer who have estrogen-receptor-positive tumors, N. Engl. J. Med. 320 (1989) 479–484.

[24] V.C. Jordan, Tamoxifen: a personal retrospective, Lancet Oncol. 1 (2000) 43–49.

[25] R. Peto, J. Boreham, M. Clarke, C. Davies, V. Beral, UK and USA breast cancer deaths down 25% in year 2000 at ages 20-69 years, Lancet 355 (2000) 1822.

[26] B. Fisher, J.P. Costantino, D.L. Wickerham, C.K. Redmond, M. Kavanah, W.M. Cronin, et al., Tamoxifen for prevention of breast cancer: report of the National Surgical Adjuvant Breast and Bowel Project P-1 Study, J. Natl. Cancer Inst. 90 (1998) 1371–1388.

[27] R.J.T. Cote, C.R.: immunohistochemical detection of steroid hormone receptors, in: W.B. Saunders (Ed.), Immunomicroscopy: A Diagnostic Tool For the Surgical Pathologist, Philadelphia, Pa, USA, 1994.

[28] J.M. Harvey, G.M. Clark, C.K. Osborne, D.C. Allred, Estrogen receptor status by immunohistochemistry is superior to the ligand-binding assay for predicting response to adjuvant endocrine therapy in breast cancer, J. Clin. Oncol. 17 (1999) 474–1481.

[29] J.L. Wittliff, Steroid hormone receptors in breast cancer, Cancer 53 (1984) 630–643.

[30] T. Yamamoto, S. Ikawa, T. Akiyama, Similarity of protein encoded by the human c-erb-B-2 gene to epidermal growth factor receptor, Nature 319 (1986) 230–234.

[31] A.L.A. Eisenberg, S. Koifman, Cancer de mama: marcadores tumorais, Revista Brasileira De Cancerologia 47 (2001) 11.

[32] A. Citri, K.B. Skaria, Y. Yarden, The deaf and the dumb: the biology of ErbB-2 and ErbB-3. Exp, Cell Res. 284 (2003) 54–65.

[33] D.J. Slamon, B. Leyland-Jones, S. Shak, H. Fuchs, V. Paton, A. Bajamonde, et al., Use of chemotherapy plus a monoclonal antibody against HER2 for metastatic breast cancer that overexpresses HER2. The New Engl, J. Med. 344 (2001) 783–792.

[34] M.A. Molina, J. Codony-Servat, J. Albanell, F. Rojo, J. Arribas, J. Baselga, Trastuzumab (Herceptin), a humanized anti-HER2 receptor monoclonal antibody, inhibits basal and activated HER2 ectodomain cleavage in breast cancer cells, Cancer Res. 61 (2001) 4744–4749.

[35] T. Vu, F.X. Claret, Trastuzumab: Updated mechanisms of action and resistance in breast cancer, Front. Oncol. 2 (2012) 62.

[36] T. Narita, H. Funahashi, Y. Satoh, H. Takagi, C-erbB-2 protein in the sera of breast cancer patients, Breast Cancer Res. & Treat. 24 (1992) 97–102.

[37] L.D. Tan, Y.Y. Xu, Y. Yu, X.Q. Li, Y. Chen, Y.M. Feng, Serum HER2 level measured by dot blot: a valid and inexpensive assay for monitoring breast cancer progression, PLoS One 6 (2011) e18764 Article ID.

[38] R. Dent, M. Trudeau, K.I. Pritchard, W.M. Hanna, H.K. Kahn, C.A. Sawka, et al., Triple-negative breast cancer: clinical features and patterns of recurrence, Clin. Cancer Res. 13 (2007) 4429–4434.

[39] E.Y. Cho, M.H. Chang, Y.L. Choi, J.E. Lee, S.J. Nam, J.H. Yang, et al., Potential candidate biomarkers for heterogeneity in triple-negative breast cancer (TNBC), Cancer Chemother. & Pharmacol. 68 (2011) 753–761.

[40] T. Sørlie, R. Tibshirani, J. Parker, T. Hastie, J.S. Marron, A. Nobel, et al., Repeated observation of breast tumor subtypes in independent gene expression data sets, PNAS, United States of America 100 (2003) 8418–8423.

[41] P. Vielh, S. Chevillard, V. Mosseri, B. Donatini, H. Magdelenat, Ki67 index and S-phase fraction in human breast carcinomas. Comparison and correlations with prognostic factors, Am. J. Clin. Pathol. 94 (1990) 681–686.

[42] A.V. Patil, R. Singhai, R.S. Bhamre, V.W. Patil, Ki-67 biomarker in breast cancer of Indian women, North Am. J. Med. Sci. 3 (2011) 119–128.

[43] K. Tawfik, B.F. Kimler, M.K. Davis, F. Fan, O. Tawfik, Ki-67 expression in axillary lymph node metastases in breast cancer is prognostically significant, Human Pathol 44 (2013) 39–46.

[44] E. Luporsi, F. Andre, F. Spyratos, P-M. Martin, J. Jacquemier, F. Penault-Llorc, et al., Ki-67: level of evidence and methodological considerations for its role in the clinical management of breast cancer: Analytical and critical review, Breast Cancer Res. & Treat. 132 (2012) 895–915.

[45] R.M. Elledge, G.M. Clark, S.A.W. Fuqua, Y.Y. Yu, D.C. Allred, p53 protein accumulation detected by five different antibodies: Relationship to prognosis and heat shock protein 70 in breast cancer, Cancer Res. 54 (1994) 3752–3757.

[46] A. Thathiah, D.D. Carson, MT1-MMP mediates MUC1 shedding independent of TACE/ADAM17, Biochem. J. 382 (2004) 363–373.

[47] M.J. Duffy, S. Shering, F. Sherry, E. McDermott, N.N. O'Higgins, CA, 15-3: A prognostic marker in breast cancer, Int. J. Biol. Markers 4 (2000) 330–333.

[48] J.A. Thompson, F. Grunert, W. Zimmermann, Carcinoembryonic antigen gene family: Molecular biology and clinical perspectives, J. Clin. Laboratory Analysis 5 (1991) 344–366.

[49] R. Molina, J.M. Auge, B. Farrus, G. Zanón, J. Pahisa, M. Muñoz, et al., Prospective evaluation of Carcinoembryonic Antigen (CEA) and carbohydrate antigen 15.3 (CA 15.3) in patients with primary locoregional breast cancer, Clin. Chem. 56 (2010) 1148–1157.

[50] M. Uehara, T. Kinoshita, T. Hojo, S. Akashi-Tanaka, E. Iwamoto, T. Fukutomi, Long-term prognostic study of carcinoembryonic antigen (CEA) and carbohydrate antigen 15-3 (CA 15-3) in breast cancer, Int. J. Clin. Oncol. 13 (2008) 447–451.

[51] J.G. Geraghty, E.C. Coveney, F. Sherry, N.J. Ohiggins, M.J. Duffy, Ca-15-3 in patients with locoregional and metastatic breast-carcinoma, Cancer 70 (1992) 2831.

[52] F.G. Ebeling, P. Stieber, M. Untch, D. Nagel, G.E. Konecny, U.M. Schmitt, et al., Serum CEA and CA15-3 as prognostic factors in primary breast cancer, Br. J. Cancer 86 (2002) 1217–1222.

[53] M.T. Sandri, M. Salvatici, E. Botteri, R. Passerini, L. Zorzino, N. Rotmensz, Prognostic role of CA15.3 in 7942 patients with operable breast cancer, Breast Cancer Res. Treat 132 (2012) 317–326.

[54] G.A. Mendes, F.C.R. Sturmer, D.L. Basegio, Utilizac̃aodosmarcadores, CA15. 3 e CEA no seguimento de pacientes comneoplasia mam´aria, News Lab 102 (2010) 7.

[55] W. Jager, K. Eibner, B. Loffler, S. Gleixner, S. Kramer, Serial CEA and CA 15-3 measurements during follow-up of breast cancer patients, Anticancer Res. 20 (2000) 5179–5182.

[56] P.L. Welcsh, K.N. Owens, M.C. King, Insights into the functions of BRCA1 and BRCA2, Trends in Genet 16 (2000) 69–74.

[57] N. Senst, M. Llacuachaqui, J. Lubinski, H. Lynch, S. Armel, S. Neuhausen, et al., Parental origin of mutation and the risk of breast cancer in a prospective study of women with a BRCA1 or BRCA2 mutation, Clin. Genet. 84 (2013) 43–46.

[58] M.E. Robson, Treatment of hereditary breast cancer, Semin. in Oncol. 34 (2007) 384–391.

[59] M. Vanstone, W. Chow, L. Lester, P. Ainsworth, J. Nisker, M. Brackstone, Recognizing BRCA gene mutation risk subsequent to breast cancer diagnosis in south western Ontario, Can. Fam. Physician 58 (2012) E258–E266.

[60] T.R. Rebbeck, T. Friebel, H.T. Lynch, S.L. Neuhausen, L.van 't Veer, J.E. Garber, et al., Bilateral prophylactic mastectomy reduces breast cancer risk in BRCA1 and BRCA2 mutation carriers: The PROSE study group, J. Clin. Oncol. 22 (2004) 1055–1062.

[61] P. Apostolou, F. Fostira, Hereditary breast cancer: The era of new susceptibility genes, Biomed. Res. Int. (2013) 747318 Article ID.

[62] P. Malik, T.K. Mukherjee, Recent Progress in gold and silver nanoparticles biomedical attributes towards lung and breast cancer treatment, Int. J. Pharm. 553 (2018) 483–509.

[63] L. Wang, L.Shao C.Hu, The antimicrobial activity of nanoparticles: Present situation and prospects for the future, Int. J. Nanomed. 12 (2017) 1227–1249.

[64] T.C. Dakal, A. Kumar Majumdar Rs, V. Yadav, Mechanistic basis of antimicrobial actions of silver nanoparticles, Front. Microbiol. 7 (2016) 1831 Article.

[65] P. Malik, R. Shankar, V. Malik, N.K. Sharma, T.K. Mukherjee, Green chemistry based benign routes for nanoparticle synthesis, J. Nanopart. (2014) 302429 Article ID.

[66] M. Wuithschick, A. Birnbaum, S. Witte, M. Sztucki, U. Vainio, N. Pinna, Turkevich in new robes: Key questions answered for the most common gold nanoparticle synthesis, ACS Nano 9 (2015) 7052–7071.

[67] M. Brust, M. Walker, D. Bethell, D.J. Schiffrin, R. Whyman, Synthesis of thiol-derivatised gold nanoparticles in a two-phase liquid-liquid system, J. Chem. Soc. Chem. Commun. 7 (1994) 5–7.

[68] S. Iravani, H. Korbekandi, S.V. Mirmohammadi, B. Zolfaghari, Synthesis of silver nanoparticles: Chemical, physical and biological methods, Res. Pharm. Sci. 9 (2014) 385–406.

[69] V. Kumar, S. Yadav, Plant-mediated synthesis of silver and gold nanoparticles and their applications, J. Chem. Technol. & Biotechnol. 84 (2008) 151–157.

[70] M. Kitching, M. Ramani, E. Marsili, Mini-review: Fungal biosynthesis of gold nanoparticles: Mechanism and scale up, Microbial Biotechnol 8 (2014) 904–917.

[71] N. Bertrand, J. Wu, X. Xu, N. Kamaly, O.C. Farokhzad, Cancer Nanotechnology: The impact of passive and active targeting in the era of modern cancer biology, Adv. Drug Deliv. Rev. 66 (2014) 2–25.

[72] C. Bernard-Marty, F. Lebrun, A. Awada, M.J. Piccart, Monoclonal antibody-based targeted therapy in breast cancer. Current status and future directions, Drugs 66 (2006) 1577–1591.

[73] M. Liu, X. Yu, Z. Chen, T. Yang, D. Yang, Q. Liu, K. Du, B. Li, Z. Wang, S. Li, Y. Deng, N. He, Aptamer selection and applications for breast cancer diagnostics and therapy, J. Nanobiotechnol. 15 (2017) 81.

[74] Y. Gilad, M. Firer, G. Gellerman, Recent innovations in peptide based targeted drug delivery to cancer cells, Biomedicines 4 (2016) 11.

[75] C.S. Kue, A. Kamkaew, K. Burgess, L.V. Kiew, L.Y. Chung, H.B. Lee, Small molecules for active targeting in cancer, Med. Res. Rev. 36 (2016) 494–575.

[76] P.G. Calavia, M.J. Martin, I. Chambrier, M.J. Cook, D.A. Russell, Towards optimisation of surface enhanced photodynamic therapy of breast cancer cells using gold nanoparticle-photosensitizer conjugates, Photochem. Photobiol. Sci. 17 (2018) 281–289.

[77] Y. Li, D. He, J. Tu, R. Wang, C.C. Zu, C. You, et al., Comparative effect of wrapping solid gold nanoparticles and hollow gold nanoparticles with doxorubicin loaded thermosensitive liposomes for cancer thermo-chemotherapy, Nanoscale 10 (2018) 8628–8641.

[78] H. Banu, N. Renuka, S.M. Faheem, R. Ismail, V. Singh, Z. Saadatmand, et al., Gold and silver nanoparticles biomimetically synthesized using date palm pollen extract induce apoptosis and regulate p53 and Bcl-2 expression in human breast adenocarcinoma cells, Biol. Trace. Elem. Res. 186 (2018) 122–134.

[79] R.J.C. Bose, S.U. Kumar, Y. Zeng, R. Afjei, E. Robinson, K. Lau, et al., Tumor cell-derived extracellular vesicle-coated nanocarriers: An efficient theranostic platform for the cancer-specific delivery of anti-miR-21 and imaging agents, ACS Nano 12 (2018) 10817–10832.

[80] S. Mahalunkar, A.S. Yadav, M. Gorain, V. Pawar, R. Braathen, S. Weiss, et al., Functional design of pH-responsive folate-targeted polymer-coated gold nanoparticles for drug delivery and in vivo therapy in breast cancer, Int. J. Nanomed. 14 (2019) 8285–8302.

[81] P. Ruenraroengsak, D. Kiryushko, I.G. Theodorou, M.M. Klosowski, E.R. Taylor, T. Niriella, et al., Frizzled-7-targeted delivery of zinc oxide nanoparticles to drug-resistant breast cancer cells, Nanoscale 11 (2019) 12858–12870.

[82] B. Shrestha, L. Wang, H. Zhang, C.Y. Hung, L. Tang, Gold nanoparticles mediated drug-gene combinatorial therapy for breast cancer treatment, Int. J. Nanomed. 15 (2020) 8109–8119.

[83] M. Khoobchandani, K.K. Katti, A.R. Karikachery, V.C. Thipe, D. Srisrimal, D. Kumar, et al., New approaches in breast cancer therapy through green nanotechnology and nano-ayurvedic medicine-pre-clinical and pilot human clinical investigations, Int. J. Nanomed. 15 (2020) 181–197.

[84] G.M. Sulaiman, H.M. Waheeb, M.S. Jabir, S.H. Khazaal, Y.H. Dewir, Y. Naidoo, Hesperidin loaded on gold nanoparticles as a drug delivery system for a successful biocompatible, anti-cancer, anti-inflammatory and phagocytosis inducer model, Sci. Rep. 10 (2020) 9362.

[85] M. Ganji, F. Dashtestani, H.K. Neghab, M.H. Soheilifar, F. Hakimian, F. Haghiralsadat, Gold nanoparticles conjugated L-lysine for improving cisplatin delivery to human breast cancer cells, Curr. Drug Deliv. (2020), doi:10.2174/1567201818666201203150931.

[86] J. Swanner, C.D. Fahrenholtz, I. Tenvooren, B.W. Bernish, J.J. Scars, A. Hooker, et al., Silver nanoparticles selectively treat triple-negative breast cancer cells without affecting non-malignant breast epithelial cells in vitro and in vivo, FASEB BioAdv 1 (2019) 639–660.

[87] M.K. Gopisetty, D. Kovacs, N. Igaz, A. Ronavari, P. Belteky, Z. Razga, et al., Endoplasmic reticulum stress: major player in size-dependent inhibition of P-glycoprotein by silver nanoparticles in multidrug-resistant breast cancer cells, Nanobiotechnol 17 (2019) 9.

[88] A.A.H. Abdellatif, Z. Rasheed, A.H. Alhowail, A. Alqasoumi, M. Alsharidah, R.A. Khan, et al., Silver citrate nanoparticles inhibit PMA-induced TNFα deactivation of NF-κB activity in human cancer cell lines, MCF-7, Int. J. Nanomed. 15 (2020) 8479–8493.

[89] Y. Chong, J. Huang, X. Xu, C. Yu, X. Ning, S. Fan, et al., Hyaluronic acid-modified Au-Ag alloy nanoparticles for radiation/nanozyme/Ag$^+$ multimodal synergistically enhanced cancer therapy, Bioconjug. Chem. (2020), doi:10.1021/acs.bioconjchem.0c00224.

[90] E.M. Halawani, A.M. Hassan, S.M.F.G. El-Rab, Nanoformulation of biogenic cefotaxime-conjugated-silver nanoparticles for enhanced antibacterial efficacy against multidrug-resistant bacteria and anticancer studies, Int. J. Nanomed. 15 (2020) 1190–1889.

[91] S. Dinparvar, M. Bagirova, A.M. Allahverdiyev, E.S. Abamor, T. Safarov, M. Aydogdu, et al., A nanotechnology-based new approach in the treatment of breast cancer: Biosynthesized silver nanoparticles using *Cuminum cyminum* L. seen extract, J. Photochem. & Photobiol. B: Biology 208 (2020) 111902.

[92] K.O. Gonclaves, D.P. Vieira, D. Levy, S.P. Bydlowski, L.C. Courrol, Uptake of silver, gold and hybrids silver-iron, gold-iron and silver-gold aminolevulinic acid nanoparticles by MCF-7 breast cancer cells, Photodiag. & Photodyn. Ther. 32 (2020) 102080.

[93] S. Ulagesan, T.J. Nam, Y.H. Choi, Cytotoxicity against human breast carcinoma cells of silver nanoparticles biosynthesized using *Capsosiphon fulvescens* extract, Bioprocess & Biosyst. Engg. 44 (2021) 901–911.

CHAPTER 10

Silver and gold nanoparticles: Promising candidates as antimicrobial nanomedicines

Anjana K. Vala[a], Nidhi Andhariya[b], Bhupendra Kumar Chudasama[b]

[a]Department of Life Sciences, Maharaja Krishnakumarsinhji Bhavnagar University, Bhavnagar, Gujarat, India [b]Thapar Institute of Engineering and Technology, Patiala, Punjab, India

10.1 Introduction

In recent years, it is getting challenging to beat micro-organism with traditional antibiotics [1]. The irrational use of antibiotics might be the cause of this situation; the clinical world is facing right now world-wide [2]. In view of this scenario, the development of new antibiotics which can work efficiently on these stubborn micro-organisms is of utmost requirement. While developing a new drug molecule is a very lengthy, time consuming, and costly process, one effective alternative is to use existing metals such as silver and gold in their nano or colloidal form as their antibacterial activities are well known from old days [3,4]. According to Rai et al., AgNPs are the new generation of antimicrobials [5]. AgNPs, AuNPs, and bimetallic nanoparticles can be synthesized using physicochemical and biological route for potential applications as antimicrobial nanomedicines. In this chapter, an effort has been made to give insight of the synthesis of AgNPs and AuNPs with different routes involving chemical and biosynthesis approaches [6,7]. Further, the antibacterial and antifungal effects of these metal nanoparticles have been reviewed on different gram-positive and gram-negative organisms and on different fungi [8–10].

10.2 Antimicrobial potential of nanoparticles fabricated through physicochemical route

The chemical route is the simplest way for the synthesis of AgNPs and AuNPs that are carried out mostly in the presence of different small organic molecules, natural and synthetic biopolymers as capping or stabilizing agents. The different size, shape, and morphology of AgNPs and AuNPs can work differently on the antibacterial activity to different pathogens [1]. Conventional antibiotics can work on few pathogens, while studies have been reported that AgNPs can work on over hundreds of microbes without having any issues related to development of resistance [11]. Comparative studies have been done by tagging molecules of traditional antibiotics with AgNPs and interesting results have been achieved where ineffective

Gold and Silver Nanoparticles.
DOI: https://doi.org/10.1016/B978-0-323-99454-5.00013-5

traditional antibiotics showed promising good results on inhibition of microorganism, when combined with AgNPs. Studies also have proven the additive silver effect of conventional antibiotics on microorganism [12]. Some important studies showing antibacterial effect of AgNPs over different pathogens are discussed below.

The simple and environment friendly synthesis approach was adopted to prepare polystyrene/silver nanospheres embedded in polyvinylpyrrolidone (PVP) [13]. This synthesis can be used as a template for synthesizing core-shell structure. Nanospheres can be used as a catalyst for the reduction of organic dyes and antibacterial agents against gram negative bacteria *Salmonella* and *Escherichia coli*. A comparative study of antibacterial activity of glucosamine functionalized AgNPs and pure AgNPs have been studied against eight gram negative and eight gram positive bacteria [14]. It was concluded that *Klebsiella pneumoniae* and *Bacillus cereus* expressed high level of inhibition compared to other organisms in the presence of glucosamine-AgNPs.

The citrate route is the most common approach to obtainAgNPsby using tri-sodium citrate [15]. No other capping agents are required during synthesis. The citrate capped AgNPs showed antibacterial and antifungal activity against two bacterial strains of *Bacillus subtilis* (gram-positive) and *E. coli* (gram-negative), and two yeasts *Saccharomyces cerevisiae* and *Candida albicans*, respectively. The study was carried out with four different concentrations of AgNPs and observed a linear relationship between the concentration of silver and the antibacterial and antifungal activity. Most significant inhibitions have been seen in the case of *E. coli* and *Saccharomyces cerevisiae*.

Effect of different ligands on surface of AgNPs have been explained in details by Padmos et al. [16]. PVP and cystein has been attached to AgNPs via chemical rout. The bacterial strains used to carry out antibacterial activity study were *Staphylococcus aureus* and *E. coli*. The comparative study of two different ligands suggested that cystein coated AgNPs were significantly effective on both the gram positive and gram negative strains compared to PVP coated AgNPs. Similarly, one more study on AgNPs coated with three different polymers POEM, PVA, and SMA has been carried out by Li and his coworkers [17]. The cytotoxicity and antibacterial activity have been well explained in the study. Antibacterial study was performed on *E. coli*. The in vitro study has been performed on mouse skin fibro blast and human heptacarcinoma (HePG2), and mouse monocyte macrophages. Overall study concluded that AgNPs coated with SMA polymer is better, safe, and effective. The study shows that polymer stabilizer played an important role in determining the toxicity of AgNPs. In another study, the AgNPs was synthesized by chemical reduction using sodium borohydrate in the presence of three different capping agents namely citrate, SDS and PVP [18]. The study also carried out to examine the combined antimicrobial effect of silver with three different conventional antibiotics (streptomycin, ampicillin, and tetracycline) on *E. coli* and *Staphylococcus aureus*. The study shown that PVP capped AgNPs combined with all the three antibiotics have more profound resistance compared to the citrate and SDS coated AgNPs.

The polymeric micelle has been used for coating AgNPs and natural polymer curcumin has been embedded in it to enhance antibacterial activity on *Pseudomonas aeruginosa* and *S. aureus*. Curcumin has effectively increased the AgNPs efficiency [19]. Gozdziewska et al. described a simple synthetic approach to obtain nitroxide coated AgNPs [20]. Antimicrobial activity of nitroxide coated AgNPs was tested against *P. aeruginosa* and *S. aureus* showed significantly lower values of minimum inhibitory concentrations, that is, 4 µg mL^{-1} and 12 µg mL^{-1}, respectively. The study proved that nitroxide favored the antibacterial activity of AgNPs because of the ability of oxidation of nitroxides by reactive oxygen species (ROS) to positively charged oxoammonium ions which can interact strongly with the bacterial membrane.

The microwave assisted synthesis was adopted to obtain eco-friendly AgNPs by using a carboxymethylated gum kondagogu (CMGK) as capping and reducing agent [21]. Bacterial strains used in the study are *P. aeruginosa*, *B. subtilis*, *Bacillus cereus* and *E. coli*. The strong antibacterial effect has been observed in the case of *B. subtilis*, *B. cereus* and *E. coli*, while the moderate level of antimicrobial activity has been observed in the case of *Pseudomonas aeruginosa*. The study showed the scavenging activity of AgNPs, which is carried out using DPPH radical scavenging.

Chudasama et al. has adopted a one pot synthesis of magnetite nanoparticles coated with AgNPs [22]. Themagnetitecore-shell nanostructures are well-described to show antibacterial activity and recyclability of AgNPs by the virtue of its magnetic core. Antibacterial study had been carried out for the core-shell structure and it has shown good activity against *B. subtilis* (gram-positive) and *E. coli* (gram-negative) bacteria. Similarly, more study has been reported on the potential antibacterial applications of magnetic core-shell nanostructures coated with PEG along with the advantage of optical labels for cellular imagining [23–26].The Fe_3O_4 core provides an easy magnetic separation and targeting and magnetic resonance imaging (MRI) contrast ability, and the Ag nanocrystals provide stable strong fluorescence and antibacterial activity. This type of core-shell structures provides a great opportunity for magnetic drug targeting and optical labeling. Development of hybrid nanocomposite to give combined effect against microorganism is a promising way to fight effectively against stubborn bacterial strains like *K. pneumoniae* and *S. aureus*. Biopolymer chitosan coated AgNPs is one of such successful effort towards that path. Antibacterial activity studies of the hybrid composite have been carried out on gram positive and gram negative bacteria [23–26].

Roe et al. carried out coating of AgNPs on the surface of catheters [27]. They tested antimicrobial activity and also evaluated the risk of systemic toxicity of the catheters. Significant in vitro antimicrobial activity was displayed by the coated catheters. The authors reported sustained release of silver over a period of 10 days. Elution of nearly 15% of the coated silver was obtained in 10 days. Organ accumulation of silver was not found, whereas 3% accumulation was found at transplantation site and 8% excretion in feces was observed.

In recent years, restorative dental materials with antimicrobial properties are in great demand. In order to mitigate the emergence of antibiotic resistance and to reduce side effects, antibiotics formulated in nanostructures with localized drug release could be more promising [28]. AgNPs have been employed in various areas of dentistry including dental biomaterials, dental implants, and orthodontic adhesives [28–31]. Enan et al. evaluated effect of Ag nanobiotics incorporated into glass-ionomer cement (GIC) against pathogens associated with dental caries [28]. The AgNPs were synthesized using *Cupressusmacrocarpa* extract. The authors observed synergistic effect of AgNPs with amoxicillin and GIC dental restorative material against the tested microbes. Synergistic effect against biofilm activity was also observed. The authors also reported that addition of the agents did not adversely affect the mechanical properties of the cement, hence, its use was found to be safe. Clinical trials have been suggested for validating the in vitro observations. Fig. 10.1 showed the SEM images of *Staphylococcus aureus* after exposure to various tested agents for 24 h.

In recent times, severe fungal infections have significantly contributed to the increasing morbidity of immunocompromised patients who need intensive treatment including

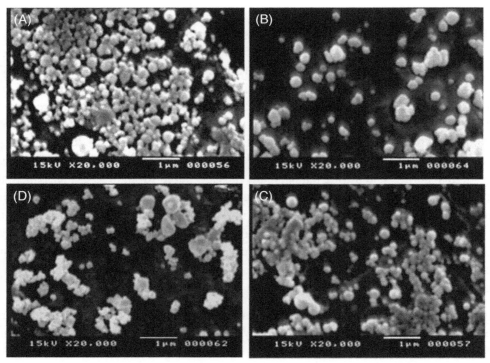

Fig. 10.1: SEM images of S. aureus on group A (GIC), group B (GIC with AgNPs) group C (GIC with amoxicillin) group D (GIC with amoxicillin and AgNPs) after 24 hours of immersion in bacterial culture. *(Adapted with permission from ref. [28]).*

broad-spectrum antibiotic therapy [32,33]. Currently most of the available effective antifungal agents are based on polyenes (amphotericin B), triazoles (fluconazole, itraconazole, voriconazole, posaconazole) or echinocandins (caspofungin, micafungin, and anidulafungin). However, administration of these antifungals is often accompanied by various complications such as amphotericin B toxicity and adverse effects of some azoles including toxicity and drug interactions [34–36] and yeast resistance to antifungal therapy [37]. Due to that, other options for effective antifungal therapy must be found to avoid the above-mentioned adverse effects. AgNPs could be utilized as potential candidates as an antifungal agent.

In case of fungal infection, *Candida* spp. represent one of the most common pathogens which are responsible for fungal infections often causing hospital-acquired sepsis with an associated mortality rate of up to 40%. Panecaek et al. [37] has studied antifungal activity of AgNPs synthesized by modified Tollens' process. Minimum inhibitory concentrations have been carried out for *Candida* spp. The results revealed that the SDS coated AgNPs worked better in inhibiting the yeast. Most importantly, it inhibits the growth below its cytotoxic limit against the tested human fibroblasts. Ionic silver on the other hand works at cytotoxic level. Therefore, the polymer coating shows good inhibition as well as kept cytotoxicity of silver below its limit for human fibroblasts [38].

Kim et al. have studied the antifungal activity of AgNPs synthesized by chemical route against *C. albicans* [39]. *Saccharomyces cerevisiae* and *Trichosporon beigelci* and *C. albicans* were used to examine the antifungal activity of AgNPs. Further, hemolytic activity against human erythrocytes has also been carried out. Flow cytometric analysis has been carried out to understand plasma membrane potential. AgNPs showed antifungal activity by disrupting membrane integrity of fungus.

Pinto et al. has studied antifungal activity of nanocomposite thin films of pullulan and silver against *Aspergillus niger* [40]. Transparent cast film around 66–74 micro meter has been prepared from silver hydrosol. As a result of antifungal activity, disruption of the spores of *Aspergillus niger* was observed by SEM. In another work, the antifungal activity of PVP coated AgNPs with the average size of 5 nm, 10 nm, and 60 nm were tested against *Candida albicans* and *Candida glabrata* [41]. Results of MIC (0.4–3.3 µg mL^{-1}) shows that nanoparticles are very effective on *Candida glabrata* biofilms. Study indicated that the particle size and stabilizing agent did not interfere in the antifungal activity against *Candida* biofilms. This study can contribute to the development of improvement in oral health.

Pereira et al. [42] studied the comparative antifungal activity of AgNPs prepared from different rout. Chemical and biosynthesis route has been employed to synthesis AgNPs, and tested against *Penicillium chrysogenum* and *Aspergillus oxyzae*. Chemical route showed higher antifungal activity against the tested fungus. In another study, the comparative antifungal activity of chitosan nanoparticles and chitosan coated AgNPs was performed against *F. oxysporum* species complex [43]. Both the nanoparticles have been synthesis by sol-gel technique. Chitosan and chitosan coated AgNPs, both of them could be successful in damaging and disrupting wall of *F. oxysporum*. Both the nanoparticles were able to reduce the fungal growth, and caused morphological and ultra-structural changes to *F. oxysporum*.

Ifukua et al. [44] has carried out antifungal activity of chitin coated AgNPs on 10 different plant pathogenic fungi, such as Alternariaalternata, Alternaria brassicae, Alternaria brassicicola, Bipolarisoryzae, Botrytis cinerea, and Penicillium digitatum, Colletotrichum higginsianum, Colletotrichum orbiculare, Fusarium oxysporum, and Pyriculariaoryzae. Chitin coated AgNPs has shown higher antifungal activity. The effect of quantum size of PVP coatedAgNPs on Candidaalbicans have been studied by Selveraj et al. [45]. The comparative antifungal study has been carried out in the presence of well-known antifungal drugs such as amphotericin B, fluconazole, and ketoconazole. PVP coated AgNPs showed 80% of inhibitions against *C. albicans* and its MIC value are much lower than all the three antifungal drugs used in the study. XTT assay, FE-SEM, and TEM measurements of the AgNPs treated fungi shows that silver has succeeded to lose the structural integrity of a fungi to a greater extent.

The antifungal activity of multifunctional core-shell Fe_3O_4-Ag nanocolloids was studied against *Aspergillus glaucus* isolates [46]. Facile one-pot chemical reduction formula was employed to prepare core-shell structure of silver coated magnetite nanoparticles. The Fe_3O_4-Ag nanocolloids displayed severe toxicity against the multidrug resistant fungus *Aspergillus glaucus*. Further investigations in the field can lead to the development of a medical methodology for the treatment of infections caused in immune-compromised patients. The antibacterial and antifungal activities of soda lime glass containing AgNPs have been carried out by Tejeda et al. [47]. The commercial soda lime glass was milled in the presence of isoropyle alcohol with silver sepiolite in it to achieve less than 30 micrometer particle size. Antibacterial and antifungal study have been carried out on three different organisms named *E. coli* (gram-negative bacteria), *Micrococcus luteus* (gram-positive bacteria) and *Issatchenkia orientalis* (yeast). Important outcome of this study was that glass, that is, calcium ions favor silver to be safe and fast disinfectant.

Gold nanoparticles (AuNPs) have also been studied for antimicrobial activities. A detailed size and shape dependent study on antifungal activity of AuNPs have been carried out by Wani and Ahmad [48]. The study indicated that gold disc of 30 nm size showed strong antifungal activities against *Candida* in comparison to polyhedral AuNPs. Similar to AgNPs, AuNPs also showed excellent antimicrobial activity against gram positive and gram negative bacteria [49]. AuNPs coated with mesoporous silica possessed dual enzyme activities similar to peroxidase and oxidase. Oxidase generates reactive oxygen species (ROS) which gives good antibacterial property against *E. coli* and *S. aureus*.

Armelao et al. described comparative study of pure titanium oxide and gold coated titanium oxide nanoparticles [50]. Synthetic approach involves the radio frequency (RF)-Sputtering to get TiO_2 nanoparticles, whereas the sol-gel route to get core-shell structure of Au TiO_2^{-1} nanoparticles. The introduction of Au on the TiO_2 matrix enhanced the photocatalytic performance to degrade azo dye, but the antibacterial activity of Au TiO_2^{-1} films is reduced compared to pure TiO_2. The antibacterial activity of AuNPs and AuNPs combined with conventional antibiotic gentamycin was investigated against *E. coli* [51]. Results showed that no major difference can be achieved in antibacterial activity between pure AuNPs and antibiotics coated AuNPs presumably because of the very small amount of drug have attached to the AuNPs.

The in vitro and in vivo study of antimicrobial peptides (PEP) coated AuNPs against bacterial strain of *S. aureus* revealed good antibacterial effect [52]. Also, the PEP coated AuNPs can be used as a carrier for *in vivo* gene activation in tissue regeneration. The combined effect of well-known antibiotic kanamycin plus AuNPs against bacterial strains *Streptococcus* epidermidise and *Enterobacter aerogenes* have been studied by Payne et al. [53]. The results showed excellent antibacterial effect of kanamycin and AuNPs. Perni et al. designed the light activated silica encapsulated methylene blue encapsulated AuNPs of size 5 nm–20 nm [54]. The antibacterial activity against *E. coli* and *S. Epidermidis*. Study indicated that as size of AuNPs increases the antibacterial effect decreases.

Zhu et al. [55] has studied the photothermal-based bacterial activity of gold nanorods array on *E. coli*. Gold nanorods (2D and 3D) array have been synthesized using confined convective arraying technique. The results showed promising bactericidal effect. The severity of pathogen destruction was measured and quantified using fluorescence microscopy, bioatomic force microscopy (Bio-AFM) and flow cytometry techniques. The results indicated that the fabricated gold nanorods arrays at higher concentrations were highly capable of complete bacterial destruction by photothermal effect compared to the low concentration gold nanorods arrays. Subsequent laser irradiation of the gold nanorods arrays resulted in rapid photoheating with remarkable bactericidal activity, which could be used for water treatment to produce microbe-free water.

Cytotoxicity and antibacterial activity of gold supported cerium oxide nanoparticles have been studied by Babu et al. [56]. The microorganisms, such as *B. subtilis*, *Salmonella enterritidis, E. coli,* and *S. aureus* are studied on both mono and co-cultured systems. For co-cultured studies, a lactic acid bacterium, *L. plantarum*, reported to have antibacterial activity. Therefore, co-cultured with each set of pathogens. Cytotoxicity have been studied towards normal RAW264.7 macrophage cells and A549 lung cancer cells. The outcome demonstrated better selectivity of Au CeO_2^{-1} nanoparticles towards cancerous cells. In comparative studies between pure CeO_2, Au, and Au CeO_2^{-1} on in vivo as well as antibacterial studies, Au CeO_2^{-1} shows better and promising results as compare to CeO_2 and Au nanoparticles.

Castillo-Martínez et al. showed antibacterial activities of seven different bacterial strains by using photothermal therapy of gold nanorods [57]. The plasmonic photothermal therapy (PPT) has been used against seven oral micro-organisms, such as *Enterococcusfaecalis*, *Staphylococcusaureus*, *Streptococcusmutans*, *Streptococcussobrinus*, *Streptococcus oralis*,

Streptococcussalivarius, and *E. coli*. MIC and MBC have been reported for all the seven bacterial strains. Study concludes that PTT of gold nanorods have enhanced its antimicrobial effect on microorganism than using gold nanorods alone.

Mikhailova reviewed the synthesis and potentials of AuNPs [58]. AuNPs have been reported to weaken biofilm formation by inhibiting synthesis of virulence factors and metabolic activity [59]. Geethalakshmi and Sarada [60] demonstrated the involvement of bacterial cell surface attachment, loss of flagella, clumping inside biofilms etc. as various antibacterial processes. Generally, the enhanced antimicrobial activities have been observed in case of AuNPs conjugates, however, the underlying mechanisms is not completely clear. As small diameter particles display surface effects and large surface area to mass ratio, absorption of large quantities of molecules occurs [61].

Skin burns are among the main reasons for morbidity and mortality in burn patients. Emergence of multidrug resistant bacteria has resulted in severe opportunistic infections which are very difficult to cure. Hence, new antimicrobial agents are required [62–64].While systemically administered antibiotics may encounter difficulties reaching damaged skin tissues, topically applied antimicrobials could be better due to their low systemic toxicity, ease of application, controlled effects etc. [64–66]. Arafa et al. developed thermoresponsive gels containing AuNPs that displayed more prolonged and sustained effects than an AuNPs suspensions [64]. The authors also examined in vivo burn healing and antibacterial properties of the formulations. The AuNPs thermoresponsive gels exhibited improved skin permeation, bioavailability, anti-inflammatory as well as antibacterial activities (Fig. 10.2 and Fig. 10.3). Besides, the gels did not exert skin irritation (in rabbits) or sensitization (in guinea pigs) suggesting their applicability as topical antibacterial drug delivery systems during skin inflammation and wound healing.

10.3 Antimicrobial activity of nanoparticles synthesized through biological route

The biological route for the synthesis of nanoparticles offers some additional advantages compared to the physicochemical route, especially, when the synthesized nanoparticles are planned for biomedical applications [67]. The nanoparticles should be biocompatibility with less-toxic/nontoxic in nature for biomedical applications. Microorganisms (bacteria and fungi), algae, and plants are the useful biogenic sources for the synthesis of nanoparticles. Various metabolites produced by diverse biota help in fabricating stable metal nanoparticles (Fig. 10.4). Biosynthesis of nanoparticles could be intracellular and/or extracellular [68,69]. When the biosynthesis is extracellular, easy recovery could be possible. Compared to intracellularly synthesized particles, extracellularly synthesized nanoparticles are easier to produce on a large scale, further, their production is more cost-effective also [70,71].

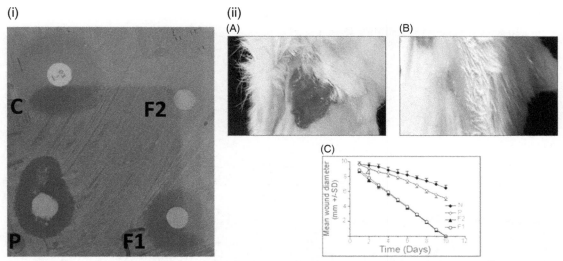

Fig. 10.2: (I) Antibacterial activity of the prepared formulations (F1 and F2) against *Staphylococcus aureus,*determined by disc diffusion antibiotic sensitivity method. C: control; Oxacillin antibiotic disc, P: commercial product; silver sulphadiazine, F1: (AuNPs-PF127) and F2: (AuNPs-PF127-HPMC). (II) Burn wound position on mouse dorsal skin treated with F1 formula, mouse before application; day 1 (A), mouse after end of treatment; day 9 (B), mean wound diameter (mm ± SD) daily application (C) of F1(AuNPs-PF127), F2 (AuNPs-PF127-HPMC) and P (silver sulphadiazine cream) compared with N (normal saline). *(Adapted with permission from ref. [64]).*

The gram positive and gram negative groups of bacteria have been employed for the biosynthesis of AgNPs and AuNPs.Klaus et al. [72] reported the biosynthesis of silver-based crystals (size upto 200 nm) using *Pseudomonasstutzeri* AG259 isolated from a silver mine. The synthesis was observed in the periplasmic space of the cell. A number of bacilli have also showed nanoparticles biosynthesis ability [73,74]. Saravanan et al. [75] have reported the rapid synthesis of AgNPs using *Bacillusmegaterium*. Priyadarshini et al. [75] fabricated anisotropic AgNPs using novel strain, *Bacillus flexus*. The particles were stable for 5 months and proved to be effective against multidrug resistant bacteria. Reddy et al. [73] studied the biosynthesis of AgNPs and AuNPs by *B. subtilis*. While AgNPs were synthesized extracellularly, AuNPs were found to be synthesized extra- and intracellularly. Based on SDS-PAGE analysis, the authors speculated the involvement of 25–66 kDa protein and 66–116 kDa protein in synthesizing AuNPs and AgNPs, respectively. Selvakumar et al. [76]reported that *Streptomycesrochei* as a good source for synthesizing extracellular AgNPs with antimicrobial activity.

Nanoparticles like AgNPs and AuNPs can be employed to cure a variety of ailments. Use of biogenic nanoparticles could be more advantageous [77]. Biosynthesized nanoparticles are biocompatible. Rossi-Bergmann et al. [78] observed IC_{50} values of mycosynthesized AgNPs to be more than three times less than that observed for chemically synthesized AgNPs. Alti et

(A)

(B)

(C)

(D)

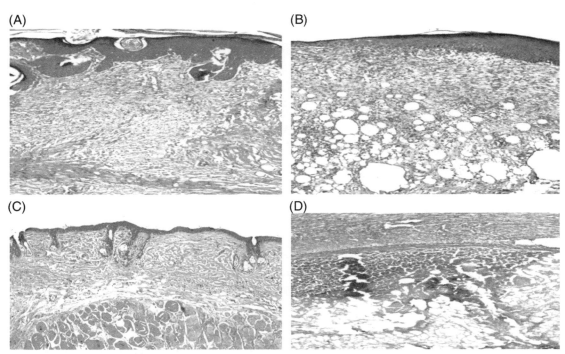

Fig. 10.3: Photomicrographs of tissue sections stained with H and E (16x mag), represent skin of mice treated with silver sulphadiazine, showing focal hyperkeratosis and acanthosis in the epidermis as well as focal granulation tissue formation in the underlying dermis (A), skin of mice treated with F2(AuNPs-PF127-HPMC), showing hyperkeratosis and acanthosis in the epidermis with granulation tissue formation and inflammatory cells infiltration in the underlying dermis (B), skin of mice treated with F1(AuNPs-PF127), showing intact normal histopathological structure of epidermis and dermis with hyalinosis in underlying skeletal muscle and few inflammatory cells infiltration (C) and skin of mice treated with saline showing focal suppuration withliquefaction and pus formation in the subcutaneous tissue (D). *(Adapted with permission from ref. [64]).*

al. reported reduced IC_{50} values of biogenic nanoparticles [79]. Further, it was reported that the IC_{50} values vary with the change in biomaterial sources. Rodrigues et al. [80] reported extracellular synthesis of spherical 35 ± 10 nm AgNPs by fungi *Aspergillus tubingensis* and *Bionectria ochroleuca*. The AgNPs were extremely effective (MIC 0.11–1.75 µg/mL) against *Candida* sp. Interestingly, AgNPs produced by *A. tubingensis* had high positive surface potential, which was unlike all known fungi. These AgNPs proved more effective antibacterial inhibiting 98% of *Pseudomonasaeruginosa*.

Depending on their size, shape, and morphology, nanoparticles exhibit unique properties that enable them to interact with various biota including animals, microbes, and plants [71]. Biogenically synthesized AgNPs and AuNPs have been reported to have antimicrobial activities against plant pathogenic bacteria and fungi also. However, such potential was lacking in the conventionally synthesized nanoparticles [81]. The biosynthesized AgNPs exhibit bactericidal activity against gram positive as well as gram negative bacteria. Drug resistance among microbes is one

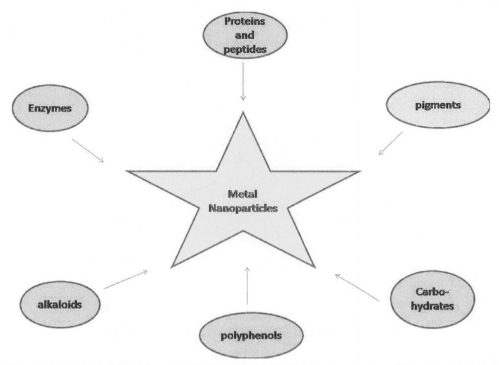

Fig. 10.4: Various biomolecules playing role in the formation and stabilization of metal nanoparticles.

of the serious problem these days, coupling of AgNPs with antibiotics could be quite useful in such cases. The culture supernatants of *K. pneumonia* can be used to form AgNPs within 5 minutes. The authors evaluated combined effect of the AgNPs and certain antibiotics against pathogenic bacteria. The highest enhancing effects in efficacy of antibiotics in the presence of AgNPs were observed in case of vancomycin, amoxicillin, and penicillin G against *S. aureus* [82].

The exact mechanism of antimicrobial activity of AgNPs and AuNPs is not completely known, however, a number of reports suggesting various ways for the attack of nanoparticles to microbial cells. Cellular components are released upon damage to cell wall due to the reaction of nanoparticles with sulfhydryl and phosphorus groups. The ATP depletion and interruption of respiratory chain occurs due to the generation of reactive oxygen species (ROS), when nanoparticles penetrate the bacterial cell membrane and attach to NADH dehydrogenase. Replication of DNA is disrupted and proteins are inactivated [67,81,83–85]. Recently, Tian et al. [86] has reviewed the antibacterial applications of AuNPs and the underlying mechanisms. Fig. 10.5 summarizes the effects of AuNPs on microbial cell.

Manikprabhu et al. [87] reported that due to their large surface area, AgNPs can attach with microbial cell membrane and cause damage to it. AgNPs have been reported to be effective against bacteria causing nosocomial infection [88]. Asghar et al. [89] compared the antifungal

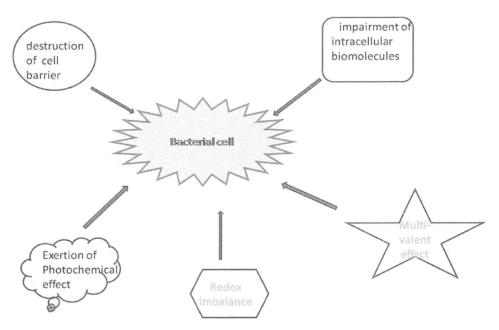

Fig. 10.5: Various bactericidal effects exerted by AuNPs.

efficacy of iron (Fe), copper (Cu), and silver nanoparticles (AgNPs) synthesized using green/black tea leaf extract and observed the AgNPs as the most effective antifungal agents followed by CuNPs and FeNPs. Further, the authors suggested inactivation of protein and damage to DNA leading to the replication failure. The authors further suggested easy penetration of nanoparticles (due to small size) into the cell wall and induction of cell lysis.

In recent years, the marine bionanotechnology has gained much attention. Marine biota ranging from bacteria to soft corals, invertebrates and plants have been observed to fabricate various nanoparticles with diverse shape and size, and hence, diverse applications [90,91]. Marine biota is endowed with ability to produce a battery of natural products including enzymes, carbohydrates, polyphenols, alkaloids, polymers, pigments, and other metabolites, in order to survive in extremes of conditions in marine environments. Such products could be harnessed as bioreducing, capping, and stabilizing agents for successful one-step biofabrication of nanostructures materials [67, 91–93].

Micro-organisms (including bacteria and fungi) from marine habitats have been reported to contribute significantly to biosynthesis of AgNPs and AuNPs [68,69,94–96]. Maharani et al. [97] reported marine *E. coli* synthesized AgNPs to display cytotoxic effects. Marine fungi have also been observed to produce Ag-Au bimetallic nanoparticles with cytotoxic effects on cancer cell lines [98]. Marine-derived fungi have been observed to be more efficient in producing nanoparticles compared to their bacterial counterparts. Upon screening many marine bacteria from various sea cost in India, Sharma et al. [95] obtained bacterium *Marinobacterpelagius* having ability to synthesize AuNPs.

Not only marine microorganisms, but even marine invertebrates and other marine animals have been observed to be sources of biofabrication of nanoparticles [91]. Aqueous extracts of polychaete *Marphysa moribidii* have been reported to synthesize AgNPs (40 nm) with antibacterial activities [99]. AuNPs with antibacterial activity were also synthesized using the extracts of the same species [100]. A giant jellyfish *Nemopilema nomurai* was used to synthesize 35.2 nm AuNPs with cytotoxic activities. Inbakandan et al.[101] harnessing marine sponge extract, biosynthesized flower like silver nanocolloids and reported their effect on oral biofilm bacteria and oral cancer cell lines. For biosynthesis of nanomaterial (100–120 nm), the water-soluble organic amines were suggested to be useful. Table 10.1 summarizes the biosynthesis of AgNPs and AuNPs and their potential applications.

Basu et al. [116] reported the biosynthesis of protein-coated AuNPs (5–25 nm) by the edible, mycorrhizal fungus *Tricholoma crassum* (Berk.) Sacc. A 40 kDa protein was observed to cap the nanoparticles. The nanoparticles of different size could be obtained by altering the synthetic parameters. The as produced nanoparticles exhibited antifungal and antibacterial activities. Neethu et al. [117] reported rapid and extracellular formation of AgNPs (10–15 nm) by an algicolous endophytic fungus *Penicilliumpolonicum*. The authors suggested possible role of protein components present in the fungal extract in synthesizing the nanoparticles. The as synthesized particles showed antibacterial activity against biofilm forming, multidrug-resistant with remarkable MIC and MBC values. Complete killing of the bacterial cells within 6 hours was achieved when killing kinetic assay was performed. The authors also studied effect of AgNPs on ultrastructure of the bacteria.

Very recently, Noman et al. [104] utilized secondary metabolic products of fungus Penicillium pedernalens 604 EAN for biosynthesizing AgNPs. The as synthesized nanoparticles were examined for their activity against spores of airborne fungi. Upon optimization using RSM and ANN, the authors could achieve remarkable inactivation of fungal spores. Damage to spore morphology due to the penetration of AgNPs was observed as revealed by FESEM analysis. Complete damage to spore surface was confirmed by AFM analysis. Based on Raman spectroscopic data, deformation of lipids was reported.

Yeasts are unicellular, nonfilamentous fungi, like their filamentous counter parts, yeasts have also been observed to be good resource for the green synthesis of nanoparticles. AgNPs and AuNPs have been reported to be synthesized by using yeasts. Apte et al. [118] reported the synthesis of AgNPs using psychrotrophic yeast *Yarrowialipolytica* NCYC 789. The AgNPs were fabricated using cell-associated melanin and exhibited antimicrobial properties. Pimprikar et al. [119] reported the biofabrication of AuNPs by tropical marine yeast *Yarrowialipolytica* NCIM 3589. Based on the experimental data, the authors suggested that by varying yeast biomass and supplied gold concentration, a variety of

Table 10.1: Biosynthesized AgNPs and AuNPs and their potential applications.

Sources	Name	Nanoparticles	Size/shape	Applications	Refs.
Bacteria					
	Bacillus subtilis	Ag Au		NR*	[73]
	B. cereus		4–5 nm; spherical	NR	[74]
	Pseudomonas sp. ef1	Ag	20–100 nm	antimicrobial	[102]
	Streptomyces sp. VITSTK7	Ag	20–60 nm	antifungal	[103]
	Marinobacter pelagius	Au	<10 nm	NR	[95]
	Streptomycesalbidoflavus	Ag	10–40 nm	antibacterial	[96]
Fungi					
	Penicillium pedernalens 604 EAN	Ag		antifungal	[104]
	Aspergillus niger (marine-derived)	Au		optics	[105]
	Aspergillus niger (marine-derived)	Ag		optics	[106]
	Penicillium sp	Ag	25–30 nm	Antibacterial	[107]
	Curvularia lunata	Ag	10–50 nm	Antibacterial	[108]
	Fusarium solani	Au	40–50 nm	anticancer	[109]
Plants					
	Gloriosasuperba	Ag, Au, Ag-Au		Antibacterial	[110]
	Halophila stipulacea	Ag (and Fe₂O₃)	17.7–25.0 nm	Antialgae	[111]
	Rhizophora mucronata	Ag	12 nm	antifungal	[112]
	Cinnamomum camphora	Ag, Au	55–80 nm	NR	[113]
	Citrus limetta	Ag	18 nm	Antibacterial, antifungal	[114]
	Olaxscandens	Ag	20–60 nm	Antibacterial	[115]

(*NR Not reported).

nanoparticles and nanoplates could be obtained. Further, the authors proposed the possible role of hydroxyl, carboxyl, and amide groups present on the cell surfaces in formation of nanoparticles.

Algae, the photosynthetic organisms produce a range of bioactive components and polymers including alginates, carrageenan, fucoidan, laminaran, lipids, fatty acids, and phenolic compounds, which play important role in synthesizing various nanoparticles [91,120]. AgNPs with average size 26.5 nm could be obtained using a microalga

Trichodesmium erythraeum. These AgNPs showed antibacterial, antioxidant, and antiproliferative properties [121]. When a red marine alga *Gelidium corneum*, was employed to synthesize AgNPs, 20–50 nm nanoparticles with antimicrobial and antibiofilm activities were obtained. Based on the FTIR analysis, azo compounds and oxidation of the aldehyde group were suggested to play a key role in the bio-reduction [122]. Algotiml et al. [123] compared the biosynthesis AgNPs by brown, red, and green algae and observed the extracts of green algae *Ulva rigida* to be the most effective. The algae mediated AgNPs could be efficient alternative antidermatophytes and anticancer agents especially against the MCF-7 cell line.

Plant-mediated synthesis of nanoparticles has also gained much attention as it offers cost-effective, environmentally benign, and rapid single-step procedure [124]. More than half of the publications on nanoparticles synthesis by plants were on AgNPs, this could be due to their strong microbicidal properties as well as easy reduction of silver salts [67]. Anjana et al. reported the synthesis of AgNPs using rapid microwave-assisted green method (with *Cyanthillium cinereum* as a reducing as well as capping agent) to exhibit antibacterial activity against *Staphylococcus aureus* and *Klebsiella pneumonia* [125]. Botteon et al. [126] examined the antibacterial and antifungal activities of AuNPs biosynthesized using Brazilian red propolis (BRP) extract and its fractions. Significant antimicrobial activities were exhibited by AuNPs synthesized by BRP crude extract and its fractions. The authors suggested the presence of medicarpin, vestitol, and benzophenones (Guttiferone E and Oblongifolin B) in the BRP crude extract and the fractions responsible for the antimicrobial activities besides heterogeneous shape of the nanoparticles. The authors also observed high cytotoxicity when bladder and prostate cancer cell lines were tested.

Huang et al. [113] exploited the sundried biomass of *Cinnamomum camphora* leaf for biosynthesizing AgNPs and AuNPs and obtained 55–80 nm AgNPs and triangular and spherical shaped AuNPs. The authors suggested polyol components and the water-soluble heterocyclic components to be responsible for the reduction of metal ion and stabilization of nanoparticles. Ahila et al. [127] obtained extracellular synthesis of stable AgNPs by seagrass *Syringodiumisoetifolium* with the proteins as reducing and capping agents. The 2–50 nm AgNPs exhibited antibacterial and cytotoxic potentials. Distinctive properties of AgNPs viz. small size, high stability, diversity in shape, remarkable conductivity etc. play important role in imparting antimicrobial properties to AgNPs. Unique physicochemical properties of AuNPs, biocompatibility, good scattering, and absorption characteristics, chemical inertness, and oxidation resistance play important role in conferring antimicrobial properties to them [91]. Bimetallic nanoparticles exhibit unique traits compared to their monometallic counterparts. Amina et al. [128] carried out *Asparagusracemosus* mediated biosynthesis of Ag, Au, and Ag-Au bimetallic nanoparticles (BNP) (Fig. 10.6). Upon examining the biomedically relevant properties of the particles, the authors found that compared to individual Ag and Au particles, BNP exhibited better antibacterial and immunomodulatory potentials. The authors

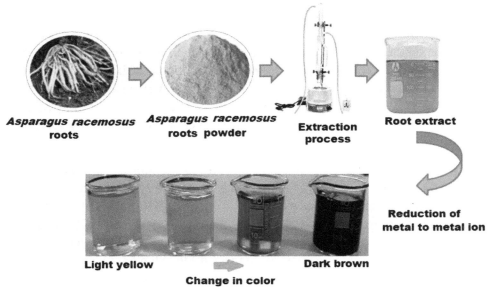

Asparagus racemosus roots   **Asparagus racemosus roots powder**   **Extraction process**   **Root extract**

Reduction of metal to metal ion

Light yellow → **Dark brown**

Change in color

Fig. 10.6: Schematic representation of the green synthesis of Ag-Au alloy nanoparticles by the *Asparargus racemosus* root extract in an ionic liquid medium and sequential color change from light yellow to dark brown. *(Adapted with permission from ref. [128]).*

explained mechanism of antibacterial and immunomodulatory potentials of the nanoparticles using a schematic diagram (Fig. 10.7). The authors suggested possible exploitation of the Ag-Au bimetallic alloy nanoparticles for the treatment of anti-inflammatory ailments. Botha et al. also reported that BNP to be more potent than individual AgNPs and AuNPs. The authors suggested that increased toxicity of the BNP could be due to the presence of Au functional portion in the BNP that aid in uptake of nanoparticles across the cell membrane [129]. Au-AgNPs have been reported as potential candidates for bacterial imaging and antibacterial activity [130].

10.4 Conclusions and future perspectives

In last one or two decades, attention has been given to antimicrobial activities of AgNPs, AuNPs, and Ag-Au alloy NPs. Promising reports are available narrating efficacy of individual as well as conjugated nanoparticles against pathogens including MDR ones. Size and dose dependent effects of nanoparticles also have been studied. Nontoxic behavior of AgNPs, AuNPs, and Ag-Au alloy NPs to healthy mammalian cells make them more advantageous as antimicrobials. However, majority of the reports available so far are confined to laboratory scale experiments. Clinical studies involving animal models, targeted delivery and studies pertaining to in vivo stability of particles would give a better view of applicability

(A)

Antibacterial

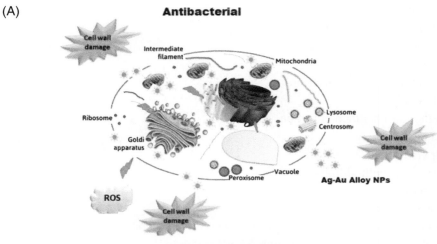

Immunomodulatory

(B)

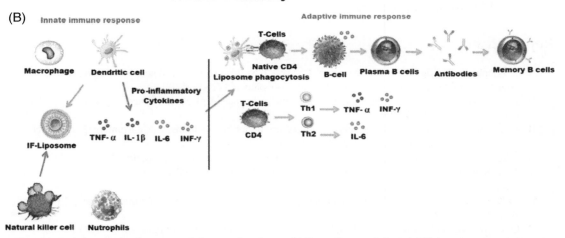

Fig. 10.7: Schematic diagram of the mechanism of (A) antibacterial and (B) immunomodulatory potentials of Ag-Au alloy NPs. *(Adapted with permission from ref. [128]).*

of nanoparticles. Further, exact mechanisms of antimicrobial effects of AgNPs, AuNPs, and Ag-Au alloy NPs need to be worked out. This chapter would definitely give a momentum to their biomedical applications.

Abbreviations

AgNPs   silver nanoparticles
ANN   artificial neural network
AuNPs   gold nanoparticles

BNP	bimetallic nanoparticles
DPPH	2,2-Diphenyl-1-(2,4,6-trinitrophenyl)hydrazyl
MBC	minimum bactericidal concentration
MDR	multi drug resistance
MIC	minimum inhibitory concentration
PEG	poly ethylene glycol
POEM	poly(oxyethylene)-segmented imide
PVA	Poly vinyl alcohol
PVP	polyvinylpyrrolidone
RSM	response surface methodology
SDS	sodium dodecyl sulfate
SEM	Scanning electron microscope
SMA	poly(styrene-*co*-maleic anhydride)-grafting poly(oxyalkylene)
XTT	2,3-bis(2-methoxy-4-nitro-5-sulphophenyl)-5-carboxanilide-2H-tetrazolium,monosodium salt

References

[1] J. Talapko, T. Matijevic, M. Juzbasic, A.A. Pozgain, I. Skrlec, Antibacterial activity of silver and its application in dentistry, Cardiol. Dermatol., Microorganism 8 (2020) 1400.

[2] F. Prestinaci, P. Pezzotti, A. Pantosti, Antimicrobial resistance: A global multifaceted phenomenon, Pathog. Glob. Health 109 (2015) 309–318.

[3] E.A. Mboya, L.A. Sanga, J.S. Ngocho, Irrational use of antibiotics in the Moshi Municipality Northern Tanzania: A cross sectional study, Pan Afr. Med. J. 31 (2018) 135.

[4] J. Talapko, D. Drenjancevic, M. Stupnisek, M. Tomic Paradzik, I. Kotris, D. Belic, M. Bogdan, P. Sikiric, *In vitro* evaluation of the antibacterial activity of pentadecapeptide BPC 157, Acta Clin. Croat. (2018) In press.

[5] M.K. Rai, A.P. Yadav, A.K. Gade, Silver nanoparticles as a new generation of antimicrobials, Biotechnol. Adv. 27 (2009) 76–82.

[6] G. Carotenuto, G.P. Pepe, L. Nicolais, Preparation and characterization of nano-sized Ag/PVP composites for optical applications, Eur. Phys. J. B 16 (2000) 11–17.

[7] E. Stathatos, P. Lianos, Photocatalytically deposited silver nanoparticles on mesoporous TiO_2 films, Langmuir 16 (2000) 2398–2400.

[8] S.V. Kyriacou, W.J. Brownlow, X.H.N. Xu, Using nanoparticle optics assay for direct observation of the function of antimicrobial agents in single live bacterial cells, Biochemistry 43 (2004) 140–147.

[9] X. Feng, H. Ma, S. Huang, W. Pan, X. Zhang, F. Tian, C. Gao, Y. Cheng, J. Luo, Aqueous-organic phase-transfer of highly stable gold, silver, and platinum nanoparticles and new route for fabrication of gold nanofilms at the oil/water interface and on solid supports, J. Phys. Chem. B 110 (2006) 12311–12317.

[10] S. Choi, K. Kim, S. Yeon, J. Cha, H. Lee, C. Kim, I. Yoo, Fabrication of silver nanoparticles via self-regulated reduction by 1-(2-hydroxyethyl)-3-methylimidazolium tetrafluoroborate, Korean J. Chem. Eng. 24 (2007) 856–859.

[11] L. Wang, C. Hu, L. Shao, The antimicrobial activity of nanoparticles: present situation and prospects for the future, Int. J. Nanomed. 12 (2017) 1227.

[12] A.H. Abo-Shama, H. El-Gendy, W.S. Mousa, R.A. Hamouda, W.E. Yousuf, H.F. Hetta, E.E. Abdeen, Synergistic and antagonistic effects of metal nanoparticles in combination with antibiotics against some reference strains of pathogenic microorganisms, Drug Resist 13 (2020) 351–362.

[13] Z. Deng, H. Zhu, B. Peng, H. Chen, Y. Sun, X. Gang, P. Jin, J. Wang, Synthesis of PS/Ag nanocomposite spheres with catalytic and antibacterial activities, ACS Appl Mater & interfaces 4 (10) (2012) 5625–5632.

[14] M. Veerapandian, S.K. Lim, H.M. Nam, G. Kuppannan, K.S. Yun, Glucosamine-functionalized silver glyconanoparticles: characterization and antibacterial activity, Anal. Bioanal. Chem. 398 (2) (2010) 867–876.

[15] U.T. Khatoon, G.N. Rao, K.M. Mohan, A. Ramanaviciene, A. Ramanavicius, Antibacterial and antifungal activity of silver nanospheres synthesized by tri-sodium citrate assisted chemical approach, Vacuum 146 (2017) 259–265.

[16] J.D. Padmos, R.T. Boudreau, D.F. Weaver, P. Zhang, Impact of protecting ligands on surface structure and antibacterial activity of silver nanoparticles, Langmuir 31 (12) (2015) 3745–3752.

[17] J. Lin, W. Lin, R. Dong, S. Hsu, The cellular responses and antibacterial activities of silver nanoparticles stabilized by different polymers, Nanotechnology 23 (2012) 065102.

[18] A.J. Kora, L. Rastogi, Enhancement of antibacterial activity of capped silver nanoparticles in combination with antibiotics, on model gram-negative and gram-positive bacteria, Bioinorg. Chem. Appl. (2013) 871097 Article ID.

[19] F. Huang, Y. Gao, Y. Zhang, T. Cheng, H. Ou, L. Yang, J. Liu, L. Shi, J. Liu, Silver-decorated polymeric micelles combined with curcumin for enhanced antibacterial activity, ACS Appl. Mater. Interfaces 9 (2017) 16880–16889.

[20] M. Gozdziewska, G. Cichowicz, K. Markowska, K. Zawada, E. Megiel, Nitroxide-coated silver nanoparticles: Synthesis, surface physicochemistry and antibacterial activity, RSC Adv. 5 (72) (2015) 58403–584015.

[21] K. Seku, B.R. Gangapuram, B. Pejjai, K.K. Kadimpati, N. Golla, Microwave-assisted synthesis of silver nanoparticles and their application in catalytic, antibacterial and antioxidant activities, J. Nanostr. Chem. 8 (2018) 179–188.

[22] B. Chudasama, A.K. Vala, N. Andhariya, R.V. Upadhyay, R.V. Mehta, Enhanced antibacterial activity of bifunctional Fe_3O_4-Ag core-shell nanostructures, Nano Res. 2 (12) (2009) 955–965.

[23] D. Wei, W. Sun, W. Qian, Y. Ye, X. Ma, The synthesis of chitosan-based silver nanoparticles and their antibacterial activity, Carbohydr. Res. 344 (2009) 2375–2382.

[24] R.J.B. Pinto, S.C.M. Fernandes, P. Sadocco, C.S.R. Freirea, J. Causio, C.P. Neto, T. Trindade, Antibacterial activity of optically transparent nanocomposite films based on chitosan or its derivatives and silver nanoparticles, Carbohydr. Res. 348 (2012) 77–83.

[25] H.V. Tran, L. Dai Tran, C.T. Ba, H.D. Vu, T.N. Nguyen, D.G. Pham, PX. Nguyen, Synthesis, characterization, antibacterial and antiproliferative activities of monodisperse chitosan-based silver nanoparticles, Colloids Surf. A 360 (1-3) (2010 May 5) 32–40.

[26] K. Shameli, M.B. Ahmad, M. Zargar, W.M. Zin W Y, N.A. Ibrahim, P. Shabanzadeh, M.G. Moghaddam, Synthesis and characterization of silver/montmorillonite/chitosan bionanocomposites by chemical reduction method and their antibacterial activity, Int. J. Nanomed. 6 (2011) 271–284.

[27] D. Roe, B. Karandikar, N. Bonn-Savage, B. Gibbins, J.B. Roullet, Antimicrobial surface functionalization of plastic catheters by silver nanoparticles, J. Antimicrob. Chemoth. 61 (2008) (2008) 869–876.

[28] E.T. Enan, A.A. Ashour, S. Basha, N.H. Felemban, S.M.F. Gad El-Rab, Antimicrobial activity of biosynthesized silver nanoparticles, amoxicillin, and glass-ionomer cement against *Streptococcusmutans* and *Staphylococcusaureus*, Nanotechnol 32 (2021) 215101, doi:10.1088/1361-6528/abe577.

[29] M. Azarsina, S. Kasraei, R. Yousef-Mashouf, N. Dehghani, M. Shirinzad, The antibacterial properties of composite resin containing nanosilver against Streptococcusmutans and Lactobacillus, J. Contemp. Dent. Pract 14 (2013) 1014–1018.

[30] R.A. Bapat, et al., An overview of application of silver nanoparticles for biomaterials in dentistry, Mater. Sci. Eng. C. 91 (2018) 881–898.

[31] A. Sodagar, et al., Evaluation of the antibacterial activity of a conventional orthodontic composite containing silver/hydroxyapatite nanoparticles Prog, Orthod 17 (2016) 40.

[32] M.A. Pfaller, D.J. Diekema, Epidemiology of invasive candidiasis: A persistent public health problem, Clin. Microbiol. Rev. 20 (2007) 133–163.

[33] G.S. Martin, D.M. Mannino, S. Eaton, M. Moss, The Epidemiology of Sepsis in the United States from 1979 through 2000, New England J. Med. 348 (2003) 1546–1554.

[34] M.D. Levin, J.G. Hollander, B. Holt, B.J. Rijnders, M. Vliet, P. Sonneveld, R.H.N. Schaik, Hepatotoxicity of oral and intravenous voriconazole in relation to cytochrome P450 polymorphisms, J. Antimicrobial Chemotherapy 60 (2007) 1104–1107.

[35] C. Taxvig, U. Hass, M. Axelstad, M. Dalgaard, J. Boberg, H.R. Andeasen, A.M. Vinggaard, Endocrine-disrupting activities in vivo of the fungicides tebuconazole and epoxiconazole, Toxicol. Sci. 100 (2) (2007) 464–473.

[36] T.C. White, S. Holleman, F. Dy, L.F. Mirels, D.A. Stevens, Resistance mechanisms in clinical isolates of *Candidaalbicans*, 46 (2002) 1704–1713

[37] A. Panacek, L. Kvitek, R. Prucek, M. Kolar, R. Vecerova, N. Pizurova, V.K. Sharma, T. Nevecna, R. Zboril, Silver colloid nanoparticles: Synthesis, characterization, and their antibacterial activity, J. Phys. Chem. B 110 (2006) 16248–16253.

[38] L. Kvitek, R. Prucek, A. Panacek, R. Novotny, J. Hrbac, R. Zboril, The influence of complexing agent concentration on particle size in the process of SERS active silver colloid synthesis, J. Mater. Chem. 15 (2005) 1099–1105.

[39] K.J. Kim, W.S. Sung, B.K. Suh, S.K. Moon, J.S. Choi, J.G. Kim, D.G. Lee, Antifungal activity and mode of action of silver nano-particles on *Candidaalbicans*, Biometals 22 (2009) 235–242.

[40] R.J.B. Pinto, A. Almeida, S.C.M. Fernandes, C.S.R. Freirea, A.J.D. Silvestrea, C.P. Netoa, T. Trindadea, Antifungal activity of transparent nanocomposite thin films of pullulan and silver against *Aspergillusniger*, Coll. Surf. B: Biointerfaces 103 (2013) 143–148.

[41] D.R. Monteiro, S. Silva, M. Negri, L.F. Gorup, E.R. Camargo, R. Oliveira, D.B. Barbosa, M. Henriques, Silver nanoparticles: influence of stabilizing agent and diameter on antifungal activity against *Candidaalbicans* and *Candidaglabrata* biofilms, Lett. Appl. Microbiol. 54 (2012) 383–391.

[42] L. Pereira, N. Dias, J. Carvalho, S. Fernandes, C. Santos, N. Lima, Synthesis, characterization and antifungal activity of chemically and fungal-produced silver nanoparticles against Trichophyton rubrum, J. Appl. Microbiol. 117 (2014) 1601–1613.

[43] S.H.S. Dananjaya, W.K.C.U. Erandani, C.H. Kim, C. Nikapitiyac, J. Leec, M.D. Zoysa, Comparative study on antifungal activities of chitosan nanoparticles and chitosan silver nano composites against *Fusarium oxysporum* species complex, J. Biol. Macromole 105 (2017) 478–488.

[44] S. Ifukua, Y. Tsukiyamaa, T. Yukawaa, M. Egusaa, H. Kaminaka, H. Izawaa, M. Morimotoc, H. Saimoto, Facile preparation of silver nanoparticles immobilized on chitin nanofiber surfaces to endow antifungal activities, Carbohydr. Polym. 117 (2015) 813–817.

[45] M. Selvaraj, P. Pandurangan, N. Ramasami, S.B. Rajendran, S.N. Sangilimuthu, P. Perumal, Highly potential antifungal activity of quantum-sized silver nanoparticles against *Candidaalbicans*, Appl. Biochem. Biotechnol. 173 (2014) 55–66.

[46] B. Chudasama, A.K. Vala, N. Andhariya, R.V. Upadhyay, R.V. Mehta, Antifungal activity of multifunctional Fe_3O_4–Ag nanocolloids, J. Magn. Magn. Mater. 323 (2011) 1233–1237.

[47] L.E. Tejeda, F. Malpartida, A.E. Cubillo, C. Pecharroman, J.S. Moya, The antibacterial and antifungal activity of a soda-lime glass containing silver nanoparticles, Nanotechnology 20 (2009) 085103.

[48] I.A. Wani, T. Ahmad, Size and shape dependant antifungal activity of gold nanoparticles: a case study of *Candida*, Colloids Surf B: Biointerfaces 101 (2013) 162–170.

[49] Yu Tao, E. Ju, J. Ren, X. Qu, Bifunctionalized mesoporous silica-supported gold nanoparticles: Intrinsic oxidase and peroxidase catalytic activities for antibacterial applications, Adv. Mater. 27 (2015) 1097–1104.

[50] L. Armelao, D. Barreca, G. Bottaro, A. Gasparotto, C. Maccato, C. Maragno, E. Tondello, U. Stangar, M. Bergant, D. Mahne, Photocatalytic and antibacterial activity of TiO_2 and Au/TiO_2 nanosystems, Nanotechnology 18 (2007) 375709.

[51] G.L. Burygin, B.N. Khlebtsov, A.N. Shantrokha, L.A. Dykman, V.A. Bogatyrev, N.G. Khlebtsov, On the enhanced antibacterial activity of antibiotics mixed with gold nanoparticles, Nanoscale Res. Lett. 4 (2009) 794–801.

[52]   L.H. Peng, Y.F. Huang, C.Z. Zhang, J. Niu, Y. Chen, Y. Chu, Z.H. Jiang, J.Q. Gao, Z.W. Mao, Integration of antimicrobial peptides with gold nanoparticles as unique non-viral vectors for gene delivery to mesenchymal stem cells with antibacterial activity, Biomaterials 103 (2016) 137–149.

[53]   J.N. Payne, H.K. Waghwani, M.G. Connor, W. Hamilton, S. Tockstein, H. Moolani, F. Chavda, V. Badwaik, M.B. Lawrenz, R. Dakshinamurthy, Novel synthesis of kanamycin conjugated gold nanoparticles with potent antibacterial activity, Front. Microbiol. 7 (2016) 607 Article.

[54]   S. Perni, C. Piccirillo, A. Kafizas, M. Uppal, J. Pratten, M. Wilson, I. Parkin, Antibacterial activity of light-activated silicone containing methylene blue and gold nanoparticles of different sizes, J. Clust. Sci. 21 (2010) 427–438.

[55]   Y. Zhu, M. Ramasamy, D.K. Yi, Antibacterial activity of ordered gold nanorod arrays, ACS Appl. Mater. Interfaces 6 (2014) 15078–15085.

[56]   K.S. Babu, M. Anandkumar, T.Y. Tsai, T.H. Kao, B.S. Inbaraj, BH. Chen, Cytotoxicity and antibacterial activity of gold-supported cerium oxide nanoparticles, Int. J. Nanomed. 9 (2014) 5515–5531.

[57]   J.C. Castillo-Martínez, G.A. Martínez-Castañón, F. Martínez-Gutierrez, N.V. Zavala Alonso, N. Patiño-Marín, N. Niño-Martinez, V. Zaragoza-Magaña, C. Cabral-Romero, Antibacterial and antibiofilm activities of the photothermal therapy using gold nanorods against seven different bacterial strains (2009) Article ID783671

[58]   E.O. Mikhailova, Gold nanoparticles: Biosynthesis and potential of biomedical application, J.Funct. Biomater. 12 (2021) 70, https://doi.org/10.3390/jfb12040070.

[59]   A. Samanta, R. Gangopadhyay, C.K. Ghosh, M. Ray, Enhanced photoluminescence from gold nanoparticle decorated polyaniline nanowire bundles, RSC Adv. 7 (2017) 27473–27479.

[60]   R. Geethalakshmi, D.V.L. Sarada, Gold and silver nanoparticles from *Trianthemadecandra*: Synthesis, characterization, and antimicrobial properties, Int. J. Nanomed. 7 (2012) 5375–5384.

[61]   C. Tao, Antimicrobial activity and toxicity of gold nanoparticles: Research progress, challenges and prospects, Lett. Appl. Microbiol. 67 (2018) 537–543.

[62]   D. Church, S. Elsayed, O. Reid, B. Winston, B.R. Lindsay, Burn wound infections, Clini. Microbiol. Rev. 19 (2006) 403–434.

[63]   R.F. El-Kased, Natural antibacterial remedy for respiratory tract infections, Asian Pacific J. Trop. Biomed. 6 (2016) 270–274.

[64]   M.G. Arafa, R.F. El-Kased, M.M. Elmazar, Thermoresponsive gels containing gold nanoparticles as smart antibacterial and wound healing agents, Sci Rep. 8 (2018) 13674, https://doi.org/10.1038/s41598-018-31895-4.

[65]   S. Honari, Topical therapies and antimicrobials in the management of burn wounds, Crit. Care Nurs. Clin. North. Am. 16 (2004) 1–11.

[66]   M. G.Arafa, B.M. Ayoub, DOE optimization of nano-based carrier of pregabalin as hydrogel: new therapeutic and chemometric approaches for controlled drug delivery systems, Sci. Rep. 7 (2017) 41503.

[67]   A.K. Vala, H. Trivedi, H. Gosai, H. Panseriya, B. Dave, Biosynthesized silver nanoparticles and their therapeutic applications, in: S K Verma, A K Das (Eds.), Biosynthesized silver nanoparticles and their therapeutic applications, Compr. Anal. Chem. 94 (2021) 547–584.

[68]   A.K. Vala, Intra- and extracellular biosynthesis of gold nanoparticles by a marine-derived fungus *Rhizopus oryzae*, Synth. React Inorg. Met. Org. Nano Met. Chem. 44 (9) (2014) 1243–1246.

[69]   A.K. Vala, Exploration on green synthesis of gold nanoparticles by a marine derived fungus *Aspergillus sydowii*, Env. Prog. Sustain Energy 34 (1) (2015) 194–197.

[70]   N. Durán, D. Priscyla, P.D. Marcato, O. Alves, G. De Souza, E. Esposito, Mechanistic aspects of biosynthesis of silver nanoparticles by several *Fusarium oxysporum* strains, J. Nanobiotechnol. 3 (2005) 1–7.

[71]   K.S. Siddiqi, A. Husen, R.A.K R.A.K. Rao, A review on biosynthesis of silver nanoparticles and their biocidal properties, J. Nanobiotechnol. 16 (14) (2018), https://doi.org/10.1186/s12951-018-0334-5.

[72]   T. Klaus, R. Joerger, E. Olsson, C.G. Granqvist, Silver-based crystalline nanoparticles, microbially fabricated, Proc Natl Acad Sci USA. 96 (1999) 13611–13614.

[73]   A.S. Reddy, C.Y. Chen, C.C. Chen, J.S. Jean, H.R. Chen, M.J. Tseng, C.W. Fan, JC. Wang, iological synthesis of gold and silver nanoparticles mediated by the bacteria *Bacillus subtilis*, J. Nanosci. Nanotechnol. 10 (2010) 6567–6574.

[74] M.M. Ganesh Babu, P. Gunasekaran, Production and structural characterization of crystalline silver nanoparticles from *Bacillus cereus* isolate, Coll Surf B 74 (2009) 191–195.

[75] M. Saravanan, H. Vahidi, D.M. Cruz, A. Vernet-Crua, E. Mostafavi, R. Stelmach, T.J. Webster, M.A. Mahjoub, M. Rashed, H. Barabadi, Emerging antineoplastic biogenic gold nanomaterials for breast cancer therapeutics: A systematic review, Internat. J. Nanomed. 15 (2020) 3577.

[76] P. Selvakumar, S. Viveka, S. Prakash, S. Jasminebeaula, R. Uloganathan, Antimicrobial activity of extracellularly synthesized silver nanoparticles from marine derived *Streptomycesrochei*, Int. J. Pharm. Biol. Sci. 3 (2012) 188–197.

[77] S. Iravani, Green synthesis of metal nanoparticles using plants, Green Chem. 13 (2011) 2638–2650.

[78] B. Rossi-Bergmann, W. Pacienza-Lima, P.D. Marcato, R. De Conti, N. Durán, Therapeutic potential of biogenic silver nanoparticles in murine cutaneous leishmaniasis, J. Nano Res. 20 (2012) 89–97.

[79] M. Dayakar Alti, D. Veeramohan Rao, N. Rao, R. Maurya, SK. Kalangi, Gold–silver bimetallic nanoparticles reduced with herbal leaf extracts induce ros-mediated death in both promastigote and amastigote stages of Leishmaniadonovani, ACSOmega 5 (26) (2020) 16238–16245, doi:10.1021/acsomega.0c02032.

[80] A.G. Rodrigues, L.Y. Ping, P.D. Marcato, O.L. Alves, M.C.P. Silva, R.C. Ruiz, I.S. Melo, L. Tasic, A.O. De Souza, Biogenic antimicrobial silver nanoparticles produced by fungi, Appl. Microbiol. Biotechnol. 3 (2013) 775–782, doi:10.1007/s00253-012-4209-7.

[81] L. Castillo-Henríquez, K. Alfaro-Aguilar, J. Ugalde-Álvarez, L. Vega-Fernández, G. Montes de Oca-Vásquez, J.R. Vega-Baudrit, Green synthesis of gold and silver nanoparticles from plant extracts and their possible applications as antimicrobial agents in the agricultural area, Nanomaterials 10 (9) (2020) 1763.

[82] A.R. Shahverdi, A. Fakhimi, H.R. Shahverdi, S. Minaian, Synthesis and effect of silver nanoparticles on the antibacterial activity of different antibiotics against *Staphylococcusaureus* and *Escherichiacoli*, Nanomed: Nanotechnol, Biol. Med. 3 (2) (2007) 168–171.

[83] A. Ahmad, F. Syed, M. Imran, A.U. Khan, K. Tahir, Z.U.H. Khan, Q. Yuan, Phytosynthesis and antileishmanial activity of gold nanoparticles by *Maytenus royleanus*, J. Food Biochem. 40 (2016) 420–427.

[84] Y. Cui, Y. Zhao, Y. Tian, W. Zhang, X. Lü, X. Jiang, The molecular mechanism of action of bactericidal gold nanoparticles on *Escherichiacoli*, Biomaterials 33 (2022) 2327–2333.

[85] G. Franci, A. Falanga, S. Galdiero, L. Palomba, M. Rai, G. Morelli, M. Galdiero, Silver nanoparticles as potential antibacterial agents, Molecules. 20 (2015) 8856–8874.

[86] E.K. Tian, Y. Wang, R. Ren, W. Zheng, W. Liao, Gold nanoparticle: Recent progress on its antibacterial applications and mechanisms, J. Nanomater. 2021 (2021) 2501345. Article ID, https://doi.org/10.1155/2021/2501345.

[87] D. Manikprabhu, J. Cheng, W. Chen, A.K. Sunkara, S.B. Mane, R. Kumar, M. Das, W.N. Hozzein, Y.Q. Duan, W.J. Li, Sunlight mediated synthesis of silver nanoparticles by a novel actinobacterium (*Sinomonasmesophila* MPKL 26) and its antimicrobial activity against multi drug resistant *Staphylococcus aureus*, J. Photochem. Photobiol. B: Biol. 158 (2016) 202–205.

[88] M. Hekmati, S. Hasanirad, A. Khaledi, D. Esmaeili, Green synthesis of silver nanoparticles using extracts of *Allium rotundum* l, *Falcaria vulgaris* Bernh, and *Ferulago angulate* Boiss, and their antimicrobial effects in vitro, Gene Reports 19 (2020) 100589.

[89] M.A. Asghar, E. Zahir, S.M. Shahid, M.N. Khan, M.A. Asghar, J. Iqbal, G. Walker, Iron, copper and silver nanoparticles: Green synthesis using green and black tea leaves extracts and evaluation of antibacterial, antifungal and aflatoxin B1 adsorption activity, LWT 90 (2018) 98–107.

[90] S. AlNadhari, N.M. Al-Enazi, F. Alshehrei, F. Ameen, A review on biogenic synthesis of metal nanoparticles using marine algae and its applications, Environ. Res. 194 (2021) 110672, https://doi.org/10.1016/j.envres.2020.110672.

[91] N. Yosri, S.A. Khalifa, Z. Guo, B. Xu, X. Zou, H.R. El-Seedi, Marine organisms: Pioneer natural sources of polysaccharides/proteins for green synthesis of nanoparticles and their potential applications, Int. J. Biol. Macromole 193 (2021) 1767–1798.

[92] S.A.M. Khalifa, N. Elias, M.A. Farag, L. Chen, A. Saeed, M.-E.F. Hegazy, M.S. Moustafa, A. Abd El-Wahed, S.M. Al-Mousawi, S.G. Musharraf, F.-R. Chang, A. Iwasaki, K. Suenaga, M. Alajlani, U.

Gˮoransson, H.R. El-Seedi, Marine natural products: A source of novel anticancer drugs, Mar. Drugs. 17 (2019) 491, https://doi.org/10.3390/md17090491.

[93] H.R. El-Seedi, R.M. El-Shabasy, S.A.M. Khalifa, A. Saeed, A. Shah, R. Shah, F.J. Iftikhar, M.M. Abdel-Daim, A. Omri, N.H. Hajrahand, J.S.M. Sabir, X. Zou, M.F. Halabi, W. Sarhan, W. Guo, Metal nanoparticles fabricated by green chemistry using natural extracts: Biosynthesis, mechanisms, and applications, RSC Adv. 9 (2019) 24539–24559, https://doi.org/10.1039/c9ra02225b.

[94] R.S. Prakasham, B.S. Kumar, Y.S. Kumar, G.G. Shankar, Characterization of silver nanoparticles synthesized by using marine isolate *Streptomycesalbidoflavus*, J. Microbiol Biotechnol. 22 (5) (2012) 614–621.

[95] N. Sharma, A.K. Pinnaka, M. Raje, et al., Exploitation of marine bacteria for production of gold nanoparticles, Microb. Cell Fact. 11 (2012) 86, https://doi.org/10.1186/1475-2859-11-86.

[96] A.K. Vala, S. Shah, R. Patel, Biogenesis of silver nanoparticles by marine-derived fungus *Aspergillus flavus* from Bhavnagar Coast, Gulf of Khambhat, India, J. Mar. Biol. Oceanogr. 3 (2014) (2014) 1.

[97] V. Maharani, A. Sundaramanickam, T. Balasubramanian, In vitro anticancer activity of silver nanoparticle synthesized by *Escherichiacoli* VM1 isolated from marine sediments of Ennore southeast coast of India, Enzym. Microb. Technol. 95 (2016) 146–154, https://doi.org/10.1016/j.enzmictec.2016.09.008.

[98] A.K. Vala, H.B. Trivedi, B.P. Dave, Marine-derived fungi: Potential candidates for fungal nanobiotechnology, in: R. Prasad (Ed.), Advances and Applications Through Fungal Nanobiotechnology. Fungal Biology, Springer, Cham, 2016, https://doi.org/10.1007/978-3-319-42990-8_3.

[99] N.S.R. Rosman, N.A. Harun, I. Idris, W.I.W. Ismail, Eco-friendly silver nanoparticles (AgNPs) fabricated by green synthesis using the crude extract of marine polychaete Marphysa moribidii: biosynthesis, characterisation, and antibacterial applications 6 (2020) e05462, https://doi.org/10.1016/j.heliyon.2020. e05462.

[100] A.U.E. Pei, P.C. Huai, M.A.A. Masimen, W.I.W. Ismail, I. Idris, N.A. Harun, Biosynthesis of gold nanoparticles (AuNPs) by marine baitworm Marphysa moribidii idris, hutchings and arshad 2014 (A*nnelida*: *Polychaeta*) and its antibacterial activity, Adv. Nat. Sci. Nanosci. Nanotechnol. 11 (2020) 15001.

[101] D. Inbakandan, C. Kumar, M. Bavanilatha, D.N. Ravindra, R. Kirubagaran, S.A. Khan, Ultrasonic-assisted green synthesis of flower like silver nanocolloids using marine sponge extract and its effect on oral biofilm bacteria and oral cancer cell lines, Microb. Pathog. 99 (2016) 135–141, https://doi.org/10.1016/j. micpath.2016.08.018.

[102] M.S. John, J.A. Nagoth, K.P. Ramasamy, A. Mancini, G. Giuli, A. Natalello, P. Ballarini, C. Miceli, S. Pucciarelli, Synthesis of bioactive silver nanoparticles by a *Pseudomonas* strain associated with the Antarctic psychrophilic protozoon *Euplotesfocardii*, Mar. Drugs 18 (2020) 38, https://doi.org/10.3390/ md18010038.

[103] M. Thenmozhi, K. Kannabiran, R. Kumar, V.G Khanna, Antifungal activity of Streptomyces sp. VITSTK7 and its synthesized Ag2O/Ag nanoparticles against medically important *Aspergillus* pathogens, Journal de Mycologie Médicale 23 (2) (2013) 97–103. 156-5233, https://doi.org/10.1016/j.mycmed.2013.04.005.

[104] E. Noman, A. Al-Gheethi, R.M.S.R. Mohamed, B. Talip, N. Othman, S. Hossain, et al., Inactivation of fungal spores from clinical environment by silver bio-nanoparticles; optimization, artificial neural network model and mechanism, Environ. Res. 204 (A) (2022) 111926, https://doi.org/10.1016/j. envres.2021.111926.

[105] V. Dave, A.K. Vala, R. Patel, Observation of weak localization of light in gold nanofluids synthesized using the marine derived fungus Aspergillus niger, RSC Adv. 5 (22) (2015) 16780–16784.

[106] A.K. Vala, B. Chudasama, R.J. Patel, Green synthesis of silver nanoparticles using marine-derived fungus *Aspergillus niger*, Micro Nano Lett 7 (2012) 859–862.

[107] S. D, V. Rathod, S. Ninganagouda, J. Hiremath, A.K. Singh, J. Mathew, Optimization and Characterization of Silver Nanoparticle by Endophytic Fungi Penicillium sp. Isolated from Curcuma longa (Turmeric) and Application Studies against MDR E. coli and S. aureus, Bioinorg. Chem. Appl. 2014 (2014) 408021, doi:10.1155/2014/408021 Article ID.

[108] P. Ramalingmam, S. Muthukrishnan, P. Thangaraj, Biosynthesis of silver nanoparticles using an endophytic fungus, *Curvularialunata* and its antimicrobial potential, J. Nanosci. Nanoeng. 1 (4) (2015) 241–247.

[109] P. Clarance, B. Luvankar, J. Sales, A. Khusro, P. Agastian, J.-C. Tack, M.M. Al Khulaifi, H.A. AL-Shwaiman, A.M. Elgorban, A. Syed, H.-J. Kim, Green synthesis and characterization of gold nanoparticles using endophytic fungi *Fusariumsolani* and its in-vitro anticancer and biomedical applications, Saudi. J. Biol Sci. 27 (2) (2020) 706–712.

[110] K. Gopinath, S. Kumaraguru, K. Bhakyaraj, S. Mohan, K.S. Venkatesh, M. Esakkirajan, et al., Green synthesis of silver, gold and silver/gold bimetallic nanoparticles using the *Gloriosasuperba* leaf extract and their antibacterial and antibiofilm activities, Microb. Pathog. 101 (2016) 1–11.

[111] H.Y. El-Kassas, M.G. Ghobrial, Biosynthesis of metal nanoparticles using three marine plant species: Anti-algal efficiencies against "*Oscillatoriasimplicissima*, Environ. Sci. Pollut. Res. 24 (2017) 7837–7849, https://doi.org/10.1007/s11356-017-8362-5.

[112] M. Singh, M. Kumar, R. Kalaivani, S. Manikandan, A.K. Kumaraguru, Metallic silver nanoparticle: A therapeutic agent in combination with antifungal drug against human fungal pathogen, Bioprocess Biosyst. Eng. 36 (2013) 407–415, https://doi.org/10.1007/s00449-012-0797-y.

[113] J. Huang, Q. Li, D. Sun, Y. Lu, Y. Su, Y. Yang, X. Chen, C., Biosynthesis of silver and gold nanoparticles by novel sundried *Cinnamomumcamphora* leaf, Nanotechnol. 18 (10) (2007) 105104.

[114] T. Dutta, N.N. Ghosh, M. Das, R. Adhikary, V. Mandal, A.P. Chattopadhyay, Green synthesis of antibacterial and antifungal silver nanoparticles using *Citruslimetta* peel extract: Experimental and theoretical studies, J. Environ. Chem. Eng. 8 (4) (2020) 104019, https://doi.org/10.1016/j.jece.2020.104019.

[115] S. Mukherjee, D. Chowdhury, R. Kotcherlakota, S. Patra, V. Binothkumar, M.P. Bhadra, B. Sreedhar, C.R. Patra, Potential theranostics application of bio-synthesized silver nanoparticles (4-in-1 system), Theranostics 4 (2014) 316–335, doi:10.7150/thno.7819.

[116] A. Basu, S. Ray, S. Chowdhury, et al., Evaluating the antimicrobial, apoptotic, and cancer cell gene delivery properties of protein-capped gold nanoparticles synthesized from the edible mycorrhizal fungus *Tricholoma crassum*, Nanoscale Res. Lett. 13 (2018) 154, https://doi.org/10.1186/s11671-018-2561-y.

[117] S. Neethu, S.J. Midhun, E.K. Radhakrishnan, M. Jyothis, Green synthesized silver nanoparticles by marine endophytic fungus Penicillium polonicum and its antibacterial efficacy against biofilm forming, multidrug-resistant Acinetobacter baumanii, Microb. Pathog. 116 (2018) 263–272, https://doi.org/10.1016/j.micpath.2018.01.033.

[118] M. Apte, D. Sambre, S. Gaikawad, S. Joshi, A. Bankar, A.R. Kumar, S. Zinjarde, Psychrotrophic yeast Yarrowia lipolytica NCYC 789 mediates the synthesis of antimicrobial silver nanoparticles via cell-associated melanin, AMB Express 3 (2013) 32, https://doi.org/10.1186/2191-0855-3-32.

[119] P.S. Pimprikar, S.S. Joshi, A.R. Kumar, S.S. Zinjarde, S.K. Kulkarni, Influence of biomass and gold salt concentration on nanoparticle synthesis by the tropical marine yeast Yarrowia lipolytica NCIM 3589, Colloids Surf. B Biointerfaces 74 (2009) 309–316, https://doi.org/10.1016/j.colsurfb.2009.07.040.

[120] M. J¨onsson, L. Allahgholi, R.R.R. Sardari, G.O. Hreggviðsson, E. Nordberg Karlsson, Extraction and modification of macroalgal polysaccharides for current and next-generation applications, Molecules 25 (2020) 930, https://doi.org/10.3390/molecules25040930.

[121] R.S. Sathishkumar, A. Sundaramanickam, R. Srinath, T. Ramesh, K. Saranya, M. Meena, P. Surya, Green synthesis of silver nanoparticles by bloom forming marine microalgae Trichodesmium erythraeum and its applications in antioxidant, drug-resistant bacteria, and cytotoxicity activity, J. Saudi Chem. Soc. 23 (2019) 1180–1191, https://doi.org/10.1016/j.jscs.2019.07.008.

[122] B. Yılmaz Öztürk, B. Yenice Gürsu, ˙I. Dağ, Antibiofilm and antimicrobial activities of green synthesized silver nanoparticles using marine red algae Gelidium corneum, Process Biochem. 89 (2020) 208–219, doi:10.1016/j. procbio.2019.10.027.

[123] R. Algotiml, A. Gab-Alla, R. Seoudi, H.H. Abulreesh, M. Zaki El Readi, K. Elbanna, Anticancer and antimicrobial activity of biosynthesized Red Sea marine algal silver nanoparticles, Sci. Rep. 12 (2022) 2421, doi:10.1038/s41598-022-06412-3.

[124] N. Ahmad Kuthi, S. Chandren, N. Basar, M.S.S. Jamil, Biosynthesis of gold nanoisotrops using carallia brachiata leaf extract and their catalytic application in the reduction of 4-nitrophenol, Front. Chem. 9 (2022) 800145, doi:10.3389/fchem.2021.800145.

[125] V.N. Anjana, M. Joseph, S. Francis, A. Joseph, EP. Koshy, B. Mathew, Microwave assisted green synthesis of silver nanoparticles for optical, catalytic, biological and electrochemical applications, Artificial Cells, Nanomedicine, and Biotechnology 49 (1) (2021) 438–449, doi:10.1080/21691401.2021.1925678.

[126] C.E.A. Botteon, L.B. Silva, G.V. Ccana-Ccapatinta, et al., Biosynthesis and characterization of gold nanoparticles using Brazilian red propolis and evaluation of its antimicrobial and anticancer activities, Sci. Rep. 11 (2021) 1974, doi:10.1038/s41598-021-81281-w.

[127] N.K. Ahila, V.S. Ramkumar, S. Prakash, B. Manikandan, J. Ravindran, P.K. Dhanalakshmi, E. Kannapiran, Synthesis of stable nanosilver particles (AgNPs) by the proteins of seagrass Syringodium isoetifolium and its biomedicinal properties, Biomed. Pharmacother. 84 (2016) 60–70, doi:10.1016/j. biopha.2016.09.004.

[128] M. Amina, N.M. Al Musayeib, N.A. Alarfaj, M.F. El-Tohamy, G.A. Al-Hamoud, Antibacterial and immunomodulatory potentials of biosynthesized Ag, Au, Ag-Au bimetallic alloy nanoparticles using the *Asparagus racemosus* root extract, Nanomaterials (Basel) 10 (12) (2020) 2453, doi:10.3390/nano10122453.

[129] T.L. Botha, E.E. Elemike, S. Horn, D.C. Onwudiwe, J.P. Giesy, V. Wepener, Cytotoxicity of Ag, Au and Ag-Au bimetallic nanoparticles prepared using golden rod (*Solidago canadensis*) plant extract, Sci Rep. 9 (2019) 4169, doi:10.1038/s41598-019-40816-y.

[130] X. Ding, P. Yuan, N. Gao, H. Zhu, Y. Yan Yang, Q.H. Xu, Au-Ag core-shell nanoparticles for simultaneous bacterial imaging and synergistic antibacterial activity, Nanomed: Nanotech Biol. Med. 13 (1) (2017) 297–305.

CHAPTER 11

Covalent organic framework-functionalized Au and Ag nanoparticles: Synthesis and applications

Ritambhara Jangir

Department of Chemistry, Sardar Vallabhbhai National Institute of Technology, Surat, Gujarat, India

11.1 Introduction

Nanoparticles (NPs) have taken a lot of attention due to their unique physical and chemical properties such as high adsorption ability, rapid diffusion rate, changeable surface features [1]. In the class of nanoparticles, gold and silver nanoparticles (AuNPs and AgNPs) are well known nanoparticles due to their widespread applications in nanobiotechnology, biosensor studies, visualization of cell structures, and targeted drug delivery [2,3]. However, limited applicability has been observed with NPs due to strong tendency to aggregate, thus the effective utilization is getting affected [4]. To rectify this issue, NPs have been anchored with surfactants, chelating ligands, activated carbons, mesoporous silica, porous coordination polymers, metal oxides, graphene, metal-organic frameworks (MOFs), etc. Protection of NPs not only imparts stability against aggregation but also many other advantages such as improved solubility, controlled shape, ordered assembling of nanoparticles, improved electron-transfer efficiency etc. [5–8].

Inorganic porous materials have been widely used to get nanocomposites of desirable applications due to lesser reactivity in the given conditions. The lager surface area and porous nature of MOFs attract them to combine with NPs but practical usability will be limited due to lower stability especially in acidic conditions [8]. A new class of porous crystalline materials that is, covalent organic frameworks (COFs) has been introduced by Yaghi et al. which has shown tremendously high surface area, defined pore structure and desirable functionality [9–12]. Due to these outstanding properties, COFs have been widely employed in gas storage, catalyst supports, semiconductors, and optoelectronic devices. Desirable building blocks having nitrogen-rich and oxygen-rich functional groups can easily be introduced into COFs. Controlling in growth and minimization in aggregation of NPs can be done by growing them in regulated pores of COFs [13,14]. Despite low crystalline nature, these materials have high thermal

Gold and Silver Nanoparticles.
DOI: https://doi.org/10.1016/B978-0-323-99454-5.00001-9

355

and chemical stability including strong acidic and basic conditions which are the foremost requirements of many fields such as heterogeneous catalysis, remediation, and biomedicine where COFs have proven to excellent carriers for NPs [15,16].

Three main synthesis approaches have been used to get COF@NPs: (1) growth of NPs on already synthesized COFs, letting controlled the growth of NPs by the COF network. This results in NPs with relatively smaller sized due to COF pore dimensions [17,18]. (2) Growth of COFs on pre-synthesized NPs, where larger size of NPs can be expected but the metal NPs should be resistant to COF synthesis conditions [19]. (3) Simultaneously growing COFs and NPs where the synthesis conditions should be optimum for both [20].

In 2012, Roland A. Fischer et al. have reported first COF composites with NPs [21]. In the solvent-free conditions, gas-phase infiltration of volatile organometallic precursor has been done. Under UV-light irradiation, the precursor diffuses onto the pores of the COF. In this case the crystallinity of the COF was maintained, the average size of the NPs reported relatively larger than pore size of COF due to not compromising the structure. However, gas phase infiltration was not common due to less volatile nature of metal precursors. The most common method is solution infiltration of NPs precursors onto already prepared COF. Banergee et al. have grown AuNPs on TpPa-1 using reducing agent $NaBH_4$. COF composites of AuNPs have also been prepared via pre-functionalization with polyvinylpyrrolidone (PVP) followed by COF growth [17,18].

Silver NPs have been grown on pre-synthesized COF in a single step by addition of silver metal precursor. Zhong-Min Su et al. in 2019 were the first to report combination of AgNPs with COF, which show high efficiency of degradation of organic pollutants such as various nitroaromatic compounds and rhodamine B (RhB), methylene blue (MB), methyl orange (MO) and congo red (CR) from waste water [22].

Using metal NPs for several light-driven applications may not result desirably since metal NPs are generally having weak light-emitting capabilities. However, plasmon-mediated conversion of light into heat and/or charge transfer and plasmon enhancement approach can be applied for the preservation of the desired plasmonic behavior. The plasmonic properties of metal nanoparticles can change significantly with changes in particle size, shape, composition, and arrangement. COF functionalized NPs show interesting light-driven applications such as photodynamic therapy, photorelease, and photocatalysis [23]. This chapter briefly discussed various synthetic strategies for AuNPs, AgNPs, and COFs, and their characterization techniques, properties, and applications. Emphasizing on some literature reported synthesis of COF-AgNPs and COF-AuNPs based nanocomposites has been done along with applications such as degradation and removal of hazardous pollutants from water, drug delivery, and small molecules detections. COF@NPs composites have also been compared with individual COF and NPs.

11.2 Gold and silver nanoparticles (AuNPs and AgNPs): Synthesis methods

Gold nanoparticles are widely used in biotechnology and biomedical field because of their large surface area and high electron conductivity. Silver nanoparticles are also getting attracted due to distinctive properties, such as good conductivity, chemical stability, catalytic activity, and antimicrobial activity. Both the NPs show significant biological applications, for example AuNPs proved to be the safest and much fewer toxic agents for drug delivery whereas AgNPs have individual plasmon optical spectra properties which allow being silver nanoparticles used in biosensing applications. These properties of NPs depend on the preparation method used. Various methods such as chemical, physical, and biological methods have been employed for their synthesis (Fig. 11.1).

11.2.1 Chemical methods

This is the most commonly used methods for the synthesis of spherical NPs. "Top-Down" and "Bottom-Up" approaches are generally used for the synthesis of AuNPs and AgNPs. In the "Top-Down" approach, bulk materials are used for NPs synthesis while in the "Bottom-Up" approach, NPs are synthesized based on the packaging of atoms, molecules or clusters. Synthesis procedures that involve the top-down approach comprise laser ablation [24], ion sputtering [25], UV and IR irradiation [26,27], and aerosol technology [28], whereas the reduction of metal ion to metal is the bottom-up approach.

Turkevich et al. have reported first chemical method for the synthesis of AuNPs and obtained size in the range of 1–2 nm [29]. The basic principle of this technique involves the reduction of gold ions (Au^{3+}) to produce gold atoms (Au^0) by using some reducing agents like amino acids [30], ascorbic acid [31], UV light, citrate [32,33], etc. Stabilization of AuNPs is carried out by using different capping/stabilizing agents. In 1994, a new method called "The Brust method" has been reported which involves a two-phase reaction using phase transfer reagent such as tetraoctylammonium bromide and reduction has been done using sodium borohydride along with an alkanethiol [34]. Further, for the synthesis of rod-shaped NPs, seed-mediated growth has been used which is based on the basic principle of first synthesizing seed particles by reducing gold salts [35,36]. The next step involves the transferring of the seed particles to a metal salt. Weak reducing agent like ascorbic acid has been used which prevents further nucleation and speeds up the synthesis of AuNPs of rod shape. For the preparation of monodispersed gold nanoparticles, digestive ripening method has been used which involves higher temperature heating of the colloidal solution in the presence of excessive ligands [37].

AgNPs can easily be synthesized using chemical methods. Monodisperse silver nanocubes of particle size 20 nm have been prepared using polyol process in which polyvinylpyrrolidone (PVP) polymer with $AgNO_3$ in the presence of ethylene glycol as reducing agent have been used [38]. The production of AgNPs by the chemical method depends on stabilizing agent, reducing agents, and Ag precursor. However, the shape of nanoparticles also depends on other experimental parameters like pH and temperature. The chemical reduction using organic and inorganic reducing agents such as sodium citrate, ascorbate, sodium borohydride ($NaBH_4$), N,N-dimethylformamide (DMF), and poly(ethylene glycol)-block copolymers, elemental hydrogen, polyol process, and Tollens reagent are common for the synthesis of AgNPs.

11.2.2 Physical methods

Uniform sized AuNPs of range 5–40 nm can be synthesized by γ-irradiation technique in the presence of suitable stabilizer [39]. Cetyltrimethylammonium bromide as binding agent and citric acid as reducing agent have been used in microwave irradiation for the synthesis of AuNPs [40]. Photochemical reduction and heating method have also been used for getting 10–50 nm AuNPs [40]. Various physical methods such as condensation, evaporation, thermal decomposition, heating, arc discharge technique, and sputtering have been used for to obtain uniform shape and controlled size of AgNPs [41,42].

11.2.3 Biological methods

A new clean, dependable, and bio-friendly method has been introduced without the use of any stabilizer and reductant in the presence of citrus fruit juice extracts (Citrus limon, Citrus reticulate, and Citrus sinensis) for the synthesis of AuNPs [43–45]. Biological resources such as bacteria, plants, algae, fungi, and biomolecules have proven to be excellent candidates for the synthesis of both AuNPs and AgNPs. *Shewanella one Dennis*, *Trichoderma ride fungus*, *Bacillus sp.* and *Cadaba Fruticosa* leaves have been utilized as reducing agent (a metal reducing agent) for the biosynthesis of AgNPs [46–50]. These are extra cellular synthesis approaches which produce pure nanoparticles as compared to the intracellular synthesis process which requires additional purification steps. Contrarily, these methods of synthesis are slow and tedious.

To determine various properties such as the size, geometry, shape, crystallinity and surface area, these NPs can be characterized using transmission electron microscopy (TEM), scanning electron microscopy (SEM), UV-Vis spectroscopy, X-ray photoelectron spectroscopy (XPS), dynamic light scattering (DLS) atomic force microscopy (AFM), powder X-ray diffractometry (XRD), and Fourier transform infrared spectroscopy (FTIR). TEM, AFM, and SEM are used for shape and particle size of nanoparticles, particle height, and volume can be measured in three-dimensional images by AFM, dynamic light scattering is used for the determination of particle size distribution and the crystalline nature can be determined by XRD.

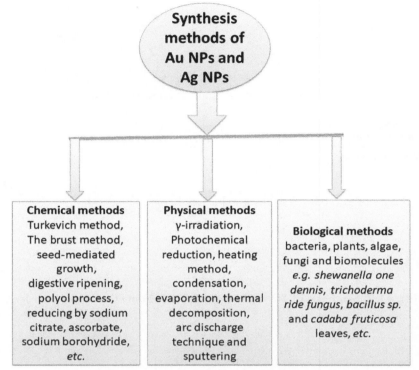

Fig. 11.1: Various synthesis methods of gold and silver nanoparticles.

11.3 COFs: Synthesis, characterization, and applications

11.3.1 Synthesis methods

It is very important to find out an appropriate construction of COFs. The first COF was successfully developed by Yaghi and coworkers using solvothermal method [9], afterwards several research groups adapted the method with few modifications [51–53]. The other conventional methods are ionothermal, microwave, and room-temperature synthesis. Prior knowledge of construction units, synthetic routes, and regulated thermodynamic equilibrium during covalent bond formation is vital to construct highly ordered and crystalline COFs. Development of proficient methods for synthesizing COFs at large scale is still challenging. Many other simple and useful technical methods such as ionothermal method [54–56], micro-wave method [57,58], mechanochemical method [59,60] room-temperature method [61,62], and interface method [63] have been used to synthesize COFs.

11.3.1.1 Solvothermal synthesis

This is the most widely accepted method for the synthesis of COFs. In this method, first, activation of materials is done in the Pyrex tube by Freezing Pump-Thawing Circulator

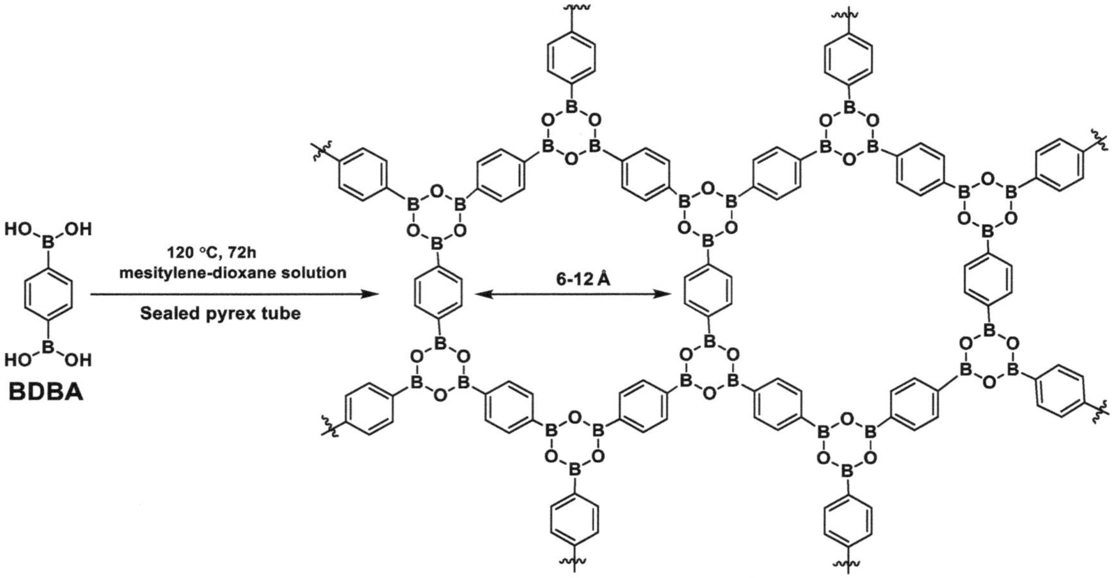

Fig. 11.2: Synthesis of 2D COFs (A) COF-1 (obtained as white powder).

which is a degassing treatment. Second, the tube is sealed and heated to ~80–120°C (temperature depends on the solvents used), this practice takes 2–9 days. The obtained crude solid is purified by washing with common organic solvents like acetone. Adjusting pressure inside the sealed vessel is very important, as it has the potential to significantly affect reaction yields. Yaghi and co-workers were the first to synthesize two dimensional COFs using the solvothermal process. COF-1 was produced using a self-molecular assembly method as a layered hexagonal structure by dehydration reaction of 1,4-benzenediboronic acid (BDBA) (Fig. 11.2). COF-5 was prepared as a coplanar extended sheet structure by condensation of BDBA and 2,3,6,7,10,11-hexahydroxytriphenylene (HHTP) (Fig. 11.3). Despite the fact that the solvothermal method is commonly used in the synthesis of COFs, maintaining high temperature and pressure and longer reaction time make it unsuitable for large-scale production.

11.3.1.2 Ionothermal synthesis

Ionic liquids or molten salts are used as a solvent or catalyst in an ionothermal process. This method requires very high temperature to complete the reaction. Amorphous materials with the absence of long-range molecular orderings are the most commonly used monomers in ionothermal methods. In 2008, Thomas and co-workers have used this method for synthesizing covalent triazine-based frameworks (CTFs) by cyclotrimerizing of nitrile building

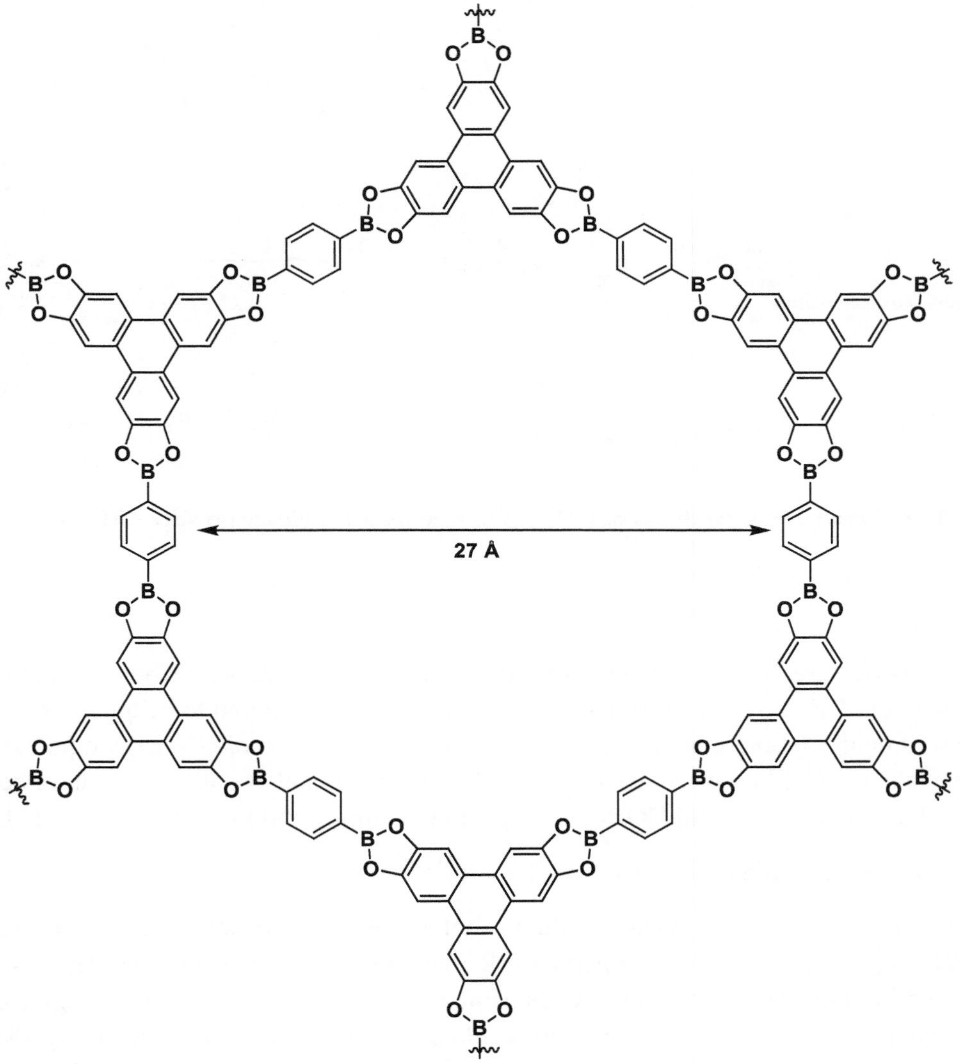

Fig. 11.3: Structure of COF-5 obtained as grey-purple solid by condensation of 1,4-benzenediboronic acid with 2,3,6,7,10,11-hexahydroxytriphenylene.

units in molten $ZnCl_2$ at 400°C (Fig. 11.4) [54]. It has been observed that only a few starting materials are suitable for this process due to less solubility in ionic liquids/molten salts and tendency to decompose at higher temperature. Hence, ionothermal synthesis method is not a feasible for mass production.

11.3.1.3 Microwave synthesis

Microwaves have been extensively studied as an alternative energy source for chemical reaction since last few years. Starting materials were dissolved in a mixture of solvents and

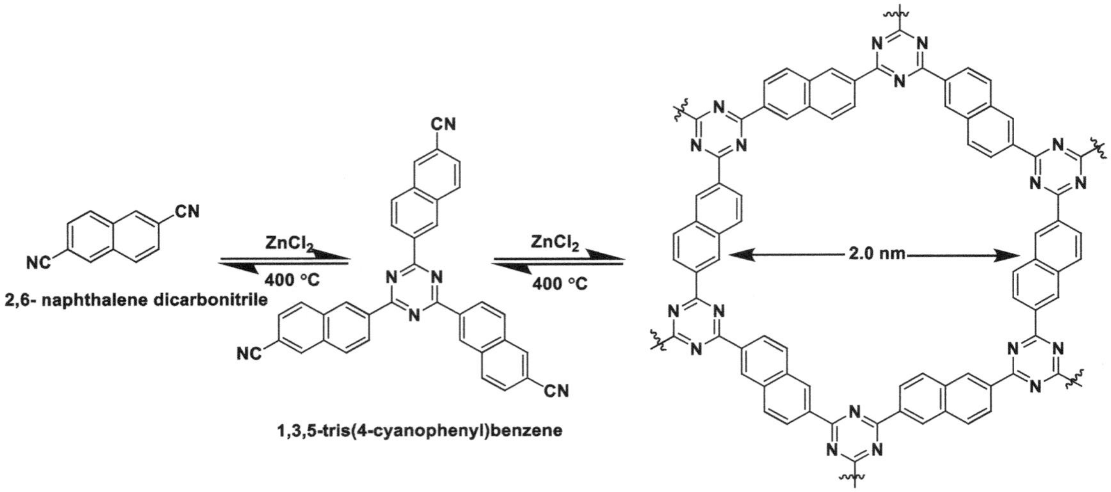

Fig. 11.4: Ionothermal synthesis of CTF-2 obtained as black glossy powder with 81% yield.

nitrogen gas was used to seal the glass tube. After that, the machine was heated with microwaves and kept 100°C for less than half an hour. This method is found to be quite faster than the conventional solvothermal synthesis method and the formation of cleaner compounds has been taken [57]. Microwave-assisted solvothermal and ionothermal methods have been employed to get various useful COFs with higher crystallinity and high-stability [64,65].

11.3.1.4 Mechanochemical synthesis

The method which requires manual grinding to achieve easy, rapid, solvent-free, and room temperature synthesis and has an important role in modern synthetic chemistry. The first β-ketoenamine-linked COFs was developed using a mortar and pestle where the degree of poly-condensation and visual color changes could be observed [66]. Unlike COFs synthesized through the solvothermal process, the obtained same COF had a graphene-like layered structure which could be due to the exfoliation of COF layers during the mechanochemical synthesis. Hence, this method is user and environmental friendly and time-saving method for mass-production of COFs. This method is a powerful tool for the development of COFs at industrial scale.

11.3.1.5 Room temperature synthesis

This method is found to be easiest and economical method. TpBD was synthesized at room temperature conditions, and its BET surface area (885 $m^2.g^{-1}$) was found higher than that of TpBD [537 $m^2.g^{-1}$] synthesized using solvothermal method [67]. The majority of commonly used building monomers are not stable at high temperatures, emphasizing the importance of using a room-temperature approach to prepare COFs. It has been recently discovered that imine-based COFs can be easily produced at room temperature and in the presence of ambient atmosphere. Wang and co-workers were the first to discuss room-temperature synthesis of COFs [68]. The precursor monomers were dissolved in the same solvents used in the solvothermal process with catalytic amount of acetic acid. The imine-based COFs synthesized and precipitated out from solution after long standing for three days. Zamora et al. have prepared RT-COF-1 at room temperature using *m*-cresol or DMSO as a solvent rather than the conventional mixture solvent [62,69]. Since this synthesis method sidesteps harsh reaction conditions, it has only been used to make –C=N– based COFs.

11.3.1.6 Interfacial synthesis

Interface synthesis is a bottom-up technique for producing COF films at the solid-liquid, solid-vapor, liquid-liquid, and liquid-air interfaces. Graphene nanosheets and polyamide nanofilms have been prepared using interfacial synthesis method [70]. In this strategy, the interface is the place where the monomers react with each other, thus COF growth is restricted to a small area at the interface, resulting in the formation of thin films. This synthesis method is an excellent technique for preparing well-structured COFs, and it has made considerable progress in recent years, but the process has a disadvantage of low production capacity.

11.3.2 Characterization of COFs

Characterization of COFs includes the structural regularity, atomic connectivity, porosity, and morphology. Unlike structure identification of MOFs using single crystal X-ray diffraction, structure determination of COFs cannot be done due to difficulty in obtaining single crystals of COFs. Powder X-ray diffraction (PXRD) technique is used for structure determination of microcrystalline COF materials. Pore size determination is being done using gas absorption isotherm data. Solid-state NMR spectroscopy has been utilized to know atomic connectivity in the COF materials, especially for the formation of new covalent bonds. IR and UV-V is spectroscopies have also been used to other complementary information. The porosity and surface areas of COF materials are normally assessed nitrogen or argon gas adsorption-desorption measurements. Surface morphology of COF materials can be examined using SEM. X-ray photoelectron spectroscopy (XPS) is supportive to investigate the state of metal ions incorporated into the COF materials. The theoretical investigations such as molecular dynamics and density functional theory (DFT) calculations can provide structure factors and other important information.

11.3.3 Applications of COFs

Due to their unique properties, such as wide surface area, orderly passages, high porosity, adjustable structure by preselection of the building wedges, simple operationalization, and strong thermal and chemical stability, they have gained a lot of interest [71]. COF materials have attracted a lot of attention in recent years for their contributions to sample treatment, separation, and sensing processes, providing highly structured resources for analytical applications such as gas storage, photoelectricity, catalysis and so on. The application progress has been summarized below:

11.3.3.1 Gas storage

1. *Hydrogen storage.* Due to its clean combustion and high chemical energy density, hydrogen has been pursued as an ideal substitute for traditional fossil fuels, especially in the automotive applications [72,73]. It has been observed that COFs having larger surface areas hold higher hydrogen uptake capacities when compared under the same conditions. For example, COF-18A° with a BET surface area of 1263 $m^2 g^{-1}$ shows the highest hydrogen uptake (1.55 wt% at 1 bar, 77 K) among similar 2D COFs with different alkyl chain lengths [74]. However, most of the hydrogen uptakes reported for porous materials were measured at 77 K, far away from the ideal temperature range required for the practical applications. Theoretical predictions indicate that the metal-doped COFs can show hydrogen storage at 298 K and 100 bar.

2. *Methane storage.* Methane is abundant and inexpensive in comparison with conventional fossil fuels and contributing as main component of natural gas. The capability of methane storage in certain COF materials has been examined and it has been found that the capacities of methane storage in 3D COFs are higher than those of 2D COFs [75].

3. *Carbon dioxide storage.* It is major contributing gas to the global warming. Among all the COFs, COF-102 has shown maximum uptake of 1200 mg g^{-1} [75]. CO_2 capture capacity of COFs can be enhanced by incorporating other suitable functional moieties into COFs.

4. *Ammonia storage.* For the production of nitrogen fertilizers ammonia is widely used in industries. It is difficult to handle toxicity and corrosiveness of liquid ammonia, so it is wise choice to collect it in gaseous form in porous materials. Boron containing COF-10 shows maximum ammonia uptake capacity (15 mol kg^{-1} at 298 K and 1 bar) which is found to highest among all the reported porous compounds so far [76]. The exceptionally high uptake capacity is due to ammonia-borane coordination bond.

11.3.3.2 Photoelectric applications

COF materials functionalized with photoelectric moieties can have unique optical and electrical properties. For example, pyrene-functionalised COFs that is, PPy-COF was obtained via the self-condensation of pyrene-2,7-diboronic acid (PDBA) [77], and TP-COF was obtained by co-condensation of 2,3,6,7,10,11-hexahydroxytriphenylen (HHTP) and PDBA [78]. Both PPy-COF and TP-COF have 2D eclipsed structures with the BET surface areas of 868 m^2 g^{-1} and 932 m^2 g^{-1}, respectively. TP-COF is found to be highly luminescent and exhibits the p-type semiconductive characters due to the eclipsed arrangement of triphenylene and pyrene units. Similarly, PPy-COF also shows a fluorescence shift comparable to the PDBA. Many other COFs having 2D eclipsed structure show beneficial optoelectronic properties.

11.3.3.3 Catalysis

Suitably functionalized COF materials can work as highly efficient and robust catalysts for many useful reactions. The ideal catalytic activity can be explained by the easy accessibility to the catalytic sites and the efficient mass transport inside the porous catalytic materials. Pd(II)-coordinated COF-LZU1 that is, Pd/COF-LZU1 show high catalytic performance for Suzuki-Miyaura coupling reaction with high stability and easy recyclability of the catalysts [79]. The larger activity of the Pd/COF-LZU1 catalyst can be explained by its unique eclipsed layered-sheet arrangement. The distance between nitrogen atoms in adjacent layers make COF-LZU1 a robust scaffold for incorporating active catalytic sites. However, it is still challenging to use COFs as catalysts at industrial scale for practical applications.

11.4 Synthesis and applications of COF@AuNPs and COF@AgNPs

Ultrafine silver nanoparticles were successfully supported on the predesigned conjugated microporous polymer (CMP) by using simple liquid impregnation and light-induced reduction method [80] (Fig. 11.5). CMP is added to $AgNO_3$ solution stirred overnight in the dark. The grayish yellow sample is collected by centrifugation and washed three times with distilled water, and the solution is added to 20 mL of distilled water to be stirred for another 4 hours under visible light irradiation. Upon light-induced reduction from Ag^+ to Ag^0, Ag^0@ CMP composite materials forms and are recovered by centrifugation and further washed three times with distilled water. The silver content was determined by ICP with a mean loading of ca. 0.73 wt %. These composites work as high-performance nanocatalysts for the reduction of nitrophenols, a family of priority pollutants, at various temperatures and ambient pressure. The composite nanocatalysts also feature convenient recovery and excellent reusability [80].

A synthetic strategy to couple COFs with materials having good biocompatibility, high conductivity and unique catalytic activity is essential. Combining AuNPs can be good idea to develop electrochemical sensors due to extraordinary catalytic activity, excellent

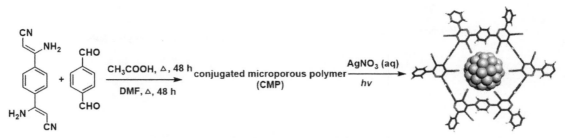

Fig. 11.5: Silver nanoparticle immobilization in CMP. *Reproduced with permission from [ref. 80]. Copyright 2017 American Chemical Society.*

conductivity and outstanding stability. The grouping of AuNPs and COFs material was expected to show outstanding performance as a novel electrode material, which can improve the sensitivity, stability and reproducibility. A successful synthesis of a COFs composite material TAPB-DMTPCOF@Au NPs (TAPB: 1,3,5-tris(4-aminophenyl)benzene; DMTP:2,5-dimethoxyterephaldehyde) with imino linkage using solution infiltration method have been done by T. Zhang et al. in 2018 [81]. In this method, TAPB-DMTP-COF was first treated with hydrochloric acid to tune the pH about 4. $HAuCl_4 \cdot 4H_2O$ was added dropwise into the above mixture under vigorous stirring, and freshly prepared $NaBH_4$ in methanol was added (Fig. 11.6). The COF@AuNPs has been obtained as precipitate. The obtained precipitate was centrifuged and washed with methanol. Finally, the yellow powder was dried in vacuum oven at 60°C for 12 hours [81].

Further to obtain modified electrode, chlorogenic acid was first polished to a mirror-like surface with alumina slurry, then TAPB-DMTP-COFs/AuNPs was dropped on the surface of GCE and dried under refrigerator. 0.50% chitosan solution is used to fix the TAPB-DMTP-COFs/AuNPs layer. The characterization of these COF composite materials has been done using PXRD where additional distinguishable peaks of AuNPs have been present. Further confirmation was done FTIR and X-ray photoelectron spectroscopy, high-resolution

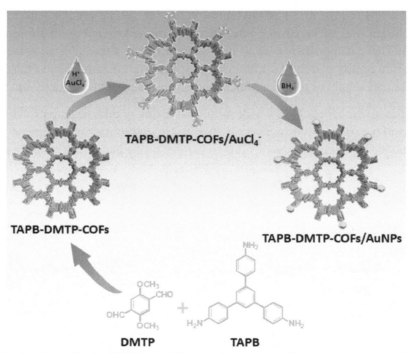

Fig. 11.6: Synthesis method of TAPB-DMTPCOF@Au NPs. *Adapted with permission from [ref. 81]. Copyright 2018 Elsevier.*

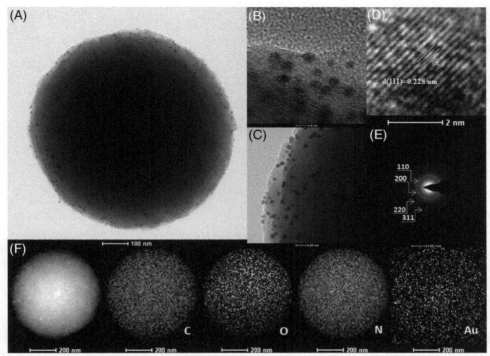

Fig. 11.7: HRTEM images of TAPB-DMTP-COFs/AuNPs with different magnified folds (A–C). HRTEM and SAED patterns of TAPB-DMTP-COFs/AuNPs (D and E). EDS elemental mapping picture of TAPB-DMTP-COFs/AuNPs (F). *Adapted with permission from [ref. 81]. Copyright 2018 Elsevier.*

transmission electron microscopy (Fig. 11.7). Electrochemical behavior of bare glassy carbon electrode (GCE) and APB-DMTP-COFs/AuNPs/GCE was compared where the peak current of later is found several-folds higher. This is attributed due to the effective electroactive surface area, due to the addition of conductive material Au NPs on the surface. These composites further can be functional in biosensors, energy storage devices and electrocatalysis with enriched properties [81].

A postmodification strategy is used to synthesize sulphur-containing COF (S-COF) [82] (Fig. 11.8). First, propenyl-functionalized COF is synthesized by condensation of 2,5-bis(allyloxy)terephthalohydrazide and 1,3,5-triformylbenzene. S-COF is prepared by a click reaction of propenyl-functionalized COF and 2,2-azobisisobutyronitrile and 1,2-ethanedithiol. Furthermore, the obtained S-COF can be used as matrix to successful preparation of S-COF-supported AuNPs (Au-S-COF) with narrow size distribution (4.2 ± 1.2 nm) and high distribution density. This composite material shows excellent catalytic activities in reduction

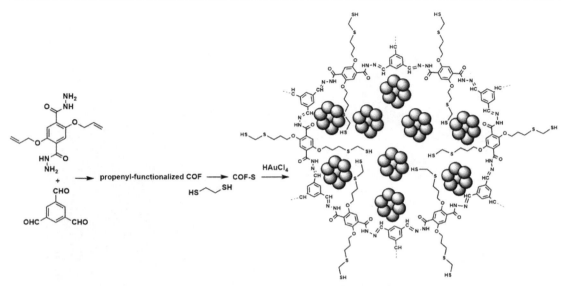

Fig. 11.8: Synthesis of Au-S-COF [82].

of 4-nitrophenol with significant stability and recyclability. In the Au-S-COF heterogeneous catalytic system, the strong anchoring ability between AuNPs and sulphur-containing groups over COF's surface enhance the binding of AuNPs with COFs to maintain the dispersity of NPs and their catalytic activity for long-term performance [82].

Due to higher reactivity and atom utilization with respect to pollutants in aqueous environments, NPs can be used for water purification. Nevertheless, the aggregation and instability in acidic solution limit their practical applications. Removal of toxic metal like mercury from acidic solution is quite problematic. Hence, a tunable porous covalent organic framework can be employed to enhance the spatial dispersion of NPs and their stability in acidic solution [83]. The COF supports in-situ growth of AgNPs which show reasonably good mercury adsorption performance. More importantly, the AgNPs@COF composites have high selectivity and stability, and also reusability for Hg(II) removal from acidic aqueous solutions (Fig. 11.9) [83].

Magnetic nanoparticles [84–86] for example, $CuFe_2O_4$ and Fe_3O_4 mixed with AgNPs and AuNPs can be used for several water-treatment applications like reduction of 4-nitrophenols. Due to the outstanding stability and chemical robustness of COF materials, a facile, rapid and easy separation is possible. The COF coating with higher surface area ensures high loading capacity of the reaction substrates. The large number of affinity sites and strong magnetic property are due to the combination of the unique π-π electron structure of COF and strong magnetic property of magnetic of magnetic nanoparticles and COFs. These magnetic composites of COFs can be prepared by fabricated by using a seed deposition method. In this

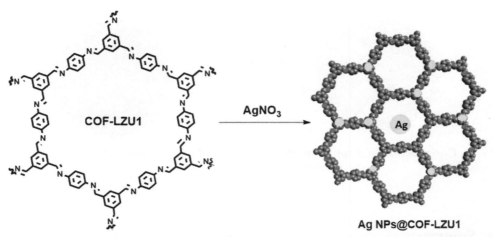

Fig. 11.9: Schematic representation of the synthesis of AgNPs@COF-LZU1. *Reproduced with permission from [ref. 83]. Copyright 2020 Elsevier.*

method, KOH and polyethyleneimine (PEI) is added to COF solution with stirring. This solution is added dropwise to weighed number of $CuFe_3O_4/Fe_3O_4$ nanoparticles in methanol. Finally, the noble metal sol (Ag/Au) has been added to get composites materials.

Nanoparticles supported by a COF exhibit an unusual catalytic performance towards organic pollutants, however the applications are limited by the recovery and reuse from the mixture due to the size of NPs@COF at the nanoscale. Adding up magnetic carriers can be a good choice where magnetic separation can be used to recover these prepared catalysts but the use of an excessive number of magnetic equipment is not in favor of the application. It was studied that the porous micro-size carrier assembled by nanomaterials provides a large surface area for supporting noble metal nanoparticles. The catalytic alteration arisen on the surface of the catalyst including a complex adsorption-transformation-desorption procedure, where the design of heterogeneous catalyst structure regulates the interaction between the substrates and catalysts, can influence the catalytic performance significantly. In this progress ultrafine gold metal nanoparticles (AuNPs) are in-situ reduced on the surface of sulfhydryl groups functionalized new COF with $NaBH_4$, followed by the deposition on polydopamine modified micronsized porous SMt (spherical montmorillonite) to construct a spherical SMt@COF@AuNPs heterostructure (Fig. 11.10) [87]. The resultant heterostructure is employed as heterogeneous catalyst for catalytic reduction of methylene blue (MB) in the presence of $NaBH_4$. The catalytic reduction process can be explored in various reaction conditions that is, concentration of catalyst, reaction temperature, and pH. Recyclability was further studied. The granular SMt@COF@Au NPs heterostructure simplifies recovery from the mixture without using any devices [87].

Fig. 11.10: Construction of a spherical SMt@COF@AuNPs heterostructure. *Adapted with permission from [ref. 87]. Copyright 2021 Elsevier.*

11.5 Summary

Gold and silver nanoparticles can be prepared using different methods and based on their morphology, these are used in different fields including drug delivery, sensing, and catalysis (Table 11.1 and Table 11.2). The widespread applicability is due to their extraordinary chemical and physical properties, high surface area, tunable optical, stability, properties small size, and noncytotoxicity. Incorporating inorganic NPs into materials having tremendously high surface area, defined pore structure, and desirable functionality that is, COFs by well-designed synthetic protocols, remains in its infancy and has tremendous potential for further advancement in the properties. It has been studied that generally imine-based COFs having nitrogen richness are used to obtain COF@MNPs with desired properties. Au and Ag nanoparticle-loaded porous crystalline materials show very narrow size distribution without any larger agglomerates even at high loadings.

The alliance of AuNPs and COFs material shows outstanding performance as a novel electrode material, with improved sensitivity, stability, and reproducibility. Suitably modified electrodes can be employed in simultaneous electrochemical sensing of many biochemicals of interest such as dopamine and uric acid from human urine. These sensors have superior performance, low cost, simple structure, wide linear range, and high sensitivity than some traditional methods. Sulphur supported COF combined with AuNPs shows excellent catalytic property in the reduction of environmental pollutants such as nitoaromatics with exceptional stability, separability, and reusability. COF@AgNPs and COF@AuNPs are commonly employed in degradation and removal of hazardous pollutants from water, drug delivery and small molecules detections.

Table 11.1: Various reported COF@AuNPs/COF@AgNPs employed for different applications.

Sr no.	COF@AuNPs/ COF@AgNPs	Applications	Refs.
1.	TAPB-PDACOFs@AuNPs	Electrochemical determination of enrofloxacin	[88]
2.	TAPB-DMTP-COF@AuNPs	Reduction of 4-nitrophenol	[89]
3.	P6-Au-COF (P6:pillar[6]arene)	Reduction of nitroaromatics	[90]
4.	HRP-pSC4-AuNPs@COFs	Detection of colorectal cancer-derived exosomes	[91]
5.	DP-Py COF@ AuNPs	Sensing of theophylline and caffeine	[92]
6.	TB-Au-COFs	Detection of cardiac troponin I (cTnI)	[93]
7.	o-aminothiophenol functionalized AuNPs (o-ATP@AuNPs)	Detection of aflatoxin B1	[94]
8.	PTAzo@ AuNPs	Detection of Hg^{2+}	[95]
9.	COF-NH_2-MWCNT/Au/GCE	Detection of dopamine and uric acid	[96]
10.	2HP6@Au@CP6@COF	Electrochemical sensing of sodium picrate	[97]
11.	Fe_3O_4@COF@Au-β-CD	Magnetic sorbent for solid phase extraction and determination of sulfonamides	[98]
12.	Ag@TATF-COM	Hydrogenation of nitroaromatics	[99]
13.	MCOF-AuNPs (MCOF: melamine based COF)	Detection of riboflavin in Pharmaceutical and human fluids samples	[100]
14.	Au/COF/MnO_2 composites	Immunosensors for human chorionic gonadotropin	[101]
15.	AuPt@MnO_2@COF	Detection of prostate-specific antigen (PSA)	[102]
16.	TpPa-1@Au@GSH	Extraction of N^1-methyladenosine (m^1A)	[103]
17.	AuNPs/carboxylated COFs/ Poly(fuchsinbasic) film	Detection of ascorbic acid, dopamine, and uric acid	[104]
18.	COF-Ag NPs@Sand catalyst	Reduction of 4-nitrophenol, Congo red, and methylene blue in the waste water	[105]
19.	AgNPs-COF/GCE (GCE: glassy carbon electrode)	Determination of DNA bases in human blood serum, urine and saliva samples	[106]
20.	CuTAPc-MCOF@AgNPs	Detection of nitric oxide released from cancer cells	[107]
21.	AgNPs@NCOF	4-nitrophenol reduction	[108]
22.	TpPa-1@Ag@GSH	Rapid and highly efficient enrichment of N-linked glycopeptides	[109]

Table 11.2: List of abbreviations.

Acronyms	Expansion
TAPB	3,5-tris(4-aminophenyl)benzene
PDA	p-phthalaldehyde
DMTP	2,5-dimethoxyterephthaldehyde
P6	P6:pillar[6]arene
HRP	horseradish peroxidase
pSC_4	*para*-sulfocalix [4] arene hydrate
DP	2,6-diaminopyridine
Py	pyrene
TB	toluidine blue
o-ATP	*o*-aminothiophenol
PT	1,3,5-Tris-(4-formyl-phenyl)triazine
Azo	4, 4'-azodianiline
MWCNT	Multi-walled carbon nanotubes
GCE	Glassy Carbon Electrode
2HP6	dihydroxylatopillar [6]arene
CP6	cationic pillar [6]arene
β-CD	β-Cyclodextrin
COM	covalent organic microspheres
MCOF	melamine based COF
Tp	1,3,5-Triformylphloroglucinol
Pa-1	Paraphenylenediamine
GSH	tripeptide glutathione
CuTAPc-MCOF	metallo-copper phthalocyanine-based covalent-organic framework
NCOF	nitrogen-containing covalent organic framework
GSH	glutathione
SMt	spherical montmorillonite

References

[1] I. Gopalakrishnan, R. Sugaraj Samuel, K. Sridharan, Nanomaterials-based adsorbents for water and wastewater treatments, in: K. Sridharan (Ed.), Emerging Trends of Nanotechnology in Environment and Sustainability. SpringerBriefs in Environmental Science, Springer, Cham, 2018.

[2] L. Zhang, E. Wang, Metal nanoclusters: New fluorescent probes for sensors and bioimaging, Nano Today 9 (2014) 132–157.

[3] P. Ghosh, G. Han, M. De, C.K. Kim, V.M. Rotello, Gold nanoparticles in delivery applications, Adv. Drug Deliv. Rev. 60 (2008) 1307–1315.

[4] H. Wei, K. Rodriguez, S. Renneckar, W. Leng, P.J. Vikesland, Preparation and evaluation of nanocellulose–gold nanoparticle nanocomposites for SERS applications, Analyst 140 (16) (2015) 5640–5649.

[5] S.H. Joo, S.J. Choi, I. Oh, J. Kwak, Z. Liu, O. Terasaki, R. Ryoo, Ordered nanoporous arrays of carbon supporting high dispersions of platinum nanoparticles, Nature 412 (6843) (2001) 169–172.

[6] X. Lan, C. Du, L. Cao, T. She, Y. Li, G. Bai, Ultrafine Ag Nanoparticles Encapsulated by Covalent Triazine Framework Nanosheets for CO2 Conversion, ACS Appl. Mater. Interfaces 10 (45) (2018) 38953–38962.

[7] B. Li, Y. Zhang, D. Ma, Z. Shi, S. Ma, Mercury nano-trap for effective and efficient removal of mercury(II) from aqueous solution, Nat. Commun. 5 (1) (2014) 1–7.

[8] H. Li, H. Liu, J. Zhang, Y. Cheng, C. Zhang, X. Fei, Y. Xian, Platinum Nanoparticle Encapsulated Metal–Organic Frameworks for Colorimetric Measurement and Facile Removal of Mercury(II), ACS Appl. Mater. Interfaces 9 (46) (2017) 40716–40725.

[9] A.P. Cote, A.I Benin, N.W Ockwig, A.JMatzger M.O'Keeffe, O.M Yaghi, Porous, Crystalline, Covalent Organic Frameworks, Science 310 (2005) 1166.

[10] N. Huang, P. Wang, D. Jiang, Covalent organic frameworks: a materials platform for structural and functional designs, Nat. Rev. Mater. 1 (2016) 1–19.

[11] A.G. Slater, A.I. Cooper, Function-led design of new porous materials, Science 348 (2015) 6238.

[12] H.M. El-Kaderi, J.R. Hunt, J.L. Mendoza-Cortes, A.P. Cote, R.E. Taylor, M. O'Keeffe, O.M. Yaghi, Designed Synthesis of 3D Covalent Organic Frameworks, Science 316 (2007) 268.

[13] D. Sun, S. Jang, S.J. Yim, L. Ye, D.P. Kim, Metal Doped Core–Shell Metal-Organic Frameworks@ Covalent Organic Frameworks (MOFs@COFs) Hybrids as a Novel Photocatalytic Platform, Adv. Funct. Mater. 28 (13) (2018) 1707110.

[14] M. Bhadra, H.S. Sasmal, A. Basu, S.P. Midya, S. Kandambeth, P. Pachfule, E. Balaraman, R. Banerjee, Predesigned Metal-Anchored Building Block for In Situ Generation of Pd Nanoparticles in Porous Covalent Organic Framework: Application in Heterogeneous Tandem Catalysis, ACS Appl. Mater. Interfaces 9 (15) (2017) 13785–13792.

[15] N. Huang, L. Zhai, H. Xu, D. Jiang, Stable Covalent Organic Frameworks for Exceptional Mercury Removal from Aqueous Solutions, J. Am. Chem. Soc. 139 (2017) 2428–2434.

[16] A. Halder, S. Karak, M. Addicoat, S. Bera, A. Chakraborty, S.H. Kunjattu, P. Pachfule, T. Heine, R. Banerjee, Ultrastable Imine-Based Covalent Organic Frameworks for Sulfuric Acid Recovery: An Effect of Interlayer Hydrogen Bonding, Angew. Chem. Int. Ed. Engl. 57 (2018) 5797–5802.

[17] P. Pachfule, S. Kandambeth, D. Díaz Díaz, R. Banerjee, Highly stable covalent organic framework–Au nanoparticles hybrids for enhanced activity for nitrophenol reduction, Chem. Commun. 50 (2014) 3169–3172.

[18] P. Pachfule, M.K. Panda, S. Kandambeth, S.M. Shivaprasad, D.D. Díaz, R. Banerjee, Multifunctional and robust covalent organic framework–nanoparticle hybrids, J. Mater. Chem. A 2 (2014) 7944–7952.

[19] X. Shi, Y. Yao, Y. Xu, K. Liu, G. Zhu, L. Chi, G. Lu, Imparting Catalytic Activity to a Covalent Organic Framework Material by Nanoparticle Encapsulation, ACS Appl. Mater 9 (2017) 7481–7488.

[20] M. Bhadra, H.S. Sasmal, A. Basu, S.P. Midya, S. Kandambeth, P. Pachfule, E. Balaraman, R. Banerjee, Predesigned Metal-Anchored Building Block for In Situ Generation of Pd Nanoparticles in Porous Covalent Organic Framework: Application in Heterogeneous Tandem Catalysis, ACS Appl. Mater. 9 (2017) 13785–13792.

[21] S.B. Kalidindi, H. Oh, M. Hirscher, D. Esken, C. Wiktor, S. Turner, G. Van Tendeloo, R.A. Fischer, Metal@COFs: Covalent Organic Frameworks as Templates for Pd Nanoparticles and Hydrogen Storage Properties of Pd@COF-102 Hybrid Material, Chem. Eur. J. 18 (2012) 10848–10856.

[22] R.-L. Wang, D.-P. Li, L.-J. Wang, X. Zhang, Z.-Y. Zhou, J.-L. Mu, Z.-M. Su, The preparation of new covalent organic framework embedded with silver nanoparticles and its applications in degradation of organic pollutants from waste water, Dalton Trans. 48 (2019) 1051–1059.

[23] H. Kang, J.T. Buchman, R.S. Rodriguez, H.L. Ring, J. He, K.C. Bantz, C.L. Haynes, Stabilization of Silver and Gold Nanoparticles: Preservation and Improvement of Plasmonic Functionalities, Chem. Rev. 119 (2019) 664–699.

[24] B. Tangeysh, K.M. Tibbetts, J.H. Odhner, B.B. Wayland, R.J. Levis, Gold Nanoparticle Synthesis Using Spatially and Temporally Shaped Femtosecond Laser Pulses: Post-Irradiation Auto-Reduction of Aqueous $[AuCl_4]^-$, J. Phys. Chem. C 117 (36) (2013) 18719–18727.

[25] R.C. Birtcher, M.A. Kirk, K. Furuya, G.R. Lumpkin, In situ Transmission Electron Microscopy Investigation of Radiation Effects, J. Mater. Res. 20 (7) (2005) 1654–1683.

[26] M. Sakamoto, M. Fujistuka, T. Majima, Light as a construction tool of metal nanoparticles: Synthesis and mechanism, J Photochem Photobiol C 10 (1) (2009) 33–56.

[27] Y. Zhou, C.Y. Wang, Y.R. Zhu, Z.Y. Chen, A novel ultraviolet irradiation technique for shape-controlled synthesis of gold nanoparticles at room temperature, Chem. Mater. 11 (9) (1999) 2310–2312.

[28] T.J. Krinke, K. Deppert, M.H. Magnusson, F. Schmidt, H. Fissan, Microscopic aspects of the deposition of nanoparticles from the gas phase, J. Aerosol Sci. 33 (10) (2002) 1341–1359.

[29] J. Turkevich, P.H.J. Cooper, A study of the nucleation and growth processes in the synthesis of colloidal gold, Discuss. Faraday Soc. 55 (c) (1951) 55–75.

[30] N. Wangoo, K.K. Bhasin, S.K. Mehta, C.R. Suri, Synthesis and capping of water-dispersed gold nanoparticles by an amino acid: Bioconjugation and binding studies, J. Colloid Interface Sci. 323 (2) (2008) 247–254.

[31] Y. Niidome, K. Nishioka, H. Kawasaki, S. Yamada, Localized Surface Plasmonic Properties of Au and Ag Nanoparticles for Sensors: a Review, Chem. Comm. 18 (18) (2003) 2376–2377.

[32] A. Pal, K. Esumi, T. Pal, Preparation of nanosized gold particles in a biopolymer using UV photoactivation, J. Colloid Interface Sci. 288 (2) (2005) 396–401.

[33] S. Kumar, K.S. Gandhi, R. Kumar, Modeling of Formation of Gold Nanoparticles by Citrate Method, Ind. Eng. Chem. Res. 46 (10) (2007) 3128–3136.

[34] M. Brust, M. Walker, D. Bethell, D.J. Schiffrin, R. Whyman, Gold nanoparticle research before and after the Brust–Schiffrin method, J Chem Soc Chem Commun 7 (1994) 5–7.

[35] T.K. Sau, C.J. Murphy, Room Temperature, High-Yield Synthesis of Multiple Shapes of Gold Nanoparticles in Aqueous Solution, J. Am. Chem. Soc. 126 (28) (2004) 9–10.

[36] Y. Chen, X. Gu, C-G. Nie, Z-Y. Jiang, Z-X. Xie, C-J. Lin, Shape controlled growth of gold nanoparticles by a solution synthesis, ChemComm 1 (33) (2005) 4181.

[37] P. Sahu, B.L.V. Prasad, Time and Temperature Effects on the Digestive Ripening of Gold Nanoparticles: Is There a Crossover from Digestive Ripening to Ostwald Ripening? Langmuir 30 (34) (2014) 10143–10150.

[38] K.M. Koczkur, S. Mourdikoudis, L. Polavarapu, S.E. Skrabalak, Polyvinylpyrrolidone (PVP) in nanoparticle synthesis, Dalton Trans. 44 (2015) 17883–17905.

[39] T.S. Rezende, G.R.S. Andrade, L.S. Barreto, JrN.B. Costa, I.F. Gimenez, L.E. Almeida, Facile preparation of catalytically active gold nanoparticles on a thiolated chitosan, Mater. Lett. 64 (2010) 882–884.

[40] W. Guo, Y. Pi, H. Song, W. Tang, J. Sun, Layer-by-layer assembled gold nanoparticles modified anode and its application in microbial fuel cells, Colloid Surfac A: Physicochem Engineer Aspects 415 (2012) 105–111.

[41] D.K. Lee, Y.S. Kang, Synthesis of Silver Nanocrystallites by a New Thermal Decomposition Method and Their Characterization, ETRI J. 26 (2004) 252–256.

[42] J.H. Jung, O.H. Cheol, N.H. Soo, J.H. Ji, K.S. Soo, Metal nanoparticle generation using a small ceramic heater with a local heating area, J. Aerosol Sci. 37 (2006) 1662–1670.

[43] H. Namazi, A.M.P. Fard, Preparation of gold nanoparticles in the presence of citric acid-based dendrimers containing periphery hydroxyl groups, Mater. Chem. Physic. 129 (2011) 189–194.

[44] R. Tarnawski, M. Ulbricht, Amphiphilic gold nanoparticles: Synthesis, characterization and adsorption to PEGylated polymer surfaces, Colloid Surfac A: Physicochem Engineer Aspects 374 (2011) 13–21.

[45] P. Nalawade, T. Mukherjee, S. Kapoor, High-yield synthesis of multispiked gold nanoparticles: Characterization and catalytic reactions, Colloid Surfac A: Physicochem Engineer Aspects 396 (2012) 336–340.

[46] L. Sintubin, W. Verstraete, N. Boon, Biologically produced nanosilver: Current state and future perspectives, Biotechnol. Bioeng. 109 (2012) 2422–2436.

[47] K. Suresh, D.A. Pelletier, W. Wang, Silver Nanocrystallites: Biofabrication using Shewanella oneidensis, and an Evaluation of Their Comparative Toxicity on Gram-negative and Gram-positive Bacteria, Environ. Sci. Technol. 44 (2010) 5210–5215.

[48] M. Fayaz, K. Balaji, M. Girilal, R. Yadav, P.T. Kalaichelvan, Biogenic synthesis of silver nanoparticles and their synergistic effect with antibiotics: a study against gram-positive and gram-negative bacteria, Nanomed.: Nanotechnol., Biol., and Med. 6 (2010) 103–109.

[49] A.V. Thirumalai, D. Prabhu, M. Soniya, Stable silver nanoparticle synthesizing methodsand its applications, J. Biosci. Res. 1 (2010) 259–270.

[50] S.K.P. Venkata, N. Savithramma, Synthesis of silver nanoparticles and anti microbial activity from Cadaba Fruticosa – an important ethnomedicinal plant to treat vitiligo of Kurnool district, Andhra Pradesh, India, Indo Am. J. Pharma. Res. 3 (2013) 1285–1292.

[51] X. Feng, L. Chen, Y. Dong, D. Jiang, Porphyrin-based two-dimensional covalent organic frameworks: synchronized synthetic control of macroscopic structures and pore parameters, Chem. Commun. 47 (2011) 1979–1981.

[52] J.W. Colson, A.R. Woll, A. Mukherjee, M.P. Levendorf, E.L. Spitler, V.B. Shields, M.G. Spencer, J. Park, W.R. Dichtel, Oriented 2D Covalent Organic Framework Thin Films on Single-Layer Graphene, Science 332 (2011) 228–231.

[53] A. Halder, M. Ghosh, A. Khayum M, S. Bera, M. Addicoat, H.S. Sasmal, S. Karak, S. Kurungot, R. Banerjee, Interlayer Hydrogen-Bonded Covalent Organic Frameworks as High-Performance Supercapacitors, J. Am. Chem. Soc. 140 (2018) 10941–10945.

[54] P. Kuhn, M. Antonietti, A. Thomas, Porous, Covalent Triazine-Based Frameworks Prepared by Ionothermal Synthesis, Porous, Chem. Int. Ed. 47 (2008) 3450–3453.

[55] M.J. Bojdys, J. Jeromenok, A. Thomas, M. Antonietti, Rational Extension of the Family of Layered, Covalent, Triazine-Based Frameworks with Regular Porosity, Adv. Mater. 22 (2010) 2202–2205.

[56] H. Ren, T. Ben, E. Wang, X. Jing, M. Xue, B. Liu, Y. Cui, S. Qiu, G. Zhu, Targeted synthesis of a 3D porous aromatic framework for selective sorption of benzene, Chem. Commun. 46 (2010) 291–293.

[57] N.L. Campbell, R. Clowes, L.K. Ritchie, A.I. Cooper, Rapid Microwave Synthesis and Purification of Porous Covalent Organic Frameworks, Chem. Mater. 21 (2009) 204–206.

[58] L.K. Ritchie, A. Trewin, A. Reguera-Galan, T. Hasell, A.I. Cooper, Synthesis of COF-5 using microwave irradiation and conventional solvothermal routes, Microporous Mesoporous Mater. 132 (2010) 132–136.

[59] B.P. Biswal, S. Chandra, S. Kandambeth, B. Lukose, T. Heine, R. Banerjee, Constructing Ultraporous Covalent Organic Frameworks in Seconds via an Organic Terracotta Process, J. Am. Chem. Soc. 135 (2013) 5328–5331.

[60] S. Karak, S. Kandambeth, B.P. Biswal, H.S. Sasmal, S. Kumar, P. Pachfule, R. Banerjee, Constructing Ultraporous Covalent Organic Frameworks in Seconds via an Organic Terracotta Process, J. Am. Chem. Soc. 139 (2017) 1856–1862.

[61] T. Shiraki, G. Kim, N. Nakashima, Room-temperature Synthesis of a Covalent Organic Framework with Enhanced Surface Area and Thermal Stability and Application to Nitrogen-doped Graphite Synthesis, Chem. Lett. 44 (2015) 1488–1490.

[62] A. de la Peña Ruigómez, D. Rodríguez-San-Miguel, K.C. Stylianou, M. Cavallini, D. Gentili, F. Liscio, S. Milita, O.M. Roscioni, M.L. Ruiz-González, C. Carbonell, Direct On-Surface Patterning of a Crystalline Laminar Covalent Organic Framework Synthesized at Room Temperature, Chem. Eur. J. 21 (2015) 10666–10670.

[63] E.L. Spitler, J.W. Colson, F.J. Uribe-Romo, A.R. Woll, M.R. Saldivar, R. Dichtel William, Lattice Expansion of Highly Oriented 2D Phthalocyanine Covalent Organic Framework Films, Angew. Chem. 124 (2012) 2677–2681.

[64] W. Zhang, L.G. Qiu, Y.P. Yuan, A.J. Xie, Y.H. Shen, J.F. Zhu, Microwave-assisted synthesis of highly fluorescent nanoparticles of a melamine-based porous covalent organic framework for trace-level detection of nitroaromatic explosives, J. Hazard. Mater. 147 (2012) 221.

[65] W. Zhang, F. Liang, C. Li, L.G. Qiu, Y.P. Yuan, F.M. Peng, X. Jiang, A.J. Xie, Y.H. Shen, J.F. Zhu, Microwave-enhanced synthesis of magnetic porous covalent triazine-based framework composites for fast separation of organic dye from aqueous solution, J. Hazard. Mater. 186 (2011) 984.

[66] B.P. Biswal, S. Chandra, S. Kandambeth, B. Lukose, T. Heine, R. Banerjee, Mechanochemical Synthesis of Chemically Stable Isoreticular Covalent Organic Frameworks, J. Am. Chem. Soc. 135 (2013) 5328–5331.

[67] C.X. Yang, C. Liu, Y.M. Cao, X.P. Yan, Facile room-temperature solution-phase synthesis of a spherical covalent organic framework for high-resolution chromatographic separation, Chem. Commun. 51 (2015) 12254–12257.

[68] S.Y. Ding, W. Wang, Covalent organic frameworks (COFs): from design to applications, Chem. Soc. Rev. 42 (2013) 548–568.

[69] C. Montoro, D. Rodriguez-San-Miguel, E. Polo, R. Escudero-Cid, M.L. Ruiz-Gonzalez, J.A. Navarro, P. Ocon, F. Zamora, Ionic Conductivity and Potential Application for Fuel Cell of a Modified Imine-Based Covalent Organic Framework, J. Am. Chem. Soc. 139 (2017) 10079–10086.

[70] X. Cai, Y. Luo, B. Liu, H.M. Cheng, Preparation of 2D material dispersions and their applications, Chem. Soc. Rev. 47 (2018) 6224–6266.

[71] S. Kumar, V.V. Kulkarni, R. Jangir, Covalent-organic framework composites: A review report on synthesis methods, Chem. Select 6 (41) (2021) 11201–11223.

[72] L. Schlapbach, A. Zuttel, Hydrogen-storage materials for mobile applications, Nature 414 (2001) 353.

[73] J. Graetz, New approaches to hydrogen storage, Chem. Soc. Rev. 38 (2009) 73.

[74] R.W. Tilford, S.J. Mugavero III, P.J. Pellechia, J.J. Lavigne, Tailoring Microporosity in Covalent Organic Frameworks, Adv. Mater. 20 (2008) 2741.

[75] H. Furukawa, O.M. Yaghi, Storage of Hydrogen, Methane, and Carbon Dioxide in Highly Porous Covalent Organic Frameworks for Clean Energy Applications, J. Am. Chem. Soc. 131 (2009) 8875.

[76] C.J. Doonan, D.J. Tranchemontagne, T.G. Glover, J.R. Hunt, O.M. Yaghi, Exceptional ammonia uptake by a covalent organic framework, Nat. Chem. 2 (2010) 235.

[77] S. Wan, J. Guo, J. Kim, H. Ihee, D. Jiang, A Belt-Shaped, Blue Luminescent, and Semiconducting Covalent Organic Framework, Angew. Chem., Int. Ed. 47 (2008) 8826.

[78] S. Wan, J. Guo, J. Kim, H. Ihee, D. Jiang, A Photoconductive Covalent Organic Framework: Self-Condensed Arene Cubes Composed of Eclipsed 2D Polypyrene Sheets for Photocurrent Generation, Angew. Chem., Int. Ed. 48 (2009) 5439.

[79] F. Xamena, A. Abad, A. Corma, H. Garcia, MOFs as catalysts: Activity, reusability and shape-selectivity of a Pd-containing MOF, J. Catal. 250 (2007) 294.

[80] H-L. Cao, H-B. Huang, Z. Chen, B. Karadeniz, J. Lü, R. Cao, Ultrafine Silver Nanoparticles Supported on a Conjugated Microporous Polymer as High-Performance Nanocatalysts for Nitrophenol Reduction, ACS Appl. Mater. Interfaces 9 (2017) 5231–5236.

[81] T. Zhang, Y. Chen, W. Huang, Y. Wang, X. Hu, A novel AuNPs-doped COFs composite as electrochemical probe for chlorogenic acid detection with enhanced sensitivity and stability, Sensors & Actuators: B. Chemical 276 (2018) 362–369.

[82] Q-P. Zhang, Y.-l. Sun, G. Cheng, Z. Wang, H. Ma, S-Y. Ding, B. Tan, J-h. Bu, C. Zhang, Highly dispersed gold nanoparticles anchoring on post-modified covalent organic framework for catalytic application, Chem. Eng. J. 391 (2020) 123471.

[83] L. Wang, H. Xu, Y. Qiu, X. Liu, W. Huang, N. Yan, Z. Qu, Utilization of Ag nanoparticles anchored in covalent organic frameworks for mercury removal from acidic waste water, J. Hazard. Mater. 389 (2020) 121824.

[84] Y. Yang, K. Jiang, J. Guo, J. Li, X. Peng, B. Hong, X. Wang, H. Ge, Facile fabrication of Au/Fe3O4 nanocomposites as excellent nanocatalyst for ultrafast recyclable reduction of 4-nitropheol, Chem. Eng. J. 381 (2020) 122596.

[85] Y. Xu, X. Shi, R. Hua, R. Zhang, Y. Yao, B. Zhao, T. Liu, J. Zheng, G. Lu, Remarkably catalytic activity in reduction of 4-nitrophenol and methylene blue by Fe3O4@COF supported noble metal nanoparticles, Appl. Catal. B 260 (2020) 118142.

[86] C. Hou, D. Zhao, W. Chen, H. Li, S. Zhang, C. Liang, Covalent Organic Framework-Functionalized Magnetic CuFe2O4/Ag Nanoparticles for the Reduction of 4-Nitrophenol, Nanomaterials 10 (2020) 426.

[87] F. Wang, F. Pan, S. Yu, D. Pan, P. Zhang, N. Wang, Towards mass production of a spherical montmorillonite@covalent organic framework@gold nanoparticles heterostructure as a high-efficiency catalyst for reduction of methylene blue, Appl. Clay Sci. 203 (2021) 106007.

[88] S. Lu, S. Wang, P. Wu, D. Wang, J. Yi, L. Li, P. Ding, H. Pan, A composite prepared from covalent organic framework and gold nanoparticles for the electrochemical determination of enrofloxacin, Adv. Powder Technol. 32 (2021) 2106–2115.

[89] X. Shi, Y. Yao, Y. Xu, K. Liu, G. Zhu, L. Chi, G. Lu, Imparting Catalytic Activity to a Covalent Organic Framework Material by Nanoparticle Encapsulation, ACS Appl. Mater. Interfaces 9 (2017) 7481–7488.

[90] X. Tan, W. Zeng, Y. Fan, J. Yan, G. Zhao, Covalent organic frameworks bearing pillar[6]arene-reduced Au nanoparticles for the catalytic reduction of nitroaromatics, Nanotechnology 31 (2020) 135705.

[91] M. Wang, Y. Pan, S. Wu, Z. Sun, L. Wang, J. Yang, Y. Yin, G. Li, Detection of colorectal cancer-derived exosomes based on covalent organic frameworks, Biosens. Bioelectron. 169 (2020) 112638.

[92] Q. Guan, H. Guo, R. Xue, M. Wang, N. Wu, Y. Cao, X. Zhao, W. Yang, Electrochemical sensing platform based on covalent organic framework materials and gold nanoparticles for high sensitivity determination of theophylline and caffeine, Microchim. Acta 188 (2021) 85.

[93] T. Zhang, N. Ma, A. Ali, Q. Wei, D. Wu, X. Ren, Electrochemical ultrasensitive detection of cardiac troponin I using covalent organic frameworks for signal amplification, Biosens. Bioelectron. 119 (2018) 176–181.

[94] Y. Gu, Y. Wang, X. Wu, M. Pan, N. Hu, J. Wang, S. Wang, Quartz crystal microbalance sensor based on covalent organic framework composite and molecularly imprinted polymer of poly(o-aminothiophenol) with gold nanoparticles for the determination of aflatoxin B1, Sensors & Actuators: B. Chem. 291 (2019) 293–297.

[95] W. Li, Y. Li, H-L. Qian, X. Zhao, C-X. Yang, X–P. Yan, Fabrication of a covalent organic framework and its gold nanoparticle hybrids as stable mimetic peroxidase for sensitive and selective colorimetric detection of mercury in water samples, Talanta 204 (2019) 224–228.

[96] Q. Guan, H. Guo, R. Xue, M. Wang, X. Zhao, T. Fan, W. Yang, M. Xu, W. Yang, Electrochemical sensor based on covalent organic frameworks-MWCNT-NH$_2$/AuNPs for simultaneous detection of dopamine and uric acid, J. Electroanal. Chem. 880 (2021) 114932.

[97] X. Tan, Y. Fan, S. Wang, Y. Wu, W. Shi, T. Huang, G. Zhao, Ultrasensitive and highly selective electrochemical sensing of sodium picrate by Dihydroxylatopillar[6]arene-Modified gold nanoparticles and cationic Pillar[6]arene functionalized covalent organic framework, Electrochim. Acta 335 (2020) 135706.

[98] Y. Yang, G. Li, D. Wu, A. Wen, Y. Wu, X. Zhou, β-Cyclodextrin-/AuNPs-functionalized covalent organic framework-based magnetic sorbent for solid phase extraction and determination of sulfonamides, Microchim. Acta 187 (2020) 278.

[99] Subodh, K. Prakash, D.T. Masram, Silver Nanoparticles Immobilized Covalent Organic Microspheres for Hydrogenation of Nitroaromatics with Intriguing Catalytic Activity, ACS Appl. Polym. Mater. 3 (2021) 310–318.

[100] P. Arul, N.S.K. Gowthaman, E. Narayanamoorthi, S.A. John, S-T. Huang, Synthesis of homogeneously distributed gold nanoparticles built-in metal free organic framework: Electrochemical detection of riboflavin in pharmaceutical and human fluids samples, J. Electroanal. Chem. 887 (2021) 115143.

[101] H. Liang, G. Ning, L. Wang, C. Li, J. Zheng, J. Zeng, H. Zhao, C-P. Li ACS Appl, Covalent Framework Particles Modified with MnO$_2$ Nanosheets and Au Nanoparticles as Electrochemical Immunosensors for Human Chorionic Gonadotropin, Nano Mater 4 (2021) 4593–4601.

[102] J. Zheng, H. Zhao, G. Ning, W. Sun, L. Wang, H. Liang, H. Xu, C. He, H. Zhao, C-P. Li, A novel affinity peptide–antibody sandwich electrochemical biosensor for PSA based on the signal amplification of MnO2-functionalized covalent organic framework, Talanta 233 (2021) 122520.

[103] Y-F. Ma, F. Yuan, F. Yu, Y-L. Zhou, X-X. Zhang, Synthesis of a pH-Responsive Functional Covalent Organic Framework via Facile and Rapid One-Step Postsynthetic Modification and Its Application in Highly Efficient N1-Methyladenosine Extraction, Anal. Chem. 92 (2020) 1424–1430.

[104] Y. He, X. Lin, Y. Tang, L. Ye, A selective sensing platform for the simultaneous detection of ascorbic acid, dopamine, and uric acid based on AuNPs/carboxylated COFs/Poly(fuchsin basic) film, Anal. Methods 13 (2021) 4503–4514.

[105] F. Pan, F. Xiao, N. Wang, Towards application of a covalent organic framework-silver nanoparticles@sand heterostructure as a high-efficiency catalyst for flow-through reduction of organic pollutants, Appl. Surf. Sci. 565 (2021) 150580.

[106] P. Arul, S-T. Huang, N.S.K. Gowthaman, S. Shankar, Simultaneous electrochemical determination of DNA nucleobases using AgNPs embedded covalent organic framework, Microchim. Acta 188 (2021) 358.

[107] M. Wang, L. Zhu, S. Zhang, Y. Lou, S. Zhao, Q. Tan, L. He, M. Du, A copper(II) phthalocyanine-based metallo-covalent organic framework decorated with silver nanoparticle for sensitively detecting nitric oxide released from cancer cells, Sensors & Actuators: B. Chem. 338 (2021) 129826.

[108] A. Shen, R. Luo, X. Liao, C. He, Y. Li, Highly dispersed silver nanoparticles confined in a nitrogen-containing covalent organic framework for 4-nitrophenol reduction, Mater. Chem. Front. 5 (2021) 6923–6930.

[109] Y-F. Ma, L-J. Wang, Y.-L. Zhou, X-X. Zhang, A facilely synthesized glutathione-functionalized silver nanoparticle-grafted covalent organic framework for rapid and highly efficient enrichment of N-linked glycopeptides, Nanoscale 11 (2019) 5526–5534.

CHAPTER 12

Recent advancements in designing Au/Ag based plasmonic photocatalysts for efficient photocatalytic degradation

Samriti[a], Maneet[a], Tripti Ahuja[b], Jai Prakash[a]

[a]Department of Chemistry, National Institute of Technology Hamirpur, Hamirpur, India [b]Department of Chemistry, Indian Institute of Technology Delhi, Hauz Khas, New Delhi, India

12.1 Introduction

Plasmonic photocatalyst nanoparticles (PPNPs) are growing rapidly as innovative materials for various advanced engineering applications. The design, fabrication, and manipulation of nanostructured materials to gain various physicochemical properties particular to a specific range of sizes is a prominent subject of research in Materials Science [1,2]. Noble metal nanoparticles (NMNPs) and metal-oxide semiconducting nanostructures have a range of capabilities that could benefit future societies, including optoelectronics, sensing, bioimaging, catalysis, etc. In recent decades, much research has been done to create nanocomposites for application in sensors, solar cells, optical devices, photocatalysis, and antibacterial activities. [3–5]. In this book chapter, we have outlined the recent achievements in photocatalytic degradation of organic pollutants by utilizing the nanomaterials belonging to the category of NMNPs and metal oxides, with a focus on TiO_2.

TiO_2 is a promising photocatalyst among the other metal-oxide semiconducting nanomaterials because not only it possesses long-term physical and chemical stability, but also have low manufacturing costs. Because of its large surface area and ease of production, TiO_2 NPs have shown amazing effectiveness as photocatalysts, adsorbents, and sensors in early studies [4,6,7]. However, intrinsic drawbacks such as quick recombination, poor charge carrier transfer rate, and high recycling cost restrict their photocatalytic effectiveness. Despite its attractiveness as a photocatalyst nanomaterial, the broad band gap of TiO_2 in the UV range and quick recombination of charge carriers limits its possible applicability in the visible regime of light irradiation. [2,4,5]. As a result, limiting photogenerated charge carrier recombination in TiO_2 nanomaterials, as well as expanding their optical absorption to the visible range of sunlight, are critical stages in improving photocatalytic efficiency. Considerable efforts have been made in the recent decades to produce TiO_2-based efficient visible light active photocatalysts.

Gold and Silver Nanoparticles.
DOI: https://doi.org/10.1016/B978-0-323-99454-5.00011-1

These efforts include modification or functionalization of TiO_2 nanomaterials by producing functional groups on the TiO_2 surface (hydroxyl groups, oxygen deficits), metal or nonmetal doping or by making nanocomposites with carbon nanotubes (CNTs), and noble metal, other semiconductors etc [1,2,8,9].

Modification of TiO_2 based nanomaterials with NMNPs has been extensively studied as NMNPs can absorb light in visible region because of their unique property that is, surface plasmon resonance (SPR). This SPR can be tuned by varying their size, shape, and surrounding medium. In hybrid nanocomposites of TiO_2 with NMNPs, that is, PPNPs, NMNPs can also behave as an electron trap and active reaction sites. Due to these factors, these PPNPs act as an efficient visible-light active photocatalyst. In case of PPNPs, the photogenerated electrons and holes in TiO_2, may be efficiently separated by the metal-semiconductor interface along with photogenerated electrons are generated due to the SPR of NMNPs under the visible light. General characteristics of NMNPs, their plasmonic behavior, and TiO_2 NPs followed by synthesis and tailoring of PPNPs are discussed in the next sections.

12.2 General characteristics of plasmonic and metal oxide semiconductor photocatalysts

The plasmonic and metal oxide semiconductor nanomaterials have unique physicochemical properties that can be tailored for multifunctional applications. Plasmonic, a relevant subject of materials science that deals with the creation, amplification, and transfer of optical signals by generating surface driven oscillations of free electrons in thin layers or NPs of noble metals, is bringing new possibilities in the world of nanotechnology [10,11]. On the other side, TiO_2 semiconducting NPs have become increasingly popular for usage in technological applications such as in green hydrogen production, sensors, LED, and in waste water treatment etc. among various commonly used semiconductors. [4,5,9]. These plasmonic NMNPs and TiO_2 semiconductor photocatalysts are briefly discussed below.

12.2.1 Plasmonic (Au/Ag) nanomaterials

In plasmonic photocatalysis, the word "plasmonic" define the strong oscillation of metal free electrons in phase with the incident electromagnetic radiations [10,11]. In this system, mainly noble metals (Ag/Au) were used because their resonance wavelength which falls in the region of UV-visible-IR [12,13]. The major objective behind the plasmonic photocatalysis was to make metal oxide semiconductors that is, TiO_2 nanomaterial, highly photoreactive under visible light so that maximum photocatalysis efficiency and stability under sunlight irradiation could be achieved. Plasmonic photocatalysis have two different features w.r.t other photocatalytic systems; first is Schottky junction and second is LSPR [10]. Schottky junction is formed near the metal-semiconductor interface which creates an internal electric field. This electric field helps in the movement of free electrons and holes to move in different directions. Due to

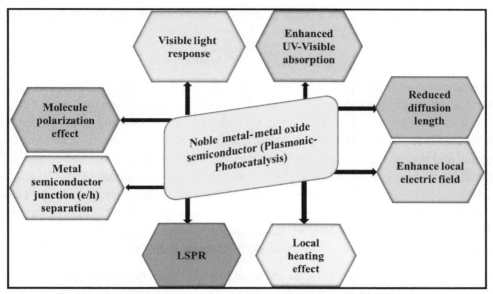

Fig. 12.1: Major properties of plasmonic photocatalysts [2].

this charge separation occurs and the chances of recombination of photoexcited charge carriers is decreased. LSPR enhances the UV-Visible absorption. [10,14,15]. Some of the important properties of plasmonic nanomaterials have been described in Fig. 12.1.

Various forms of NMNPs like prisms, spheres, nano shells, etc. [10] exist but mainly deal with spherical NMNP in plasmonic photocatalysis because it has the simplest structure. And, more importantly, it produces strict analytical solution in response to an electromagnetic field. When visible light interacts with NMNPs, the free conduction electrons in the NMNPs are displaced by external electric field. Due to this phenomenon charge separation and formation of dipole occur. The dipole contribution goes falls down as we move away from the NMNP surface and enhancement in electric field is also maximum near the noble metal surface [10,16]. LSPR is explained by Mie's theory on solving Maxwell Equation:-The extinction spectrum ($E_{xtinction}$) in case of dispersed NMNP's (where the size of NMNP is much smaller than wavelength of interacting EMR) is the product of two terms:-Wavelength dependent term which is related to scattering and the dielectric constants of the NMNPs and dispersing medium as shown in Eq. (12.1) [2,17–20].

$$\text{Sextinction} = \frac{24^2 R_a^3 \varepsilon_{dm}^{3/2}}{\lambda} \left[\frac{\varepsilon_{ic}}{\left(\varepsilon_{rc} + 2\varepsilon_{dm}\right)^2 + \varepsilon_{ic}^{~2}} \right] \tag{12.1}$$

Where R_a is the radius of spherical NMNP, ϵ_{dm} is the dispersing medium dielectric constant where in NP is dispersed ϵ_{rc} and ϵ_{ic} are the real and imaginary components of dielectric function of NMNP. Mie resonance occur at LSPR frequency under the following condition depicted in Eq. (12.2): -

$$\left(e_{rc} + 2e_{dm}\right)^2 + e_{ic}^2 = minimum \tag{12.2}$$

When size of NMNP is approximately equal to wavelength of light used or greater than, then scattering is dominant over absorption [21,22]. When size of NMNP increases from 20 to 80 nm resonating wavelength also increases from 520 to 540 nm. Also, change the shape of NMNP shows variations resonating wavelength. When shape of NMNPs changes from spherical to pentagonal and pentagonal to triangular resonant wavelength shift from 445 to 520 and 520 to 670 nm. Resonating wavelength is also shifted towards higher wavelength (350–430 nm) as the surface contact area of spherical NMNP with photocatalyst TiO_2 increases. As the value of dielectric constant increases, we get red shift in the resonating wavelength [2,3]. It was observed that change in shape and size of NMNPs enhances their visible light activity. Due to which the hybrid of NMNPs with TiO_2 shows visible light stimulated photocatalytic activity [2,23].

The morphologies of NMNPs other than spherical like nanorods, nanowires etc. also show multiple resonance peaks due to multipolar resonances in different directions [24]. In the absorption spectra, two peaks at different wavelengths are observed which are useful for plasmonic photocatalysis as it utilizes a broader spectral range [25]. In the plasmonic photocatalysis due to LSPR, the collective oscillation of free electrons helps in more excitation of electrons and holes by creating a space charge region in TiO_2 near Au surface. When TiO_2 comes in contact with Au, the electrons from TiO_2 goes to the Au side by creating a positive charge region in TiO_2. At equilibrium condition, equal number of electrons are trapped by Au and an internal electric field is created from TiO_2 to Au side. Due to this force of local internal electric field, electrons go towards Au and holes in TiO_2 direction and recombination is decreased. After that these electrons and holes are captured by acceptors and donors and participate in redox reactions [26–29].

Apart from many advantages, such plasmonic photocatalysts possess drawbacks too due to the shading effect that is, reduction in light receiving area of semiconductors surface due to deposition of NMNPs on the semiconductor. This shading effect decreases the photocatalytic activity, because the photo absorption is decreased. This effect is more when we increase the metal loading because NMNPs present on the surface of semiconductor blocks some part of semiconductor for upcoming light absorption. Moreover, due to the metal corrosion and leaching, photocatalytic performance also decreases [10,30].

12.2.2 TiO₂ photocatalysts

TiO_2 is one of the most efficient photocatalyst because of its physical and chemical stability, low toxicity, high catalytic activity, etc. Its unique qualities, such as high energy conversion efficiency and recycling ability, make it the most promising semiconductor nanomaterial. It is the most extensively used photocatalyst with its potential uses in other industries such as sensor, antibacterial agent, solar cell component, and so forth [4,5,31,32]. Fujishima et al.

[33] discovered photocatalytic water splitting using TiO_2 NPs under UV light for the first time in 1972. The photocatalytic capabilities of a semiconductor are dependent on its electronic structure, that is, its band gap. Light absorption and photo excitation of electron (e^-) and hole (h^+) pairs occur during the photocatalytic event if the incoming photon energy meets or exceeds the semiconductor's bandgap.[4,34] When UV or visible light is directed on TiO_2 nanomaterials, photo-excited e^- excites from the valence band (VB) to the conduction band (CB), resulting in a positively charged h^+ in the VB (particularly near-ultraviolet (315–400 nm) [4,34]. These photoexcited e^- h^{+-1} pairs can participate in reduction or oxidation reactions. When a compound is adsorbed on the surface of the photocatalysts. In response to photoexcitation, the generated h^+ facilitates an oxidation reaction whereas the e^- promotes a reduction process by reacting with O_2, H_2O, or OH^-. These oxidation and reduction reactions produce reactive oxygen species (ROS), such as superoxide anion radicals ($\cdot O^{2-}$), hydroxyl radicals ($\cdot OH$) and hydrogen peroxide (H_2O_2) molecules. The origin of ROS species can be understood clearly from the following Fig. 12.2A–B [2,35,36].

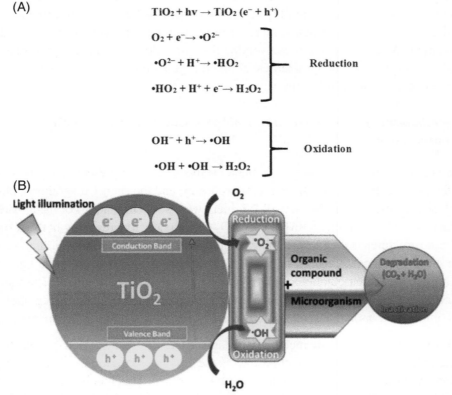

(A)

$$TiO_2 + hv \rightarrow TiO_2\ (e^- + h^+)$$

$$O_2 + e^- \rightarrow \cdot O^{2-}$$
$$\cdot O^{2-} + H^+ \rightarrow \cdot HO_2$$
$$\cdot HO_2 + H^+ + e^- \rightarrow H_2O_2$$

Reduction

$$OH^- + h^+ \rightarrow \cdot OH$$
$$\cdot OH + \cdot OH \rightarrow H_2O_2$$

Oxidation

Fig. 12.2: (A) Complete equation showing mechanism of photocatalytic degradation [2]. (B) Photocatalytic mechanism of TiO_2 nanoparticle under UV light irradiation. *Reproduced with permission from [2].*

Wide band gap (3.2 eV) which lies in UV region of the electromagnetic spectrum, greater combination rate of electrons and holes, delayed charge carrier transfer, and visible light inactivity are all the major drawbacks of TiO_2 nanomaterials. These drawbacks restrict its proper implementation for visible light activity. The fast recombination of $e^- h^{+-1}$ pair results in low quantum yield [5]. Furthermore, sunlight only includes around 5% UV radiation and that is the reason TiO_2-based nanomaterials are less efficient under solar radiation. Hence, there is a need for the proper utilization of TiO_2 nanomaterials. Much efforts have been made to overcome these barriers [2,4,5]. As discussed in the last section, there are several ways to modify and improve the functional properties of TiO_2 nanomaterial such as surface functionalization, doping, composite formation with plasmonic metals, semiconductors doping, 2D functional materials, dye sensitization, and so on [19,23,37–41]. These modifications not only improve the visible light activity but also reduces the recombination rate of charge carriers which play a major role in the enhancement of photocatalytic activity.

12.3 Plasmonic photocatalysts

Plasmonic photocatalysis has lately emerged as a very promising approach for high-performance photocatalytic activity [42]. It entails dispersing NMNPs into semiconductor photocatalysts, resulting in a dramatic increase in photo-reactivity when irradiated with UV and visible light. When compared to conventional semiconductor photocatalysis, plasmonic photocatalysis has two significant features: a Schottky junction and LSPR; each of the feature benefits photocatalysis in its different way.

12.3.1 Mechanism of TiO_2 based plasmonic photocatalysts and tailoring of optical properties

The fundamental mechanism of plasmonic photocatalysis has been presented in this section, covering the types, general benefits, the major processes, and the energy diagrams of plasmonic photocatalysis. As discussed in Section 12.2.2, there were some limitations of using bare TiO_2 as photocatalysts. The addition of NMNPs to TiO_2 semi-conductors resulting in the formation of hybrid nanostructures is the most suitable technique for overcoming those limitations of TiO_2. These hybrid nanostructures show absorption of light in a range of the solar light spectrum resulting in the increased photocatalytic activity. Because of the unique LSPR features of NMNPs, enhanced visible light absorption might be accomplished by suppressing photo-excited $e^- h^{+-1}$ recombination or charge separation by electron trapping via Schottky junction.

Under UV irradiation, plasmonic metals-TiO_2 nanostructures are energized, primarily the TiO_2 semiconductor, but the dominant process is the charge separation mechanism of plasmonic metals to produce a Schottky junction at the interface. This junction leads to the generation of an internal electric field inside the photocatalyst. Once produced inside or near the Schottky junction, the e- h^{+-1} would be forced to migrate in opposite directions. The Schottky

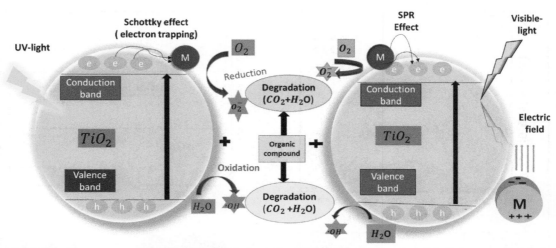

Fig. 12.3: Charge separation methods of photocatalysis under UV and visible light are depicted schematically [2].

junction traps the photoexcited e^- and prevents the recombination of e^- h^{+-1} pairs and facilitates the rate of ROS formation as shown in Fig. 12.3 [2].

Under visible light, plasmonic-TiO_2 nanostructures shows charge transfer (CT) mechanism due to the LSPR of plasmonic metals. The LSPR property of plasmonic metals has been discussed in detail in Section 12.2.1. When visible light strikes the plasmonic photocatalyst, the LSPR of the plasmonic metals excites the electrons on the metal surface, which are then transported to the TiO_2 CB. The excited electrons in TiO_2 CB engage in subsequent chemical processes to create ROS, which increases the photocatalytic activity. Tian et al. [43] proposed the mechanism of CT in plasmonic metal based TiO_2 hybrid nanostructures.

As discussed in Section 12.2.1, there are various factors that affect the optical properties of the plasmonic metals. Similarly, the sizes, morphology, composition, surrounding media, etc. play an important role in tuning the optical properties of the hybrid nanostructures contained NMNPs. For example, Lin et al. [44] studied the effect of AuNP size on photocatalytic activity of Au-TiO_2-x. Because of their stronger contacts with TiO_{2-x}, smaller AuNPs provide more reactive sites, wider regions for light absorption, and hot electron transport at the interface. Small size AuNPs shows larger adsorption energy which indicates stronger interaction of Au and TiO_2. Fig. 12.4A PL spectra shows that intensity of the signals decreased with the decreasing of the Au NPs sizes. It reflects that the e^- h^{+-1} pair recombination was suppressed with the smaller size of AuNPs.

Similarly, Bamola et al. [45] prepared plasmonic NPs-TiO_2 nanorods (NRs) hybrids, (Ag and Au plasmonic NPs of various sizes were adorned over TiO_2 NRs. Using an inert gas

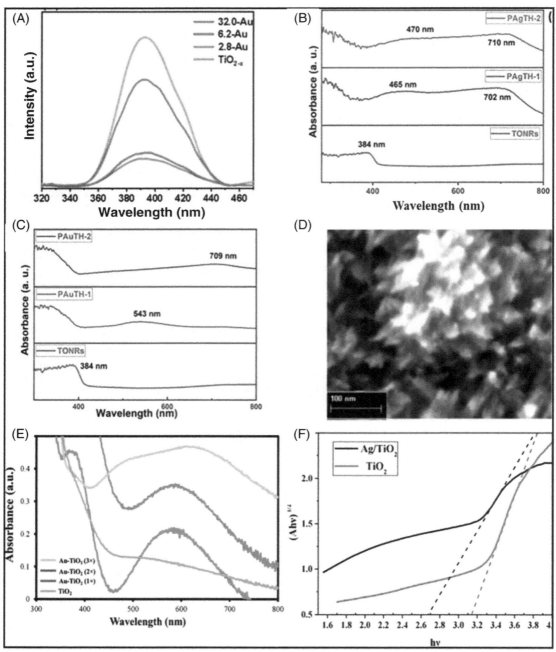

Fig. 12.4: (A) PL spectra of Au/TiO$_2$ nanocomposite with varying size. *Reproduced with permission from [44].* (B–C) UV spectra of Au-TiO$_2$ coral-needle nanoparticles (PAgTH-1 plasmonic Ag- TiO$_2$ NRs hybrid with Ag size 3 nm and PAgTH-2 plasmonic Ag-TiO$_2$ NRs hybrid with Ag size 5 nm and similarly with PAuTH-1 and PAuTH-2 respectively). *Reproduced with permission from [45].* (D) SEM image of Au decorated TiO$_2$ surface. (E) UV spectra of Au-TiO$_2$ coral-needle nanoparticles. *Reproduced with permission from [46]* (F) UV spectra of Ag/TiO$_2$ Nanoparticles. *Reproduced with permission from [25].*

evaporation process at various vapor pressures, the size of Ag and Au PNPs were controlled. In Figs. 12.4(B–C), red-shift in UV spectra was seen with the condition of resonance wavelength that strongly related to the increasing size of Ag and Au NPs. Au-TiO_2 nanocomposite showed higher zeta potential values indicating that plasmonic NPs possessed higher periphery surface charge, which could promote repulsion of NPs, avoiding formation of aggregates and improve the stability. The use of Au NPs with tiny size to adjust the interface and plasmonic activity of TiO_2 NRs were proven to be more effective than Ag NPs. Wibowo et al. [46] studied that Au-TiO_2 coral-needle nanoparticles (CNPs) have high-response of absorption activity in the visible light area in a range of 555–600 nm. The successful decoration of Au exhibited the shining morphological on TiO_2 surface shown in Fig. 12.4D The Au-TiO_2 CNPs produced the glittering morphological effect and drastic plasmonic photocatalytic activity after successful decoration of AuNPs. UV spectra of Fig. 12.4E show the high-response of Au-TiO_2 (3x) on visible range absorption from 555 to 600 nm.

Zheng et al. [25] synthesized cubic Ag TiO_2^{-1} nanocomposite via combination of hydrothermal and photoreduction process. The Ag TiO_2^{-1} nanocomposite showed absorption in visible region which was due to the SPR property of Ag. By using Tauc plot equation they calculated the band gap energies such as 3.15 eV and 2.7 eV for pure TiO_2 and Ag TiO_2 composites, respectively, as shown in Fig. 12.4F. Similarly, Shoueir et al. [47] designed a plasmonic fiber which act as an efficient visible-light photocatalyst for the breakdown of organic and inorganic pollutants. The degrading activity, kinetic behavior, catalytic reduction, and stability of the plasmonic fiber were good as compared to Au-TiO_2. Such fabricated photocatalysts are expected to aid in energy conservation and environmental applications. Figs. 12.5A–B TiO_2, Au@TiO_2, and plasmonic fiber samples had band gap energies of 3.2, 3.11, and 3.13 eV, respectively. This was due to the creation of an isolated defect energy level from Ti^{3+} below the CB of TiO_2 [47].

Concentration of the precursor solution of plasmonic metals showed tremendous changes in its photocatalytic activity of the TiO_2-plasmonic hybrid photocatalysts. For example, in Fig. 12.5C, Ling et al. [48] showed the TEM results where the Ag NPs were attached on the surface of TiO_2 NPs. Fig. 12.5D demonstrated that TiO_2 sample loaded with Ag NPs of different concentration had a visible light absorption range of 400–800 nm. The binding energies of the Ti 2p and O 1s peaks in Ag TiO_2^{-1} was found to be considerably displaced as studied by X-ray photoelectron spectroscopy (XPS) which indicating the strong binding of Ag and TiO_2. It was seen that as the concentration of Ag NPs increased on TiO_2, the degradation of dye molecule also increased showing increased photodegradation capability. The sample of 1.4 wt% Ag TiO_2^{-1} showed the best photocatalytic activity. Similarly, Wu et al. [50] studied the effect of different concentration of Ag dopant on the photocatalytic activity of Ag-TiO_2 nanocomposite. The absorbance in visible light increased with increasing concentration of Ag doping attributed to the typical absorption of Ag_2O and SPR by Ag depositing on TiO_2 surface.

Scarisoreanu et al. [49] observed the effect of synthesis method that is, laser pyrolysis (LP) and different plasmonic metals (Ag, Au, and Pt) on the optical and photocatalytic activity of

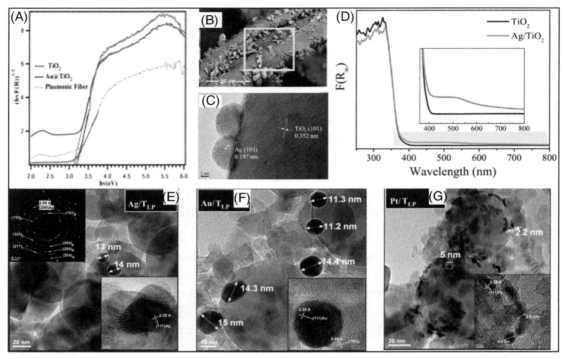

Fig. 12.5: (A) UV spectra of Au@TiO$_2$ plasmonic nanofibers. (B) SEM of Au@TiO$_2$. *Reproduced with permission from [47]* (C). TEM images of Ag TiO$_2^{-1}$. (D) UV spectra of Ag TiO$_2^{-1}$ nanoparticle. *Reproduced with permission from [48]* (E–G) TEM images of samples of Ag, Au, and Pt composite with TiO$_2$. *Reproduced with permission from [49].*

TiO$_2$-plasmonic metal hybrid. Reduction in the band gap energy of hybrid was observed due to synergistic effect of near atmospheric pressure in LP synthesis and metal addition on TiO$_2$ NPs. Morphological analyses by TEM images showed that ellipsoidal Au, Ag, and Pt NPs were dispersed onto TiO$_2$ NPs surface as shown in Figs. 12.5E–G.

12.3.2 Synthesis of TiO$_2$ based plasmonic photocatalysts

There are several promising physical and chemical methods which have been used for the synthesis of plasmonic photocatalysts [2,42]. The synthesis of PPNSs by optimizing the experimental conditions and parameters in order to produce highly stable, reproducible and sensitive nanomaterials [42]. These methods are briefly discussed in this section along with the recent advancements carried out in fabrication strategies by the researchers.

For plasmonic photocatalysts materials (TiO$_2$-Ag Au^{-1}), sol–gel synthesis is a widely utilized synthesis process. A colloidal solution is made via this approach, which involves polycondensation of precursor in a liquid. [51] The sol-gel methodology, in particular, allows for precise control of parameters, resulting in the synthesis of a wide range of nanocomposites. The steps for the production of TiO_2-Ag Au^{-1} nanocomposite in sol–gel are as follows; [52] (1) In the beginning, a uniform solution is prepared by dissolving noble metal (Ag Au^{-1}) and/or titanium organic precursors in a solvent, that could be water or an organic solvent. [2] (2) The prepared solution is then treated with an adequate acid or basic reagent to convert it to a sol, which is then molded into the desired form using thermal heating or sintering to produce the final product Fig. 12.6A. For example, Prakash et al. [53] used the sol-gel method to produce Ag–TiO_2 nanocomposite NPs and studied their multifunctional applications in photocatalysis, antimicrobial activity, and surface-enhanced Raman scattering (SERS). The lack of control over reaction rates is the fundamental disadvantage of the sol-gel synthesis process. In some metal-oxide precursors, reactions occur more quickly, resulting in the uncontrollable morphological and structural outcomes. It has also been discovered that the varied reactivity of metal alkoxides in the sol–gel process does not allow good control over the composition and uniformity of the product [2,54].

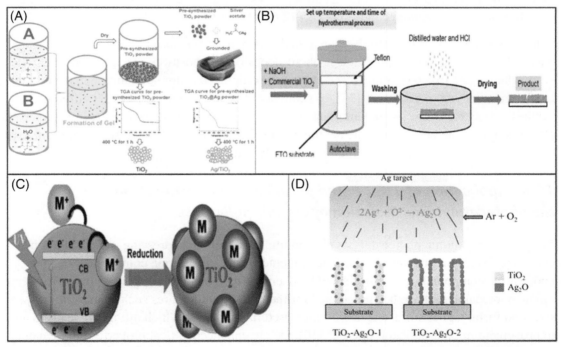

Fig. 12.6: Various methods of the synthesis of plasmonic photocatalysts, (A) Sol-gel method. *Reproduced with permission from [54]* (B) hydrothermal. *Reproduced with permission from [55]* (C) photoreduction method. *Reproduced with permission from [2]* and (D) sputtering. *Reproduced with permission from [56].*

Hydrothermal synthesis route is another most promising methods which provide the possibility of synthesizing all kinds of plasmonic photocatalysts nanostructures by varying the experimental parameters Fig. 12.6B [55]. A quartz autoclave is commonly used for the hydrothermal process [4,5]. Precursors of plasmonic metals and TiO_2 were first dissolved in either concentrated aqueous or alkaline solutions, then the final solution was transferred into the autoclave and the experimental parameters were maintained [25,57]. When the reaction finishes, the system is cooled down to normal temperature and desired product was filtered and dried [58]. The solvothermal method is similar to hydrothermal taking place in an autoclave, but the precursor solutions are usually not aqueous but organic solvents [59].

Photoreduction is another effective method for the synthesis of plasmonic photocatalysts by reduction of plasmonic metal ions adsorbed on TiO_2 surface Fig. 12.6C [48]. This approach is based on the photocatalysis reaction discussed in Section 12.2.2. This technique is very useful for investigating photoinduced interactions in metal-semiconductor systems and the interfacial CT mechanism for photocatalytic enhancement [2,25].

Sputtering is an effective method for producing thin films on a substrate. In this method, atoms are ejected from a solid target because of bombardment by energetic particles. Because they are in the gas phase and not in an equilibrium state, sputtered atoms are easily deposited on substrates under vacuum conditions Fig. 12.6D [56,60,61].

Other approaches in addition to the ones stated above, have been used to manufacture TiO_2-based plasmonic photocatalysts such as photo-deposition, deposition–precipitation, and ion-exchange, etc [62–64]. Some researchers have employed a simple biomimetic technique for the manufacture of TiO_2 NPs [37]. In summary, it is possible to deduce that different synthesis processes have significant impacts on morphologies, crystallinity, and thus on the photocatalytic characteristics of the nanomaterials. The choice of targeted shape and synthesis technique is critical in generating high photocatalytic activity.

12.4 Application of TiO_2 based plasmonic photocatalysts in the photocatalytic degradation of organic pollutants

TiO_2-based plasmonic photocatalysts play an important role in addressing water pollution challenges by degrading various pollutants, including organochlorine compounds, organophosphorus compounds, polychlorinated bisphenols, and phenolic compounds [65]. Plasmonic photocatalysts have been regulated for their structural features such as morphology, size, and form, higher surface area with increased functionality, which ultimately leads to an improvement in their optical properties [10]. Due to enhanced optical properties, these nanocomposites can be used in many fields such as in catalysis, sensing, energy storage etc. In this section, we have mainly focused on the effect of Ag and Au on the photocatalytic activity of TiO_2 nanomaterial specifically photodegradation of organic pollutants.

12.4.1 Ag-TiO$_2$ based plasmonic photocatalysts

Combining TiO$_2$ nanostructures with Ag NPs is a recent trend in the development of effective photocatalysts [66]. Ag is the most attractive element, and its integration with TiO$_2$ NPs opens a wide range of potential applications [67]. Various transition metals (Ag, Au, Co, Cr, Cu, Fe, Ni, Pd, Pt, Y, and Zn) doped TiO$_2$ were synthesized and followed by the study of their photocatalytic activity. Moreover, it was seen that among all, Cu, Pd, Pt, Zn, and Ag presented the excellent photodegradation phenomenon under UV light Fig. 12.7A [57]. Doped samples show better photodegradation activity as compared to bare TiO$_2$ and commercial TiO$_2$-P25. It was observed that Ag doped TiO$_2$ composite showed highest photocatalytic activity as studied by Wu et al. [57] For the detailed study of photocatalytic activity of such nanocomposite, doping concentration was optimized. It was found that at higher doping concentration above 5.00 mol%, Ag$_2$O, and TiO$_2$ rutile phase were both present and further increased in doping concentration caused phase transformation from anatase to rutile. Similarly, UV spectra showed visible light absorption due to the SPR effect of Ag. It was also seen that apart from high photocatalytic activity, it had excellent disinfection property. Similar, study was done by Hariharan et al. [68] where they studied anticancer and photocatalytic activity of the Ag TiO$_2$$^{-1}$ nanocomposite. In comparison to the other concentrations, 0.015M of Ag@TiO$_2$ had superior optical properties, including significant visible light absorption, resulting in an excess of electron transport to cell lines. Because of their UV absorption, the photocatalytic activity of Ag@TiO$_2$ had been explored for the breakdown of picric acid (an example of explosive) in the visible light area Figs. 12.7B–C.

Zheng et al. [25] described that the photocatalytic performance of Ag-TiO$_2$ composites towards the breakdown of aqueous solution of methyl orange (MO) was higher than that of pure nano-TiO$_2$. As discussed in above in Fig. 12.4F that Ag doping decreased the band gap energy and facilitated the visible light absorption. Fig. 12.7D showed the photodegradation efficiency of aqueous solution of MO in pure TiO$_2$ and Ag TiO$_2$$^{-1}$ composite under exposure to UV light. It was calculated that pure TiO$_2$ showed degradation efficiency of 56%, and Ag TiO$_2$$^{-1}$ composite showed increased Photodegradation up to 65.4%. Also, the photocatalytic study was done under visible light, and it was seen that photocatalytic activity of Ag TiO$_2$$^{-1}$ in UV–Visible region was higher than that of in UV region. It was due to synergic effect of both morphology and Ag deposition which reduced the charge recombination rate and enhanced the photocatalytic activity under UV-visible light. The involved mechanism has been shown in Fig. 12.7E.

Ling et al. [48] prepared Ag TiO$_2$$^{-1}$ nanocomposite by using photo-reduction method and studied the photocatalytic activity and adsorption of MO under microwave (MW) irradiation. As discussed in Fig. 12.5D, UV spectra showed the reduction of band gap energy due to the Ag NPs addition with TiO$_2$. Using these Ag TiO$_2$$^{-1}$ nanocomposite, four different sources of irradiation were employed to study the photocatalytic activity that is, MW, UV, combined UV-MW and combined UV with heating method (UV+ CH). It was concluded that only MW radiation does not have observable effect on photocatalytic activity of Ag-TiO$_2$. From Fig. 12.7F, it was

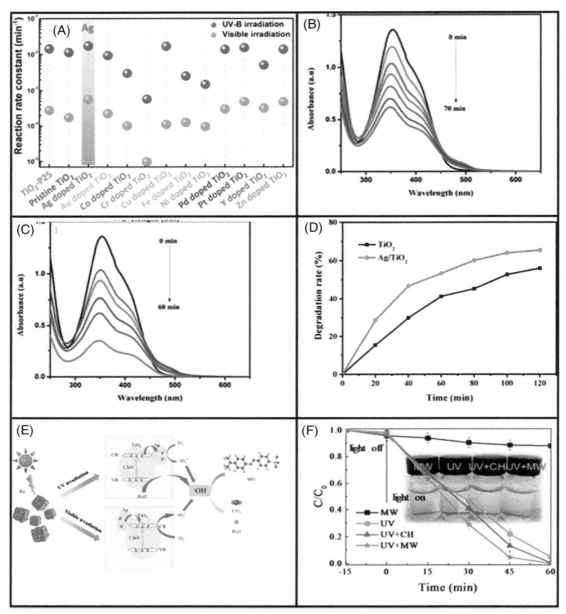

Fig. 12.7: (A) The photodegradation activity of TiO_2 P25, pristine TiO_2, and various transient metal doped TiO_2 NFs under UV-B irradiation and visible light irradiation. *Reproduced with permission from [57]* (B–C) Photocatalytic study of picric acid by Ag@TiO_2–0.005 M and Ag@TiO_2–0.015 M. *Reproduced with permission from [68]* (D) The degradation efficiency of MO solution in the presence of pure TiO_2 and Ag TiO_2 composites under UV light. (E) Schematic of catalytic mechanism of Ag TiO_2^{-1}. *Reproduced with permission from [25]* (F) MO under different reaction conditions (50 mL 10 ppm 4-CP, or MO, 50 mg catalyst). *Reproduced with permission from [48].*

clearly observed that degradation efficiency was enhanced with the increase of reaction system temperature (UV + CH), which was mainly due to the acceleration of reaction rate by thermal effect. In Raman spectra, before UV light irradiation only three peaks were shown but after UV light illumination characteristic peaks appears at 750 cm^{-1} and 1587 cm^{-1}, which confirmed that analyte molecules were adsorbed on the surface of TiO_2. Strongest peaks were seen in the case of MW + CH indicating the intense adsorption of analyte molecules on TiO_2 surface.

Sun et al. [69] fabricated Ag-TiO_2 nanowires by one-step hydrothermal process. The results indicated that Ag treatment considerably increased TiO_2 nanowire absorption in the visible region. When compared to other samples, the Ag-TiO_2 nanowire composites with 3 mol percent Ag displayed the highest photocatalytic activity which might be attributed to improve light harvesting and decreased charge recombination. Similarly, Zarepour et al. [59] employed solvothermal method for the synthesis of TiO_2 nanorices and study the effect of deposition of Ag NPs on the photocatalytic behavior of TiO_2. The photocatalytic activity of samples containing varying quantities of Ag NPs was thoroughly studied. The greatest photocatalytic activity was demonstrated by TiO_2 samples containing 0.125 percent Ag in the breakdown of Acid Blue 92 and Crystal Violet as pollutants.

Singh et al. [70] have performed CTAB (cetyl trimethyl ammonium bromide) assisted wet chemical synthesis of TiO_2 with citrate reduction of Ag^+ ions resulting Ag–TiO_2 hybrid PPNPs. Under 60 minutes of solar light irradiation, Ag–TiO_2 hybrid NPs demonstrated enhanced photocatalytic efficiency of 89.2 %, which was 1.7 times greater than TiO_2 NPs Fig. 12.8A. The photoluminescence (PL) results depicted in Fig. 12.8B showed that the addition of Ag NPs inhibited charge recombination in TiO_2 NPs.

PPNPs as heterojunctions of plasmonic Ag NPs on Ti^{3+}-doped TiO_2 (Ag TiO_{2-x}^{-1}) were fabricated by Lu et al. [71]. The visible-light photodegradation of Rhodamine B (RhB) on 3%-Ag TiO_{2-x}^{-1} catalyst was 14-fold higher than those of bare TiO_2. It was explained based on the synergistic effects of Ag plasmonic and Ti^{3+-}doping in Ag TiO_{2-x}^{-1} heterojunction. High photocatalytic decomposition of RhB under visible light was due to the coupling between the SPR of the Ag and oxygen vacancy created by Ti^{3+} in TiO_{2-x}. There was a formation of a Schottky barrier at the junction shown in Fig. 12.8C which reduced the charge carrier recombination rate and generated more ROS which enhanced the photodegradation activity. Presence of Ti^{3+} in nanocomposite was confirmed by XPS spectra Fig. 12.8D.

The activity of varied weight percent of Ag co-catalyst over TiO_2 surface under varying UV illuminations (30, 60, and 90 minutes) for photodegradation of salicylic acid was investigated [62]. Considerable color change was observed as the illumination time was increased. It was observed that Ag deposited nanocatalysts (1, 3, and 5 wt%) prepared at 90 minutes showed enhanced activity for the photodegradation of salicylic acid under UV-light irradiation as shown in Figs. 12.9A–C. The greatest activity of the 5 wt% Ag@TiO_2 catalyst was due to the larger size, number, and uniform distribution of Ag co-catalyst NPs over the TiO_2 surface. [62].

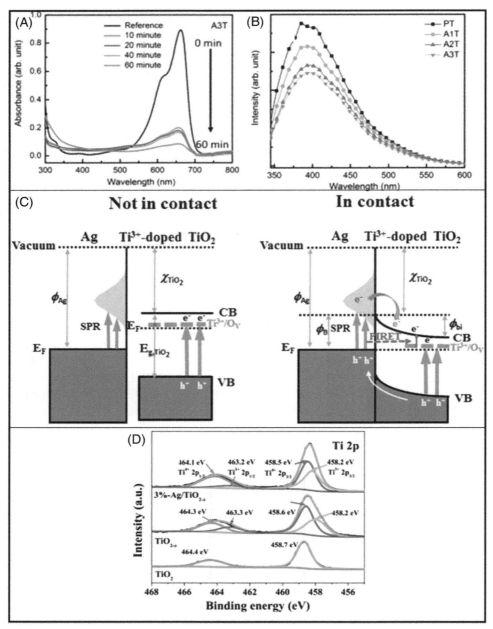

Fig. 12.8: (A) Optical absorption spectra of methylene blue (MB) by A3T as photocatalysts (B) PL spectra of pristine TiO$_2$ sample and Ag–TiO$_2$ nanohybrid samples. (5.5 mM, 11 Mm, and 22 mM of AgNO3 are hereafter denoted to as PT, A1T, A2T, and A3T, respectively.) *Reproduced with permission from [70]* (c) Schematic energy diagram of the Ag micro- nano-particles^{-1} and TiO$_2$-x (Ti$_3$+-doped TiO$_2$) framework before and after forming the heterojunction. *Reproduced with permission from [71]* (D) XPS spectra of pure TiO$_2$, TiO$_2$-x and 3%-Ag TiO$_2$-x^{-1}. *Reproduced with permission from [23].*

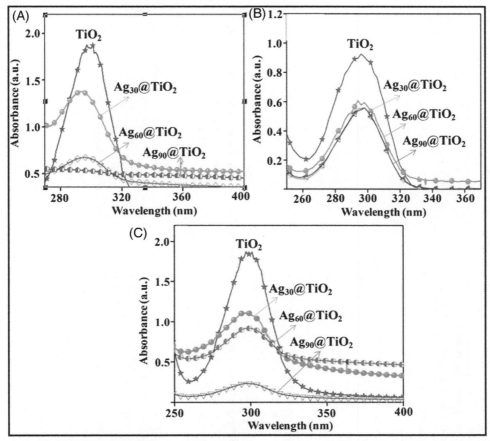

Fig. 12.9: (A–C) Absorption spectra of salicylic acid during its degradation (1 wt%, 3 wt%, 5 wt%) Ag@TiO$_2$ samples (obtained after 30–90 minutes Ag photo deposition) under 80 minutes UV irradiation. *Reproduced with permission from [62].*

Moreover, the recent Ag based TiO$_2$ plasmonic photocatalysts that is, PPNPs, their synthesis, various optical properties, photocatalytic efficiency, mechanisms etc. have been summarized in Table 12.1

12.4.2 Au-TiO$_2$ based plasmonic photocatalysts

Au NPs-TiO$_2$ photocatalyst nanostructures have been widely exploited as plasmonic photocatalysts in a wide range of processes. Similar to Ag based TiO$_2$ plasmonic photocatalysts, Au based plasmonic photocatalysts have also been extensively investigated from all aspect of size, concentration, morphology, etc. point of view in order to enhance the photocatalytic activity of the hybrid nanostructures. For example, Lin et al. studied the Au-TiO$_2$ system by

Table 12.1: Various plasmonic TiO$_2$ nanocomposites systems including their synthesis methods, morphologies, improved important properties, mechanism of action, and corresponding results in photocatalytic degradation of organic pollutants are summarized.

Nanocomposites TiO$_2$-Ag/Au	Synthesis methodology	Morphology	Improved optoelectronic and structural properties	Mechanism	Efficiency	Refs.
Ag TiO$_2^{-1}$	Hydrothermal and Photoreduction	Cubic	Decrease in band gap energy after Ag deposition	SPR and Schottky barrier help in generation of ROS	Degradation rate of Ag TiO$_2^{-1}$ composites was up to 65.4%	[25]
Au-TiO$_2$	Sol gel method and chemical deposition	Plasmonic nanohybrids	Decrease in band gap and absorption in the visible region.	SPR and Schottky junction which produce more ROS	Degradation efficiency of Au-TiO$_2$ nanohybrids is 94%with MB,85%with MO,87% with mix. Of MB and MO under 20 minutes irradiation	[51]
Bimetallic Au Ag^{-1} TiO$_2$	Facile phase inversion method	Nanorods	Broad absorption band in visible range	Due to synergetic effect of SPR of bimetallic Au and Ag NPs.	Degradation rate of optimal Au0.1Ag0.9 TiO$_2^{-1}$ CA^{-1} membrane up to 90%	[38]
Ag decorated TiO$_2$	Solvothermal method	Nano-rice	Band gap is reduced and strong absorption in visible region.	Plasmonic effect of Ag NPs and reduce electron-hole recombination make photocatalyst highly reactive under visible range	Degradation of acid blue 92 and crystal violet are 72% and 53.16%	[59]
Ag doped TiO$_2$	Hydrothermal method	Nanofibers	Ag ions increase the absorption band in visible region on reducing band gap	Due to synergistic effect of Ag2O and SPR cause generation of more photoinduced ROS	Photodegradation of MO with 5 mol% Ag doped TiO$_2$-600 gives highest rate.	[57]
Au TiO$_2$-X^{-1}	Hydrothermal method	Nanobelts	Smaller Au NPs have more reactive sites, large area for light absorption and hot e-transfer LSPR intensity increase with increase size of Au NPs, narrow the band gap of TiO$_2$-X	Due to LSPR, e- transfer mechanism,	High catalytic performance rate	[58]

Nanocomposites TiO$_2$-Ag/Au	Synthesis methodology	Morphology	Improved optoelectronic and structural properties	Mechanism	Efficiency	Refs.
Ag TiO$_2$$^{-1}$	Photo-reduction method	Nanoparticles	Band gap is reduced and absorption in visible region	SPR and Schottky effect	Rate of photocatalytic degradation with 1.4 wt% Ag TiO$_2$$^{-1}$ of 4-CP and MO enhanced	[48]
Ag-TiO$_2$	CTAB-assisted facile wet chemical method	Hybrid nanoparticles	Decrease in band gap due to modification of metal oxide semiconductor with Ag NPs	Due to Schottky junction and plasmonic the amount of ROS increase	Enhancement in photo-catalytic activity	[70]
Au-TiO$_2$	Sol-gel method, Hydrothermal method	Nano composites	PVDF provide mechanical strength, chemical stability, resistance to acid and alkali corrosion, resistance to UV-radiation etc.	Broad absorption peak around 550 nm due to Au's SPRE	Degradation rate of -TiO$_2$ PVDF^{-1} composite membrane is 75%	[74]
Ag TiO$_2$-X^{-1}	Microwave-assisted process and photoreduction	Fabricated	Ti+3 doped TiO$_2$ gives strong absorption in visible light on creating oxygen vacancy. Band gap is also reduced	Ti+3 doping, Schottky junction and plasmonic of Ag NPs enhance the catalytic performance	Highest degradation rate of 3%-Ag TiO$_2$-X^{-1} was 97.6%	[23]
Ag TiO$_2$$^{-1}$	Sol-gel, Hydrothermal etc.	Nanoparticles	Decrease in band gap energy on doping of Ag	due to SPR and Schottky barrier generates ROS	Degradation rate of Ag-TiO$_2$/r-GO is 99.5% towards RhB dye	[83]
Au-TiO$_2$	-	Composite film	Reduction in band gap energy on loading Au NPs	Due to large surface area of pristine TiO$_2$ and surface plasmon effect high absorption in visible region	Efficiency of PC and dissolved N2 are 96.7% and 59% using UV/Au-TiO$_2$ process	[72]
Ag and Au TiO$_2$$^{-1}$	Hydrothermal method	Nanorods	Band gap is reduced	Ag, Au NPs with small size creates Schottky, LSPR effects	Enhancement in photo-catalytic activity of MB dye with 95% reduction	[45]
Ag@TiO$_2$	Photo deposition	Spherical	Color intensity of Ag@TiO$_2$ change with deposition time	Strong SPR effect cause e- transfer	1, 3, and 5 wt% Ag deposited TiO$_2$ exhibit highest activity at 90 minutes UV irradiation	[62]

(Continued)

Table 12.1: (Cont'd).

Nanocomposites TiO$_2$-Ag/Au	Synthesis methodology	Morphology	Improved optoelectronic and structural properties	Mechanism	Efficiency	Refs.
Au TiO$_2^{-1}$	Spin coating	Asymmetric nanostructures	Band gap is reduced and exhibit absorption in visible region	Schottky barrier and SPR generates high energy hot electrons and holes	Maximum water oxidation at polarization angle 90o	[84]
Au and Ag decorated TiO$_2$	Sol-gel method	Nanospheres	Au-Ag decorated TiO$_2$ gives SPR over broad range of visible light(415–525nm) and enhancement in specific area	Due to SPR more ROS	Au0.4Ag0.6TiO$_2$ gives best result for degradation of MB	[82]
Au@TiO$_2$ NPs	Hydrothermal method	Yolk shell	LSPR of Au, Schottky junction, high cavity of yolk-shell structure, large specific area, more mesoporous, Ti+3 facilitate separation of charge carriers	Light reflected, refracted Charge separation provide more active sites and more ROS	57.3% of gaseous toluene removed over 0.14% yolk-shellAu@TiO$_2$ in 3 hours	[75]
Ag-TiO$_2$	Sol gel method	Nanocomposites	Decrease in band gap with increase in Ag concentration enhance optical, anti-microbial properties	SPR and Schottky barrier at interface helps to produce more ROS	Ag conc.(1–5 mol%) in Ag-TiO$_2$ show higher efficiency	[53]
Ag Au^{-1} TiO$_2$ NPs	Photopolymerization	NPs	Visible light absorption	LSPR and Schottky junction	Degradation efficiency toward azo dye was high	[85]

varying Au loading amounts and found that with the rise in the size of Au NPs, the electron transfer mechanism and the LSPR exhibited decay mostly switches from nonradiative damping to radiative damping, Fig. 12.10A. These transitions result in significant reduction of the carriers participating in photocatalytic reaction. On the other side, smaller AuNPs provide more reactive sites, larger areas for light absorption, and hot electron transfer at the interface and show enhanced photocatalytic activity Fig. 12.10B [58].

Similar study on the effect of size was done by Bamola et al. [45] as discussed in Section 12.3.1. It was seen that the tiny size Au NPs has been found to be effective in modifying the Schottky effect and plasmonic activity. The small sized Au NPs based nanocomposites showed better photocatalytic activity than Ag based hybrids nanocomposite Figs. 12.10C–D. This was due to the good plasmonic activity and low recombination rate of charge carriers at the interface. Small size Au NPs TiO_2^{-1} nanocomposite showed excellent photocatalytic activity against methylene blue (MB) dye with 95% photodegradation efficiency.

The uniqueness of morphology of the plasmonic photocatalysts also affects its photocatalytic activity. For example, Fig. 12.10E showed the thin films morphology of plasmonic photocatalysts. Cao et al. [72] studied the photodegradation of phycocyanin (PC) by Au-TiO_2 thin film in detail. The Au-TiO_2 composite film had a decreased electron-hole recombination rate, wider optical response, and a high electron transfer rate, as shown in Fig. 12.10F. The removal efficiency of PC was recorded approximately to 96.7%. Wibowo et al. [46] prepared Au-TiO_2 coral-needles. They observed that Au-TiO_2 showed excellent photodegradation ability against MO under visible light irradiations. The photocatalytic activity of Au plated on TiO_2 was repeated for high-sensitivity destruction of MO under visible light irradiation as shown in Fig. 12.10G.

Core/shell nanostructures, which consist of a noble metal core surrounded by a semiconductor oxide shell, have also received a lot of attention by scientists owing to their stimulating photocatalytic properties. This arrangement creates a physical barrier with significantly higher calcination temperatures and a larger contact area. Wu et al. [73] synthesized core shell morphology of Au@TiO_2 NPs by using a hydrothermal route. The Fig. 12.11A shows TEM image of Au@TiO_2 core-shell NPs. The synthesized plasmonic photocatalyst (Au@TiO_2) showed better photocatalytic activity than bare TiO_2 as shown in Figs. 12.11B–C. The increased photocatalytic activity of Au@TiO_2 was attributed to the Au core acting as an electron trap to promote the separation of photogenerated charge carriers (Fig. 12.11D). Similar study was performed by Yan et al. [74] synthesized imitated core-shell nanostructures of Au-TiO_2 $PVDF^{-1}$ membrane via sol-gel method. The photodegradation rate of membranes under visible light was investigated with tetracycline. It was seen that Au-TiO_2 $PVDF^{-1}$ membrane showed up to 75% degradation in 120 min.

Wang et al. [75] synthesized Au@TiO_2 yolk-shell which showed excellent photocatalytic activity against removal of volatile organic compounds. The yolk shell morphology of Au@TiO_2

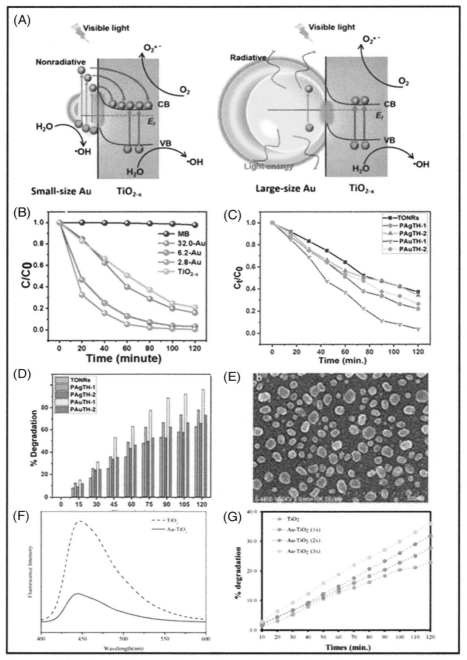

Fig. 12.10: (A) Schematic of size effect on charge transfer in Au-TiO$_2$-x. *Reproduced with permission from [58]* (B) Photocatalytic degradation of MB solution under visible-light irradiation of TiO$_2$-x, 2.8 -Au, 6.2-Au, and 32.0-Au. *Reproduced with permission from [58]* (C) Photodegradation of Ag TiO$_2$$^{-1}$ hybrid (D) Bar graph showing % degradation of MB in the presence of TONRs, PAgTH-1, PAgTH-2, PAuTH-1, and PAuTH-2. *Reproduced with permission from [45]* (E) SEM image of Au-TiO$_2$ composite film. *Reproduced with permission from [72]* (F) UV spectra of Au-TiO$_2$ composite film. *Reproduced with permission from [72]* (G) Degradation of MO aqueous solution under different catalysts. *Reproduced with permission from [46].*

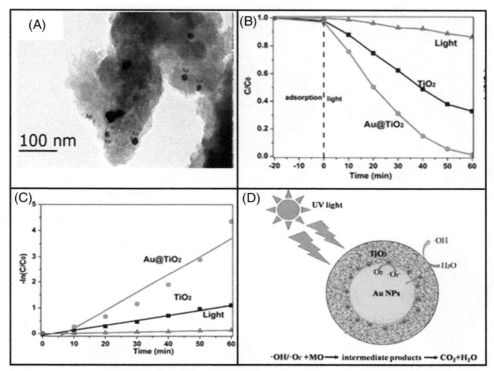

Fig. 12.11: (A) High-resolution TEM (HR-TEM) images of Au@TiO$_2$. *Reproduced with permission from [73]* (B–C) Photocatalytic degradation of MO over only light, TiO$_2$ and Au@TiO$_2$ under UV light irradiation and kinetic traces according to the Langmuir-Hinshelwood model. *Reproduced with permission from [73]* (D) Schematic illustration of photocatalytic mechanism of Au@TiO$_2$ under UV irradiation. *Reproduced with permission from [73].*

exhibited good stability and high photodegradation ability. Over 57.3% gaseous toluene was eliminated by 0.14% yolk-shell Au@TiO$_2$ in 3 hours. Furthermore, the photocatalytic process benefited from the comprehensive effect of Au LSPR response, high cavity volume of yolk-shell structure, bigger specific surface area, more mesoporous, and the presence of Ti^{3+} [75].

As an efficient visible-light photocatalyst for the breakdown of organic and inorganic contaminants, a plasmonic fiber was logically constructed. Shoueir et al. [47] used the photo deposition approach to create the nanostructured Au@TiO$_2$ with tunable sizes. Further, to create the photoactive plasmonic fiber the Au@TiO$_2$ was subsequently soaked in chitosan fiber. In the presence of citric acid, the active plasmonic fiber could convert poisonous chromium, Cr (VI), to biocompatible Cr (III) (98.9%) in 21 minutes. The plasmonic fiber demonstrated improved degrading activity, consistent kinetic behavior, catalytic reduction, and stability.

Nano Au-TiO$_2$ heterostructures with two unique TiO$_2$ morphologies, namely nanospindles and nanocubes as shown in Figs. 12.12A–B were effectively produced and used for MB

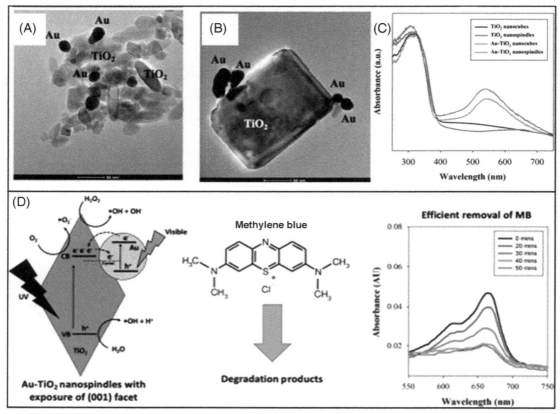

Fig. 12.12: (A–B) TEM images of the corresponding nano Au-TiO$_2$ heterostructures. (C) UV–vis spectra of Au-TiO$_2$ (D) The proposed schematic mechanism for the photodegradation of MB using the as-prepared nano catalyst. *Reproduced with permission from [76].*

photodegradation. It was discovered that TiO$_2$ nanospindles outperformed TiO$_2$ nanocubes in photocatalytic performance, owing mostly to the exposure of the (001) crystal facet. It was found that exposing the (001) crystal facet increased the photocatalytic reaction rate four-fold. Furthermore, due to the existence of SPR (as shown in UV-visible spectra of Fig. 12.12C), integration with Au NPs were found to have a synergistic effect in photodegradation activity. According to the results, Au-TiO$_2$ nanospindles were discovered to be the most efficient catalyst for photocatalytic MB degradation, with a pseudo-first order reaction rate of 0.1570 min^{-1}. Fig. 12.12D shows the compete mechanism of photodegradation of MB by Au-TiO$_2$ nanospindles [76].

The spread of antibiotics in nature is one of the most serious environmental concerns as it affects biological metabolism. It also causes the presence of bacterial resistance in drinking water sources. Au TiO$_2$$^{-1}$ NPs showed good photodegradation rate against various drugs that has been discovered. Martin et al. reported 91% and 49% higher photocatalytic degradation of ciprofloxacin (5 mg L^{-1}) under UV and visible radiation respectively. The reason behind this

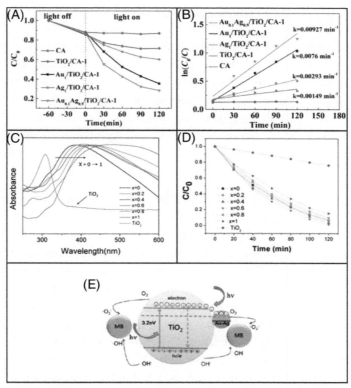

Fig. 12.13: (A) Au0.1Ag0.9 TiO_2^{-1} CA-1^{-1} membrane; TC degradation rate of different photocatalytic membranes supported with the same content of NRs (1 wt%) at a TC concentration of 5 mg L–1 (B) and corresponding rate constant k. *Reproduced with permission from [38]* (C) UV–vis spectra of AuxAg(1−x) TiO_2^{-1} composites (x = 0, 0.2, 0.4, 0.6, 0.8, and 1) and the pure TiO_2 spheres. (D) Concentration changes of MB in the presence of pure TiO_2 particles, and AuxAg(1−x) TiO_2^{-1} composite with x = 0.2, 0.4, 0.6, 0.8, and 1 under sunlight irradiation (E) Mechanism of MB photodegradation by Au-Ag decorated TiO_2 composites. *Reproduced with permission from [82].*

was the good binding ability of ciprofloxacin towards Au TiO_2^{-1} surface and enhanced charge transfer property due to deposition of Au NPs. Additionally, Au NPs also shifted the absorption from UV to visible light of the Au TiO_2^{-1} nanocomposite [77]. Similar study was performed by Do et al. [78] who studied the photocatalytic activity of Au TiO_2^{-1} hybrid against the degradation of amoxicillin, ampicillin, doxycycline, oxytetracycline, lincomycin, vancomycin, sulfamethazine, and sulfamethoxazole etc. It was seen that most of antibiotics were degraded within 20 min. The excellent photocatalytic activity was due to synergistic effects of both the larger surface area of TiO_2 and SPR of Au NPs. In Table 12.1, Au based TiO_2 photocatalysts their properties and photocatalytic activities are summarized.

12.4.3 Ag-Au TiO$_2$$^{-1}$ based plasmonic photocatalysts

Researchers are more interested in bimetallic nanoparticles (NPs) than monometallic nanoparticles (NPs) from both a scientific and technological standpoint. They differ from monometallic counterparts in terms of catalytic, electronic, and magnetic properties [79]. Because of the strong absorption and SPR effect in the visible region, the presence of Ag Au^{-1} in Ag-Au-TiO$_2$ NPs aids in the generation of hot electrons. The higher energetic hot-electron density raises the Fermi energy level of Au and Ag, making it more energetically advantageous to provide the SPR hot-electron into the TiO$_2$ conduction band of Ag-Au-TiO$_2$ NPs [80,81].

Integration of plasmonic photocatalysts with membranes is the most promising strategy for water treatment. For example, Li et al. [38] synthesized Au/Ag fabricated TiO$_2$ nano-composite membrane for improved photocatalytic degradation and bactericidal efficiency. Au$_{0.1}$Ag$_{0.9}$ NPs adorned TiO$_2$ nanorods were embedded in cellulose acetate to produce membrane (Au0.1Ag0.9 TiO$_2$$^{-1}$ CA^{-1}). The produced membrane shows enhanced photocatalytic breakdown of antibiotics (tetracycline, TC) for eliminating harmful bacteria. It was seen that Au0.1Ag0.9 TiO2^{-1} CA^{-1} membrane shows high crystallinity and pore accessibility. Furthermore, the fabrication of Au0.1Ag0.9 nanoparticles considerably increased the visible light consumption rate and charge separation. The photodegradation rate of tetracycline could reach up to 90% after 120 minutes under visible light irradiation. It was observed that the degradation rate was significantly higher than the rates of the monometallic Au1 TiO$_2$$^{-1}$ CA^{-1} and Ag1 TiO$_2$$^{-1}$ CA^{-1} membrane Figs. 12.13A–B.

One of the most effective solutions for enhancing visible light absorption was to decorate TiO$_2$ NPs with noble metals that have SPR effect. The exact match of the SPR absorption band with the incident light spectrum is critical for plasmonic enhancement. Au-Ag co-decoration exhibits a highly tunable, composition-dependent SPR agent over a wide range of visible light wavelengths (about from 415 to 525 nm). Yang et al. [82] synthesized Au-Ag TiO$_2$$^{-1}$ nanocomposite as most promising hybrid for tuning optical properties of TiO$_2$. The Au-Ag TiO$_2$$^{-1}$ nanocomposite has excellent stability and large specific surface area. UV spectra Fig. 12.13C shows that the Au0.4Ag0.6 TiO$_2$$^{-1}$ have SPR at 460 nm which was better than other samples such as bare TiO$_2$, Au-TiO$_2$ and Ag-TiO$_2$ nanocomposites. The SPR produced by Ag-Au co-decoration assisted TiO$_2$ particles helps in absorbing light from near-ultraviolet to visible light. Fig. 12.13D showed the photodegradation rate of Au$_{0.4}$Ag$_{0.6}$ TiO$_2$$^{-1}$ nanocomposite which possess excellent photocatalytic activity. The mechanism of whole process has been shown in Fig. 12.13E that Au-Ag TiO$_2$$^{-1}$ nanocomposite reduces the photogenerated electron-hole recombination via modifying the bandgap and trapping the charge carrier.

Thus, various Au, Ag, and Au-Ag based TiO$_2$ plasmonic photocatalysts were discussed for their synthetic approaches, their photocatalytic efficiency and their potential applications in the photodegradation of environmental pollutants. The synergistic plasmonic and photolytic characteristics of these noble metals and TiO_2 nanostructures, respectively, have already been realized, resulting in increased photocatalytic properties. A complete detail about such plasmonic photocatalysts are shown in Table 12.1.

12.5 Summary

This book chapter provides a complete overview of recent developments of Ag and Au based TiO_2 nanostructures by presenting several synthetic procedures, their physicochemical properties, and their capacity of photodegradation. These multimaterial nanocomposites are made to improve the properties of the parent material and to create multifunctional applications. It was discovered that modifying the TiO_2 lattice with noble metals such as Ag and Au was more favorable for improving TiO_2 photocatalytic performances and shifting its photo-response to the visible light range. Furthermore, it plays a crucial function in decreasing TiO_2 charge recombination. The photocatalytic activity of Ag Au^{-1} based TiO_2 nanocomposites, along with their synthesis, optical properties, dependent factors, have been briefly discussed, with their fundamental mechanisms for a better understanding. These NMNPs serve as an electron acceptor material, aiding electron–hole separation via the creation of the Schottky barrier, resulting in charge carrier recombination suppression, the generation of more ROS, and hence greater photocatalytic efficacy. It is also recognized that more research should be conducted in order to improve the functions, as well as greater stability and reproducibility. In comparison to single NMNPs, nanocomposites comprising bimetallic NMNSs also serve as superior candidates in this regard. To maximize novel functional properties, enhanced stability and protection of embedded NMNSs and TiO_2 nanomaterials from the environment could be provided.

Acknowledgment

Authors gratefully acknowledge the MHRD and Department of Chemistry, National Institute of Technology, Hamirpur for financial support throughout their PhD/MSc research work. The author (JP) acknowledges DST, India for providing the prestigious INSPIRE Faculty award [INSPIRE/04/2015/002452(IFA15-MS-57)] and research grant.

References

[1] J. Prakash, R. Harris, H. Swart, Embedded plasmonic nanostructures: Synthesis, fundamental aspects and their surface enhanced Raman scattering applications, Int. Rev. Phys. Chem. 35 (2016) 353–398.

[2] J. Prakash, S. Sun, H.C. Swart, R.K. Gupta, Noble metals-TiO_2 nanocomposites: From fundamental mechanisms to photocatalysis, surface enhanced Raman scattering and antibacterial applications, Applied Mater. Today 11 (2018) 82–135.

[3] J. Prakash, J. Pivin, H. Swart, Noble metal nanoparticles embedding into polymeric materials: From fundamentals to applications, Adv. in colloid interface sci. 226 (2015) 187–202.

[4] T. Gupta, J. Cho, J. Prakash, Hydrothermal synthesis of TiO_2 nanorods: Formation chemistry, growth mechanism, and tailoring of surface properties for photocatalytic activities, Mater. Today Chem. 20 (2021) 100428.

[5] J. Prakash, A. Kumar, H. Dai, B.C. Janegitz, V. Krishnan, H.C. Swart, S. Sun, Novel rare earth metal doped one dimensional TiO_2 nanostructures: Fundamentals and multifunctional applications, Mater. Today Sustainability (2021) 100066.

[6] S. MiarAlipour, D. Friedmann, J. Scott, R. Amal, TiO_2/porous adsorbents: Recent advances and novel applications, J. Hazard. Mater. 341 (2018) 404–423.

[7] J. Prakash, B. Kaith, S. Sun, S. Bellucci, H.C. Swart, Recent Progress on Novel Ag–TiO 2 Nanocomposites for Antibacterial Applications, Microbial Nanobionics (2019) 121–143.

[8] N. Singh, J. Prakash, R.K. Gupta, Design and engineering of high-performance photocatalytic systems based on metal oxide–graphene–noble metal nanocomposites, Molecular Systems Design Eng. 2 (2017) 422–439.

[9] N. Singh, J. Prakash, M. Misra, A. Sharma, R.K. Gupta, Dual functional Ta-doped electrospun TiO_2 nanofibers with enhanced photocatalysis and SERS detection for organic compounds, ACS applied mater. interfaces 9 (2017) 28495–28507.

[10] X. Zhang, Y.L. Chen, R.-S. Liu, D.P. Tsai, Plasmonic photocatalysis, Rep. Prog. Phys. 76 (2013) 046401.

[11] M. Liu, X. Jin, S. Li, J.-B. Billeau, T. Peng, H. Li, L. Zhao, Z. Zhang, J.P. Claverie, L. Razzari, Enhancement of scattering and near field of TiO_2–Au nanohybrids using a silver resonator for efficient plasmonic photocatalysis, ACS Applied Mater. Interfaces 13 (2021) 34714–34723.

[12] P. Zhang, T. Wang, J. Gong, Mechanistic understanding of the plasmonic enhancement for solar water splitting, Adv. Mater. 27 (2015) 5328–5342.

[13] J. Prakash, R. Harris, H. Swart, Embedded plasmonic nanostructures: Synthesis, fundamental aspects and their surface enhanced Raman scattering applications, Int. Rev. Phys. Chem. 35 (2016) 353–398.

[14] Q. Huang, T.D. Canady, R. Gupta, N. Li, S. Singamaneni, B.T. Cunningham, Enhanced plasmonic photocatalysis through synergistic plasmonic–photonic hybridization, ACS Photonics 7 (2020) 1994–2001.

[15] L. Lin, X. Feng, D. Lan, Y. Chen, Q. Zhong, C. Liu, Y. Cheng, R. Qi, J. Ge, C. Yu, Coupling effect of Au nanoparticles with the oxygen vacancies of TiO_2–x for enhanced charge transfer, J. Phys. Chem. C 124 (2020) 23823–23831.

[16] S.Y. Lee, H.T. Do, J.H. Kim, Microplasma-assisted synthesis of TiO_2–Au hybrid nanoparticles and their photocatalytic mechanism for degradation of methylene blue dye under ultraviolet and visible light irradiation, Appl. Surf. Sci. 573 (2022) 151383.

[17] P.K. Jain, X. Huang, I.H. El-Sayed, M.A. El-Sayed, Noble metals on the nanoscale: Optical and photothermal properties and some applications in imaging, sensing, biology, and medicine, Acc. Chem. Res. 41 (2008) 1578–1586.

[18] Y. Fu, Z. Yin, L. Qin, D. Huang, H. Yi, X. Liu, S. Liu, M. Zhang, B. Li, L. Li, Recent progress of noble metals with tailored features in catalytic oxidation for organic pollutants degradation, J. Hazard. Mater. 422 (2022) 126950.

[19] T.K. Pathak, R. Kroon, V. Craciun, M. Popa, M. Chifiriuc, H. Swart, Influence of Ag, Au and Pd noble metals doping on structural, optical and antimicrobial properties of zinc oxide and titanium dioxide nanomaterials, Heliyon 5 (2019) e01333.

[20] J. Prakash, J. Pivin, H. Swart, Noble metal nanoparticles embedding into polymeric materials: From fundamentals to applications, Advances in colloid interface sci. 226 (2015) 187–202.

[21] V. Juvé, M.F. Cardinal, A. Lombardi, A. Crut, P. Maioli, J. Pérez-Juste, L.M. Liz-Marzán, N. Del Fatti, F. Vallée, Size-dependent surface plasmon resonance broadening in nonspherical nanoparticles: Single gold nanorods, Nano Lett. 13 (2013) 2234–2240.

[22] A. Lombardi, M. Loumaigne, A. Crut, P. Maioli, N. Del Fatti, F. Vallée, M. Spuch-Calvar, J. Burgin, J. Majimel, M. Tréguer-Delapierre, Surface plasmon resonance properties of single elongated nano-objects: Gold nanobipyramids and nanorods, J Langmuir 28 (2012) 9027–9033.

[23] L. Lu, G. Wang, Z. Xiong, Z. Hu, Y. Liao, J. Wang, J. Li, Enhanced photocatalytic activity under visible light by the synergistic effects of plasmonics and Ti3+-doping at the Ag/TiO_2-x heterojunction, Ceram. Int. 46 (2020) 10667–10677.

[24] J. Cao, T. Sun, K.T. Grattan, Gold nanorod-based localized surface plasmon resonance biosensors: A review, Sensors actuators B: Chem. 195 (2014) 332–351.

[25] X. Zheng, D. Zhang, Y. Gao, Y. Wu, Q. Liu, X. Zhu, Synthesis and characterization of cubic Ag/TiO_2 nanocomposites for the photocatalytic degradation of methyl orange in aqueous solutions, Inorg. Chem. Commun. 110 (2019) 107589.

[26] K. Awazu, M. Fujimaki, C. Rockstuhl, J. Tominaga, H. Murakami, Y. Ohki, N. Yoshida, T. Watanabe, A plasmonic photocatalyst consisting of silver nanoparticles embedded in titanium dioxide, J. Am. Chem. Soc. 130 (2008) 1676–1680.

[27] T. Zhang, M. Xu, J. Li, Enhanced photocatalysis of TiO_2 by aluminum plasmonic, Catal. Today 376 (2021) 162–167.

[28] A. Wang, S. Wu, J. Dong, R. Wang, J. Wang, J. Zhang, S. Zhong, S. Bai, Interfacial facet engineering on the Schottky barrier between plasmonic Au and TiO_2 in boosting the photocatalytic CO2 reduction under ultraviolet and visible light irradiation, Chem. Eng. J. 404 (2021) 127145.

[29] J. Prakash, Silver nanostructures, chemical synthesis methods, and biomedical applications. In: Applications of Nanotechnology for Green Synthesis, Springer, 2020, pp. 281–303.

[30] A. Primo, A. Corma, H. García, Titania supported gold nanoparticles as photocatalyst, Phys. Chem. Chem. Phys. 13 (2011) 886–910.

[31] V. Rajput, R.K. Gupta, J. Prakash, Engineering metal oxide semiconductor nanostructures for enhanced charge transfer: Fundamentals and emerging SERS applications, J. Mater. Chem. C (2022).

[32] J. Prakash, J. Cho, Y.K. Mishra, Photocatalytic TiO_2 nanomaterials as potential antimicrobial and antiviral agents: Scope against blocking the SARS-COV-2 spread, Micro Nano Eng. (2021) 100100.

[33] A. Fujishima, K. Honda, Electrochemical photolysis of water at a semiconductor electrode, Nature 238 (1972) 37–38.

[34] S. Peiris, H.B. de Silva, K.N. Ranasinghe, S.V. Bandara, I.R. Perera, Recent development and future prospects of TiO_2 photocatalysis, J. Chin. Chem. Soc. 68 (2021) 738–769.

[35] O. Al-Madanat, Y. AlSalka, W. Ramadan, D.W. Bahnemann, TiO_2 photocatalysis for the transformation of aromatic water pollutants into fuels, Catalysts 11 (2021) 317.

[36] O. Al-Madanat, M. Curti, C. Günnemann, Y. AlSalka, R. Dillert, D.W. Bahnemann, TiO_2 photocatalysis: Impact of the platinum loading method on reductive and oxidative half-reactions, Catal. Today 380 (2021) 3–15.

[37] D. Bhardwaj, R. Singh, Green biomimetic synthesis of Ag–TiO 2 nanocomposite using Origanum majorana leaf extract under sonication and their biological activities, Bioresources Bioprocess. 8 (2021) 1–12.

[38] W. Li, B. Li, M. Meng, Y. Cui, Y. Wu, Y. Zhang, H. Dong, Y. Feng, Bimetallic Au/Ag decorated TiO_2 nanocomposite membrane for enhanced photocatalytic degradation of tetracycline and bactericidal efficiency, Appl. Surf. Sci. 487 (2019) 1008–1017.

[39] L. Zani, M. Melchionna, T. Montini, P. Fornasiero, Design of dye-sensitized TiO_2 materials for photocatalytic hydrogen production: Light and shadow, J. Phys.: Energy 3 (2021) 031001.

[40] P. Hajipour, A. Bahrami, M.Y. Mehr, W.D. van Driel, K. Zhang, Facile synthesis of ag nanowire/TiO_2 and Ag nanowire/TiO_2/go nanocomposites for photocatalytic degradation of rhodamine b, J. Phys.: Energy 14 (2021) 763.

[41] M.C. Joshi, R.K. Gupta, J. Prakash, Hydrothermal synthesis and Ta doping of TiO_2 nanorods: Effect of soaking time and doping on optical and charge transfer properties for enhanced SERS activity, Mater. Chem. Phys. (2021) 125642.

[42] E. Kowalska, Plasmonic Photocatalysts. In: Catalysts 11, 2021, p. 410.

[43] Y. Tian, T. Tatsuma, Plasmon-induced photoelectrochemistry at metal nanoparticles supported on nanoporous TiO 2, Chem. Commun. (2004) 1810–1811.

[44] L. Lin, Q. Zhong, Y. Zheng, Y. Cheng, R. Qi, R. Huang, Size effect of Au nanoparticles in Au-TiO_2-x photocatalyst, Chem. Phys. Lett. 770 (2021) 138457.

[45] P. Bamola, M. Sharma, C. Dwivedi, B. Singh, S. Ramakrishna, G.K. Dalapati, H. Sharma, Interfacial interaction of plasmonic nanoparticles (Ag, Au) decorated floweret TiO_2 nanorod hybrids for enhanced visible light driven photocatalytic activity, Mater. Sci. Eng.: B 273 (2021) 115403.

[46] D. Wibowo, M.Z. Muzakkar, S.K.M. Saad, F. Mustapa, M. Maulidiyah, M. Nurdin, A.A. Umar, Enhanced visible light-driven photocatalytic degradation supported by Au-TiO_2 coral-needle nanoparticles, J. Photochem. Photobiol. A: Chem. 398 (2020) 112589.

[47] K. Shoueir, S. Kandil, H. El-hosainy, M. El-Kemary, Tailoring the surface reactivity of plasmonic Au@ TiO$_2$ photocatalyst bio-based chitosan fiber towards cleaner of harmful water pollutants under visible-light irradiation, J. Cleaner Prod. 230 (2019) 383–393.

[48] L. Ling, Y. Feng, H. Li, Y. Chen, J. Wen, J. Zhu, Z. Bian, Microwave induced surface enhanced pollutant adsorption and photocatalytic degradation on Ag/TiO$_2$, Appl. Surf. Sci. 483 (2019) 772–778.

[49] M. Scarisoreanu, A.G. Ilie, E. Goncearenco, A.M. Banici, I.P. Morjan, E. Dutu, E. Tanasa, I. Fort, M. Stan, C.N. Mihailescu, Ag, Au and Pt decorated TiO$_2$ biocompatible nanospheres for UV & vis photocatalytic water treatment, J Applied Surface Sci. 509 (2020) 145217.

[50] M.-C. Wu, T.-H. Lin, K.-H. Hsu, J.-F. Hsu, Photo-induced disinfection property and photocatalytic activity based on the synergistic catalytic technique of Ag doped TiO$_2$ nanofibers, Appl. Surf. Sci. 484 (2019) 326–334.

[51] J. Singh, K. Sahu, B. Satpati, J. Shah, R. Kotnala, S. Mohapatra, Facile synthesis, structural and optical properties of Au-TiO$_2$ plasmonic nanohybrids for photocatalytic applications, J. Phys. Chem. Solids 135 (2019) 109100.

[52] C. Wang, D. Astruc, Nanogold plasmonic photocatalysis for organic synthesis and clean energy conversion, Chem. Soc. Rev. 43 (2014) 7188–7216.

[53] J. Prakash, P. Kumar, R. Harris, C. Swart, J. Neethling, A.J. van Vuuren, H. Swart, Synthesis, characterization and multifunctional properties of plasmonic Ag–TiO$_2$ nanocomposites, Nanotechnology 27 (2016) 355707.

[54] R. Saravanan, D. Manoj, J. Qin, M. Naushad, F. Gracia, A.F. Lee, M.M. Khan, M. Gracia-Pinilla, Mechanothermal synthesis of Ag/TiO$_2$ for photocatalytic methyl orange degradation and hydrogen production, Process Safety Environ. Protection 120 (2018) 339–347.

[55] P. Van Viet, C.M. Thi, The directed preparation of TiO 2 nanotubes film on FTO substrate via hydrothermal method for gas sensing application, AIMS Mater. Sci. 3 (2016) 460–469.

[56] Y.-C. Liang, Y.-C. Liu, Design of nanoscaled surface morphology of TiO$_2$–Ag$_2$O composite nanorods through sputtering decoration process and their low-concentration NO$_2$ gas-sensing behaviors, Nanomaterials 9 (2019) 1150.

[57] M.-C. Wu, T.-H. Lin, K.-H. Hsu, J.-F. Hsu, Photo-induced disinfection property and photocatalytic activity based on the synergistic catalytic technique of Ag doped TiO$_2$ nanofibers, Appl. Surf. Sci. 484 (2019) 326–334.

[58] L. Lin, Q. Zhong, Y. Zheng, Y. Cheng, R. Qi, R. Huang, Size effect of Au nanoparticles in Au-TiO$_2$-x photocatalyst, Chem. Phys. Lett. 770 (2021) 138457.

[59] M.A. Zarepour, M. Tasviri, Facile fabrication of Ag decorated TiO$_2$ nanorices: Highly efficient visible-light-responsive photocatalyst in degradation of contaminants, J. Photochem. Photobiol. A: Chem. 371 (2019) 166–172.

[60] S. Veziroglu, M. Ullrich, M. Hussain, J. Drewes, J. Shondo, T. Strunskus, J. Adam, F. Faupel, O.C. Aktas, Plasmonic and non-plasmonic contributions on photocatalytic activity of Au-TiO$_2$ thin film under mixed UV–visible light, Surface Coatings Technol. 389 (2020) 125613.

[61] J. Singh, K. Sahu, A. Pandey, M. Kumar, T. Ghosh, B. Satpati, T. Som, S. Varma, D. Avasthi, S. Mohapatra, Atom beam sputtered Ag-TiO$_2$ plasmonic nanocomposite thin films for photocatalytic applications, Appl. Surf. Sci. 411 (2017) 347–354.

[62] S. Bhardwaj, D. Sharma, P. Kumari, B. Pal, Influence of photodeposition time and loading amount of Ag co-catalyst on growth, distribution and photocatalytic properties of Ag@ TiO$_2$ nanocatalysts, Opt. Mater. 106 (2020) 109975.

[63] A. Sandoval, A. Aguilar, C. Louis, A. Traverse, R. Zanella, Bimetallic Au–Ag/TiO$_2$ catalyst prepared by deposition–precipitation: High activity and stability in CO oxidation, J. Catal. 281 (2011) 40–49.

[64] Y. Ren, S. Xing, J. Wang, Y. Liang, D. Zhao, H. Wang, N. Wang, W. Jiang, S. Wu, S. Liu, Weak-light-driven Ag–TiO$_2$ photocatalyst and bactericide prepared by coprecipitation with effective Ag doping and deposition, Opt. Mater. 124 (2022) 111993.

[65] A. Kumar, P. Choudhary, A. Kumar, P.H. Camargo, V. Krishnan, Recent advances in plasmonic photocatalysis based on TiO$_2$ and noble metal nanoparticles for energy conversion, environmental remediation, and organic synthesis, Small 18 (2022) 2101638.

[66] N. Tasić, J. Ćirković, V. Ribić, M. Žunić, A. Dapčević, G. Branković, Z. Branković, Effects of the silver nanodots on the photocatalytic activity of mixed-phase TiO_2, J. Am. Ceram. Soc. 105 (2022) 336–347.

[67] Y. Shi, J. Ma, Y. Chen, Y. Qian, B. Xu, W. Chu, D. An, Recent progress of silver-containing photocatalysts for water disinfection under visible light irradiation: A review, Sci. Total Environ. 804 (2022) 150024.

[68] D. Hariharan, P. Thangamuniyandi, A.J. Christy, R. Vasantharaja, P. Selvakumar, S. Sagadevan, A. Pugazhendhi, L. Nehru, Enhanced photocatalysis and anticancer activity of green hydrothermal synthesized Ag@ TiO_2 nanoparticles, J. Photochem. Photobiol. B: Biol. 202 (2020) 111636.

[69] Y. Sun, Y. Gao, J. Zeng, J. Guo, H. Wang, Enhancing visible-light photocatalytic activity of Ag-TiO_2 nanowire composites by one-step hydrothermal process, Mater. Lett. 279 (2020) 128506.

[70] J. Singh, N. Tripathi, S. Mohapatra, Synthesis of Ag–TiO_2 hybrid nanoparticles with enhanced photocatalytic activity by a facile wet chemical method, Nano-Structures Nano-Objects 18 (2019) 100266.

[71] L. Lu, G. Wang, Z. Xiong, Z. Hu, Y. Liao, J. Wang, J. Li, Enhanced photocatalytic activity under visible light by the synergistic effects of plasmonics and Ti3+-doping at the Ag/TiO_2-x heterojunction, Ceram. Int. 46 (2020) 10667–10677.

[72] Z. Cao, C. Liu, D. Chen, J. Liu, Preparation of an Au-TiO_2 photocatalyst and its performance in removing phycocyanin, Sci. Total Environ. 692 (2019) 572–581.

[73] L. Wu, S. Ma, P. Chen, X. Li, The mechanism of enhanced charge separation and photocatalytic activity for Au@ TiO_2 core-shell nanocomposite, Int. J. Environ. Anal. Chem. (2020) 1–11.

[74] M. Yan, Y. Wu, X. Liu, Photocatalytic nanocomposite membranes for high-efficiency degradation of tetracycline under visible light: An imitated core-shell Au-TiO_2-based design, J. Alloys Compounds 855 (2021) 157548.

[75] Y. Wang, C. Yang, A. Chen, W. Pu, J. Gong, Influence of yolk-shell Au@ TiO_2 structure induced photocatalytic activity towards gaseous pollutant degradation under visible light, Appl. Catal. B 251 (2019) 57–65.

[76] M. Khalil, E.S. Anggraeni, T.A. Ivandini, E. Budianto, Exposing TiO_2 (001) crystal facet in nano Au-TiO_2 heterostructures for enhanced photodegradation of methylene blue, Appl. Surf. Sci. 487 (2019) 1376–1384.

[77] P. Martins, S. Kappert, H. Nga Le, V. Sebastian, K. Kühn, M. Alves, L. Pereira, G. Cuniberti, M. Melle-Franco, S.J.C. Lanceros-Méndez, Enhanced photocatalytic activity of Au/TiO_2 nanoparticles against ciprofloxacin, Catalysts 10 (2020) 234.

[78] T.C.M.V. Do, D.Q. Nguyen, K.T. Nguyen, P.H. Le, TiO_2 and Au-TiO_2 nanomaterials for rapid photocatalytic degradation of antibiotic residues in aquaculture wastewater, Materials 12 (2019) 2434.

[79] V. Fauzia, A. Yudiana, Y. Yulizar, M.A. Dwiputra, L. Roza, I. Soegihartono, The impact of the Au/Ag ratio on the photocatalytic activity of bimetallic alloy AuAg nanoparticle-decorated ZnO nanorods under UV irradiation, J. Phys. Chem. Solids 154 (2021) 110038.

[80] K.J. Rao, S. Paria, Phytochemicals mediated synthesis of multifunctional Ag-Au-TiO_2 heterostructure for photocatalytic and antimicrobial applications, J. Cleaner Prod. 165 (2017) 360–368.

[81] A. Zielińska-Jurek, Progress, challenge, and perspective of bimetallic TiO_2-based photocatalysts, J. Nanomaterials 2014 (2014).

[82] X. Yang, Y. Wang, L. Zhang, H. Fu, P. He, D. Han, T. Lawson, X. An, The use of tunable optical absorption plasmonic Au and Ag decorated TiO_2 structures as efficient visible light photocatalysts, Catalysts 10 (2020) 139.

[83] H. Chakhtouna, H. Benzeid, N. Zari, R. Bouhfid, Recent progress on Ag/TiO_2 photocatalysts: Photocatalytic and bactericidal behaviors, Environ. Sci. Pollution Res. 28 (2021) 44638–44666.

[84] Y. Gao, W. Nie, Q. Zhu, X. Wang, S. Wang, F. Fan, C. Li, The polarization effect in surface-plasmon-induced photocatalysis on Au/TiO_2 nanoparticles, Angew. Chem. 132 (2020) 18375–18380.

[85] V. Melinte, T. Buruiana, I. Rosca, A.L.J.C. Chibac, TiO_2-based photopolymerized hybrid catalysts with visible light catalytic activity induced by in situ generated Ag/Au NPs, Chem. Select 4 (2019) 5138–5149.

CHAPTER 13

DNA functionalized gold and silver nanoparticles

Subrata Dutta

Department of Chemistry, Sardar Vallabhbhai National Institute of Technology, Surat, Gujarat, India

13.1 Introduction

Nanoparticles (NPs) are microscopic materials with dimensions ranging from 1 to 100 nm in at least one of the dimensions [1]. Because of their large surface area and quantum effect due to the nanosize, NPs exhibit distinct chemical and physical characteristics compared to solid bulk materials [2]. Nanoparticles are widely used in the biotechnology and biomedical field. Because of their distinct properties and enhanced cell permeability, they are promising candidates for various biomedical applications. Among all other nanoparticles, gold (Au) and silver (Ag) nanoparticles have attracted the interest of many researchers.

DNA oligonucleotides have a high degree of programmability, recognition, and catalytic characteristics. Attaching DNA with AuNPs/AgNPs allow the formation of diverse structures with distinct properties due to the synergistic action of both components [3]. DNA-conjugated AuNPs or AgNPs have excellent optical properties, high biocompatibility, and stability. As a result, their hybrids have enabled a wide range of applications, such as biocatalysts, sensors for the rapid detection of genes in disease diagnostics, and the detection of small molecules. At the same time, this system also possesses interesting fundamental properties, such as sharp melting transitions and tighter binding to complementary DNA. This field of DNA-conjugated NPs has expanded exponentially since the initial articles were published in 1996 by Mirkin and Alivisatos independently. Within the scope of this book chapter, a summary of the fundamental features of DNA - conjugated AuNPs/AgNPs linked to DNA-directed assembly and sensing is discussed. Then, several distinct physical features are highlighted. Finally, with specific examples, the applications of these materials in sensing, drug administration, and gene silencing are discussed.

13.1.1 Structure of DNA

Nucleotides are the building blocks of DNA, and each one consists of a deoxyribose sugar, a phosphate group, and a nitrogen-containing base. There are four bases in DNA: adenine (A), cytosine (C), guanine (G), or thymine (T). A phosphate group forms a bridge between the $5'$-carbon of a nucleotide and $3'$- carbon of another nucleotide, and they are linked together in a

Gold and Silver Nanoparticles.
DOI: https://doi.org/10.1016/B978-0-323-99454-5.00004-4

chain to form a single-stranded DNA. In a neutral state, base pairing occurs between A and T or G and C base pair due to hydrogen bonding, called Watson-Crick base-pairing interactions. So, A and T or G and C are complementary base pairs. Using this base pairing, two single-stranded DNA can be held together by twisting each other to form a helix structure with a diameter of just 2 nm. This twisted form is called a double helix. The two single strands of the DNA double helix are anti-parallel to one another, that is, one of the single strands will have the 3′ - carbon of the nucleotide in the bottom position, and the other single strand will have the 3′- carbon in the opposite position (Fig. 13.1) [4]. After the discovery of DNA synthesis on solid-phase synthesis in 1981 by Caruthers and co-workers, [5] the chemical synthesis of DNA became very easy and cheap now.

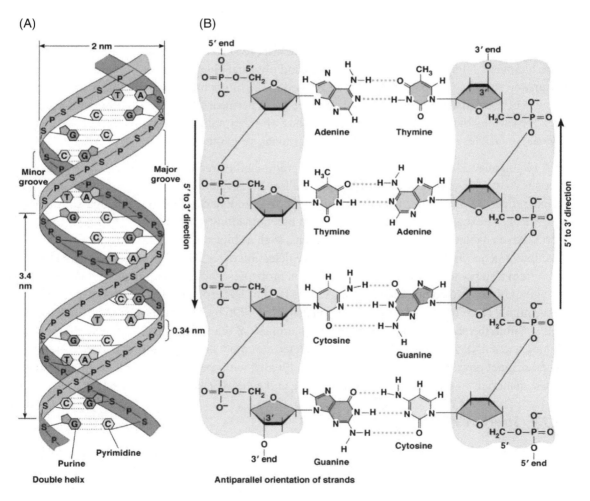

Fig. 13.1: Structure of a DNA molecule. *Adapted with permission from ref. [4].*

13.1.2 Synthesis of gold and silver nanoparticles

The reduction of Au^{3+} salt (generally $HAuCl_4$) in water by sodium citrate, introduced by Turkevich in 1951 [6], has become one of the most widely used methods for the synthesis of gold nanoparticles (AuNPs). Sodium citrate acts as a mild reducing agent as well as a capping agent, stabilizing the AuNPs. AuNPs with diameters ranging from 10 to 20 nm were produced using this approach. The ratio of citrate to $AuCl_4^-$ ions controls particle size; more citrate leads to smaller nanoparticles. The color of the reaction mixture gradually changes during synthesis from colorless to light yellow, dark blue, purple, and finally ruby red. Pong et al. in 2007 investigated this reduction process using transmission electron microscopy (TEM) [7]. When Au^{3+} was reduced by sodium citrate, it first formed nanoclusters with a diameter of around 5 nm, which subsequently self-assembled to create an extensive network of nanowires. Following that, spherical particles with dimensions ranging from 10 to 13 nm began to split from the nanowires, as confirmed by TEM (Fig. 13.2).

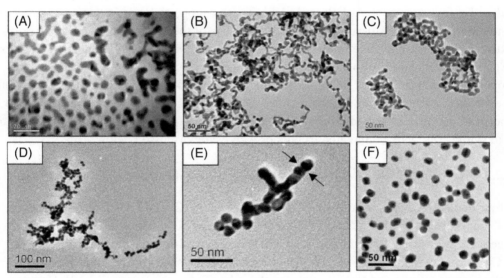

Fig. 13.2: TEM images were obtained at different reaction stages of AuNPs synthesis by citrate reduction of $AuCl_4^-$.

(A) The colorless solution formed after the addition of citrate contained nanoclusters 3–5 nm in diameter and some irregularly shaped chain-like segments. (B) As the solution turned dark blue, an extensive network of Au nanowires of 5 nm diameter was formed. (C) In the dark purple stage, small sections of the nanowire network started to break with a diameter of approximately 8 nm. (D) Spherical particles between 10 and 13 nm in diameter began to "cleave" off from the nanowires when the color of the reaction turned purple. (E) Constriction and cleavage of spherical particles from the nanowires. (F) Well-defined spherical particles of diameter 13–15 nm were finally formed in the ruby-red solution. *Adapted with permission from ref. [7].*

In 1982, Lee's group applied a reduction approach similar to that used for the preparation of AuNPs to produce silver nanoparticles (AgNPs) by reducing Ag^+ with sodium citrate [8]. However, this citrate reduction approach was less effective in controlling the size of AgNPs. Pyatenko et al. modified this procedure to produce size-controlled AgNPs by combining seeding with soft laser treatment [9]. They initially synthesized 8–10 nm AgNPs by reducing $AgNO_3$ with sodium citrate, followed by soft laser treatment. A small amount of these nanoparticles was then boiled along with $AgNO_3$ and sodium citrate to synthesize larger AgNPs with diameters ranging from 10 to 40 nm. The initial small AgNPs serve as seed particles [9]. Dong et al. later discovered a straightforward approach for producing size-controlled AgNPs by reducing $AgNO_3$ at various pH levels. At high pH, both rod and spherical AgNPs were formed; however, at low pH, triangular or polygon AgNPs were formed [10].

13.2 Synthesis of DNA conjugated AuNPs and AgNPs

For the conjugation of DNA to the AuNPs and AgNPs, various covalent and non-covalent immobilization methods have been developed. DNA can be either densely attached or decorated with a few strands on AuNPs or AgNPs. These nanoparticles are prone to aggregation due to their high surface-to-volume ratio. But because of the electrostatic and steric stabilization of the DNA molecule on the nanoparticles, a high-density DNA coated on AuNPs or AgNPs exhibits good colloidal stability. This high-density DNA on the surface of the nanoparticles has also produced many exciting properties, such as sharp melting transitions and a stronger affinity for the complementary DNA (cDNA). Also, these conjugates have higher stability against degradation by nuclease compared to nonconjugated DNA. Few methods to synthesize DNA-conjugated AuNPs or AgNPs are discussed below.

13.2.1 Attaching DNA on AuNPs/AgNPs surface via adsorption or covalent bond

All four nucleotide bases of a single-stranded DNA (A, T, G, and C) can be adsorbed on AuNPs particles through the N6 exocyclic amino and the N7 atom of adenosine, C6 keto oxygen and N7 nitrogen of guanine, the C4 keto oxygen of thymine, and the N3 nitrogen and C4 keto of cytosine (Fig. 13.3A) [11]. Several studies have revealed that adenine (A) has the highest affinity for AuNPs, whereas thymine (T) has the lowest [12,13]. However, the nucleobase's interaction with AgNPs differs from that with AuNPs. Cytosine (C) has the highest affinity for AgNPs, whereas thymine (T) has the lowest affinity, with the adsorption ranks C > G > A>T (Fig. 13.3B). Exocyclic nitrogen in the DNA base pair is thought to play a significant role in AgNPs binding. The lack of exocyclic nitrogen atoms in thymine results in the lowest interaction with AgNPs [14]. Though these adsorption techniques for producing DNA-nanoparticle conjugates are straightforward, they have a disadvantage. DNA is strongly adsorbed on the nanoparticles, and the binding energy of DNA base pair - nanoparticles, particularly for AuNPs, is greater than the energy of DNA-DNA hybridization. As a result,

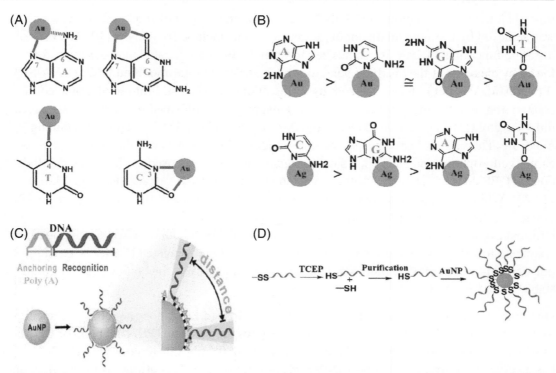

Fig. 13.3: (A) Interaction of DNA bases (adenine, guanine, thymine, and cytosine) on the gold surface (B) affinity ranking of four DNA bases on gold and silver surface. (C) Schematic of the adsorption with poly-A anchoring DNA sequence on AuNP. *Adapted with permission from ref. [13].* (D) TCEP mediated cleavage of disulfide bond before attaching on AuNPs.

complementary DNA will not hybridize with DNA adsorbed on nanoparticles, rendering it ineffective for molecular recognition [15]. However, because adenine has a strong binding affinity for gold, it is possible to produce functionalized DNA on AuNPs by employing poly-A DNA as an anchoring sequence firmly attached to the AuNPs and the remaining DNA sequence as molecular recognitions (Fig. 13.3C) [13]. Later, Zhu et al. reported a similar anchoring technique to synthesize DNA conjugate AgNPs [16]. They reported a facile approach to attached DNA containing a poly-C sequence as an anchoring sequence on AgNPs by exploiting strong silver–cytosine (Ag–C) interactions.

Thiols (-SH) have a high affinity for AuNPs or AgNPs and form Au-S or Ag-S covalent bonds [17,18]. By exploiting this interaction, DNA conjugation on AuNPs or AgNPs was efficiently accomplished using 3′ - or 5′ - thiol-modified DNA. When thiol-modified DNA was incubated with AuNPs or AgNPs, the thiol group formed a stable covalent bond with the nanoparticles, leaving the DNA base pair accessible for conjugation with the complementary DNA. Mirkin's group used this strategy to attach thiol-modified DNA to AuNPs for the first time [17]. Later, DNA-conjugated AgNPs were reported using thiol-modified DNA, which was achieved by a simple modification of the analogous method for AuNPs [19–21]. First, a disulfide functional DNA was synthesized to form the thiol-conjugated nanoparticles, followed by cleavage of the disulfide bond to produce terminal thiol-modified DNA (DNA-SH). If the disulfide is not reduced, the activity of the thiol group is lost, limiting oligonucleotide conjugation onto the nanoparticles. In the original protocol published by Mirkin et al., excess dithiothreitol (DTT) was used to cleave the disulfide bond [17]. DTT is a thiol-containing molecule. Therefore, after cleavage, the thiol-modified DNA must be isolated. Later, tris(2-carboxyethyl)phosphine (TCEP), a superior reducing agent, was used to cleave the disulfide bond. TCEP does not contain sulfur and has no affinity for AuNPs or AgNPs (Fig. 13.3D) [22,23]. While terminal mono thiol-modified DNA has been used to functionalize on AuNPs or AgNPs, the conjugates produced with such probes lack stability. DNA modified with terminal cyclic disulfide moieties, multiple thiols, or thioctic acid have also been reported for conjugate formation with these nanoparticles to get a stable conjugate [17,24,25]. Because polydentate ligands provide more stable metal-ligand complexes, the conjugates were more stable when several thiol groups were present.

Due to the negative charge, the phosphate backbone (PO) of DNA does not attach efficiently to the surface of AuNPs or AgNPs. However, the phosphorothioate (PS) backbone of DNA, in which a sulfur atom has replaced the oxygen atom in the phosphate group of the DNA, has a very strong affinity for both AuNPs and AgNPs [26–28]. This affinity is higher than the affinity for nucleobases, resulting in significantly more stable DNA adsorption than unmodified PO DNA and improved the colloidal stability of the conjugated nanoparticles. Several research groups have developed methods for functionalizing AgNPs or AuNPs using chimeric phosphorothioate modified DNA (PS-PO-DNA) [26–28]. The multivalent Ag-sulfur or Au-sulfur interaction between Ag/Au and thiol-modified DNA anchors the DNA to nanoparticle surfaces, leaving the nonmodified DNA available for hybridization.

13.2.2 Salt, pH, and surfactant-dependent conjugation

Even though there is a strong bond between thiol and gold, it is challenging to attach DNA to citrate-capped AuNPs because the negatively charged AuNPs and DNA repel each other. When thiol-modified DNA is added to AuNPs, a few DNA strands bind to the surface of the gold nanoparticles. As a result, the complex formed by this addition becomes more negatively charged, and charge repulsion between them inhibits further conjugation and seriously limits the DNA density on the AuNPs. Also, the target DNA will not be hybridized with the DNA adsorbed on AuNPs, limiting its molecular recognition properties [29]. Herne et al. found that the adsorption of thiol-modified DNA onto planar gold was significantly reduced with low salt concentration [29]. By adding salt to shield this charge repulsion and allow more DNA strands to adsorb on the nanoparticles, a high-density DNA shell can be formed. Adding salt can neutralize the charge-charge repulsion; however, adding too much salt at once causes

AuNPs to aggregate irreversibly [30,31]. Mirkin's group developed a salt aging method in which salt was added slowly over two days in modest increments of 10–50 mM each time to overcome this issue (Fig. 13.4A) [32]. Maximum DNA loading on AuNPs by this salt aging has been obtained at ~ 0.7 M salt concentration, and the loading remains relatively stable. Because of their high curvature, small AuNPs can pack DNA more densely than flat gold, up to four times more densely. However, this salt aging process was time-consuming, took more than two days to finish, and needed an excess amount of DNA. Similar salt aging methods did not work for AgNPs. As a result, better ways of loading DNA onto these nanoparticles were required.

Liu's group discovered that DNA adsorption on AuNPs or AgNPs is pH-dependent [33,34]. At acidic pH, adsorption was greatly enhanced, especially at low salt concentrations, and takes a few minutes to complete. At pH 3, a high density of DNA was attached to AuNPs or AgNPs in just 3 minutes (Fig. 13.4B) [33,34]. Later, the same group discovered that poly-A-containing DNA improves attachment to AuNPs by forming a poly-A DNA duplex at low pH (Fig. 13.4C) [35]. Placing a spacer between the thiol group and the DNA also made it easier for DNA to adhere to nanoparticles quickly and in large amounts [32].

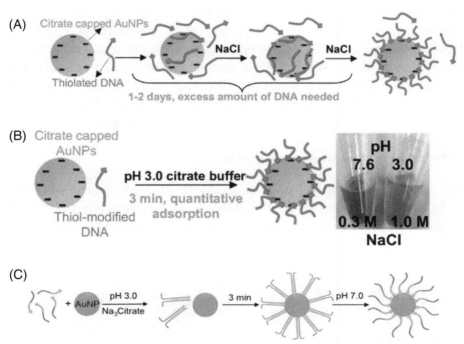

Fig. 13.4: Schematics of attaching negatively charged thiol-modified DNA on AuNPs using (A) salt aging and (B) the low pH assisted methods, and (C) pH-dependent loading of DNA containing poly-A block. *Adapted with permission from ref. [33,35].*

Although the direct salt-aging approach is straightforward, it is a time-consuming process and an inefficient procedure for bigger AuNPs. To solve this issue, Mirkin's group designed a quick salt aging procedure in which a surfactant was added before the addition of salt, and the highest salt concentration of the solution was achieved in a matter of hours [36]. Surfactants reduce the aggregation propensity of AuNPs by generating a layer on top of it. Using surfactant, they effectively functionalized AuNPs with larger sizes up to 250 nm. For this functionalization, freshly prepared thiol-modified DNA was incubated with AuNPs in phosphate buffer (10 mM, pH 7) for 20 minutes with 0.1% surfactant SDS. The salt of 50 mM was then added and incubated for 20 minutes. The salt concentration was progressively elevated to 1 M by repeating this process. Finally, the mixture was incubated at room temperature overnight. With a concentration of 1 M salt (NaCl), adding surfactants to the process increased the DNA loading capacity by 39% while simultaneously increasing the reproducibility.

13.2.3 Freezing and thaw method

Liu et al. developed a relatively straightforward approach for conjugating DNA to AuNPs utilizing freezing methods [37,38]. They discovered that DNA could be conjugated successfully by just freezing the DNA with AuNPs followed by thawing without using any additional reagent. During the freezing process, small ice crystals of pure water were formed, and the non-water components such as DNA, AuNPs, and salt concentrated in the micro pockets that exist between the ice crystals, forming a near-saturated salt concentration (Fig. 13.5). Because of this high concentration of DNA, AuNPs and salt, the conjugation process become fast. Furthermore, DNA molecules starched and aligned during the freezing procedure, making it easier to attach to AuNPs. The concentration of salt has little effect on this approach. This method requires a low concentration of salt (0.1 mM). However, a high quantity of DNA was needed to stabilize the AuNPs during this process to prevent aggregation.

13.2.4 Mononucleotide-mediated conjugation

Hsing et al. exploited the mononucleotide-coated AuNPs to rapidly functionalize high-density DNAs on AuNPs in the presence of salt [39]. Mononucleotide binds reversibly on the surface of AuNPs and stabilizes them in salt solution by producing a mononucleotide layer on the

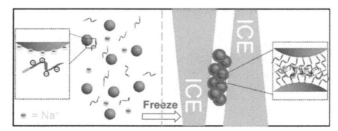

Fig. 13.5: Proposed mechanism of attaching DNA onto AuNPs by freezing. *Adapted with permission from ref. [37].*

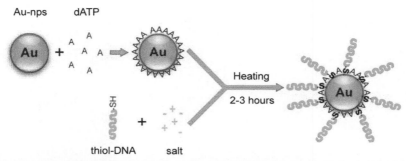

Fig. 13.6: Mononucleotide-mediated conjugation of thiol-DNA to AuNPs. *Adapted with permission from ref. [39].*

surface, which can be controlled by temperature. Reversible adsorption of mononucleotides on the surface of the AuNPs is the first step in this DNA conjugation process. Next, an incoming thiol-DNA undergoes a ligand-exchange reaction with the AuNPs' surface mononucleotides, which are facilitated by temperature. Using this approach, DNA was functionalized on AuNPs in 4 hours, and the resultant conjugate had high DNA loading (~80 DNA strands per particle) and stability (Fig. 13.6) [39].

13.2.5 Microwave-assisted heating drying method

Recently, Zhou et al. have developed a microwave-assisted heating-drying technique for conjugating thiol-modified DNA/RNA, poly-T DNA, or poly-U RNA to AuNPs using a household microwave (Fig. 13.7) [40]. This technique is quick, takes about a minute, and requires no additional reagent. In a glass container, citrate-capped AuNPs were combined with thiol-modified DNA, poly-T DNA, or poly-U RNA. The glass container was then placed in a home microwave and heated on medium power (700 W) for 3 minutes, followed by resuspension in water or buffer, resulting in DNA conjugation on AuNPs. Standard oil bath heating at 150°C was insufficient to conjugate the DNA on AuNPs; hence microwave aided heating was required. High temperatures were achieved by MW-assisted heating, which unfolded coiled DNA and evaporated water, concentrating DNA, AuNPs, and salt and thereby increasing DNA/RNA conjugation on AuNP surfaces [40].

13.2.6 Low-density DNA conjugation on nanoparticles

One limitation of DNA-AuNPs conjugation is that colloidal stability requires a relatively high density of DNA on AuNPs. Recently, Francis's group developed an enzymatic approach for conjugating DNA on AuNPs to prepare low-density DNA-AuNPs, and the process is rapid [41]. The AuNPs were initially functionalized with a phenol monolayer, inhibiting aggregation and nonspecific DNA adsorption on the AuNPs. Tyrosinase enzyme was used to activate the functionalized AuNPs, and the activated phenol on AuNPs was subsequently conjugated to thiolate DNA within 2 hours. Deng's group developed the

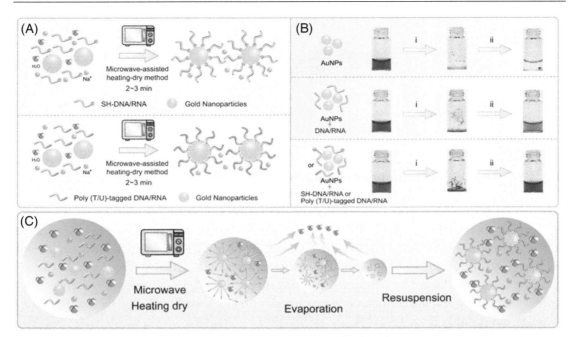

Fig. 13.7: (A) Scheme of attaching thiolated and non-thiolated DNA/RNA to AuNPs using microwave (MW) -assisted heating-dry method, in which a domestic microwave oven drives the heating-dry process for 2–3 minutes. (B) Photographs of AuNPs before and after the heating-dry process. The bare AuNPs and random DNA/RNA sequence mixed AuNPs aggregated after heating dry treatment. While poly (T/U)-tagged DNA-AuNPs retained monodispersed and seemed red after heating dry and resuspension. (1) MW-assisted heating dry; (2) resuspending with water or buffer. (C) MW-assisted heating-dry method involves two consecutive processes: (1) MW-assisted heating -induced high temperature unwinds the coiled structure of oligonucleotides. (2) MW-assisted water evaporation concentrates the DNA/RNA, AuNPs, and salt, thereby accelerating the attachment of DNA/RNA on the AuNPs surface. *Adapted with permission from ref. [40].*

DNA-AgNPs nanoconjugate with a discrete amount of DNA [42]. In this procedure, Ag^+ was initially adsorbed on short fish sperm DNA (fsDNA) before being reduced to AgNPs by $NaBH_4$. This approach yields AgNPs with a mean diameter of 2 nm, making it suitable for gel electrophoresis separation. The fsDNA stabilized AgNPs were conjugated with thiolate DNA and then purified by gel-electrophoresis to get the definite number of DNA on the surface [42].

13.3 Properties of DNA conjugated AuNPs and AgNPs

Because of the tightly packed DNA on their surface, DNA conjugated AuNPs and AgNPs exhibit several exciting properties. A few of the most intriguing properties are discussed here.

13.3.1 Sharp melting transition

Base stacking, hydrogen bonding, hydrophobic, and electrostatic are some of the many physical and chemical processes that contribute to the formation of a duplex when two DNA strands come together (Fig. 13.1). When this DNA duplex is heated above room temperature, two stands are separated gradually to generate two single-stranded DNA. This is referred to as DNA melting. The melting temperature (Tm) of a DNA duplex is the transition temperature between the two states in which half remains in duplex form while the other half is denatured into a single strand. DNA melting is measured by the absorbance at 260 nm by the DNA solution, and the absorbance increase with the melting of duplex DNA. The melting temperature of a non-modified DNA duplex is not sharp; the range is more than 20°C. But due to the high DNA density on AuNPs or AgNPs, these conjugates behave differently than non-modified DNA. They exhibit a very sharp melting transition with only a few degrees (~2–5°C) [43,44]. This sharp melting point is caused by the simultaneous cooperative or synergistic melting of all DNA strands.

13.3.2 Tighter binding to complementary DNA (cDNA)

Another important property of this DNA conjugated nanoparticle is that it strongly binds the complementary DNA (cDNA), which is why the conjugate has a higher melting temperature than free duplex DNA of the same sequence [45]. The equilibrium binding constant (K_{eq}) for 15 nm DNA-AuNPs functionalized with complementary three base pairs DNA is estimated to be three-fold greater than the K_{eq} for the identical non-modified DNA strands free in solution. It was later revealed that the very high density of DNA caused the higher binding on AuNPs [46].

13.3.3 Stability and cellular uptake of DNA conjugate AuNPs

Nature has developed an arsenal of enzymes known as nucleases. In most biological environments, to digest foreign nucleic acids that enter cells, unmodified oligonucleotides are rapidly degraded by nonspecific nucleases, rendering them less effective. However, the DNA-AuNPs conjugate is highly resistant to nuclease [47], making it appropriate for intracellular molecular diagnostics, imaging, and drug delivery.

The Au-S bond is the most commonly used linkage to achieve DNA-AuNP conjugation. But, it is critical to maintain a stable Au-S bond for this conjugation since Au-S bonds are known to be displaced by numerous thiol molecules in biological environments, such as glutathione (GSH). Chan et al. conducted considerable research on the impact of glutathione on the stability of DNA conjugated AuNPs. They discovered that thiol-modified DNA on AuNPs is resistant to glutathione displacement [48].

The Mirkin group did extensive research on the cellular uptake mechanism and intracellular trafficking of DNA-AuNP conjugates. DNA-AuNPs conjugates with high-density DNA oligonucleotide strands have been found to penetrate cells significantly and have been used for intracellular mRNA imaging and delivery of therapeutic nucleic acids. It was known that serum protein non-specifically binds to the nanoparticle surface, and also, the high concentration of DNA on the AuNPs increase this serum binding. Nonspecific adsorption of serum proteins onto the nanoparticle surface may facilitate cellular uptake. The density of the DNA loading on the surface of the particles affects uptake, with higher densities leading to increased uptake [49]. But few reports showed that serum protein act as an inhibitor for cellular uptake of DNA- AuNPs [50].

13.4 Applications of DNA conjugated AuNPs/AgNPs

Both AuNPs and AgNPs have interesting physical and chemical properties, making them suitable agents for many biomedical applications. These nanoparticles have interesting plasmons resonance absorption bands below 500 nm for AgNPs and at 500–600 nm for AuNPs. Due to their plasmon resonance, both nanoparticles can act as fluorescence quenchers. When the recognition characteristic of DNA is combined with AuNPs or AgNPs, an effective biosensor is produced. These probes can recognize a diverse range of analytics, including complimentary nucleic acids, heavy metal ions, and small molecules, and are also suitable for drug delivery and gene regulations. Several applications of DNA functionalized AuNPs or AgNPs are addressed further below.

13.4.1 Sequence-specific detection of nucleic acids

Detecting genetic mutations is a critical diagnostic emphasis, which has increased interest in nucleic acid-based detection tests for the early diagnosis of a wide range of disorders, including cancer [51]. It is feasible to identify specific DNA sequences utilizing DNA-AuNPs or DNA-AgNPs by exploiting DNA-induced nanoparticle assembly or disassembly. The color of these nanoparticles varies according to assembly or disassembly, which can be seen with the naked eye [52]. Mirkin et al. reported the first use of DNA-AuNPs to detect specific DNA [17]. They used two types of functionalized AuNPs with two different single-stranded DNA, a DNA attached to the AuNPs through 3'- end and another via 5'- end. These single-stranded sequences were complementary to a target sequence. The DNA-AuNPs color was purple to the naked eye without targeted nucleic acids. But in the presence of target nucleic acids, which were complementary to both the DNA present on AuNPs, it caused to crosslink by hybridization and aggregate the AuNPs (Fig. 13.8(I)(a)). Due to the aggregation, the color of the solution changed from purple to blue, which was readily detected visually or by UV-Vis spectroscopy (Fig. 13.8(I)(b)). Due to the aggregation of the AuNPs in the presence of target nucleic acids, the transient melting point became sharp (Fig. 13.8(I)(c)). In another similar type of study, only one type of single-stranded DNA was conjugated on AuNPs, and it was demonstrated that it is even better to detect targeted sequences because of speed and sensitivity. In the presence of perfectly matched complementary DNA at high salt concentration, complementary DNA hybridized and caused the AuNPs to aggregate because of salting out effects and the color changed from purple to blue, but in the presence of a single mismatch target, AuNPs remained in dispersed, and the color did not change [53].

DNA-AgNPs conjugates have also been reported to detect a target oligonucleotide with colorimetric analysis (Fig. 13.8(II)) [21,43]. This approach can also distinguish a single base mismatch. Two batches of DNA-AgNPs were functionalized using two complementary DNA sequences (Fig. 13.8(II)(a)). The absorbance pattern of the conjugate was not altered by DNA conjugation on AgNPs (Fig. 13.8(II)(b)). DNA-AgNPs showed a surface plasmon resonance at 410 nm and bright yellow color (Fig. 13.8(II)(c)(1)). In the presence of target nucleic acid, which was complementary to the DNA functionalized on AgNPs, plasmon resonance dampens. This dampening was caused by AgNPs aggregation as a result of hybridization, which is visible to the naked eye as a color shift from bright yellow to

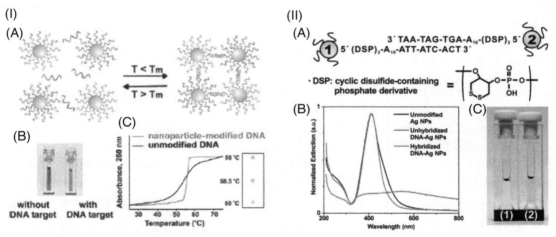

Fig. 13.8: (I) Complementary target DNA induced (A) the aggregation of AuNPs (B) leading to the color change of the solution from red to blue. (C) The aggregation process can also be monitored by using UV-vis spectroscopy. (II) (A) Schematic illustration of the hybridization of two complementary DNA-AgNPs. (B) UV-vis spectra of unmodified AgNPs (black line), unhybridized DNA-AgNPs (blue line), and hybridized DNA-AgNPs (red line). The wavelength at which the maximum of the extinction of AgNPs is obtained remains the same after DNA functionalization. After hybridization, however, the band of DNA-AgNPs broadens and redshifts significantly from 410 to 560 nm. (C) Colorimetric change was responsible for the assembly process of DNA-AgNPs. The intense yellow color of the unhybridized AgNPs (C1) turns to pale red (C2) as the particle aggregation proceeds. Heating of (C2), however, results in the return of the solution color to yellow (C1). *Adapted with permission from ref. [52 and 43].*

colorless (Fig. 13.8(II)(c)(2)). The melting profile of DNA-AgNPs aggregate was much sharper than unmodified DNA due to the presence of multiple DNA linkages and cooperative DNA melting like DNA conjugated AuNPs (Fig. 13.8(II)(a)). This method was used to detect mismatched DNAs with high sensitivity.

Storhoff and co-workers used this approach to detect genomic DNA samples of methicillin-resistant Staphylococcus pathogen [54]. They developed a colorimetric detection method called the "spot-and-read" method for detecting DNA sequences. The DNA-modified AuNPs were first spotted onto a glass surface, which was then exposed to a laser beam, and the scattered light was measured. The specific nucleic acids were then added, and the color changes caused by this process were measured. This light scattering approach allowed the detection of DNA in minimal concentrations. This method increased sensitivity by four orders of magnitude compared to absorbance-based methods.

The gene of brucella bacteria was also detected very selectively and specifically using DNA-AuNPs. A thiol-modified DNA specific to the brucella bacterial gene was conjugated to AuNPs. In the presence of the specific bacterial gene, a DNA duplex AuNPs conjugate was formed, which does not aggregate in the presence of HCl. But, in the absence of the specific gene, the DNA-AuNPs conjugates undergo aggregation, leading to a change in color from red to blue [55].

13.4.2 Detection of heavy metal ions

It has been demonstrated that DNA-conjugated nanoparticles can detect a wide range of heavy metals, from mercury to lead to cadmium and arsenic. In DNA-AuNPs, the AuNPs serve just as a color indicator, while the surface DNA molecules are responsible for recognizing a metal target. The most straightforward method for colorimetric detection of the metals using DNA-AuNPs is the crosslinking assembly of DNA-AuNPs caused by heavy metal target binding. As a result of target binding, adjacent DNA-AuNPs are brought into close contact, forming sandwich configurations that cause a color change from red to blue.

It is known that thymine (T) and cytosine (C) cannot form hydrogen bonding with the same base but form stable T-Hg^{2+}-T or C-Ag^+-C complexes. Therefore, two DNA containing mismatched T-T base pairs or C-C base-pair can form a stable duplex in the presence of Hg^{2+} or Ag^+. Mirkin et al. developed a selective and sensitive colorimetric method for Hg^{2+} detection using T mismatch DNA conjugated AuNPs [56]. They synthesized two types of DNA-AuNPs containing complementary sequences, except for an internal single T-T mismatch. In the absence of target Hg^{2+}, these particles also form aggregates and exhibit the characteristic sharp melting temperature but with a lower temperature because of the mismatch internal T in the duplex. But in the presence of Hg^{2+}, it coordinates selectively to the T-T mismatch site and the probe aggregates. Because of the aggregation, the melting temperature increase (melt at more than 46°C). Because of higher Tm, the crosslinked AuNPs remain purple at ~ 46°C.

Crosslinked AuNPs appeared purple at room temperature in the absence of Hg^{2+} but became red when heated to $\sim 46°C$. Melting point measurements of DNA-AuNPs in the presence of Hg^{2+} at various concentrations revealed a linear detection range of 100–2000 nM of Hg^{2+}, with an estimated LOD of ~ 100 nM [56].

Xue et al. further improved the detection system for Hg^{2+} by using DNA functionalized AuNPs [57]. They prepare two AuNPs functionalized with two different non-complementary DNA sequences. They also used another non-modified DNA which is complementary to the DNA present on AuNPs, except it contains few internal T-T mismatches. The melting point of the three probes was optimized by controlling the number of T–T mismatches, and the melting point of the complex was suppressed below the ambient temperature. In the presence of Hg^{2+}, the formation of T- Hg^{2+}-T complex increased the melting temperature by over $40°C$. Because of stable duplex formation, AuNPs aggregate, and the color changes from purple or colorless (Fig. 13.9A). By using this probe, 3 μM of Hg^{2+} was detected selectively in the presence of other metal ions [57]. Dong et al. designed a colorimetric sensor for Ag^+ by C- Ag^+-C complex between Ag^+ and C- mismatched DNA functionalized AuNPs (Fig. 13.9B) [58].

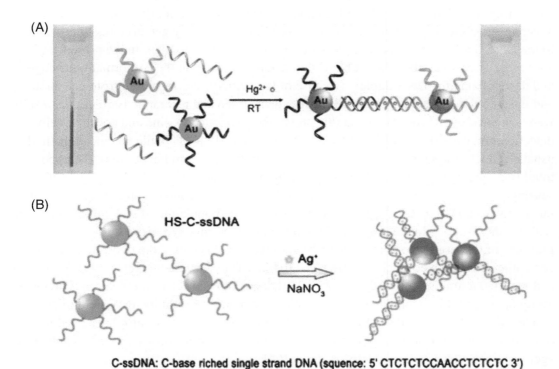

C-ssDNA: C-base riched single strand DNA (squence: 5' CTCTCTCCAACCTCTCTC 3')

Fig. 13.9: Aggregation mechanism for the DNA conjugated AuNPs in the presence of (A) Hg^{2+} and (B) Ag^+. *Adapted with permission from ref. [57,58].*

Liu et al. designed another DNA-AuNPs conjugate using a random coiled poly-T DNA adsorbed on AuNPs to detect Hg^{2+}. They took poly-T DNA and 13 nm AuNPs. The poly-T was adsorbed on AuNPs in the presence of salt and formed a highly dense negative charge on AuNPs. Due to this high-density negative charge on the AuNPs surface, it becomes dispersed and shows a purple color, but in the presence of a Hg^{2+} ion, the poly-T DNA forms a stable duplex because of T- Hg^{2+-}-T complex formation, causing the poly-T to desorb from the AuNPs surface. The tendency of the AuNPs to aggregate increases when the electrostatic repulsion that is provided by the adsorbed DNA is absent. Due to this aggregation, the color of the solution changed from purple to blue. By using this method, Hg^{2+} can be detected down to 250 nM [59].

13.4.3 Aptamer-based detection of small molecules

Aptamers are nucleic acids with excellent specificity and sensitivity for various targets, including proteins, peptides, nucleotides, antibiotics, and small compounds [60]. These aptamers are single-stranded DNA with a stem region and a flexible loop that bind to a specific target. DNA aptamers conjugated with AuNPs for the application to detect adenosine and cocaine was first reported by Li et al. [61]. They created a colorimetric sensor by connecting two kinds of AuNPs with an aptamer fragment containing a linker DNA strand. When these DNA conjugated AuNPs are mixed in solution, they get close together due to the DNA duplex formation, and the solution color changes to blue. In the presence of adenosine or cocaine, these target binds to the aptamer, causing the conformation change and destabilizing the DNA duplex. Because of that, AuNPs are separated from one another, and the solution color changes from blue to red. This sensing technique is quick; the results were achieved in minutes. By using this probe, they detect adenosine and cocaine, with detection ranges of 0.3–2 mM and 50–500 M, respectively [61]. While the above methods usually reflect disaggregation of AuNPs (i.e., color shifting from blue to red), the opposite trend (i.e., aggregation for a red-to-blue transition) has also been applied in colorimetric sensing. Li et al. used this for the detection of both adenosine and cocaine [62]. They split the aptamer into two pieces, attaching each piece to one population of AuNPs and the other with another AuNPs. In the absence of the target, they remain in dispersed form. But in the presence of the target, the two aptamers bind to the target bringing the two AuNPs close to each other, causing a color change from red to blue. They detected targets within 30 minutes and reported a limit of detection of 0.25 mM (adenosine) and 0.1 mM (cocaine). The same group later used this technique to detect cocaine using high-resolution dark-field microscopy images of latent fingerprints (LFPs) [63]. DNA aptamer AuNPs have also been used for various biomedical applications, for example, detection of bacteria and viruses, as an anticancer agent, anticoagulant and drug delivery agent [64].

13.4.4 Fluorescence-based assays

While only colorimetric sensors have been discussed so far, DNA conjugated AuNPs/ AgNPs has also been utilized to detect various analyte using fluorescence as signal output. Fluorescence methods are more sensitive than colorimetric methods and offer a low detection limit (LOD) for analytes. The high molar extinction coefficients and broad energy bandwidth of AuNPs and AgNPs make them ideal candidates for fluorescence quenchers. These nanoparticles can quench fluorophores over a longer distance than conventional molecular quenchers. When a fluorophore is placed in close proximity to the AuNPs or AgNPs, its fluorescence is quenched [65,66]. The quenching efficiency of the nanoparticle depends on the fluorophore's emission wavelength and the nanoparticle's size. Libchaber and co-workers have conducted considerable research into the quenching efficiency of 1.4 nm gold nanoparticles with regard to a variety of dyes, such as rhodamine 6G, fluorescein, texas red and Cy5 [67]. They found that all dyes are quenched efficiently. Their concept consisted of a hairpin loop of DNA that was fluorescently tagged in one end and AuNPs attached to another end, analogous to that of a molecular beacon (MB). This probe was named nanobeacon. Due to the close proximity to the gold nanoparticle and the fluorophore, the fluorescence was completely quenched. Upon binding to the target DNA, which is complementary to the loop regions, the loop of the probe opened, and the fluorescence was restored due to the increase in the distance between the AuNPs and the fluorophore (Fig. 13.10A). Maxwell et al. used a similar type of strategy using larger 2.5 nm AuNPs [68]. They were able to detect cDNA concentrations as low as 10 nM.

Wright et al. applied this gold nanoparticle-based nanobeacon to detect intracellular mRNA [69]. They found that the nanobeacon was better than a conventional molecular beacon. Because AuNPs are more efficient at quenching the fluorescence of a variety of organic dyes, the nanobeacon's fluorescence background was significantly reduced. The same group subsequently demonstrated that the nanobeacon could identify tyrosinase mRNA in melanoma cells [70]. Various other research groups also used this nanobeacon to detect mRNA or miRNA in living cells [71–73]. Tang's group improved the nanobeacon to detect several target mRNAs simultaneously [74,75]. They developed a FRET-based AuNPs-molecular beacon probe to detect and image multiple intracellular mRNA [76]. First, molecular beacons (MB) tagged with Alexa fluor 488 and Cy3 were attached to the surface of AuNPs. When two separate mRNA targets with complementary sequences bind to the molecular beacon, their hairpin structure opens, restoring the fluorophore's corresponding fluorescence and generating FRET between these fluorophores. As a result, when Alexa Fluor 488 is excited at 488 nm, the emissions of Cy3 and Alexa Fluor 488 can both be detected. Later, Mirkin's group extended this technique to detect multiple mRNAs at once and image intracellularly by incorporating four different molecular beacons barring four different fluorescence dyes (Alexa fluor 405, Alexa fluor 488, Cy3, and Cy5) on the same AuNPs (Fig. 13.10B) [77].

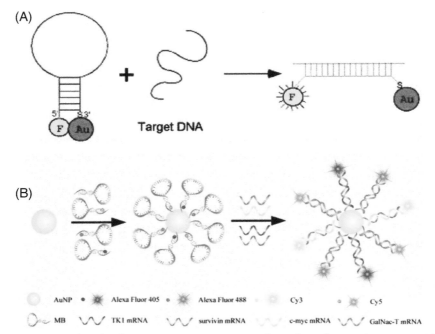

Fig. 13.10: (A) A molecular-beacon type DNA sensor based on fluorophore-tagged, hairpin oligonucleotide functionalized gold nanoparticles (B) schematic illustration of the four-color nanoprobe for detection of multiple intracellular mRNAs. *Adapted with permission from ref. [68 and 77].*

Duan et al. have reported an AgNPs-based nanobeacon similar to AuNPs nanobeacon that detects specific DNA [78]. Instead of AuNPs, they employed AgNPs as a quencher. In the presence of targeted DNA, the nanobeacon opens up, enhancing the fluorescence intensity.

13.4.5 DNA-AuNPs probes for therapy in living systems

In 2009, a team led by Mirkin and Lippard used AuNPs as a delivery vehicle for platinum (IV) complex in living cells. The complex was covalently linked to amine-functionalized DNA on AuNPs [79]. When Pt(IV) complex conjugated AuNPs entered the cell, it was reduced to Pt(II), which was then released and caused intra-strand crosslinks on nuclear DNA, leading to cytotoxicity. Mirkin's group developed a strategy for improving paclitaxel solubility and effectiveness by covalently connecting it to AuNPs using fluorophore-labeled DNA linkers. (Fig. 13.11A) [80]. These nanoconjugates showed enhanced therapeutic effectiveness compared to free drugs when tested against various cell lines, including paclitaxel-resistant cells. It is also known that some anticancer drugs, such as doxorubicin

(DOX), can intercalate into DNA or RNA, resulting in fluorescence quenching and lower drug cytotoxicity. Tang's group reported an AuNPs-based nanobeacon conjugate with a tumor-related mRNA sequence for DOX delivery [81]. In this strategy, the beacon served as drug carrier and an imaging probe. It released the drug DOX upon target mRNA binding. Fluorescence of DOX was also used to image intracellularly. Kanaras et al. later designed a DNA conjugated AuNPs dimer with intercalating two anticancer drugs, DOX and mitoxantrone (MTX) [82]. This probe can simultaneously or independently identify two mRNA targets and deliver one or two DNA-intercalating anticancer drugs (DOX or MTX) to living cells. They detected two mRNA (keratin 8 and vimentin) in A549 cells by confocal images (Fig. 13.11B).

13.4.6 Nucleic acid delivery

DNA conjugated AuNPs have been used to deliver a specific gene into cells. Mirkin's group reported DNA-AuNPs conjugates for antisense gene regulation [47]. Later, the same group created a novel antisense oligonucleotide delivery method that can monitor and regulate RNA levels in living cells [83]. It has also been shown that a nanobeacon is an effective probe for antisense gene therapy. Conde's group developed a method to detect, target, and silence gene expression in vivo by conjugating Cy3-labeled antisense DNA nanobeacon onto AuNPs [84]. Small interfering RNA (siRNA) is a double-stranded RNA that inhibits gene expression and protein synthesis [85]. However, siRNA has poor stability, low membrane permeability, and large molecular size, and it is difficult to deliver both in cells and in vivo. In 2014, Kim's group developed an AuNPs-based method to deliver siRNA in cells [86].

13.5 Summary

In summary, DNA-functionalized AuNPs and AgNPs have excellent optical and stability features. These conjugates have unique physical characteristics, such as very sharp melting transitions and higher binding affinity to complementary DNA (cDNA), which helped in the development of different biosensors. Several methods were discussed for attaching thiolated or non-thiolated DNA to AuNPs or AgNPs. The conjugates have also been utilized to create colorimetric biosensors that rely on the plasmonic properties of the nanoparticles. This chapter also discussed many fluorescence sensors by using AuNPs/AgNPs' quenching properties. Apart from biosensors, these conjugates were also used to deliver drugs and transfer nucleic acids to cells.

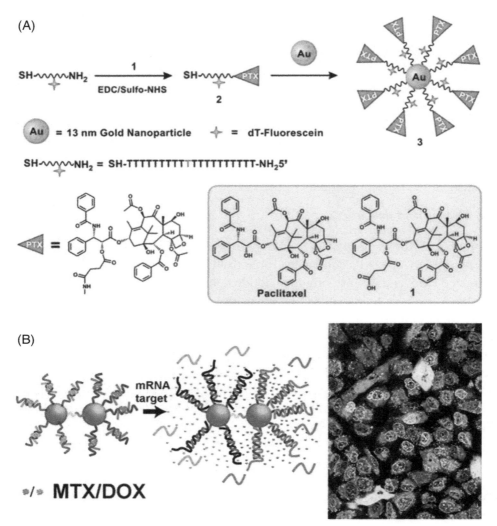

(A)

(B)

MTX/DOX

Fig. 13.11: (A) Synthesis of PTX-DNA conjugated AuNPs. (B) Schematic Illustration of AuNPs nanoparticle dimer used in drug release; (I) dimer loaded with DOX (Drug 1) and MTX (Drug 2). Both DOX and MTX have fluorescent properties, which are quenched due to the close proximity to the AuNPs core. When the target mRNA binds to the sense strand, the intercalated drugs are released (II), causing an increase in the fluorescence signal. (III) Confocal images of both drugs in A549 cells. *Adapted with permission from ref. [80, 82].*

References

[1] D. McShan, P.C. Ray, H. Yu, Molecular toxicity mechanism of nanosilver, J. Food and Drug. Analysis 22 (2014) 116–127.

[2] M.-.C. Daniel, D. Astruc, Gold nanoparticles: Assembly, supramolecular chemistry, quantum-size-related properties, and applications toward biology, catalysis, and nanotechnology, Chem. Rev. 104 (2004) 293–346.

[3] N.A. Kasyanenko, A.A. Andreeva, A.V. Baryshev, V.M. Bakulev, M.N. Likhodeeva, P.N. Vorontsov-Velyaminov, DNA integration with silver and gold nanoparticles: enhancement of DNA optical anisotropy, J. Phys. Chem. B 123 (2019) 9557–9566.

[4] T. Roy, K. Szuttor, J. Smiatek, C. Holm, S. Hardt, Conformation and dynamics of long-chain end-tethered polymers in microchannels, Polymers. 11 (2019) 488.

[5] M.D. Matteucci, M.H. Caruthers, Synthesis of deoxyoligonucleotides on a polymer support, J. Am. Chem. Soc. 103 (1981) 3185–3191.

[6] J. Turkevich, P.C. Stevenson, J. Hillier, A study of the nucleation and growth processes in the synthesis of colloidal gold, Discuss. Faraday Soc. 11 (1951) 55–75.

[7] B.-.K. Pong, H.I. Elim, J.-.X. Chong, W. Ji, B.L. Trout, J.-.Y. Lee, New insights on the nanoparticle growth mechanism in the citrate reduction of gold(III) salt: Formation of the Au nanowire intermediate and its nonlinear optical properties, J. Phys. Chem. C 111 (2007) 6281–6287.

[8] P.C. Lee, D. Meisel, Adsorption and surface-enhanced Raman of dyes on silver and gold sols, J. Phys. Chem. 86 (1982) 3391–3395.

[9] A. Pyatenko, M. Yamaguchi, M. Suzuki, Synthesis of spherical silver nanoparticles with controllable sizes in aqueous solutions, J. Phys. Chem. C 111 (2007) 7910–7917.

[10] X. Dong, X. Ji, H. Wu, L. Zhao, J. Li, W. Yang, Shape control of silver nanoparticles by stepwise citrate reduction, J. Phys. Chem. C 113 (2009) 6573–6576.

[11] J. Liu, Adsorption of DNA onto gold nanoparticles and graphene oxide: Surface science and applications, Phys. Chem. Chem. Phys. 14 (2012) 10485–10496.

[12] H. Kimura-Suda, D.Y. Petrovykh, M.J. Tarlov, L.J. Whitman, Base-dependent competitive adsorption of single-stranded DNA on gold, J. Am. Chem. Soc. 125 (2003) 9014–9015.

[13] H. Pei, F. Li, Y. Wan, M. Wei, H. Liu, Y. Su, et al., Designed Diblock oligonucleotide for the synthesis of spatially isolated and highly hybridizable functionalization of DNA–gold nanoparticle nanoconjugates, J. Am. Chem. Soc. 134 (2012) 11876–11879.

[14] S. Basu, S. Jana, S. Pande, T. Pal, Interaction of DNA bases with silver nanoparticles: assembly quantified through SPRS and SERS, J. Colloid Interface Sci. 321 (2008) 288–293.

[15] X. Zhang, M.R. Servos, J. Liu, Surface science of DNA adsorption onto citrate-capped gold nanoparticles, Langmuir 28 (2012) 3896–3902.

[16] D. Zhu, J. Chao, H. Pei, X. Zuo, Q. Huang, L. Wang, et al., Coordination-mediated programmable assembly of unmodified oligonucleotides on plasmonic silver nanoparticles, ACS Appl. Mater. Interfaces 7 (2015) 11047–118752.

[17] C.A. Mirkin, R.L. Letsinger, R.C. Mucic, J.J. Storhoff, A DNA-based method for rationally assembling nanoparticles into macroscopic materials, Nature 382 (1996) 607–609.

[18] K. Shrivas, H.F. Wu, Applications of silver nanoparticles capped with different functional groups as the matrix and affinity probes in surface-assisted laser desorption/ionization time-of-flight and atmospheric pressure matrix-assisted laser desorption/ionization ion trap mass spectrometry for rapid analysis of sulfur drugs and biothiols in human urine, Rapid Commun. Mass Spectrom. 22 (2008) 2863–2872.

[19] I. Tokareva, E. Hutter, Hybridization of oligonucleotide-modified silver and gold nanoparticles in aqueous dispersions and on gold films, J. Am. Chem. Soc. 126 (2004) 15784–15789.

[20] B.C. Vidal Jr, T.C. Deivaraj, J. Yang, H.-.P. Too, G.-.M. Chow, L.M. Gan, et al., Stability and hybridization-driven aggregation of silver nanoparticle–oligonucleotide conjugates, New J. Chem. 29 (2005) 812–816.

[21] D.G. Thompson, A. Enright, K. Faulds, W.E. Smith, D. Graham, Ultrasensitive DNA detection using oligonucleotide–silver nanoparticle conjugates, Anal. Chem. 80 (2008) 2805–2810.

[22] J. Swanner, R. Singh, Synthesis, purification, characterization, and imaging of Cy3-functionalized fluorescent silver nanoparticles in 2D and 3D tumor models, Methods Mol. Biol. 1790 (2018) 209–218.

[23] R. Wu, L.-.P. Jiang, J.-.J. Zhu, J. Liu, Effects of small molecules on DNA adsorption by gold nanoparticles and a case study of Tris(2-carboxyethyl)phosphine (TCEP), Langmuir 35 (2019) 13461–13468.

[24] Z. Li, R. Jin, C.A. Mirkin, R.L. Letsinger, Multiple thiol-anchor capped DNA–gold nanoparticle conjugates, Nucleic Acids Res 30 (2002) 1558–1562.

[25] J.A. Dougan, C. Karlsson, W.E. Smith, D. Graham, Enhanced oligonucleotide–nanoparticle conjugate stability using thioctic acid modified oligonucleotides, Nucleic Acids Res. 35 (2007) 3668–3675.

[26] W. Zhou, F. Wang, J. Ding, J. Liu, Tandem Phosphorothioate modifications for DNA adsorption strength and polarity control on gold nanoparticles, ACS Appl. Mater. Interfaces 6 (2014) 14795–14800.

[27] S. Hu, P.-J.J. Huang, J. Wang, J. Liu, Phosphorothioate DNA mediated sequence-insensitive etching and ripening of silver nanoparticles, Front. in Chem. 7 (2019) 1–9.

[28] S. Pal, J. Sharma, H. Yan, Y. Liu, Stable silver nanoparticle–DNA conjugates for directed self-assembly of core-satellite silver–gold nanoclusters, Chem. Commun. (2009) 6059–6061.

[29] T.M. Herne, M.J. Tarlov, Characterization of DNA probes immobilized on gold surfaces, J. Am. Chem. Soc. 119 (1997) 8916–8920.

[30] B. Liu, J. Liu, Methods for preparing DNA-functionalized gold nanoparticles, a key reagent of bioanalytical chemistry, Anal. Methods 9 (2017) 2633–2643.

[31] J. Liu, Y. Lu, Preparation of aptamer-linked gold nanoparticle purple aggregates for colorimetric sensing of analytes, Nat. Protoc. 1 (2006) 246–252.

[32] S.J. Hurst, A.K.R. Lytton-Jean, C.A. Mirkin, Maximizing DNA loading on a range of gold nanoparticle sizes, Anal. Chem. 78 (2006) 8313–8318.

[33] X. Zhang, M.R. Servos, J. Liu, Instantaneous and quantitative functionalization of gold nanoparticles with thiolated DNA using a pH-assisted and surfactant-free route, J. Am. Chem. Soc. 134 (2012) 7266–7269.

[34] X. Zhang, M.R. Servos, J. Liu, Fast pH-assisted functionalization of silver nanoparticles with monothiolated DNA, Chem. Commun. 48 (2012) 10114–17266.

[35] Z. Huang, B. Liu, J. Liu, Parallel polyadenine duplex formation at low pH Facilitates DNA conjugation onto gold nanoparticles, Langmuir 32 (2016) 11986–11992.

[36] S.I. Stoeva, J.-.S. Lee, C.S. Thaxton, C.A. Mirkin, Multiplexed DNA detection with biobarcoded nanoparticle probes, Angew. Chem. Int. Ed. 45 (2006) 3303–3306.

[37] B. Liu, J. Liu, Freezing directed construction of Bio/Nano interfaces: Reagentless conjugation, denser spherical nucleic acids, and better nanoflares, J. Am. Chem. Soc. 139 (2017) 9471–9474.

[38] B. Liu, J. Liu, Freezing-driven DNA adsorption on gold nanoparticles: Tolerating extremely low salt concentration but requiring high DNA concentration, Langmuir 35 (2019) 6476–6482.

[39] W. Zhao, L. Lin, I.M. Hsing, Rapid synthesis of DNA-functionalized gold nanoparticles in salt solution using mononucleotide-mediated conjugation, Bioconjugate Chem. 20 (2009) 1218–1222.

[40] M. Huang, E. Xiong, Y. Wang, M. Hu, H. Yue, T. Tian, et al., Fast microwave heating-based one-step synthesis of DNA and RNA modified gold nanoparticles, Nat. Commun. 13 (2022) 968.

[41] A.V. Ramsey, A.J. Bischoff, M.B. Francis, Enzyme activated gold nanoparticles for versatile site-selective bioconjugation, J. Am. Chem. Soc. 143 (2021) 7342–7350.

[42] Y. Zheng, Y. Li, Z. Deng, Silver nanoparticle–DNA bionanoconjugates bearing a discrete number of DNA ligands, Chem. Commun. 48 (2012) 6160–6162.

[43] J.-.S. Lee, A.K.R. Lytton-Jean, S.J. Hurst, C.A. Mirkin, Silver nanoparticle–Oligonucleotide conjugates based on DNA with triple cyclic disulfide moieties, Nano Lett. 7 (2007) 2112–2115.

[44] R. Jin, G. Wu, Z. Li, C.A. Mirkin, G.C. Schatz, What controls the melting properties of DNA-linked gold nanoparticle assemblies? J. Am. Chem. Soc. 125 (2003) 1643–1654.

[45] S.J. Hurst, H.D. Hill, C.A. Mirkin, Three-Dimensional Hybridization" with polyvalent DNA–gold nanoparticle conjugates, J. Am. Chem. Soc. 130 (2008) 12192–12200.

[46] S.J. Hurst, H.D. Hill, C.A. Mirkin, Three-dimensional hybridization" with polyvalent DNA-gold nanoparticle conjugates, J. Am. Chem. Soc. 130 (2008) 12192–12200.

[47] N.L. Rosi, D.A. Giljohann, C.S. Thaxton, A.K. Lytton-Jean, M.S. Han, C.A. Mirkin, Oligonucleotide-modified gold nanoparticles for intracellular gene regulation, Science 312 (2006) 1027–1030.

[48] K. Zagorovsky, L.Y.T. Chou, W.C.W. Chan, Controlling DNA–nanoparticle serum interactions, Proc. Natl. Acad. Sci. 113 (2016) 13600–13605.

[49] D.A. Giljohann, D.S. Seferos, P.C. Patel, J.E. Millstone, N.L. Rosi, C.A. Mirkin, Oligonucleotide loading determines cellular uptake of DNA-modified gold nanoparticles, Nano Lett. 7 (2007) 3818–3821.

[50] P.C. Patel, D.A. Giljohann, W.L. Daniel, D. Zheng, A.E. Prigodich, C.A. Mirkin, Scavenger receptors mediate cellular uptake of polyvalent oligonucleotide-functionalized gold nanoparticles, Bioconjug Chem 21 (2010) 2250–2256.

[51] T. Minamoto, Z. Ronai, Gene mutation as a target for early detection in cancer diagnosis, Crit. Rev. Oncol. Hematol. 40 (2001) 195–213.

[52] K. Saha, S.S. Agasti, C. Kim, X. Li, V.M. Rotello, Gold nanoparticles in chemical and biological sensing, Chem. Rev. 112 (2012) 2739–2779.

[53] K. Sato, K. Hosokawa, M. Maeda, Rapid aggregation of gold nanoparticles induced by non-cross-linking DNA hybridization, J. Am. Chem. Soc. 125 (2003) 8102–8103.

[54] J.J. Storhoff, A.D. Lucas, V. Garimella, Y.P. Bao, UR. Müller, Homogeneous detection of unamplified genomic DNA sequences based on colorimetric scatter of gold nanoparticle probes, Nat. Biotechnol. 22 (2004) 883–887.

[55] D. Pal, N. Boby, S. Kumar, G. Kaur, S.A. Ali, J. Reboud, et al., Visual detection of Brucella in bovine biological samples using DNA-activated gold nanoparticles, PLoS One 12 (2017) e0180919.

[56] J.S. Lee, M.S. Han, C.A. Mirkin, Colorimetric detection of mercuric ion (Hg2+) in aqueous media using DNA-functionalized gold nanoparticles, Angew. Chem. Int. Ed Engl. 46 (2007) 4093–4096.

[57] X. Xue, F. Wang, X. Liu, One-step, room temperature, colorimetric detection of mercury (Hg2+) using DNA/Nanoparticle conjugates, J. Am. Chem. Soc. 130 (2008) 3244–3245.

[58] B. Li, Y. Du, S. Dong, DNA based gold nanoparticles colorimetric sensors for sensitive and selective detection of Ag(I) ions, Anal. Chim. Acta 644 (2009) 78–82.

[59] L. Li, B. Li, Y. Qi, Y. Jin, Label-free aptamer-based colorimetric detection of mercury ions in aqueous media using unmodified gold nanoparticles as colorimetric probe, Anal Bioanal Chem 393 (2009) 2051–2057.

[60] H. Jo, C. Ban, Aptamer–nanoparticle complexes as powerful diagnostic and therapeutic tools, Exp. Mol. Med. 48 (2016) e230 -e.

[61] J. Liu, Y. Lu, Fast colorimetric sensing of adenosine and cocaine based on a general sensor design involving aptamers and nanoparticles, Angew. Chem. Int. Ed Engl. 45 (2005) 90–94.

[62] F. Li, J. Zhang, X. Cao, L. Wang, D. Li, S. Song, et al., Adenosine detection by using gold nanoparticles and designed aptamer sequences, Analyst 134 (2009) 1355–1360.

[63] K. Li, W. Qin, F. Li, X. Zhao, B. Jiang, K. Wang, et al., Nanoplasmonic imaging of latent fingerprints and identification of cocaine, Angew. Chem. Int. Ed Engl. 52 (2013) 11542–11545.

[64] C.C. Wang, S.M. Wu, H.W. Li, HT. Chang, Biomedical applications of DNA-conjugated gold nanoparticles, ChemBioChem 17 (2016) 1052–1062.

[65] K.A. Kang, J. Wang, J.B. Jasinski, S. Achilefu, Fluorescence manipulation by gold nanoparticles: From complete quenching to extensive enhancement, J Nanobiotechnology 9 (2011) 16.

[66] M. Liu, Q. Tang, T. Deng, H. Yan, J. Li, Y. Li, et al., Two-photon AgNP/DNA-TP dye nanosensing conjugate for biothiol probing in live cells, Analyst 139 (2014) 6185–6191.

[67] B. Dubertret, M. Calame, AJ. Libchaber, Single-mismatch detection using gold-quenched fluorescent oligonucleotides, Nat. Biotechnol. 19 (2001) 365–370.

[68] D.J. Maxwell, J.R. Taylor, S. Nie, Self-assembled nanoparticle probes for recognition and detection of biomolecules, J. Am. Chem. Soc. 124 (2002) 9606–9612.

[69] A. Jayagopal, K.C. Halfpenny, J.W. Perez, DW. Wright, Hairpin DNA-functionalized gold colloids for the imaging of mRNA in live cells, J. Am. Chem. Soc. 132 (2010) 9789–9796.

[70] S.R. Harry, D.J. Hicks, K.I. Amiri, DW. Wright, Hairpin DNA coated gold nanoparticles as intracellular mRNA probes for the detection of tyrosinase gene expression in melanoma cells, Chem. Commun. 46 (2010) 5557–5559.

[71] J. Xue, L. Shan, H. Chen, Y. Li, H. Zhu, D. Deng, et al., Visual detection of STAT5B gene expression in living cell using the hairpin DNA modified gold nanoparticle beacon, Biosens. Bioelectron. 41 (2013) 71–77.

[72] D. Deng, Y. Li, J. Xue, J. Wang, G. Ai, X. Li, et al., Gold nanoparticle-based beacon to detect STAT5b mRNA expression in living cells: A case optimized by bioinformatics screen, Int J Nanomedicine 10 (2015) 3231–3244.

[73] Y. Tu, P. Wu, H. Zhang, C. Cai, Fluorescence quenching of gold nanoparticles integrating with a conformation-switched hairpin oligonucleotide probe for microRNA detection, Chem. Commun 48 (2012) 10718–10720.

[74] G. Qiao, Y. Gao, N. Li, Z. Yu, L. Zhuo, B. Tang, Simultaneous detection of intracellular tumor mRNA with bi-color imaging based on a gold nanoparticle/molecular beacon, Chem. 17 (2011) 11210–11215.

[75] W. Pan, T. Zhang, H. Yang, W. Diao, N. Li, B. Tang, Multiplexed detection and imaging of intracellular mRNAs using a four-color nanoprobe, Anal. Chem. 85 (2013) 10581–10588.

[76] W. Pan, Y. Li, M. Wang, H. Yang, N. Li, B. Tang, FRET-based nanoprobes for simultaneous monitoring of multiple mRNAs in living cells using single wavelength excitation, Chem. Commun. 52 (2016) 4569–4572.

[77] W. Pan, T. Zhang, H. Yang, W. Diao, N. Li, B. Tang, Multiplexed detection and imaging of intracellular mRNAs using a four-color nanoprobe, Anal. Chem. 85 (2013) 10581–10588.

[78] J. Chen, Z. Luo, Y. Wang, Z. Huang, Y. Li, Y. Duan, DNA specificity detection with high discrimination performance in silver nanoparticle coupled directional fluorescence spectrometry, Sens. Actuators B 255 (2018) 2306–2313.

[79] S. Dhar, W.L. Daniel, D.A. Giljohann, C.A. Mirkin, SJ. Lippard, Polyvalent oligonucleotide gold nanoparticle conjugates as delivery vehicles for platinum(IV) warheads, J. Am. Chem. Soc. 131 (2009) 14652–14653.

[80] X.Q. Zhang, X. Xu, R. Lam, D. Giljohann, D. Ho, C.A. Mirkin, Strategy for increasing drug solubility and efficacy through covalent attachment to polyvalent DNA-nanoparticle conjugates, ACS Nano 5 (2011) 6962–6970.

[81] G. Qiao, L. Zhuo, Y. Gao, L. Yu, N. Li, B. Tang, A tumor mRNA-dependent gold nanoparticle–molecular beacon carrier for controlled drug release and intracellular imaging, Chem. Commun. 47 (2011) 7458–7460.

[82] M.-.E. Kyriazi, D. Giust, A.H. El-Sagheer, P.M. Lackie, O.L. Muskens, T. Brown, et al., Multiplexed mRNA sensing and combinatorial-targeted drug delivery using DNA-Gold nanoparticle dimers, ACS Nano 12 (2018) 3333–3340.

[83] A.E. Prigodich, D.S. Seferos, M.D. Massich, D.A. Giljohann, B.C. Lane, C.A. Mirkin, Nano-flares for mRNA regulation and detection, ACS Nano 3 (2009) 2147–2152.

[84] C. Bao, J. Conde, J. Curtin, N. Artzi, F. Tian, D. Cui, Bioresponsive antisense DNA gold nanobeacons as a hybrid in vivo theranostics platform for the inhibition of cancer cells and metastasis, Sci. Rep. 5 (2015) 12297.

[85] K.V. Morris, S.W. Chan, S.E. Jacobsen, D.J. Looney, Small interfering RNA-induced transcriptional gene silencing in human cells, Science 305 (2004) 1289–1292.

[86] S. Son, J. Nam, J. Kim, S. Kim, W.J. Kim, i-motif-driven Au nanomachines in programmed siRNA delivery for gene-silencing and photothermal ablation, ACS Nano 8 (2014) 5574–55784.

Index

Page numbers followed by "*f*" and "*t*" indicate, figures and tables respectively.

Lightning Source UK Ltd.
Milton Keynes UK
UKHW052310070223
416656UK00012B/156